SiC Materials and Devices

SEMICONDUCTORS
AND SEMIMETALS
Volume 52

Semiconductors and Semimetals

A Treatise

Edited by R. K. Willardson
CONSULTING PHYSICIST
SPOKANE, WASHINGTON

Eicke R. Weber
DEPARTMENT OF MATERIALS SCIENCE
AND MINERAL ENGINEERING
UNIVERSITY OF CALIFORNIA AT
BERKELEY

SiC Materials and Devices

SEMICONDUCTORS
AND SEMIMETALS

Volume 52

Volume Editor

YOON SOO PARK

OFFICE OF NAVAL RESEARCH
ARLINGTON, VIRGINIA

ACADEMIC PRESS
San Diego   London   Boston   New York
Sydney   Tokyo   Toronto

This book is printed on acid-free paper.

COPYRIGHT © 1998 BY ACADEMIC PRESS

All rights reserved.
NO PART OF THIS PUBLICATION MAY BE REPRODUCED OR TRANSMITTED IN ANY FORM OR BY ANY MEANS, ELECTRONIC OR MECHANICAL, INCLUDING PHOTOCOPY, RECORDING, OR ANY INFORMATION STORAGE AND RETRIEVAL SYSTEM, WITHOUT PERMISSION IN WRITING FROM THE PUBLISHER.

The appearance of the code at the bottom of the first page of a chapter in this book indicates the Publisher's consent that copies of the chapter may be made for personal or internal use of specific clients. This consent is given on the condition, however, that the copier pay the stated per-copy fee through the Copyright Clearance Center, Inc. (222 Rosewood Drive, Danvers, Massachusetts 01923), for copying beyond that permitted by Sections 107 or 108 of the U.S. Copyright Law. This consent does not extend to other kinds of copying, such as copying for general distribution, for advertising or promotional purposes, for creating new collective works, or for resale. Copy fees for pre-1998 chapters are as shown on the title pages; if no fee code appears on the title page, the copy fee is the same as for current chapters. 0080-8784/98 $25.00

ACADEMIC PRESS
525 B Street, Suite 1900, San Diego, CA 92101-4495, USA
1300 Boylston Street, Chestnut Hill, Massachusetts 02167, USA
http://www.apnet.com

ACADEMIC PRESS LIMITED
24–28 Oval Road, London NW1 7DX, UK
http://www.hbuk.co.uk/ap/

International Standard Book Number: 0-12-752160-7
International Standard Series Number: 0084-8784

PRINTED IN THE UNITED STATES OF AMERICA
98 99 00 01 02 BB 9 8 7 6 5 4 3 2 1

Contents

LIST OF CONTRIBUTORS ix
INTRODUCTION xi

Chapter 1 Materials Properties and Characterization of SiC
Kenneth Järrendahl and Robert F. Davis

   I. Introduction 1
  II. Structural Properties 2
      1. SiC Polytypes: Structure and Symbolic Notations 2
      2. Lattice Parameter and Related Properties 7
      3. Mechanical Properties 12
 III. Thermal Properties 14
 IV. Optical Properties 14
  V. Electrical Properties 16
 VI. Summary 17
      References 18

Chapter 2 SiC Fabrication Technology: Growth and Doping
V. A. Dmitriev and M. G. Spencer

   I. Introduction 21
  II. SiC Bulk Crystal Growth 22
      1. Bulk Growth by Physical Vapor Transport 22
      2. Bulk Growth by Chemical Vapor Deposition 34
      3. Bulk Growth from the Liquid Phase 34
 III. SiC Epitaxial Growth 35
      1. Chemical Vapor Deposition 36
      2. Liquid-Phase Epitaxy 50
      3. Sublimation Epitaxy 56
      4. Molecular Beam Epitaxy 58
 IV. SiC Doping 60
      1. Impurities in SiC 60
      2. Doping During Crystal Growth 60

3. Diffusion of Impurities in SiC	61
4. Ion Implantation	61
V. Conclusions	62
References	63

Chapter 3 Building Blocks for SiC Devices: Ohmic Contacts, Schottky Contacts, and p-n Junctions

V. Saxena and A. J. Steckl

I. Introduction	77
II. Electrical Properties of Metal–SiC Systems	81
1. The Schottky-Mott and Bardeen Limits for Barrier Height	81
2. Metal Contacts for High-Power Applications	84
III. Surface Preparation Techniques	86
IV. Metal Contacts to 6H-SiC	87
1. Ohmic Contacts to 6H-SiC	87
2. Schottky Contacts to 6H-SiC	96
3. High-Voltage Schottky Diodes on 6H-SiC	109
4. Edge Termination in 6H-SiC Schottky Diodes	113
5. Excess Leakage Current in SiC Schottky Diodes	114
V. Metal Contacts to 4H-SiC	118
1. Ohmic Contacts to 4H-SiC	118
2. Schottky Contacts and High-Voltage Schottky Diodes on 4H-SiC	118
VI. Metal Contacts to 3C-SiC	125
1. Ohmic Contacts to 3C-SiC	125
2. Schottky Contacts and Diodes on 3C-SiC	126
VII. SiC p-n Junctions Diode Rectifiers	131
1. 3C-SiC p-n Junctions	133
2. 6H-SiC p-n Junctions	138
3. 4H-SiC p-n Junctions	147
VIII. Summary and Conclusions	149
References	151

Chapter 4 SiC Transistors

Michael S. Shur

I. Introduction	161
II. SiC Field-Effect Transistors: MOSFETs, MESFETs, and JFETs	162
1. Principle of Operation	162
2. SiC MOSFETs	168
3. SiC MESFETs and JFETs	174
III. SiC Microwave Field-Effect Transistors	177
IV. SiC Digital Integrated Circuits	180
V. SiC Bipolar Transistors and Thyristors	182
VI. Two-Dimensional Modeling of SiC Transistors	185
VII. Analytical Transistor Models and Circuit Simulation (AIM-Spice)	186
VIII. Potential Performance and Applications of SiC Transistors and Integrated Circuits	188
References	189

Chapter 5 SiC for Applications in High-Power Electronics
C. D. Brandt, R. C. Clarke, R. R. Siergiej, J. B. Casady, A. W. Morse, S. Sriram, and A. K. Agarwal

I.	Introduction	195
II.	Bulk SiC Growth	198
III.	Epitaxial Growth of SiC	202
IV.	Advantages of SiC for High-Power RF Systems	204
V.	The SiC MESFET: Design Considerations	205
VI.	MESFET Fabrication	207
VII.	6H-SiC MESFET Results	209
VIII.	4H-SiC MESFET Results	211
IX.	The SiC SIT: Design Considerations	212
X.	SIT Fabrication	215
XI.	6H-SiC SIT Results	217
XII.	4H-SiC SIT Results	219
XIII.	450 W UHF SIT	221
XIV.	2.0 kW UHF Module	222
XV.	S-band SiC SITs	223
XVI.	S-band SIT Device Scale-up	224
XVII.	SiC Power Switching Devices	226
XVIII.	Conclusions	231
	References	232

Chapter 6 SiC Microwave Devices
R. J. Trew

I.	Introduction	237
II.	Background	239
III.	Semiconductor Material and Contact Properties	241
IV.	Semiconductor Device Models	247
V.	Temperature Effects	250
VI.	RF Active Devices	252
	1. MESFETs	253
	2. SiC Static Induction Transistors	264
	3. Bipolar Transistors	267
	4. IMPATT Diodes	272
VII.	Summary	279
	References	280

Chapter 7 SiC-Based UV Photodiodes and Light Emitting Diodes
J. Edmond, H. Kong, G. Negley, M. Leonard, K. Doverspike, W. Weeks, A. Suvorov, D. Waltz, and C. Carter, Jr.

I.	Introduction	283
II.	SiC Blue LEDs	286
	1. Epitaxy and Device Fabrication	286
	2. Device Performance	287
III.	SiC Green LEDs	288
	1. Epitaxy and Ion Implantation	288

	2. Device Performance	289
IV.	UV Photodiodes	290
	1. Device Fabrication	290
	2. Electrical Characteristics	292
	3. Optical Responsivity	294
	4. Applications	296
V.	Group III-Nitrides on 6H-SiC	297
	1. Epitaxial Growth, Characterization, and Device Fabrication	297
	2. Hall Effect	298
	3. Photoluminescence	300
	4. GaN: SiC Blue LEDs	302
	5. Electrical Static Discharge Survivability	305
VI.	Summary	305
	References	306

Chapter 8 Beyond Silicon Carbide! III–V Nitride-Based Heterostructures and Devices

Hadis Morkoç

I.	Introduction	307
II.	Strain and Structural Defects	309
	1. Effect of Strain and Lattice Mismatch on Crystal Structure	309
	2. Dislocations	310
	3. Stacking Fault Defects	311
	4. Point Defects	313
	5. P-Type Doping by Mg vis-à-vis Defects	320
	6. Defect Analysis by Deep-Level Transient Spectroscopy	330
	7. Defect-Aided Current in Unintentionally Doped GaN	333
III.	Optical Manifestation of Defects	335
	1. Yellow Band	335
	2. Defects Caused by P-Type Doping in GaN	339
IV.	Applications	342
	1. Field-Effect Transistors	342
	2. Blue, Green, and Yellow LEDs	373
	3. LEDs by MBE	378
	4. Lasers in Semiconductor Nitrides	380
	5. UV Detectors	385
V.	Conclusions	389
	References	390

INDEX	395
CONTENTS OF VOLUMES IN THIS SERIES	405

List of Contributors

Numbers in parenthesis indicate the pages on which the authors' contribution begins.

S. AGARWAL (195), *Northrop Grumman Science and Technology Center, Pittsburgh, PA 15235-5080*

C. D. BRANDT (195), *Northrop Grumman Science and Technology Center, Pittsburgh, PA 15235-5080*

C. CARTER, JR., (283), *Cree Research Center, Durham, NC 27713*

J. B. CASADY (195), *Northrop Grumman Science and Technology Center, Pittsburgh, PA 15235-5080*

R. C. CLARKE (195), *Northrop Grumman Science and Technology Center, Pittsburgh, PA 15235-5080*

ROBERT F. DAVIS (1), *Department of Materials Science and Engineering, North Carolina State University, Raleigh, NC 27695-7970*

V. A. DMITRIEV (21), *A. F. Ioffe, Institute, St. Petersburg, 194021 Russia.*

K. DOVERSPIKE (283), *Cree Research Inc., Durham, NC 27713*

J. EDMOND (283), *Cree Research Inc., Durham, NC 27713*

KENNETH JÄRRENDAHL (1), *Department of Physics and Measurement Technology, Linköping University, 5-581 83 Linköping, Sweden*

H. KONG (283), *Cree Research Inc., Durham, NC 27713*

M. LEONARD (283), *Cree Research Inc., Durham, NC 27713*

HADIS MORKOÇ (307), *Department of Electrical Engineering and Physics, Virginia Commonwealth University, Richmond, VA 23284-3072*

A. W. Morse (195), *Northrop Grumman Electronic Sensors and Systems Division, Baltimore, MD 21203-7319*

G. Negley (283), *Cree Research Inc., Durham, NC 27713*

V. Saxena (77), *Department of Electrical and Computer Engineering and Computer Science, University of Cincinnati, Cincinnati, OH 45221-0030*

Michael S. Shur (161), *Center for Integrated Electronics and Electronics Manufacturing, Department of Electrical, Computer, and Systems Engineering, Rensselaer Polytechnic Institute, Troy, NY 12180*

R. R. Siergiej (195), *Northrop Grumman Science and Technology Center, Pittsburgh, PA 15235-5080*

S. Sriram (195), *Northrop Grumman Science and Technology Center, Pittsburgh, PA 15235-5080*

M. G. Spencer (21), *School of Engineering, Materials Science Research Center of Excellence, Howard University, Washington, DC 20059*

A. J. Steckl (77), *Department of Electrical and Computer Engineering and Computer Science, University of Cincinnati, Cincinnati, OH 45221-0030*

A. Suvorov (283), *Cree Research Inc., Durham, NC 27713*

R. J. Trew (237), *ODDR & E(R), U.S. Department of Defense, Washington, DC 20301-3080*

D. Waltz (283), *Cree Research Inc., Durham, NC 27713*

W. Weeks (283), *Cree Research Inc., Durham, NC 27713*

Introduction

First synthesized in the 19th century, SiC has been the subject of research for semiconductor applications for over 40 years. The wide bandgap, high thermal conductivity, and robust mechanical and properties make SiC attractive for many applications including high temperature, high power, or high frequency devices that are not possible using Si or GaAs. SiC is also resistant to high radiation doses, which makes it potentially the ideal technology for nuclear power applications. Also, although it is an indirect wide bandgap material, SiC can be applied for various optical device applications such as blue and ultraviolet (UV) light emitting devices and photodetectors. Even though a significant amount of good, fundamental research was performed during the early years of SiC research and development, the development of commercially available SiC-based devices was retarded by low-quality bulk materials and inadequate epitaxial processes.

The first procedure for making synthetic SiC was developed by Acheson late in the 19th century. Single crystal SiC platelets were synthesized by Lely in 1955. A modified-Lely method proposed by Tairov and Tsvetkov in the late 1970s enabled the growth of SiC boules. Since the late 1980s, relatively large SiC wafers have been commercially available from Cree Research, Inc. The large SiC wafers produced at Northrop Grumman, Advanced Technology Materials, Inc. (ATMI), and Cree Research have created opportunities for further advancement in SiC-based device research. Improvements in epitaxial growth processes and device processing strategies were also realized during this time. Early efforts for the epitaxy of SiC with liquid phase epitaxy (LPE) was replaced with more advanced technologies such as chemical vapor deposition (CVD) and gas source molecular beam epitaxy (GSMBE). Together these factors have enabled the fabrication of high-quality device structures and have generated increased research activities in SiC devices.

Several hundred polytypes of SiC have been identified ranging from purely hexagonal 2H-SiC to purely cubic 3C-SiC. However, the polytypes which have received the most attention are 6H-SiC, 3C-SiC, and 4H-SiC. The polytype favored under most high-temperature conditions, 6H-SiC, was the first polytype to be made commercially available as single crystal wafers cut from large boules. 3C-SiC can be deposited epitaxially on Si substrates at relatively low growth temperatures, but high quality bulk crystals of this polytype are not available. High quality 4H-SiC wafers are now for sale after the recent development of their bulk crystal growth. The effort to mature the 4H-SiC synthesis was made primarily because of the lower and more isotropic effective masses compared to 6H-SiC potentially resulting in a higher mobility which is nearly independent of the crystallographic direction. Also, 4H-SiC has a wider bandgap than the 6H-SiC and 3C-SiC polytypes.

The electrical properties of SiC can now be controlled over a wide range. SiC can be doped both n- and p-type, typically using nitrogen and aluminum, respectively. Insulating SiC has also been produced by doping with vanadium.

There are many potential military and civilian applications for SiC based devices. For military aircraft, objectives are to achieve a significant improvement in aircraft flight-control system reliability, mass reduction, and reduced dependency on environmental control subsystems. Other military and commercial applications include high-frequency SiC-based devices that can deliver continuous high power at X-band (8–10GHz) at temperatures up to 500°C. These devices can be applied for radar and communications systems for unpiloted aerial vehicles and distributed satellite arrays. In the commercial market, the petroleum industry provides a major driving force for the development of high temperature devices. Sensors operating at above 300°C are needed to probe the environment around drilling equipment. In the automotive industry, the SiC devices can be used in a variety of applications such as engine control sensors. SiC technology has now advanced to where practical devices are nearly available. Therefore, basic research on materials is needed for further progress and more research effort on the device development is in great need.

In this book, we have attempted to provide an overview of the current issues relating to the development of SiC materials, electronic devices, and photonic devices. The recent development of III–V nitride materials for similar applications to those for SiC is also reviewed. In the first chapter, the physical properties and characterization of SiC materials is reviewed. Unique properties of SiC, such as high breakdown voltage and high thermal conductivity which make the SiC an ideal material for the high field and high temperature operations, are discussed in detail. The second chapter by

Dmitriev and Spencer deals with SiC fabrication technologies including growth and doping. The third chapter by Saxena and Steckl, describes the building blocks for the SiC devices such as Ohmic contacts, Schottky contacts, and p–n junctions.

Chapters 4 through 6 deal with the electronic devices fabricated from SiC materials. As SiC-based electronic devices have various applications, especially for high temperature and high power operation environments, emphases have been given on those aspects of the devices. In Chapter 4 by Shur, transistors based on SiC are reviewed. Advances in all aspects of SiC technology have enabled field-effect transistors (FETs) to be demonstrated. It is an important advancement in the SiC device technology because FETs have long been the mainstay of power switching applications due to their ease of fabrication and amenability to large-scale integration. Several types of SiC FETs have been fabricated, including MOSFETs, JFETs, and MESFETs. Bipolar transistors based on SiC have also been developed. Chapter 5 deals with SiC-based devices for power electronics. And in Chapter 6, Trew reviews various microwave devices fabricated from SiC materials.

As mentioned earlier, SiC has attracted the attention of the researchers for applications in various optical devices. Despite its indirect bandgap, SiC can be made to electroluminesce across the entire visible spectrum by the addition of various impurities. Also it can be used for the photodetector applications in the UV spectrum range. Chapter 7 by Edmond *et al.* deals with the SiC-based UV photodiodes and light emitting diodes. Recently, III–V nitride materials have attracted a lot of attention from researchers because of their potential to replace or complement SiC in both electronic and photonic device applications. Being direct bandgap materials, the III-Nitrides offer much higher quantum efficiency than SiC which makes them an ideal material system for the photonic devices in the blue and UV spectral range. Also, GaN can be alloyed with AlN and InN, permitting. The similar lattice constants of SiC and the III-Nitrides enable SiC to be used for the epitaxy of III-Nitrides that current lack large high quality substrates. In fact, Cree Research now sells blue LEDs based on III-Nitride heterostructures deposited on SiC substrates. In the final chapter, heterostructures and devices based on III-Nitrides are discussed.

YOON SOO PARK

CHAPTER 1

Materials Properties and Characterization of SiC

Kenneth Järrendahl

DEPARTMENT OF PHYSICS AND MEASUREMENT TECHNOLOGY
LINKÖPING UNIVERSITY
LINKÖPING, SWEDEN

Robert F. Davis

DEPARTMENT OF MATERIALS SCIENCE AND ENGINEERING
NORTH CAROLINA STATE UNIVERSITY
RALEIGH, NC

I.	INTRODUCTION	1
II.	STRUCTURAL PROPERTIES	2
	1. *SiC Polytypes: Structure and Symbolic Notations*	2
	2. *Lattice Parameter and Related Properties*	7
	3. *Mechanical Properties*	12
III.	THERMAL PROPERTIES	14
IV.	OPTICAL PROPERTIES	14
V.	ELECTRICAL PROPERTIES	16
VI.	SUMMARY	17
	References	18

I. Introduction

Wide energy bandgaps, high thermal conductivity, high saturated electron drift velocities, and high-breakdown electric fields make silicon carbide (SiC) a candidate of choice for high-temperature, high-speed, high-frequency, and high-power applications. SiC is also hard, chemically stable, and resistant to radiation damage. This chapter summarizes some of the basic properties of SiC.

II. Structural Properties

The symmetry and crystal structure of carborundum present an interesting and puzzling problem for which there is not as yet an satisfactory answer.
  R. W. G. Wyckoff, in the 1924 edition of *The Structure of Crystals* [51]

1. SiC Polytypes: Structure and Symbolic Notations

SiC, the only known binary compound of silicon and carbon, possesses a one-dimensional polymorphism called polytypism. In a polytypic compound, similar sheets of atoms or symmetrical variants are stacked atop each other and related according to a symmetry operator. The differences among the polytypes arise only in the direction perpendicular to the sheets (along the c-axis). The sheets can be represented as a close-packed array of spheres forming a two-dimensional pattern with sixfold symmetry. (In SiC, each sheet represents a bilayer composed of one layer of Si atoms and one layer of C atoms, as will be described in detail). Using the notations from hexagonal crystal structures, the first sheet can be defined as the basal or c-plane with Miller-indexed directions according to Fig. 1. The most stable way to stack an identical second sheet of close-packed spheres is to place the spheres atop the "valleys" in the first sheet. There are two possibilities

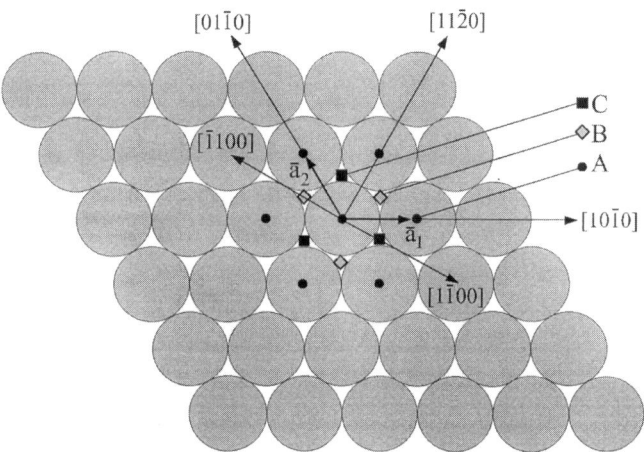

FIG. 1. A close-packed hexagonal plane of spheres, with centers at points marked A. A second and identical plane can be placed atop the first plane, with centers over either the points marked B or the points marked C.

1   MATERIALS PROPERTIES AND CHARACTERIZATION OF SiC

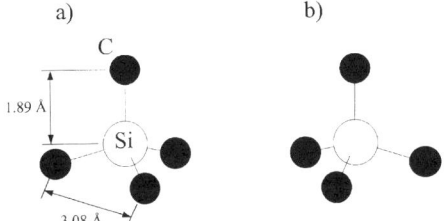

FIG. 2. (a) The basic structural unit in SiC is a tetrahedron of four carbon atoms with a silicon atom in the middle. (b) A second type rotated 180° around the stacking direction, with respect to the first type of tetrahedra, can also occur in ther SiC crystals.

for arranging the second sheet relative to the first sheet in this way. The second sheet may be displaced along, for example, [$\bar{1}$100] until the spheres lie in the valleys denoted "B" in Fig. 1, or the sheet may be displaced along, for example, [1$\bar{1}$00] until the spheres lie in the valleys denoted "C" in Fig. 1. Thus, a sheet can be denoted A, B, or C, depending on the positions of its spheres (see Fig. 1). All polytypes can be described as different stackings of A, B, and C sheets, with the restriction that sheets with the same notation cannot be stacked upon each other.

In the SiC polytypes, the basic structural unit consists of a primarily covalently bonded (88% covalent and 12% ionic) tetrahedron of four C atoms with an Si atom at the center (or four Si atoms with a C atom at the center) (Fig. 2). The approximate bond length between the Si-Si or C-C atoms is 3.08 Å, whereas the bond length between the Si-C atoms is approximately 1.89 Å. The SiC crystals are constructed with these tetrahedra joined to each other at the corners. In this way, the stacking notations (A, B, and C) for the symmetrical variants will represent Si-C bilayer sheets in the crystals. The tetrahedra have threefold symmetry around the axis parallel to the stacking direction. Thus, a second type rotated 180° around the stacking direction, with respect to the first type of tetrahedra, can also occur in SiC crystals (see Fig. 2). To distinguish between the two variants, the ABC notation can be extended with the primed letters A', B', and C' denoting bilayers belonging to rotated tetrahedra. The notation "untwinned" and "twinned" tetrahedra will here be used for the tetrahedra of the first and second types, respectively. To maintain the corner-sharing structure, the stacking of A, A', B, B', C, and C' types of bilayers must be made according to a number of rules [40]. Corner sharing is only possible if an untwinned tetrahedron is stacked upon by (*i*) an untwinned tetrahedron of the following letter (AB, BC, or CA) or (*ii*) a twinned tetrahedron of the preceding letter (AC', BA', or CB'). In the same way, a twinned tetrahedron must be stacked upon by (*iii*) an untwinned tetrahedron of the following

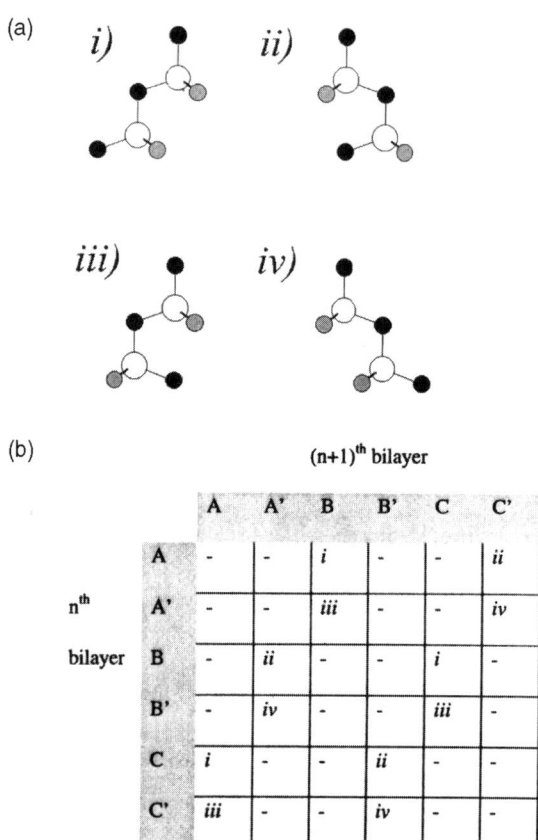

FIG. 3. Corner sharing is only possible if an untwinned tetrahedron is stacked upon by (*i*) an untwinned tetrahedron of the following letter (AB, BC, or CA) or (*ii*) a twinned tetrahedron of the preceding letter (AC', BA', or CB'). In the same way, a twinned tetrahedron must be stacked upon by (*iii*) an untwinned tetrahedron of the following letter (A'B, B'C, C'A) or (*iv*) a twinned tetrahedra of the preceding letter (A'C', B'A', or C'B') to be corner sharing. The Si atoms (white circles) and C atoms (black circles) are all in the same plane. The extra gray circles represent projections of C atoms and are added to show the twinning of the tetrahedra.

letter (A'B, B'C, C'A) or (*iv*) a twinned tetrahedra of the preceding letter (A'C', B'A', or C'B') to be corner sharing (Fig. 3).

A large number of SiC polytypes exist (some sources mention more than 250 [17]), with some having stacking sequences of several hundreds of bilayers. The crystal structures of the SiC polytypes are cubic, hexagonal, or rhombohedral. The only cubic polytype is referred to as β-SiC, whereas the hexagonal and rhombohedral polytypes are referred to collectively as α-SiC. The polytypes of SiC were originally denoted types I, II, III, and so on, in

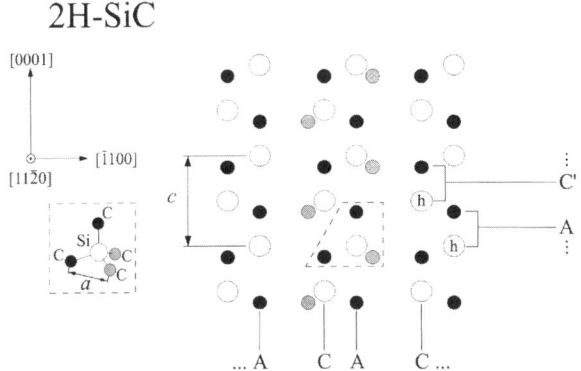

FIG. 4. The (11$\bar{2}$0) plane of 2H-SiC, showing the stacking sequence of bilayers of Si (white circles) and C (black circles). The extra gray circles represent projections of C atoms and are added to show the orientation of the untwinned and twinned tetrahedra. The tetrahedron shown in the dashed square is the same as in the dashed polygon inset but has been rotated to view all five atoms. The Jagodzinski and ABCA'B'C' notations are also shown.

the order they were found. Due to the increasing number of discovered polytypes of α-SiC, Ramsdell [41] suggested that each polytype could be named with a number according to the periodicity in the stacking direction (i.e., the number of letters A, A', B, B', C, and C' needed to define the unit cell) and the letter H, for *hexagonal*, or R, for *rhombohedral*. The common polytypes, denoted types I, II, and III, could now be referred to as 15R, 6H, and 4H, respectively. Subsequently, it became common to refer to β-SiC as 3C-SiC.

The two most simple stacking sequences for the Si-C bilayers are ...AC'AC'... and ...ABCABC..., which produce the wurtzite 2H structure (with the stacking along the [0001] direction) and the zinc-blende (or sphalerite) 3C structure (with the stacking along the [111] direction), respectively. The stacking sequences show that 2H consists of alternating twinned and untwinned types of tetrahedra (Fig. 4), whereas the 3C structure consists only of one type (Fig. 5). This makes the 2H and 3C structures differ only in the relative handedness of every fourth bond along the [0001] and [111] directions, respectively, where the 2H bonds are aligned and the 3C bonds are rotated according to Fig. 6. Using the stacking notation according to Jagodzinski [24], a bilayer can be described as having a locally (h) or locally cubic (k, from the German *kubisch*) environment, depending on the layers above and below. That is, the 2H structure with its alternating twinned and untwinned tetrahedra, and periodicity of two, can be described with the notation $(h)_2$. Accordingly, the 3C structure can be described with the notation $(k)_3$. All other polytypes can now be seen as a mix between the

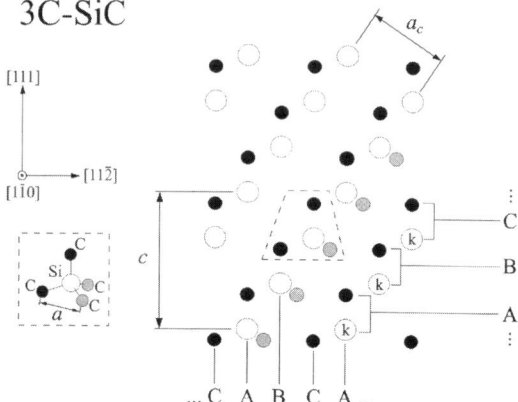

FIG. 5. The (1$\bar{1}$0) plane of 3C-SiC, showing the stacking sequence of bilayers of Si (white circles) and C (black circles). The extra gray circles represent projections of the gray C atoms seen in the inset and are added to show the orientation of the tetrahedra. The tetrahedron shown in the dashed square is the same as in the dashed polygon inset but has been rotated to view all five atoms. The Jagodzinski and ABCA'B'C' notations are also shown.

purely hexagonal and purely cubic structure, and the hexagonality of a polytype can be calculated from the fraction of h's in the Jagodzinski notation. An additional notation has been suggested by Zhadanov [53], which denotes the number of consecutive tetrahedra of each type within a period. Consequently, the 2H polytype is described with (11). The Zhadanov notation for 3C is sometimes written (∞), since the stacking is made without rotations.

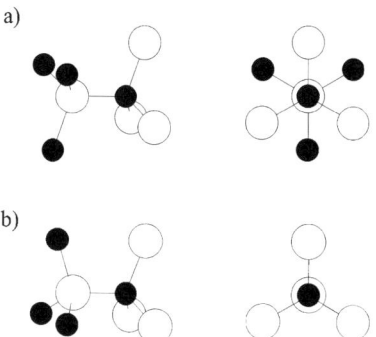

FIG. 6. Side view and view along the stacking direction for (a) the cubic type of bond, where the bonds are rotated, and (b) the hexagonal type, where the bonds are aligned.

1 MATERIALS PROPERTIES AND CHARACTERIZATION OF SiC

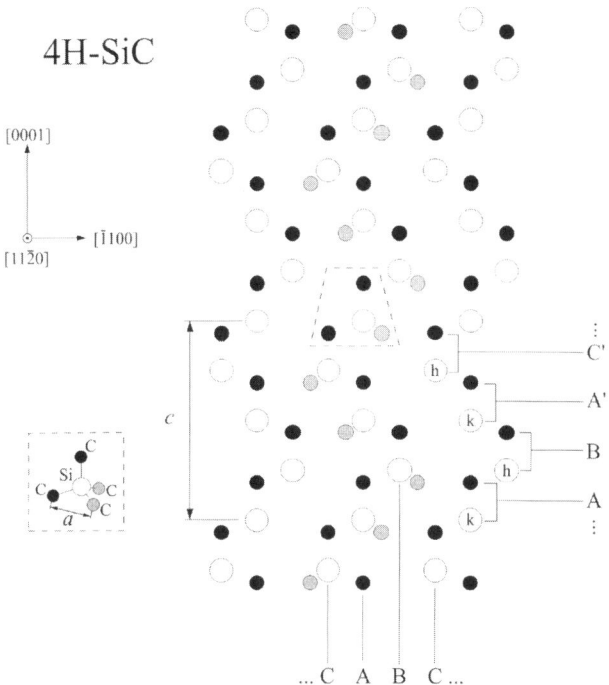

FIG. 7. The $(11\bar{2}0)$ plane of 4H-SiC, showing the stacking sequence of bilayers of Si (white circles) and C (black circles). The extra gray circles represent projections of the gray C atoms seen in the inset and are added to show the orientation of the tetrahedra. The tetrahedron shown in the dashed square is the same as in the dashed polygon inset but has been rotated to view all five atoms. The Jagodzinski and ABCA′B′C′ notations are also shown.

The wurtzite and zinc-blende are the most common crystal structures among the binary octet semiconductors. For SiC, however, the wurtzite polytype (see Fig. 4) is rarely found. The 3C (see Fig. 5), 4H (Fig. 7), 6H (Fig. 8), and 15R are the commonly found polytypes. Figures 5, 4, 7, and 8 show the stacking sequence in the $(1\bar{1}0)$ plane for 3C and in the $(11\bar{2}0)$ plane for 2H, 4H, and 6H. In Table I, the different notations for the 2H, 3C, 4H, 6H, and 15R polytypes are summarized. Information on other polytypes (e.g., 8H, 10H, and 21R) can be found elsewhere [18, 27].

2. LATTICE PARAMETER AND RELATED PROPERTIES

For complete tetrahedral symmetry, all Si-C distances must be the same, with C-S-C bond angles of $\arccos(-\frac{1}{3}) = 109.471\ldots°$. This gives a ratio between the height of the tetrahedron, c/n, and the C-C (or Si-Si) bond

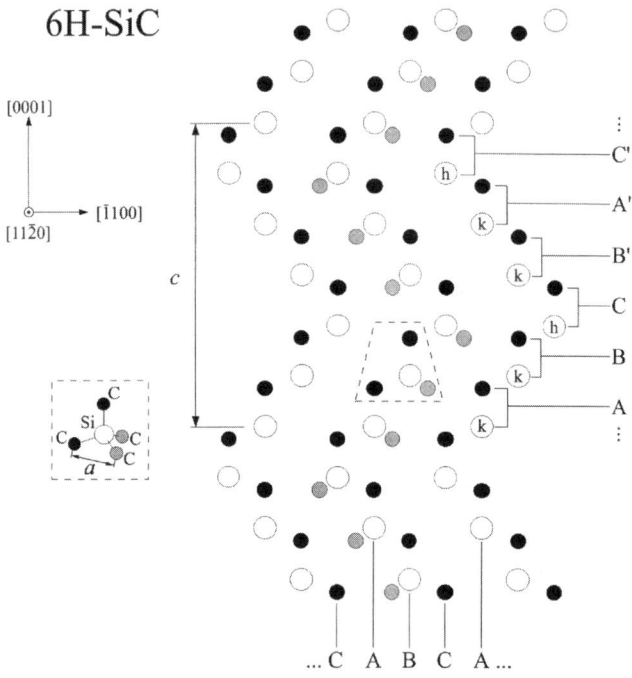

FIG. 8. The (11$\bar{2}$0) plane of 6H-SiC, showing the stacking sequence of bilayers of Si (white circles) and C (black circles). The extra gray circles represent projections of the gray C atoms seen in the inset and are added to show the orientation of the tetrahedra. The Jagodzinski and ABCA'B'C' notations are also shown. The Jagodzinski and ABCA'B'C' notations are also shown.

length, a, equal to

$$\frac{c}{na} = \sqrt{\tfrac{2}{3}} = 0.816496\ldots \qquad (1)$$

where a and c are the lattice parameters (see Figs. 4, 5, 7, and 8 and Tables II–VI), and n specifies the number of bilayers in a period of the polytype. In the 3C-SiC structure, all Si-C bond lengths are equal ($a\sqrt{\tfrac{3}{8}} = 1.88775\ldots$, using the a value reported by Tairov and Tsvetkov [46]), and the C-Si-C bond angles are 109.471°. That is, the 3C polytype can be considered to have complete tetrahedral symmetry. The c/na values for the other polytypes presented in Tables II–VI are, in general, greater than the ideal value, indicating that there must be one longer (vertical) bond and three shorter (lateral) bonds. It has been shown by DeMesquita [14] that the three

TABLE I
DIFFERENT NOTATIONS FOR THE SiC POLYTYPES

Ramsdell	3C	2H	4H	6H	15R
Structure	Zinc-blende, sphalerite	Wurtzite			
Type	β	α	α, III	α, II	α, I
Strukturbericht	B3	B4			
ABCA'B'C'	...ABC...	...AC'...	...ABA'C'...	...ABCB'A'C'...	...ABCB'A'BCAC'B'CABA'C'...
Zhadanov	(∞)	(11)	(22)	(33)	(32)$_3$
Jagodzinski	(k)$_3$	(h)$_2$	(kh)$_2$	(kh)$_2$	(kkhkh)$_3$
Hexagonality	0%	100%	50%	33.3%	40%
Space group	T_d^2	C_{6v}^4	C_{6v}^4	C_{6v}^4	C_{3v}^5
	F43m	P6$_3$mc	P6$_3$mc	P6$_3$mc	R3m
Space group no.	216	186	186	186	160
Pearson symbol	cF8	hP4	hP8	hP12	hR10
Atoms/unit cell	2	4	8	12	10
Inequivalent sites					
Hexagonal-like	0	1	1	1	2
Cubic-like	1	0	1	2	3

TABLE II

Lattice Parameters and Related Properties of 3C-SiC

a_c(Å)	a(Å)	c(Å)	c/n(Å)[b]	$c/(na)$[b]	V(Å^3)[c]	d(g/cm^3)[d]	Comment[e]	Reference to Lattice Parameters
4.3582	3.0817[a]	7.5486[a]	2.5162	0.8165	62.0837	3.2174	LT	[47]
4.3596	3.0827[a]	7.5510[a]	2.517	0.8165	62.1438	3.2143	RT	[47]
4.3589	3.0822[a]	7.5498[a]	2.5166	0.8165	62.1137	3.2158	RT	[37]
4.3596[a]	3.08269	7.55124	2.51708	0.8165	62.1453	3.2142	RT	[46]

[a] Relationship between a and c and the cubic lattice paqrameter a_c, calculated using $a = a_c/\sqrt{2}$ and $c = a_c\sqrt{3}$.
[b] $n = 3$.
[c] Unit cell volume, calculated from a and c, using $V = \dfrac{a^2 c\sqrt{3}}{2}$.
[d] Density, calculated using $d = nM/(N_A V)$, where V is the unit cell volume (expressed in cm^3), M is the Si-C molecular weight (40.09715 g mol^{-1}), $n = 3$, and N_A is the Avogadro constant ($6.0221367 \cdot 10^{23}$ mol^{-1}).
[e] LT and RT denote low-temperature (2–8 K) and room-temperature (≈ 300 K) measurements, respectively.

TABLE III

Room Temperature Lattice Parameters and Related Properties of 2H-SiC

a(Å)	c(Å)	c/n(Å)[a]	c/na[a]	V(Å^3)[b]	d(g/cm^3)[c]	Reference to Lattice Parameters
3.0763	5.0480	2.524	0.8205	41.3721	3.2187	[1, 36]

[a] $n = 2$.
[b] Unit cell volume, calculated from a and c, using $V = \dfrac{a^2 c\sqrt{3}}{2}$.
[c] Density, calculated using $d = nM/(N_A V)$, where V is the unit cell volume (expressed in cm^3), M is the Si-C molecular weight (40.09715 g mol^{-1}), $n = 2$, and N_A is the Avogadro constant ($6.0221367 \cdot 10^{23}$ mol^{-1}).

TABLE IV

Room Temperature Lattice Parameters and Related Properties of 4H-SiC

a(Å)	c(Å)	c/n(Å)[a]	c/na[a]	V(Å^3)[b]	d(g/cm^3)[c]	Reference to Lattice Parameters
3.0730	10.053	2.51325	0.8178	82.215	3.239	[52]
3.07997	10.0830	2.52075	0.8184	82.8349	3.2152	[46]
3.076	10.046	2.5115	0.8165	82.318	3.235	[15]
3.081	10.061	2.51525	0.8164	82.709	3.220	[20, 38]

[a] $n = 4$.
[b] Unit cell volume, calculated from a and c, using $V = \dfrac{a^2 c\sqrt{3}}{2}$.
[c] Density, calculated using $d = nM/(N_A V)$, where V is the unit cell volume (expressed in cm^3), M is the Si-C molecular weight (40.09715 g mol^{-1}), $n = 4$, and N_A is the Avogadro constant ($6.0221367 \cdot 10^{23}$ mol^{-1}).

TABLE V
Lattice Parameters and Related Properties of 6H-SiC

$a(\text{Å})$	$c(\text{Å})$	$c/n(\text{Å})^a$	c/na^a	$V(\text{Å}^3)^b$	$d(\text{g/cm}^3)^c$	Comment[d]	Reference to Lattice Parameters
3.080	15.1173	2.51955	0.8180	124.1956	3.217	LT	[47]
3.073	15.08	2.5133	0.8179	123.33	3.24	RT	[39, 48]
3.0806	15.1173	2.51955	0.8179	124.2440	3.2154	RT	[47]
3.08086	15.1174	2.5196	0.8178	124.2658	3.2149	RT	[46]

[a] $n = 6$.

[b] Unit cell volume, calculated from a and c, using $V = \dfrac{a^2 c \sqrt{3}}{2}$.

[c] Calculated using $d = nM/(N_A V)$, where V is the unit cll volume (expressed in cm^3), M is the Si-C molecular weight (40.09715 g mol^{-1}), $n = 6$, and N_A is the Avogadro constant (6.0221367·10^{23} mol^{-1}).

[d] LT and RT denote low-temperature (2–8 K) and room-temperature ($\approx$ 300 K) measurements, respectively.

TABLE VI
Room Temperature Lattice Parameters and Related Properties of 15R-SiC

$a_R(\text{Å})$	$\gamma(°)$	$a(\text{Å})$	$c(\text{Å})$	$c/n(\text{Å})^c$	c/na^c	$V(\text{Å}^3)^d$	$d(\text{g/cm}^3)^e$	Reference to Lattice Parameters
12.691	13.9	3.07[a,b]	37.70[a,b]	2.51	0.819	307.72	3.24	[48]
12.725[a]	13.904[b]	3.08043	37.8014	2.5201	0.8181	310.6427	3.2151	[46]
12.56[a]	14.05[b]	3.073	37.30	2.49	0.809	305.04	3.27	[15]

[a] Relationship between a and c and the rhombohedral lattice parameter a_R, calculated using $a_R = \frac{1}{3}\sqrt{3a^2 + c^2}$.

[b] Relation between a and c and the angle of the rhombohedral cell γ, calculated using, $\sin(\gamma/2) = \dfrac{3}{2\sqrt{3 + (c/a)^2}}$.

[c] $n = 15$.

[d] Unit cell volume, calculated using $V = \dfrac{a^2 c \sqrt{3}}{2}$.

[e] Calculated using $d = nM/(N_A V)$, where V is the unit cell volume (expressed in cm^3), M is the Si-C molecular weight (40.09715 g mol^{-1}), $n = 15$, and N_A is the Avogadro constant (6.0221367·10^{23} mol^{-1}).

crystallographically independent Si atoms in 6H-SiC have identical nearest-neighbor surroundings within experimental error. Each Si atom has one long (1.894 Å) Si-C bond along the stacking axis and three shorter Si-C bonds of equal (1.886 Å) length. All C-Si-C angles were between 109.4 and 109.5°. As shown in Tables II–VI, there is a tendency for the c/na ratio to increase in the polytype order 3C-6H-15R-4H-2H (i.e., with increasing hexagonality).

From the a and c lattice parameters, a unit cell volume V can be calculated using the expression

$$V = a^2 c \frac{\sqrt{3}}{2} \qquad (2)$$

For 3C-SiC, it is also natural to calculate a cubic unit cell volume from a_C^3, where a_c is the cubic lattice parameter. The density of SiC is approximately 3.2 g/cm³. In Tables II–VI, the densities, d (g/cm³), are calculated from

$$d = \frac{nM}{N_A V} \qquad (3)$$

where V is the unit cell volume according to Eq. (2) (expressed in cm³), M is the Si-C molecular weight (40.09715 g mol⁻¹), n is the number of Si-C pairs in the unit cell, and N_A is the Avogadro constant (6.0221367 · 10²³ mol⁻¹).

3. Mechanical Properties

Data related to the excellent mechanical properties of SiC are presented in Tables VII–IX.

TABLE VII

MECHANICAL PROPERTIES OF 3C-SiC

Property	Value	Reference	Comments[a]
Young's modulus (GPa)	392–448	[16, 29]	RT
Mohs hardness	≈9	[26]	
Acoustic velocity (ms⁻¹)	12,600	[4]	RT, polycrystalline

[a] RT denotes room-temperature (≈300 K) measurement.

TABLE VIII

MECHANICAL PROPERTIES OF 4H-SiC

Property	Value	Reference	Comments
Mohs hardness	≈9	[11]	
Acoustic velocity (ms⁻¹)	13,730	[25]	20 K

TABLE IX
Mechanical Properties of 6H-SiC

Property	Value	Reference	Comments[a]
Bulk modulus (GPa)	97	[43]	RT
Mohs hardness	≈9	[12]	
Acoustic velocity (ms^{-1})	13,100 ms^{-1}	[25]	5 K
	13,260 ms^{-1}	[25]	RT

[a]RT denotes room-temperature (≈300 K) measurement.

TABLE X
Thermal Properties of 3C-SiC

Property	Value	Reference	Comments[a]
Thermal conductivity (Wcm^{-1}K^{-1})	4.9	[44]	RT
Thermal expansion coefficient (K^{-1})	$2.9 \cdot 10^{-6}$	[47]	RT
Decomposition temperature (°C)	2839 ± 40	[42]	35 atm

[a]RT denotes room-temperature (≈300 K) measurement.

TABLE XI
Thermal Properties of 4H-SiC

Property	Value	Reference	Comments
Thermal conductivity (Wcm^{-1}K^{-1})	4.9	[11]	$\perp c$-axis

TABLE XII
Thermal Properties of 6H-SiC

Property	Value	Reference	Comments
Thermal conductivity (Wcm^{-1}K^{-1})	4.9	[44]	RT, $\perp c$-axis
Thermal expansion coefficient (K^{-1})	$4.2 \cdot 10^{-6}$	[47]	700 K, $\| a$-axis
	$4.68 \cdot 10^{-6}$	[47]	700 K, $\| c$-axis

III. Thermal Properties

The excellent high-temperature properties make SiC very suitable for high-temperature electronic applications. The high elastic modulus of SiC and the relatively low atomic weights of Si and C promote harmonic lattice vibrations, giving SiC a high thermal conductivity. The values at room temperature for some polytypes are given in Tables X–XII. For the variation of the thermal conductivity with doping and temperature, see also refs. [3, 30, 32].

The variation of the lattice parameters with temperature is, for instance, of importance in heteroepitaxial growth. Values of the thermal expansion coefficients for 3C- and 6H-SiC can be found in Tables X–XII. Further information about the temperature dependence on the lattice parameters for 3C- and 6H-SiC can be found in ref. [22].

Silicon carbide decomposes at high temperatures. A value of 2839°C is reported for 3C-SiC [42].

IV. Optical Properties

For many of the SiC polytypes, the optical measurements of bandgaps and dielectric functions were made more than two decades ago. Since the

TABLE XIII

OPTICAL PROPERTIES OF 3C-SiC

Property	Value	Reference	Comments[a]
Optical bandgap (eV)	2.60	[54]	LT
	2.417	[23]	LT
Optical bandgap (eV)	2.2	[35]	RT
Exciton energy gap (eV)	2.390	[5]	LT

[a]LT and RT denote low-temperature (2–8 K) and room-temperature (≈ 300 K) measurements, respectively.

TABLE XIV

OPTICAL PROPERTIES OF 2H-SiC

Property	Value	Reference	Comments[a]
Exciton energy gap	3.33	[34]	LT

[a]LT denotes low-temperature (2–8 K) measurement.

TABLE XV

OPTICAL PROPERTIES OF 4H-SiC

Property	Value	Reference	Comments[a]
Exciton energy gap (eV)	3.265	[9]	LT

[a]LT denotes low-temperature (2–8 K) measurement.

TABLE XVI

OPTICAL PROPERTIES OF 6H-SiC

Property	Value	Reference	Comments[a]
Optical bandgap (eV)	2.86	[35]	RT
Exciton energy gap (eV)	3.023	[6]	LT

[a]LT and RT denote low-temperature (2–8 K) and room-temperature (≈ 300 K) measurements, respectively.

TABLE XVII

OPTICAL PROPERTIES OF 15R-SiC

Property	Value	Reference	Comments[a]
Exciton energy gap (eV)	2.986	[33]	LT

[a]LT denotes low-temperature (2–8 K) measurement.

SiC material available today is of better quality in many respects, there should be an effort to obtain newer data.

Except for 2H- and 3C-Si-C, there are a limited number of band-structure calculations, mainly due to the rather large number of atoms in the unit cells. All polytypes investigated possess indirect bandgaps. In general, the bandgap values increase with the degree of hexagonality. Tables XIII–XVII present the values of the optical bandgaps and exciton energy gaps for the most common polytypes.

Data regarding the dielectric function (or complex refractive index) in the visible region of the electromagnetic spectrum can be found in refs. [2, 8, 21, 22, 28].

V. Electrical Properties

The SiC polytypes can be n-type doped using nitrogen or phosphorus as donors and p-type using boron, aluminum, or gallium as acceptors. In general, unintentionally doped SiC has an n-type conductivity due to nitrogen donors included as a contaminant during the production of the material. Donor and acceptor levels in the 3C, 4H, 6H, 15R, and 33R polytypes are presented for nitrogen and aluminum in ref. [22], and for nitrogen, aluminum, gallium, and boron in ref. [21].

The electron and hole mobilities in SiC are a function of the nature of the material (carrier concentration, polytype, structural quality) and temperature. There has been a tendency towards increasing mobilities and decreasing residual carrier concentrations in unintentionally doped material. A compilation of mobility data for the 3C, 4H, 6H, and 15R polytypes under various conditions can be found in ref. [21].

The high-breakdown electric field and high saturated electron drift velocity, which allow SiC to be very suitable for high-power and high-frequency applications, respectively, are presented for the 4H and 6H polytypes in Tables XVIII and XIX.

TABLE XVIII

ROOM-TEMPERATURE ELECTRICAL PROPERTIES OF 4H-SiC

Property	Value	Reference	Comments
Saturated electron drift velocity (cm s^{-1})	2.0×10^7	[13]	
Breakdown electric field (V cm^{-1})	2.2×10^6	[13]	$\|c$-axis

TABLE XIX

ROOM-TEMPERATURE ELECTRICAL PROPERTIES OF 6H-SiC

Property	Value	Reference	Comments
Saturated electron drift velocity (cm s^{-1})	2.0×10^7	[13, 49]	$\|c$-axis
Breakdown electric field (V cm^{-1})	2.5×10^6	[13]	$\|c$-axis
	$2-3 \times 10^6$	[50]	

VI. Summary

In this chapter, we have compiled some of the most important parameters of SiC. Other sources for data regarding the properties of SiC can be found in refs. [10, 18, 19, 21, 22, 28, 31, 45]. Table XX serves as a guide to other resources for more detailed studies of various SiC properties.

TABLE XX

GUIDE TO SOURCES OF VARIOUS PROPERTIES ON SiC

Property	Reference
Acceptor/donor levels	[22] (3C, 4H, 6H, 15R), [21] (3C, 4H, 6H, 15R, 33R)
Acoustic velocity	[21] (3C, 4H, 6H, 21R), [22] (3C)
Breakdown field	[21] (3C, 6H), [22] (6H)
Bulk modulus	[21] (6H), [22], [28] (α)
Carrier concentration	[21] (3C, 4H, 6H, 15R, 27R)
Carrier mobility	[21] (3C, 4H, 6H, 15R, 27R), [22] (3C, 4H, 6H, 15R), [28] (3C, 6H, 15R)
Debye temperature	[21] (3C, α), [22] (3C, 6H), [28] (3C, α)
Decomposition temperature	[28] (3C), [22]
Density	[21] (3C, 2H, 6H), [22] (3C, 6H), [28] (3C, 2H, 6H)
Dielectric constant/refractive index/absorption coefficient	[21] (3C, 2H, 4H, 6H, 15R), [22] (3C, 2H, 4H, 6H, 15R), [28] (3C, 6H), [8] (6H), [2] (3C)
Diffusion length of minority carriers	[21] (6H)
Effective mass	[21] (3C, 4H, 6H, 15R, 33R), [22] (3C, 6H, 15R), [28] (3C, 4H, 6H, 15R)
Elastic coefficient	[21] (3C, 4H, 6H), [22] (6H), [28] (3C, 6H)
Electromechanical coupling coefficient	[21] (6H)
Electro-optic coefficient	[21] (3C)
Energy bandgap	[21] (3C, 2H, 4H, 6H, 15R), [22] (3C, 6H), [28] (3C, 2H, 6H)
Energy levels of impurities	[21] (3C, 4H, 6H, 15R, 33R)
Exciton binding energy	[22] (3C)
Exciton energy bandgap	[21] (3C, 2H, 4H, 6H, 15R, 21R, 33R), [22] (3C, 2H, 4H, 6H, 8H, 15R, 21R, 24R), [28] (3C, 4H, 6H, 8H, 15R, 21R, 24R, 33R)
Gauge factor	[21] (3C, 6H)
Hardness	[28] (3C, α) [22] (3C, 6H)
Interband transitions	[22] (3C)
Lattice parameter	[21] (3C, 2H, 4H, 6H, 15R, 21R, 33R), [22] (3C, 2H, 4H, 6H, 8H, 10H, 15R, 21R, 33R), [28] (3C, 2H, 4H, 6H, 8H, 15R, 21R, 33R, 66H, 126H)
Luminescence lines	[7] (3C, 4H, 6H, 15R, 33R)
Magnetic susceptibility	[21] (6H), [22] (6H), [28] (6H)

TABLE XX. CONTINUED

Property	Reference
Phonon modes	[21] (3C, 4H, 6H, 15R), [22] (3C, 2H), [28] (3C)
Piezoelectric coefficient	[21] (6H), [28] (6H)
Piezoresistive effect	[21]
Saturated electron drift velocity	[21] (6H), [22] (6H)
Shear modulus	[28] (α)
Specific heat	[21] (3C, α), [28] (3C, α)
Spin–orbit splitting energy	[22] (3C, 6H, 15R)
Spectral emissivity	[21], [28]
Thermal conductivity	[21] (3C, 4H, 6H), [22] (6H), [28] (3C, α)
Thermal EMF	[21], [28]
Thermal expansion	[28] (3C, α)
Thermal expansion coefficient	[28] (6H), [22] (3C, 6H)
Work function	[21] (3C, 6H), [22] (6H), [28] (α)
Young's modulus	[21] (3C)

ACKNOWLEDGMENT

The authors acknowledge the support of the Office of Naval Research under contract N00014-92-J-1477. K. Järrendahl acknowledges The Swedish Foundation for International Cooperation in Research and Higher Education (STINT) for financial support.

REFERENCES

1. Adamsky, R. F. and Merz, K. M., *Zeit. Kristallogr.*, 1959. **111**, 350–356.
2. Alterowitz, S. A. and Woollam, J. A., *Cubic silicon carbide (β-SiC)*, in *Handbook of Optical Constants of Solids II*, E. D. Palik, Editor. 1991, Academic Press, Harcourt Brace Jovanovich, Publishers: Boston. pp. 705–707.
3. Burgermeister, E. A., vonMünch, W., and Pettenpaul, E., *J. Appl. Phys.*, 1979. **50**, 5790.
4. Chinone, Y., Ezaki, S., Fuijta, F., and Matsumoto, R., *Springer Proc. Phys.*, 1989. **43**, 198.
5. Choyke, W. J., Hamilton, D. R., and Patrick, L., *Phys. Rev.*, 1964. **133**, A 1163.
6. Choyke, W. J., Hamilton, D. R., and Patrick, L., *Phys. Rev.*, 1965. **139**, A 1262.
7. Choyke, W. J. and Linkov, I., *A short atlas of luminescence and absorption lines and bands in SiC, GaN, AlGaN, and AlN*, in *Silicon Carbide and Related Materials*, M. G. Spencer *et al.*, Editors. 1993, Institute of Physics Publishing: Bristol. pp. 141–146.
8. Choyke, W. J. and Palik, E. D., *Silicon Carbide (SiC)*, in *Handbook of Optical Constants of Solids*, E. D. Palik, Editor. 1985, Academic Press: Orlando. p. 587.
9. Choyke, W. J., Patrick, L., and Hamilton, D. R., in *Proc. Int. Conf. on Semiconductor Phys. Paris.*, M. Hulin, Editor. 1964, Vieweg & Sohn: Braunschweig. p. 751.
10. Choyke, W. J. and Pensl, G., MRS Bulletin, 1997. **22**, 25.

11. Cree Research Inc., "Properties and Specifications for 4H-Silicon Carbide" data sheet, Rev. 10.95, 1995.
12. Cree Research Inc., "Properties and Specifications for 6H-Silicon Carbide" data sheet, Rev. 10.95, 1995.
13. Cree Research Inc., "Physical and Electronic Properties of Silicon Carbide" data sheet, Rev. 10.95, 1995.
14. DeMesquita, A. H. G., *Acta. Cryst.*, 1967. **23**, 610.
15. Donnay, J. D. H., in *Crystal Data, Determinative Tables, 2nd ed.* 1963, American Crystallography Assoc. pp. 65–66.
16. Fekade, K., Su, Q. M., Spencer, M. G., and Wuttig, M., *Dynamic characterization of mechanical properties of 3C epitaxial SiC*, in *Silicon Carbide and Related Materials*, M. G. Spencer, *et al.*, Editors. 1993, Institute of Physics Publishing: Bristol. pp. 189–192.
17. Fisher, G. R. and Barnes, P., *Phil. Mag. B*, 1990. **61**, 217–236.
18. Freer, R., ed. *The Physics and Chemistry of Carbides, Nitrides and Borides.* E: Applied Sciences. Vol. 185. 1990, Kluwer Academic Publishers: Dordrecht.
19. Guth, J. and Petuskey, W. T., *J. Phys. Chem. Solids*, 1987. **48**, 541–549.
20. Hannam, A. L. and Schaffer, P. T. B., *J. Appl. Cryst.*, 1969. **2**, 45.
21. Harris, G. L., ed. *Properties of Silicon Carbide.* EMIS datareview series, ed. B. L. Weiss. Vol. 13. 1995, INSPEC: London.
22. Hellwege, K.-H., ed. *Landolt-Börnstein, Numerical Data and Functional Relationships in Science and Technology.* Crystal and Solid State Physics, Semiconductors, ed. M. S. O. Madelung, H. Weiss. Vol. 17a. 1982, Springer-Verlag: Berlin.
23. Humphreys, R. G., Bimberg, D., and Choyke, W. J., *J. Phys. Soc. Jpn.*, 1980. **49 Suppl. A**, 519.
24. Jagodzinski, H. V., *Acta. Cryst.*, 1949. **2**, 208.
25. Karmann, S., Helbig, R., and Stein, R. A., *J. Appl. Phys.*, 1989. **66**, 3922.
26. Kern, E. L., Hamill, D. W., Deem, H. W., and Sheets, H. D., *Mater. Res. Bull.*, 1969. **4**, S25.
27. Krishna, P., ed. *Crystal Growth and Characterization of Polytype Structures.* Progress in Crystal Growth and Characterization, ed. B. R. Pamplin. Vol. 7. 1983, Pergamon Press: Oxford.
28. Marshall, R. C., Jr., J. W. F., and Ryan, C. E., eds. *Silicon Carbide 1973.* 1974, University of South Carolina Press: Columbia.
29. Matus, L. G., Tang, L., Mehregany, M., Larkin, D. J., and Neudeck, P. G., *Silicon Carbide as a novel material for micromechanical applications*, in *Silicon Carbide and Related Materials*, M. G. Spencer *et al.*, Editors. 1993, Inswtitute of Physics Publishing: Bristol. pp. 185–188.
30. Morelli, D., Hermans, J., Beetz, C., Woo, W. S., Harris, G. L., and Taylor, C., *Carrier concentration dependence of the thermal conductivity of silicon carbide*, in *Silicon Carbide and Related Materials*, M. G. Spencer *et al.*, Editors. 1993, Institute of Physics Publishing: Bristol. pp. 313–316.
31. Morkoç, H., Strite, S., Gao, G. B., Lin, M. E., Sverdlov, B., and Burns, M., *J. Appl. Phys.*, 1994. **76**, 1363–1398.
32. Parafenova, I. I., Tairov, Y. M., and Tsvetkov, V. F., *Sov. Phys. Semicond.*, 1990. **24**(2), 158–161.
33. Patrick, L., Hamilton, D. R., and Choyke, W. J., *Phys. Rev.*, 1963. **132**, 2023.
34. Patrick, L., Hamilton, D. R., and Choyke, W. J., *Phys. Rev.*, 1966. **143**, 526.
35. Philipp, H. R. and Taft, E. A., in *Silicon Carbide — A High Temperature Semiconductor*, J. R. O'Connor and J. Smiltens, Editors. 1960, Pergamon Press: Oxford. p. 366.
36. Powder diffraction file 29-1130, JCPDS-ICDD International Center for Powder Diffraction Data, Swarthmore, PA.

37. Powder diffraction file 29-1129, JCPDS-ICDD International Center for Powder Diffraction Data, Swarthmore, PA.
38. Powder diffraction file 29-1127, JCPDS-ICDD International Center for Powder Diffraction Data, Swarthmore, PA.
39. Powder diffraction file 29-1131, JCPDS-ICDD International Center for Powder Diffraction Data, Swarthmore, PA.
40. Powell, J. A., Pirouz, P., and Choyke, W. J., *Growth and characterization of silicon carbide polytypes for electronic applications*, in *Semiconductor Interfaces, Microstructures and Devices. Properties and Applications*, Z. C. Feng, Editor. 1993, Institute of Physics Publishing: Bristol. pp. 257–293.
41. Ramsdell, L. S., *Amer. Mineral.*, 1947. **32**, 64.
42. Scace, R. I. and Slack, G. A., in *Silicon Carbide — A High Temperature Semiconductor*, J. R. O'Connor and J. Smiltens, Editors. 1960, Pergamon Press: Oxford. p. 24.
43. Shaffer, P. T. B., in *Plenum Press Handbook of High Temperature Materials No. 1*. 1964, Plenum Press. p. 107.
44. Slack, G. A., *J. Appl. Phys.*, 1964. **35**, 3460.
45. Somiya, S. and Inomata, Y., eds. *Silicon Carbide Ceramics — 1. Fundamental and Solid Reaction*. 1991, Elsevier Applied Science: London.
46. Tairov, Y. M. and Tsvetkov, V. F., *Progress in controlling the growth of polytypic crystals*, in *Crystal Growth and Characterization of Polytype Structures*, P. Krishna, Editor. 1983, Pergamon Press: Oxford. pp. 111–162.
47. Taylor, A. and Jones, R. M., in *Silicon Carbide — A High Temperature Semiconductor*, J. R. O'Connor and J. Smiltens, Editors. 1960, Pergamon Press: Oxford. pp. 147–156.
48. Thibazult, N. W., *Am. Mineral.*, 1944. **29**, 327.
49. vonMünch, W. and Pettenpaul, E., *J. Appl. Phys.*, 1977. **48**, 4823.
50. vonMünch, W. and Pfaffeneder, I., *J. Appl. Phys.*, 1977. **48**, 4831.
51. Wyckoff, R. W. G., *The Structure of Crystals*. 1924, New York: The Chemical Catalog Company, Inc.
52. Wyckoff, R. W. G., in *Crystal Structures, Vol. 1*. 1963. Wiley & Sons: New York. p. III.35.
53. Zhadanov, G. R., *Compte Rende Acad Sci. URSS*, 1945. **48**, 39.
54. Ziomek, J. S. and Pickar, P. B., *Phys. Stat. Solidi*, 1967. **21**, 271.

CHAPTER 2

SiC Fabrication Technology: Growth and Doping

V. A. Dmitriev

A. F. IOFFE INSTITUTE
ST. PETERSBURG, RUSSIA

M. G. Spencer

MATERIALS SCIENCE RESEARCH CENTER OF EXCELLENCE
HOWARD UNIVERSITY
WASHINGTON, DC

I. INTRODUCTION	21
II. SiC BULK CRYSTAL GROWTH	22
1. *Bulk Growth by Physical Vapor Transport*	22
2. *Bulk Growth by Chemical Vapor Deposition*	34
3. *Bulk Growth from the Liquid Phase*	34
III. SiC EPITAXIAL GROWTH	35
1. *Chemical Vapor Deposition*	36
2. *Liquid-Phase Epitaxy*	50
3. *Sublimation Epitaxy*	56
4. *Molecular Beam Epitaxy*	58
IV. SiC DOPING	60
1. *Impurities in SiC*	60
2. *Doping During Crystal Growth*	60
3. *Diffusion of Impurities in SiC*	61
4. *Ion Implantation*	61
V. CONCLUSIONS	62
References	63

I. Introduction

Silicon carbide (SiC) has long been a favorite material because of its combination of semiconducting and refractory properties. Unfortunately, interest in SiC waned in the seventies due to the lack of a commercial product. In the late 1980s and early 1990s, relatively high-quality large-area 6H-SiC substrates became commercially available. Devices produced on these substrates began to show performance levels that in some cases exceeded those of GaAs or Si in high-power or high-temperature applications. These successes spurred a dramatic revitalization of interest in SiC. This chapter will review SiC growth and doping technology, which has undergone tremendous development in recent years.

II. SiC Bulk Crystal Growth

SiC substrates are the key elements in the development of SiC electronics. Because of the phase equilibria in the Si and C materials system (specifically, the material sublimes before it melts) the most popular bulk growth techniques are based on physical vapor transport. These techniques were initially developed in the late fifties and have been modified and introduced for production in the early eighties. Although sublimation techniques are relatively easy to implement (at the high-growth temperatures required), these processes are difficult to control, particularly over large substrate areas. Three types of processes have been used commercially for the bulk growth of SiC: (1) the Acheson process [8]; (2) the Lely process [107]; and (3) the Modified Lely process [196]. The latter method (which was initiated at Leningrad Electrotechnical Institute for bulk SiC growth and at Ioffe Institute for epitaxial SiC growth has been adopted by many researchers, including research teams at Westinghouse Corp. [13] (now a part of Northrop Grumman Corp.), Cree Research Corp. [49], Siemens AG [190], Advanced Technology Materials Inc. [17], Sanyo Electric Co. [89], Nippon Steel Corporation [156], Kyoto University [143], Linkoping University [203], and Howard University [185]. Chemical vapor deposition (CVD) and liquid-phase techniques have also been investigated for bulk growth of SiC. Most studies of CVD and liquid-phase bulk growth were done in the late sixties and early seventies. Although these growth approaches initially produced disappointing results, both techniques are receiving renewed attention.

1. BULK GROWTH BY PHYSICAL VAPOR TRANSPORT

Physical vapor growth is accomplished by the sublimation of an SiC source placed in the hot zone of the growth furnace and the subsequent mass transport of the vapor species to a cooler region of the furnace. Single-crystal SiC material is formed from deposition of the supersaturated vapor species. Source materials may be composed of SiC powder, Si and C powders mechanically mixed, or crystalline SiC. The vapor transport is performed in either a vacuum or gas ambient. Single-crystal growth can be realized either seeded or unseeded. Typical temperature and pressure ranges for SiC sublimation growth are 1600 to 2700°C and 10^{-6} to 20 torr, respectively. Usually, the lower temperatures are employed for sublimation epitaxial SiC growth, while bulk SiC growth is performed at the higher temperatures.

a. Acheson Process

In the Acheson process, a mixture of silica and carbon with a small percentage of sawdust and common salt (e.g., coke 40%, silica 50%, sawdust 7%, and salt 3%) are heated in a trough-type electic furnace. The process temperature is about 2700°C. The result is a combination of crystallites and polycrystalline agglomerates in all states of perfection. Because of dirty starting material, crystals are heavily contaminated. Large amounts of SiC, principally for grinding applications, were produced by the Acheson process. Acheson SiC crystals have been used in research as substrates for CVD and liquid phase epitaxy (LPE).

b. Lely Process

The first technique for the production of semiconducting crystalline SiC was the Lely method, which was published in 1955 [107]. In the Lely technique, SiC powder source is placed in a porous graphite sleeve and a temperature gradient is impressed across the sleeve. The growth is conducted at a source temperature between 2200 and 2700°C. SiC crystals are nucleated on the graphite sleeve, and growth proceeds along the basal plane (which is parallel to the crystallographic *a*-axis) [88]. Nucleation of the SiC may be controlled by holes in the graphite sleeves [51]. As a result of one growth run, crystals are produced with a distribution in size and polytype. 6H-SiC crystals make up to 75 to 95% of the crystals obtained in a single growth run. Other polytypes, such as 4H and 15R, are also found in each growth but in much less quantity. Polytypes such as 8H, 21R, and 33R are rarely found. The "as-grown" Lely crystals have a pyramid shape (Fig. 1). After the growth, the top portion of the pyramids must be ground away. The resulting platelet is 0.3 to 0.35 mm thick and usually has a hexagonal shape characteristic of the Lely crystal. Because of the initial pyramid shape, one face of the crystal is always larger than another. The larger, as-grown bottom face of the pyramid is a well-oriented (0001) surface that can be either a carbon or silicon face. Doping concentration in Lely crystals ranges from 10^{16} to 10^{18} (high) cm^{-3}. Materials grown by the Lely process are of high crystal quality and often are completely free of micropipe defects (discussed

FIG. 1. Optical photograph of as-grown 6H-SiC Lely crystal (left bottom), 6H-SiC Lely substrate (left top), and 4H-SiC commercial wafer grown by modified Lely method.

later). Because of the high quality of the Lely platelets, it is possible to select crystals with dislocation-free regions extending over a few square millimeters (Fig. 2). Numerous types of SiC electronic devices have been demonstrated on Lely crystals (see reviews [7, 33] and references therein). Lely crystals are still used for laboratory research. However, due to the irregularity of the platelet shape as well as the small size (< 20 mm), this material is not viable for large-scale manufacture.

c. Modified Lely Process

Growth of SiC boules is possible using the modified, or seeded, Lely method (Fig. 3), also called the Tairov-Tsvetkov method [196, 197] or physical vapor transport technique. In this technique, a seed crystal of SiC is introduced into the Lely chamber, and growth proceeds (usually along the c-axis) by vapor transport of carbon and silicon bearing species from the source (or carbon species from the graphite walls). For a typical 6H- and 4H-SiC bulk sublimation growth process, the SiC source temperature is 2100 to 2400°C [12], growth pressure is less than 20 torr, and the temperature gradient between source and seed ranges from 20 to 35°C/cm.

2 SiC FABRICATION TECHNOLOGY: GROWTH AND DOPING 25

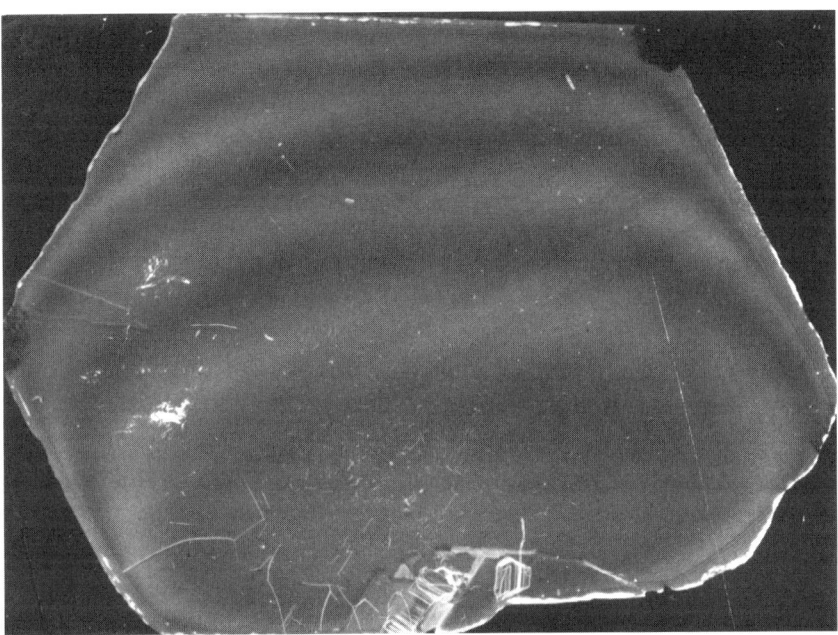

FIG. 2. Transmission x-ray topography image obtained by Lang method ($11\bar{2}0$) reflection, MoK$_\alpha$ radiation for 6H-SiC Lely substrate. Crystal size is approximately 0.7×1 cm^2.

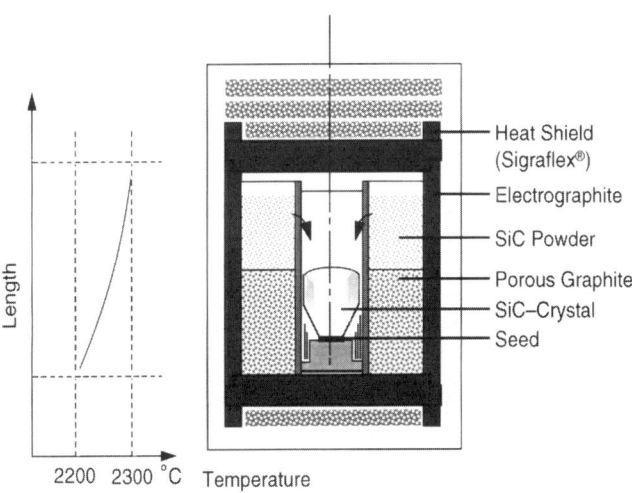

FIG. 3. Crucible assembly and temperature distribution in the growth zone for the growth of SiC single crystals by the modified Lely method. (After Glasgow [48a].)

Currently, the maximum commercially available 4H- and 6H-SiC crystals are 50 mm in diameter; however, record 3-in. boules of SiC have been grown [19]. Sublimation growth is usually performed on {0001} (Si or C face) of either 6H- or 4H-SiC seeds. Growth on faces perpendicular to the (0001) basal plane, in an effort to reduce micropipe defects, has also been reported [194].

Equipment. Sublimation growth is performed in high-temperature resistively or inductively heated furnaces. Water-cooled quartz or stainless steel chambers are used for the growth apparatus. Crucibles and other inside parts are made from high-purity graphite. Two types of crucible designs have been reported. In one version, the vapor source is separated from the growing crystal by a thin-walled cylinder of porous graphite [224]. In the second implementation, the seed crystal is placed directly above the source material [197]; this method is similar in geometry to the epitaxial "sandwich technique" [206].

The source material is (1) SiC powder, (2) SiC polycrystalline material, or (3) SiC single crystals (platelets or small pieces). The starting SiC powder may be obtained by pulverizing single or polycrystalline SiC; however, crushing of the powder introduces significant contamination from shavings of the pulverizing tool as well as residual mechanical stress in the powder (which affects the evaporation rate). An alternate approach to obtaining SiC powder is direct synthesis from high-purity Si and C powders. SiC powder sources may be synthesized with either stoichiometric, Si-rich, or C-rich compositions by adjusting the time and temperature of the synthesis reaction of Si and C. The highest purity source powder is obtained directly from gas-phase nucleation [49].

Vapor phase equilibria in SiC. The equilibrium gas species over SiC have been measured by Drowart *et al.* [31], Drowart and DeMaria [32] (Fig. 4), and Behrens and Rinehard [14]. The principal molecular species are Si, SiC_2, and Si_2C. In the SiC vapor system, carbon is transported by the Si bearing compounds. The vapor pressure of all the major species is related to the pressure of Si. In thermal equilibrium, the vapor equilibria over SiC have only one degree of freedom. If any of the species are specifid by the temperature, source powder composition, and/or crucible wall interactions, the remaining species are uniquely determined. Because Si bearing species are the most prevalent, it is usually convenient to discuss the vapor composition in terms of Si partial vapor pressure. Using the experimental data together with thermodynamic calculations, the relationship between

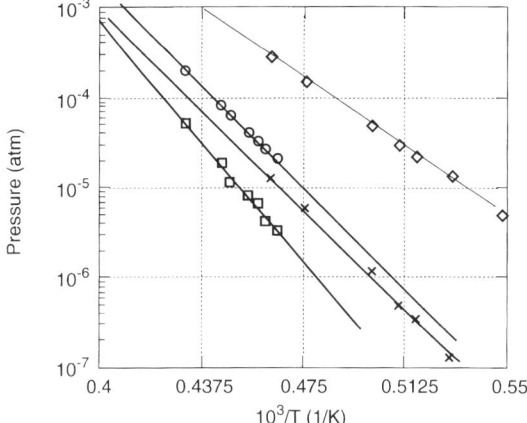

FIG. 4. Vapor pressure of the main vapor components over SiC for SiC-C and SiC-Si equilibrium systems: diamonds, Si vapor pressure for SiC-Si system; circles, Si vapor pressure for SiC-C systems; crosses, Si$_2$C vapor pressure for SiC-Si system; squares, C species vapor pressure for SiC-C system [31, 32].

the partial pressure of Si and that of Si$_2$C and SiC$_2$ for the SiC-C vapor equilibria is given by Eqs. (1) and (2) [49]:

$$P_{Si_2C} = 2.85 \times 10^2 e^{(-1.79 \times 10^4/T)} \times P_{Si} \qquad (1)$$

$$P_{SiC_2} = 9.41 \times 10^{28} e^{(-14.35 \times 10^4/T)} \frac{1}{P_{Si}} \qquad (2)$$

The partial pressure of the vapor species varies dramatically between the SiC-C equilibrium and the SiC-Si equilibrium. However, due to the large amount of carbon present in the growth chamber of most systems, the vapor interactions with the graphite walls of the crucible usually will cause the system to operate close to the SiC-C part of the curve. Since the composition of the vapor has such a large allowed existence region, the composition of the powder has important consequences for the vapor composition and the growth rate. Source graphitization is also an important problem for bulk growth. Graphitization occurs under growth conditions in which it is possible to preferentially lose Si and obtain a carbon layer over the source, which prevents further sublimation of the source. This has been treated in detail by Karpov et al. [76]. If SiC powder is used as a vapor source for bulk SiC growth, the powder composition can be adjusted (by the addition of excess Si or C to the SiC powder) to ensure a stoichiometric growth of SiC on the seed without formation of Si or C inclusions. This technique of vapor control has a number of practical problems associated with the nature of liquid Si, which can be precipitated in the system and can cause graphite parts to stick or clog. Another approach for stabilizing the Si/C ratio during the growth using gettering by refractory metals has been suggested by Konstantinov and Ivanov [92]. In their approach, a Ta getter is added to the growth system. The Ta continuously forms stable TaC, thus removing carbon from the gas phase and creating a constant Si partial pressure (making up for that which is lost through openings in the system) by providing additional Si freed from the Si_2C or SiC_2 molecule upon reaction with the Ta.

Growth rate. Typical growth rates for the bulk growth of SiC are in the range of 0.5 to 5.0 mm/hr. Figure 5 shows data for the growth of bulk SiC as a function of temperature and pressure. The general theory for vapor-phase growth in the presence of a known chemical equilibrium was developed in detail by Lever [108–111] and Mandel [115–117]. (This theory provides the basis for SiC sublimation growth.) In this development,

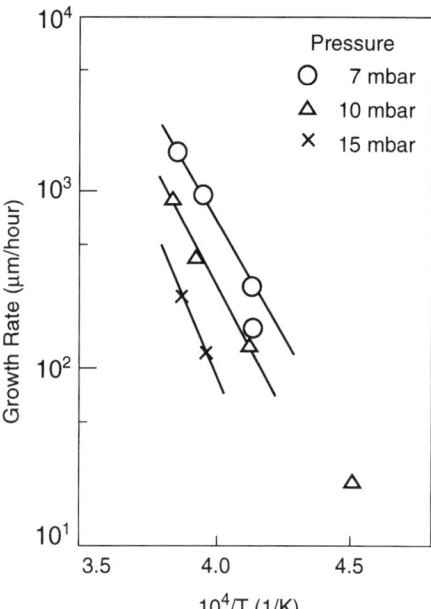

FIG. 5. Dependence of SiC growth rate on the growth temperature for modified Lely method. (After Nakata *et al.* [137a].)

the growth rate is determined by the transport of the species from the source to the seed by diffusion or the surface kinetics. Growth due to surface kinetics are treated in terms of the supersaturation of the vapor above the growth surface and "crystal impedance." The theory developed by Lever and Mandel was applied to the SiC growth system by Lilov *et al.* [114], assuming that the vapor source is not graphitized. If the growth rate is limited by the rate of diffusive transport of the gases species to the growth surface, the contribution of surface kinetics can be ignored, and the growth is determined by Fick's Law. The diffusion constant for the various gas species in the growth ambient are calculated using the kinetic theory of ideal gases. These diffusion constants are inversely proportional to the ambient pressure in the growth system. The transport limited growth rate can be controlled dynamically by changing the ambient gas pressure, which determines the coefficient of diffusion for the Si and carbon bearing species through the ambient gas. The maximum flux of the growth species transported can be estimated using the Knudsen equations.

Structural defects. There has been significant progress in the quality of material produced by seeded sublimation technology [56, 201]. Dislocation density and micropipe density in SiC bulk crystals grown by the modified Lely method currently range from 10^4 to 10^5 cm^{-2} and 10^2 to 10^3 cm^{-2}, respectively (Fig. 6). Also, record low micropipe densities of less than 1/cm^2 have been reported [201]. Other types of crystal defects found in sublimation-grown SiC crystals include basal plane tubes, cracks, and crystal domains [123, 127, 202]. However, micropipes are defects unique to the growth of SiC. These micropipes, which are physical holes, can travel large distances in the crystal. It was shown that micropipes are "killer" defects if they intersect the active regions of a device [139]. The density of the micropipes could be correlated with domains in the crystals [50]. These domains are thought to be related to growth spirals that are nucleated and interact. The relationship between micropores and growth spirals was originally suggested by Frank [45], and later papers on the subject by Krishna *et al.* [99] have expanded on this theme. More recently, x-ray studies by Fazi *et al.* [41] also have suggested this mechanism. There are, however, other views as to the origin of the micropipe defects. Some have observed that these defects are nucleated at the seed. Other investigators suggest that impurities or inclusions (or Si droplets) are responsible for micropipe formation. Support of alternate views of micropipe formation comes from the experimental observation that many of these defects are nucleated in the bulk of the materials and are not continuous. It is generally conceeded that micropipes are the most important current obstacle to the production of high-quality SiC devices. It remains to be seen whether this

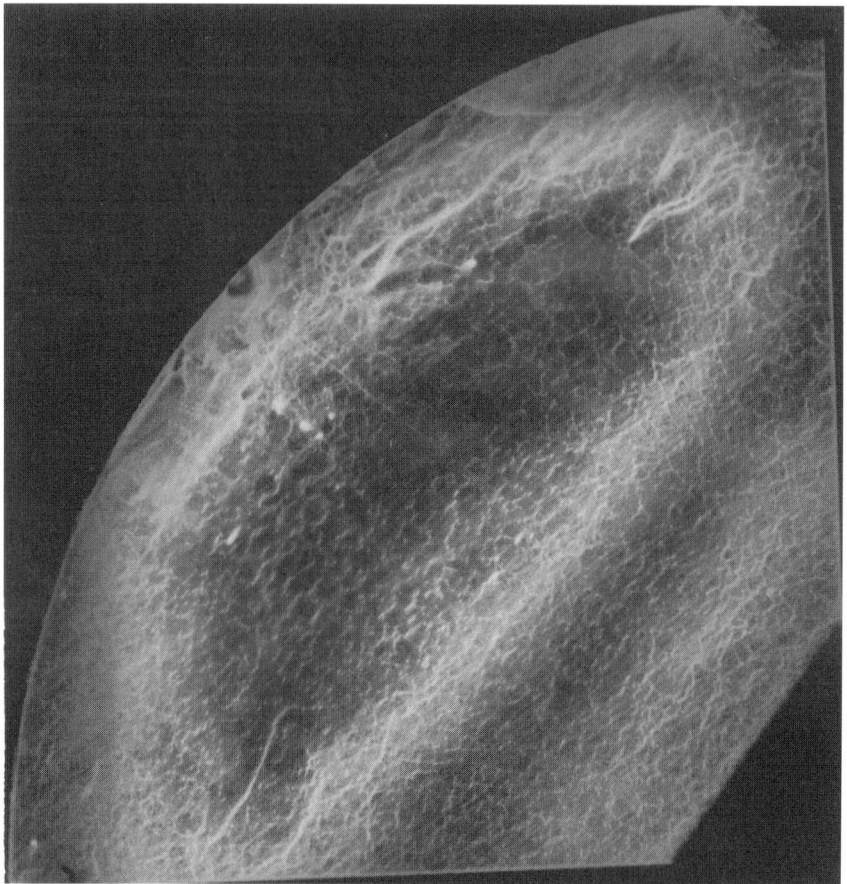

Fig. 6. Transmission x-ray topography image obtained by Lang method (11$\bar{2}$0 reflection, MoK$_\alpha$ radiation) for a part of $\frac{1}{4}$ of 30 mm diameter 6H-SiC substrate grown by modified Lely method.

is a technological problem or a fundamental property of crystal growth of large-unit cell materials. The possible mechanisms of micropipe formation are summarized by Tsvetkov *et al.* [201].

X-ray and transmission electron microscopy (TEM) techniques have been applied to the study of the quality of bulk SiC [13]. Lely platelets as well as substrates grown by the modified Lely method have been investigated. In general, the results of various investigations suggest that Lely crystals exhibit higher material quality. The full width at half maximum (FWHM) of the x-ray rocking curve is ~ 7 to 10 arc sec for Lely platelets, while for modified Lely crystals, this value ranges from 20 to 100 arc sec. For Lely crystals, a single x-ray diffraction peak is observed for each x-ray reflection. Conversely, for modified Lely crystals, several peaks are observed, indicating a mosaicity in the material.

Polytype control. A key question for the growth of bulk SiC is polytype control. If special precautions are not taken, SiC crystals grown by sublimation will contain inclusions of undesirable polytypes. Several technological parameters impact the final polytype structure of SiC crystals. These parameters include supersaturation of the vapor above growing surface, growth temperature, growth pressure, seed surface orientation and polarity, and presence of impurities.

Polytype control during the nucleation stage of the deposition was demonstrated by Tairov and Tsvetkov [198]. These researchers investigated the sublimation growth of SiC in argon and in vacuum. At the beginning of the growth process, the sublimation cell was heated up in Ar atmosphere to prevent SiC formation. When the growth temperature (1900–2000°C) was reached, ar pressure was reduced, initiating SiC growth. The polytype structure of the layer was found to be determined by pumping speed. The Ar pressure during the pumping (P) may be describe as a function of time (t) by the following expression:

$$P = P_0 e^{-t/\tau} + P_g \qquad (3)$$

where $P_0 + P_g$ is initial Ar presssure, P_g is the final Ar pressure after the pumping, and τ is pumping-speed time constant. The yield of different polytypes depended on the τ value (Fig. 7). At a small τ value, which corresponds to a low supersaturation and low initial growth rates, the 4H-SiC polytype was grown on 6H-SiC substrate. Conversely, at large values of τ, 3C-SiC was formed. Intermediate values of τ produced 6H material.

The polarity of the seed material is also important in determining whether polytype transformation will occur. All reported experiments in which the polytype was transformed during the nucleation stage were performed on carbon face substrates. Stein *et al.* [189] and Vodakov *et al.* [208] emphasized the importance of surface energy in polytype transformation. Impurities have been found to affect polytype formation as well. For example, the addition of nitrogen promotes the formation of 3C-SiC [113]. Vodakov and co-workers [206, 207] used impurities to influence the growth of 4H-SiC on 6H-SiC. The effect of growth temperature on 4H-SiC formation on 6H-SiC substrate was pointed out by Kanaya *et al.* [70]. These authors concluded that lower seed temperature (2200°C), lower growth pressure (10 torr), and

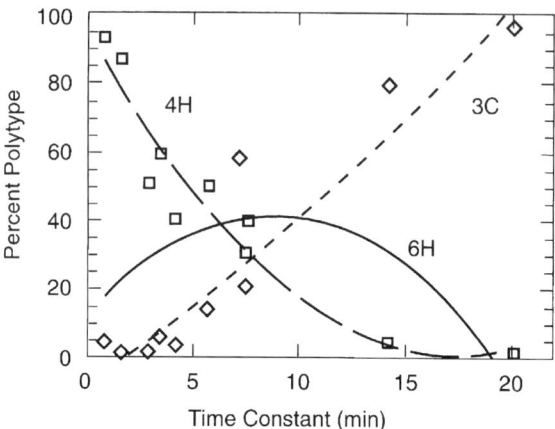

FIG. 7. Dependencies of SiC crystal yield of 3C, 6H, and 4H polytypes on the deposition kinetics in the initial stage of growth. (After Tairov and Tsvetkov [198].)

larger temperature gradient resulted in a higher yield of 4H-SiC crystals (80–85%). In commercial production, 6H- and 4H-SiC appear to be free of polytype inclusions. Polytype control for bulk 3C growth also appears possible. However, detailed investigations of the polytype purity over entire wafers have not been reported.

Electrical characteristics. Undoped SiC bulk material usually is contaminated by nitrogen, which produces n-type conductivity [17]. The typical background level of electron concentration at room temperature for undoped SiC crystals grown by the modified Lely method is 10^{16} to 10^{17} cm^{-3}. High-purity undoped SiC crystals with room temperature resistivity from 10^2 to 10^3 Ω cm have been reported by Hobgood et al. [55]. These crystals had p-type conductivity with a background carrier concentration of 10^{15} cm^{-3} due to residual boron impurities.

N-type SiC crystals with carrier concentrations up to 10^{20} cm^{-3} were produced using nitrogen doping [113, 156] (Fig. 8). The minimum reported resistivity for 6H- and 4H-SiC bulk crystals are 1.6 and 2.8 mΩ cm, respectively [49]. No information on defect density in highly doped n-type material is available. P-type SiC crystals with carrier concentrations up to 10^{20} cm^{-3} were obtained using aluminum doping [49]. Information about Al doping during bulk SiC growth is limited. No information on defect density in highly doped p-type material is available.

Semi-insulating 6H-SiC crystals were produced using vanadium doping

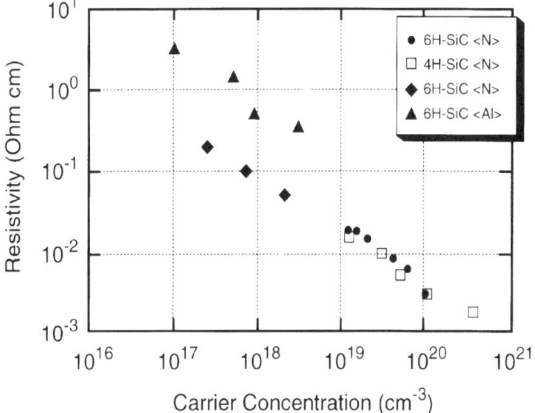

FIG. 8. Resistivity (300 K) versus doping concentration for SiC crystals grown by modified Lely method [49, 137a].

[56]. The amount of vanadium soluble in the material is limited by the precipitation of vanadium silicide. The material resistivity at room temperature can be estimated by high-temperature resistivity measurements and was determined to be in the range of $10^{15}\,\Omega$ cm. Semi-insulating 4H-SiC crystals with comparable resistivities have also been reported [49, 201].

3C-SiC bulk growth by physical vapor transport. The 3C polytype has been considered metastable and therefore difficult or impossible to grow in bulk form. However, it was found that if the quality of the 3C seed material is high, its tendency to transform to 6H during the growth is much less. Some success has been obtained in the growth of bulk 3C-SiC crystals using the modified Lely technique [67, 185, 217, 222]. Crystals of 3C-SiC betwen 100 and 400 µm thick have been grown by sublimation on 3C-SiC (100) CVD seeds at a growth rate of 100 µm/hr by Shields *et al.* [185]. 3C-SiC crystals have been grown by Furukawa *et al.* [45a] at a growth rate of 800 µm/hr. The substrates used were relatively thick 3C-SiC films previously deposited on silicon or 6H-SiC by CVD. Experiments performed on the (100) plane of 3C-SiC used the seeds produced on Si substrates by CVD, while experiments on the (111) plane utilized 3C-SiC seeds produced on 6H-SiC substrates. If SiC seeds grown by CVD on Si were used for substrates, the Si was chemically etched away before sublimation growth. The growth temperature and pressures for 3C-SiC bulk growth ranged from

1800 to 1900°C and 50 to 760 torr, respectively. 3C-SiC crystals 10 mm in diameter and ~0.7 mm thick have been grown by Nishino et al. [143]. Generally, bulk 3C-SiC grown by the modified Lely method suffers from high defect density. To have commercial process, the growth rate and crystal quality of bulk 3C-SiC must be improved.

2. BULK GROWTH BY CHEMICAL VAPOR DEPOSITION

High-quality bulk 3C-SiC crystals have been grown by CVD [52]. The crystals were deposited on resistively heated graphite rods 6 mm in diameter using a CH_3SiCl_3-H_2 gas mixture. The growth temperature was 1650 to 1750°C. The larger crystals were about 3 mm thick and $\leqslant 10$ mm wide. The material exhibited n-type conductivity with electron concentrations at room temperature between 10^{14} and 10^{17} cm^{-3}. In some growth runs, 3C-SiC crystals were doped with nitrogen and boron. Despite the high quality of these crystals [130], the method must be significantly modified to produce large-area crystals.

6H-SiC crystals have been grown by high-temperature CVD with growth rates of 0.5 mm/hr [98]. This technique is a direct adaptation of the CVD epitaxial technique for growth of SiC. The control of the Si/C ratio in such a system is excellent compared with sublimation growth; however, more research is needed to understand the possibilities of this technology for bulk SiC growth.

3. BULK GROWTH FROM THE LIQUID PHASE

Crystal growth from melt solutions is widely used for many semiconductor materials. Liquid-phase growth of SiC has not been considered promising, based on results obtained during the early stage of SiC development. Two main objections to liquid-phase technology are low solubility of SiC in the Si melt (which limits the growth rate) and inclusions incorporated in grown crystals due to parasitic phase formation. On the other hand, it has been shown that high-quality SiC can be grown from melt solutions [118, 138, 213]: undoped cubic SiC crystals grown from a Si melt at ~1500°C exhibited a mobility of 980 cm^2/V s and a carrier concentration of 4×10^{16} cm^{-3} (300 K). Significant progress has also been made in understanding the nature of SiC growth from liquids [39, 216]. It has been shown that high-quality SiC epitaxial layers with no micropipes and low dislocation density can be grown from an Si melt [173]. Epitaxial layers of 6Hand 4H-SiC were grown from Si melts on 35-mm diameter substrates, indicating the possibility of liquid-phase growth of large crystals. It was also found that the solubility of SiC in Si melt does not severely limit the growth rate, and an SiC growth rate of about ~ 0.2 mm/hr has been obtained at 1650°C.

Progress in SiC device development became possible due to improvements in the quality of bulk SiC growth. Physical vapor transport (sublimation) is the accepted technique for bulk growth of silicon carbide. The so-called modified Lely technique (seeded sublimation method) initiated by Tsvetkov and Tairov is currently the method of choice for bulk SiC growth. 4H- and 6H-SiC crystals with 50-mm diameter are available, and record crystals up to 75 mm have been demonstrated. Using nitrogen and aluminum doping, both n- and p-type conductivity can be controlled for bulk 4H and 6H crystals over a wide range. Semi-insulating SiC wafers are also available. However, for high-power devices, which require low-resistivity material combined with low defect densities, more improvement is needed. Additionally, to expand the market for SiC, large-diameter substrates are desired. Unfortunately, the defect issues (dislocations, micropipes, etc.) are expected to become more severe for larger diameter SiC crystals (3- to 4-in. diameter) [57].

3C-SiC bulk crystals have been grown by sublimation but with smaller size and inferior crystal quality. The continued progress in this polytype may result in commercialization of 3C-SiC wafers in a few years. But this will require consistent effort. Recent results on SiC bulk growth by CVD and from melt solutions look promising, but much remains to be demonstrated if these techniques are to supersede the relatively mature seeded Lely technology.

III. SiC Epitaxial Growth

To improve the quality of bulk material and produce complicated device structures, epitaxial techniques are necessary. As with other semiconductor material systems, LPE techniques and CVD were used early in the development of SiC to produce device structures. Although the material produced by LPE was of high quality, difficulties with molten Si (used as the melt) prompted the development of vapor-phase techniques such as sublimation epitaxy and CVD. Chemical vapor deposition is presently the most widely used epitaxial technique for growth of SiC device structures.

1. CHEMICAL VAPOR DEPOSITION

Chemical vapor deposition is a growth process in which gaseous compounds are transported to the substrate surface, where chemical reactions occur, resulting in formation and growth of the desired material. The growth of 6H-SiC layers on 6H-SiC {0001} substrates by CVD in the temperature range from 1500 to 1850°C have been reported since the sixties [22, 69, 79, 119, 122, 134, 135, 144, 212]. A significant lowering of the growth temperature and improvement of material quality have been achieved by using substrates that are misoriented a few degrees off the {0001} plane toward the $\langle 11\bar{2}0 \rangle$ direction. This growth on misoriented substrates has been termed *step-controlled epitaxy* and has the added advantage of stabilizing the polytype structure (see Section III.1.d).

Homoepitaxial CVD growth has been reported for 6H, 4H, and 3C polytypes of silicon carbide, while heteroepitaxy of 3C-SiC has been reported on AlN, sapphire, Si, 6H-SiC, and 15R-SiC substrates [179]. Various types of semiconductor devices have been fabricated using CVD-grown SiC structures, including high-voltage Schottky and pn diodes, light-emitting diodes, and transistors. A review of CVD growth of SiC has been published by Nishino [152] and Larkin [102]. Growth temperatures for typical SiC CVD processes range from 1200 to 1800°C, while growth pressures vary from 100 to 760 torr.

a. Growth Equipment

Three types of growth apparatus are used for SiC CVD. (1) Cold-wall horizontal atmospheric pressure reactors have been used for many years, with results reported by Siemens AG [74], Kyoto University [120], and NASA Lewis Research Center [161]. A commercial horizontal water-cooled reactor for SiC CVD, operating in the pressure range from 10 to 1000 mbar, was developed by AXITRON GmbH [183]. This system uses low pressure and a specially designed inner sleeve to maintain laminar gas flow. (2) Hot-wall horizontal atmospheric pressure reactors have been designed at Linkoping University [95]. Still another hot-wall horizontal reactor design was proposed by the Industrial Microelectronic Center in Sweden [153]. (3) Cold-wall vertical low-pressure reactors [42] have been built commercially by EMCORE Corp. and used at Howard University and Siemens AG [176]. In these reactors, laminar flow is obtained by a high-speed rotating disk (in conjunction with high gas-flow velocities), which produces a pumping action.

High temperatures in SiC epitaxial reactors can be obtained with either resistive or rf heating. In most systems, the susceptor is made from graphite.

Because of the reaction between graphite and H_2 at temperatures in excess of 1300°C, a thin SiC coating layer has been used by many researchers. Unintentional incorporation of contaminants from the susceptor during SiC CVD was studied by Karmann and co-workers [75]. In these experiments, 6H-SiC layers were grown in an atmospheric pressure reactor at 1390°C, with a growth rate of 0.7 μm/hr. Uncoated and SiC-coated (100 to 120 μm thick) graphite susceptors were used for comparison. For the uncoated susceptors, the layers were found to be contaminated with aluminum, boron, and nitrogen. Conversely, using an SiC-coated graphite susceptor in the same system, SiC layers with a concentration $N_d - N_a$ of 4×10^{15} cm^{-3} could be grown. In a low-pressure vertical reactor with high-speed substrate rotation, SiC with background concentrations in the 10^{14} cm^{-3} range was demonstrated without use of SiC-coated parts [176]. This low amount of contamination is attributed to the favorable gas-flow patterns generated in this reactor.

Temperature measurement is a major equipment issue for CVD growth at high temperatures. During growth, substrate temperature is usually measured by an optical pyrometer calibrated by melting of Si or Ge. The temperature of the susceptor typically is found to be 50 to 100°C higher than that of the SiC substrate. Power settings are often used to calibrate the substrate temperature.

b. Precursors and Reaction Chemistry

A number of precursors have been used for the growth of SiC. For transport of the Si species, the most popular choice is SiH_4 [161], but Si_2H_6 [149] and $SiCl_4$ [135] have also been used. For growth of SiC, the hydrocarbon species most reported is C_3H_8. However, there are also reports of SiC growth using C_2H_2 [112], CH_3Cl [62], CH_4 [25], CCl_4, C_7H_8, or C_6H_{14} as carbon sources. In addition to the use of individual gas species, single precursors have also been investigated, among them CH_3SiCl_3 [150] and $(CH_3)_2SiCl_2$ [25].

Gas-phase equilibrium calculations were reported for Si-H-C gas mixtures [4] in the temperature range used in CVD growth of SiC. Modeling of the gas-phase chemical processes, which occur during growth as a function of distance from the heated susceptor, was reported by Stinespring and Wormhoudt [191]. The results of the modeling study shows that the injected C species are not fully decomposed to equilibrium values at the growth interface. Gas-phase reactions in SiC CVD growth using the SiH_4-C_3H_8-H_2 gas system were experimentally investigated by Hong et al. [58] using a microcavity technique. The microcavity study suggested (1) that multiple species contribute to the film growth in the system and (2) possible precursors in the chamber contain SiH_2 (a gas-phase intermediate derived from SiH_4) and another species containing Si and C derived from SiH_2 and C_3H_8. Under certain experimental conditions, Si droplets and graphite inclusions have been observed in SiC epitaxy. Modeling of these Si droplets and graphite inclusions formation in the CVD process was done [77] considering several homogeneous and heterogeneous chemical reactions.

c. Growth Rate

Increasing the growth rate in SiC epitaxy is important because of the demands of SiC power device structures. A typical base layer in a power thyristor can be as thick as 100 μm. Since growth rates of SiC films vary from 0.1 to 6 μm/hr for a growth temperature of 1500°C, the minimum growth time for a single structure can be as long as 15 hours. Growth under normal conditions is determined by the diffusive transport of the Si species through the stagnant layer, although control of the growth by carbon species has been observed. Since the growth rate of 6 mm/hr is too low for layers greater than 10 μm, there are several ongoing studies on increasing the CVD growth rate. Growth rates of 500 μm/hr were obtained by high-temperature CVD (1800–2300°C) [68, 98] with a maximum crystal thickness of 2 mm. This result looks very promising for growth of thick layers (>100 μm) for high-power SiC devices, but many questions about this technique, including possible material contamination at high temperature, remain unanswered.

The temperature dependence of SiC CVD growth rate has been investigated by several researchers. If the growth rate is determined by surface kinetics, the activation energy of the growth is expected to be relatively large. If, on the other hand, the growth is transport-limited, the observed activation energy is expected to be small. For the $SiCl_4$-CCl_4-He gas system, an activation energy of 20 kcal/mol was reported for growth on well-oriented substrates [69]. An activation energy of 22 kcal/mol was obtained for SiC growth on the $(000\bar{1})$C face of well-oriented substrates for the C_3H_8-SiH_4-H_2 gas mixture [212]; however, the same investigators observed a complicated temperature dependence on well-oriented (0001)Si faces. The complicated temperature dependence of growth rate on the Si face was explained using a growth model in which the growth rate is limited by the adsorption-desorption of reactants at the growth interface. For step-controlled epitaxy, an activation energy of SiC growth rate was measured to be 3.0 kcal/mol (Fig. 9) [81]. The layers were grown at atmospheric pressure in the temperature range from 1200 to 1500°C on 6H-SiC {0001} substrates

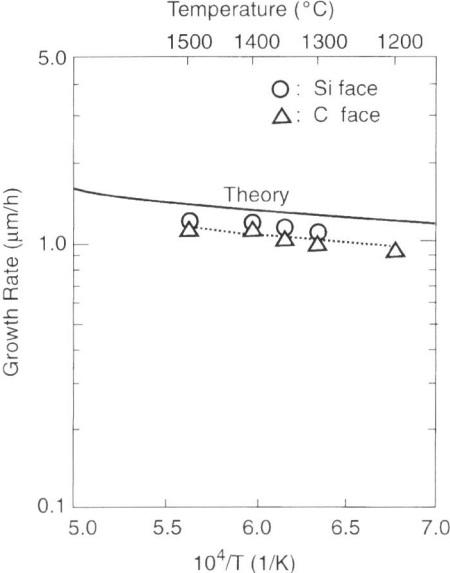

FIG. 9. Temperature dependence of growth rates on off-oriented 6H-SiC (0001)Si and (000$\bar{1}$)C faces. The result calculated based on a stagnant layer model, is shown by a solid line. (After Kimoto et al. [81].)

(with 6° off-orientation) using SiH_4-C_3H_8-H_2 precursors. The small value of activation energy is attributed to the fact that the growth in step-controlled epitaxy is mass transport–limited and the temperature dependence is due to diffusion through the stagnant layer.

d. Homoepitaxy of α-SiC

Homoepitaxial growth of α-SiC (6H-SiC, 4H-SiC, 21R-SiC) by CVD has been advanced at Kyoto University using off-oriented SiC substrates [64, 66]. This technique is called step-controlled epitaxy because the growth process is determined by the lateral growth rate of the terraces. The growth rate, substrate misorientation, and growth temperature determine whether growth will occur via the step-controlled mechanism (Fig. 10). If the growth is step-controlled, the epilayer will replicate the stacking order of the substrate. The growth mechanism for SiC homoepitaxial CVD has been discussed by a number of researchers [59, 81, 82]. Nucleation processes during SiC CVD growth were investigted by Kimoto and Matsunami [83],

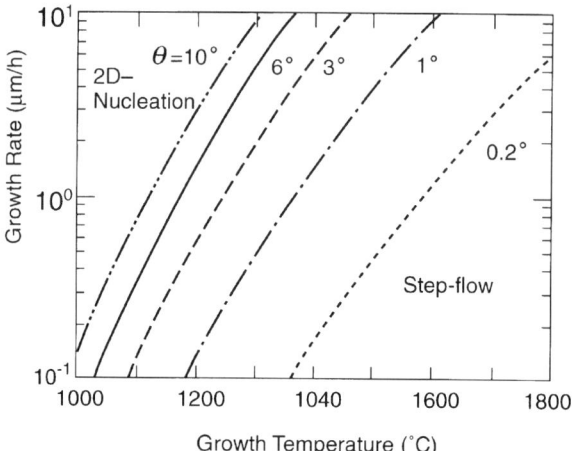

FIG. 10. Critical growth conditions as a function of growth temperature, growth rate, and substrate tilt angle. Top left and bottom right regions of each curve correspond to two-dimensional nucleation and step-flow growth conditions, respectively. (After Kimoto and Matsunami [85].)

and the surface kinetics of adatoms in CVD growth of SiC were analyzed based on Burton-Cabrera-Frank theory [85].

Typical growth of epitaxial 6H-SiC is performed on wafers that are misoriented 3.5° towards the $\langle 11\bar{2}0 \rangle$ direction. The growth temperatures are about 1500°C. 6H-SiC is usually grown with large amounts of H_2 in the chamber, and gas ratios of 1:1000 $(SiH_4 + C_3H_8):H_2$ are usual. 4H-SiC layers are typically grown (with the growth being performed at temperatures somewhat higher than those used for 6H growth) on 5 to 8° off-oriented 4H-SiC substrates using the same gas system. The (0001)Si face 4H substrates are misoriented toward the $\langle 11\bar{2}0 \rangle$ direction. A typical growth rate is about 2.5 μm/hr.

At high growth temperatures (1500°C or greater) used for 6H-SiC and 4H-SiC deposition, a wide range of Si/C ratios have been used [177]. The Si/C ratio affects not only the growth rate and crystal quality, but also the dopant incorporation. A major problem in the epitaxial growth of SiC is the unintentional incorporation of C in the gas phase, which comes from various parts of the reactor. This unintentional C is transported to the substrate via reaction with H_2 in the gas stream and the subsequent formation of hydrocarbons, making difficult to maintain a proper Si/C ratio.

Details of several epitaxial growth processes have been published. A growth process employing the SiH_4-C_2H_4-H_2 gas system was described by

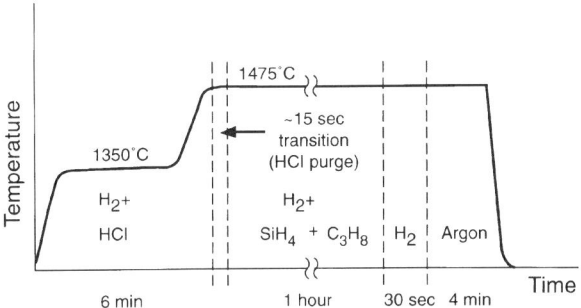

FIG. 11. Time–temperature scheme for CVD process used in growth of SiC epilayers on SiC substrates (after Powell et al. [166]). The process extends the HCl flow until the start of growth to minimize formation of morphological defects.

researchers from North Carolina State University [210]. In this process, SiC substrates were initially heated to the growth temperature (1350–1600°C) for 10 min in H_2 flow to clean the surface. In the growth procedure developed at NASA Lewis Research Center [166] for 6H-SiC and 4H-SiC CVD, the samples are initially etched by HCl at 1350°C prior to the growth (Fig. 11). The initial HCl purge reduces the density of surface defects in resulting SiC layers. The importance of pregrowth treatment of the substrates was emphasized in this study. A high-resolution x-ray diffraction study on 6H-SiC layers grown by CVD [9] also demonstrated the importance of pregrowth treatment. The layers in the study were grown in the temperature range from 1500 to 1600°C with the growth rate of 2 to 2.5 μm/hr using a C/Si ratio of 2.5 to 3.0. Prior to the growth, the substrates were etched in H_2 gas flow at 1500 to 1600°C for 10 to 30 min. A 9-arcsec FWHM x-ray rocking curve was obtained for layers grown on substrates having an FWHM of 72 arcsec (Fig. 12). Without H_2 etching, the FWHM for the layer was often observed to be larger than the FWHM of the substrate.

The role of growth termination (cool-down phase) of the CVD process on surface characteristics in grown material was studied by Rupp et al. [176]. It was shown that termination in H_2 atmosphere results in high-quality surfaces with nearly stoichiometric composition. Conversely, termination in a silane-containing atmosphere results in Si-rich, nearly amorphous surfaces. In both cases, a native oxide forms on the as-grown surface after a few hours of air exposure at room temperature. Termination in a vacuum or an inert gas atmosphere causes graphitization of thin surface layer of the grown material. These observations are very important for subsequent processing steps.

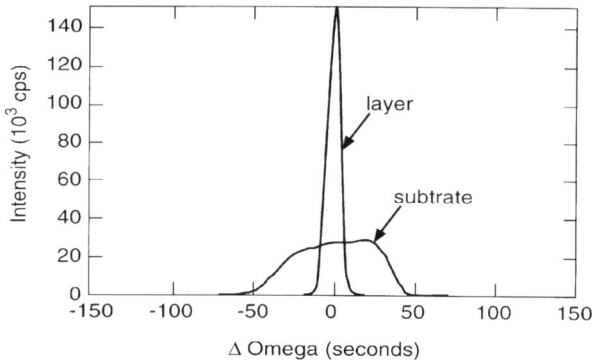

FIG. 12. X-ray rocking curves of the basal plane (0006) in the ω direction, taken from the substrate and the layer grown on this substrate. (After Bakin et al. [9].)

All the CVD processes discussed up to this point were developed for SiC deposition on polar {0001} (C or Si face) substrates. Epitaxial growth of SiC has also been investigated on the two nonpolar crystal planes: the $(10\bar{1}0)$ and $(1\bar{2}10)$ planes [18].

Structural properties. The surface of SiC epitaxial layers can contain a large number of imperfections. Surface defects observed in SiC CVD layers are growth pits, polytype inclusions (which sometimes appear as triangular features), macrosteps (often referred to as step bunching), and micropipes. Some of these defects are relatively large (tens of microns), while others have an average size of less than 1 μm.

Attempts to understand the nature of surface defects in SiC CVD layers have appeared in several studies [21, 84, 166–168, 176, 177]. A large number of factors influence the production and density of surface defects. These include substrate characteristics (orientation, face polarity, tilt angle, crystallographic direction of the misorientation), mechanical and chemical treatment of the substrate before the epitaxy, substrate pregrowth treatment in the reactor, and growth conditions such as Si/C ratio, growth rate, and growth termination procedure [94]. The best reported results thus far achieved indicate surface defect densities of 10^3 cm^{-2}. It is noteworthy that this value corresponds to the density of unknown defects in SiC pn structures, which appear to cause premature junction breakdown [26]. Investigation of surface defects in 6H-SiC and 4H-SiC layers have shown that the morphological defect density varies widely from run to run.

Growth pits are a common morphological defect observed in the growth of SiC. The relationship between growth pits on the epitaxial layer and surface imperfections in the starting substrates has been studied by Powell et al. [166]. It was concluded that the main factor in the formation of growth pits is the polishing and preparation of the substrate rather than bulk defects such as micropipes and dislocations. This conclusion has led to improved substrate polishing techniques, such as the use of colloidal silica for chemomechanical polishing [169].

Step bunching (which is the combination of atomic steps on the surface to form large macrosteps) often occurs in the CVD growth of SiC. Results on the initial investigations of step bunching on 6H-SiC epitaxy (which enhances terrace nucleation) was discussed by Kong et al. [91]. Step bunching in 6H- and 4H-SiC growth by step-controlled CVD on misoriented {0001} substrates was investigated by AFM and TEM techniques [84]. In this study, the SiC {0001} substrates were misoriented 3 to 5° toward the $\langle 11\bar{2}0 \rangle$ direction. It was observed that epitaxial growth on the (0001)Si face yielded macrosteps with an average terrace width of 280 nm and an average step height of 3 nm. On the $(000\bar{1})$C face, the surface was relatively flat and no microsteps were observed. On the (0001)Si faces, three bilayer-height steps were the most dominant type of step seen using 6H-SiC samples, while four bilayer-height steps predominated on 4H-SiC samples. Step bunching as a function of tilt angle (0.1–3.5°) was studied by Powell et al. [167], within a conclusion that step bunching in epitaxial growth can actually be reduced at higher substrate misorientations. A TEM study as well as a qualitative model of the step-bunching phenomena was reported by Chein et al. [24]. In their model, Chein and co-workers show that variations in the surface energies of the different steps that comprise the 6H unit cell are responsible for the different lateral growth velocities and consequently the step bunching.

Typical 4H- and 6H-SiC epitaxial layers also contain dislocations and micropipes. In all cases in which micropipes were observed at a layer–substrate interface, the micropipes originated in the substrate and propagated into the epitaxial layer [165]. This result implies that if micropipes in the substrate are eliminated or closed (e.g., by liquid-phase techniques), the epitaxial films will be micropipe-free.

Polytype inclusions are a common type of crystalline defect in 6H- and 4H-SiC epitaxial layers grown by CVD [94]. These polytype inclusions (usually 3C-SiC) are formed due to nucleation on terraces or dislocation sites. It was found that with proper pregrowth surface treatment with HCl etching, 6H-SiC layers without 3C-SiC inclusions can be grown by CVD on {0001} 6H-SiC substrates with small tilt angles (0.1–0.6°) [162]. Conversely, if a pregrowth HCl etch at 1375°C for 20 min were used, predominantly 6H-SiC growth was obtained. The etching process appears to be effective in removing unintentional 3C nucleation sites on 6H-SiC wafers. However, deviation from the optimal etching conditions led to 3C-SiC growth on the 6H-SiC substrates. It was also found that 4H-SiC homoepitaxial layers are more susceptible than 6H layers to 3C inclusions [166]. The mechanism of cubic SiC nucleation on off-axis 6H and 4H substrates has been investigated by Hallin et al. [54]. These investigators showed that the 3C-SiC nucleation occurs via the formation of triangular stacking faults at substrate imperfections. In 6H-SiC, these defects are usually found in on-axis material where the probability for two-dimensional nucleation of 3C-SiC is increased. In 4H-SiC epitaxial layers, 3C-SiC inclusions having a triangle shape are found even if substrates have a tilt angle of 3.5° [186]. To reduce the density of inclusions in 4H material, an 8° tilt angle was found to be necessary. In the 4H-SiC epitaxial layers grown on 8° off-substrates, the 3C-SiC inclusions are almost eliminated.

Although some progress has been made in understanding the nature and cause of structural defects such as step bunching and polytype inclusions, the origin and control of many defects in epitaxial SiC remain to be investigated.

Electrical properties. Significant progress has been achieved in producing epitaxial layers of 6H- and 4H-SiC with superior electrical properties [160]. 6H-SiC epitaxial layers having a low background doping concentration were grown and characterized [68, 96]. It was shown that, using propane as a carbon precursor, uncompensated SiC layers with donor concentrations less than 10^{15} cm^{-3} may be grown, whereas with methane, uncompensated layers can be produced with electron concentrations in the mid 10^{14} cm^{-3} range (however, with a slightly worse morphology). The optimal growth temperatures for these films were found to vary between 1550 and 1600°C. At higher temperatures, contamination from graphite parts became noticeable, and bake-out of the growth system had a significant impact on the background doping. The effect of unintentional hydrogen doping by CVD was studied by Clemen et al. [28].

4H-SiC layers with electron concentrations as low as 2×10^{14} cm^{-3} were reported [63, 86, 96]. (To obtain the low impurity concentration, the growth system was pumped for several hours prior to growth, according to some of these authors.) Electrical and optical measurements on high-quality 4H-SiC layers were reported by Kimoto and co-authors [86]. The background doping concentration in the layers was determined to be 3×10^{15} to 2×10^{16} cm^{-3}, and electron mobility in the {0001} basal plane was 600 to 720 cm^2/V s (300 K). Deep-level transient spectroscopy (DLTS) measurements on these films showed that the concentration of electron traps was approximately 10^{13} cm^{-3} independent of substrate polarity. Minority carrier lifetimes have been measured on 6H-SiC layers with $N_d - N_a$ ranging from 10^{14} to 10^{17} cm^{-3}. Lifetimes as high as 0.45 μs (300 K) have been achieved for thick low-doped samples [97]. However, the maximum reported values of minority carrier diffusion length for CVD-grown SiC pn structures do not exceed 3 μm.

Doping of SiC homoepitaxial layers grown by CVD has been reported in numerous publications. Nitrogen is commonly used as a donor, and aluminum is the acceptor of choice. Nitrogen doping has been investigated for several years as an n-type dopant (see, e.g. [73, 177]). Nitrogen doping in these studies produced donor concentrations ranging from 10^{16} to 10^{19} cm^{-3} (Fig. 13).

P-type doping has been achieved by using Al as a dopant (see, e.g. [154]). Epitaxial layers in this study were grown using SiH$_4$-C$_3$H$_8$-H$_2$-TMA (trimethylaluminum) precursors at a C/Si ratio of 2.5. The reactor pressure was 800 mbar and growth temperature was 1550°C. A growth rate was about 2 μm/hr. The atomic Al concentration in 6H-SiC was controlled from 10^{17} to 10^{21} cm^{-3} (Fig. 14). When the Al concentration exceeded 2 × 10^{20} cm^{-3}, impurity banding occurred and the Al acceptors were completely ionized, while 1% ionization was observed at lower doping levels (in keeping with the measured Al ionization energy of ~0.25 eV). Schoner and

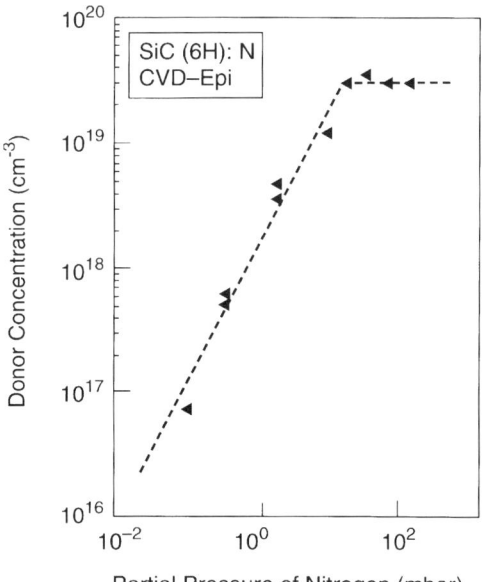

FIG. 13. Total donor concentration taken from Hall data as a function of nitrogen partial pressure during CVD growth. (After Karmann et al. [73].)

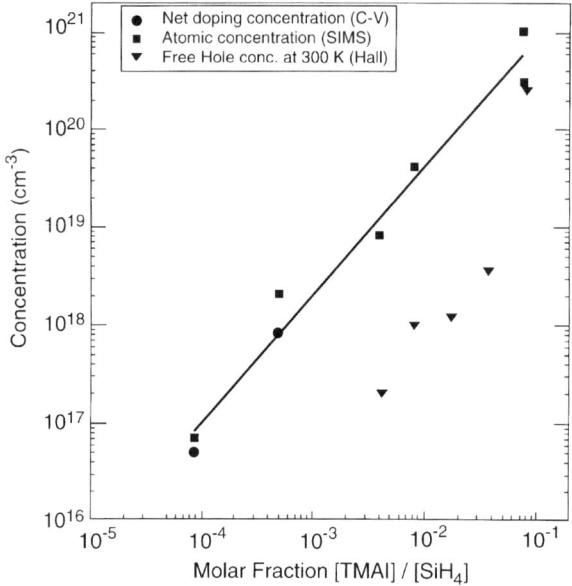

FIG. 14. Atomic and carrier concentration for Al doping in CVD-grown SiC as a function of the [TMA/SiH$_4$] molar fraction in the gas flow. (After Nordell et al. [154].)

co-workers [182] found that the Al ionization energy varied with doping concentration as well as the degree of compensation. As expected, at high doping levels, crystal quality was degraded.

A greater range of doping control is possible with site-competition epitaxy [103–105]. The method is based on varying the Si/C ratio within the CVD reactor in order to control the dopant incorporation in SiC during epitaxial growth. Site-competition epitaxy has been used for control of nitrogen, phosphorus, aluminum, and boron incorporation in 6H-SiC and 4HSiC films. The results of the site-competition epitaxy are illustrated in Fig. 15. The layers were grown using SiH$_4$-C$_3$H$_8$-H$_2$ gas precursors at 1450°C, with a typical growth rate of 3 to 4 μm/hr. Secondary ion mass spectroscopy (SIMS) determined that the Al incorporation increased as the propane flow was increased and also when the silane flow was decreased. It was found that results of site-competition doping control are similar for 6H- and 4H-SiC epitaxial layers grown on (0001)Si faces of the substrates. The effect of surface polarity and the chemistry of the particular impurity on site-competition doping control are discussed by Larkin [105]. Deep-level impurities in SiC pn structures grown by site-competition epitaxy have also been investigated by Saddow et al. [180].

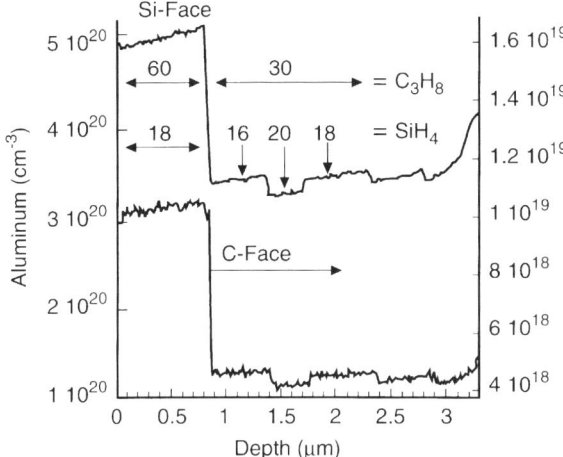

FIG. 15. SIMS depth profile of an aluminum-doped epitaxial layer simultaneously grown on a 6H-SiC (0001) Si-face and a C-face substrate. Silane flow is varied during a constant propane flow. A change in the Si/C ratio causes a change in the Al concentration. (After Larkin [105].)

Impurity memory effects on dopant concentration profiles in 4H- and 6H-SiC were investigated by SIMS [155]. It was found that dopants were absorbed by the reactor walls and re-evaporated after the dopant precursor flow was switched off. These memory effects limit the doping control range to about three orders of magnitude for aluminum and two orders of magnitude for boron. The dynamic range for Al doping was increased up to five orders of magnitude by controlling the Si/C ratio and using HCl etching during the 10-min growth interruption after gas switching. For boron, a dynamic range of more than three orders of magnitude was obtained. Doping spikes at the substrate–layer interface were also reduced by an *in situ* HCl etch [20].

e. Homoepitaxy of 3C-SiC

Studies of the homoepitaxy of 3C-SiC have been limited by the availability of bulk 3C-SiC substrates [143]. The limited literature that exists suggests that the growth of epitaxial 3C-SiC should occur at conditions similar to the growth α-SiC. Epitaxial films of (111) 3C-SiC were grown in the late sixties [10] on bulk substrates of 3C-SiC. The substrates were platelets produced by liquid-phase growth technique. The growth experiments were performed with SiH_4-C_3H_8 and CH_3SiCl_3 gas systems. The best results were obtained with the CH_3SiCl_3 gas system. Smooth transparent films were obtained at substrate temperatures of 1560°C. The growth rate was 2.5 μm/hr. Background concentration $N_d - N_a$ in the layers was measured to be about 5×10^{15} cm^{-3}. Another demonstration of 3C-SiC growth on 3C-SiC was reported by Yoshida et al. [219]. These authors compared 3C-SiC grown on 3C-SiC material initially grown on Si and separated from the Si substrate with 3C-SiC grown on 6H-SiC substrates. It was concluded that higher growth temperatures produced films with better morphological and electrical properties. It was also determined that the 3C polytype was stable on 3C-SiC substrates up to 1700°C.

f. Heteroepitaxy of SiC

Heteroepitaxial growth of SiC initially is motivated by lack of large inexpensive SiC substrates for homoepitaxy. The substrate material of choice for SiC CVD heteroepitaxy is silicon. CVD growth on Si substrates results in growth of the cubic SiC polytype. Despite intense research of 3C-SiC growth on Si and significant improvement in material properties, the quality of the grown 3C-SiC material still is not good enough for most electronic applications. To improve crystal quality of 3C-SiC layers, material was grown by CVD on 6H-SiC and 15R-SiC substrates. The main drawback to growth on all of these substrates is the occurrence of double position boundaries. (These boundaries are the result of the two allowed orientations that 3C-SiC can take on the hexagonal substrate during growth.) It was shown that CVD growth of 3C-SiC on 15R-SiC leads to reduction (but not elimination) of the density of double positional boundaries (as compared to growth on 6H-SiC). Growth of SiC on sapphire [159] and AlN has also been demonstrated.

3C-SiC CVD on silicon substrates. Heteroepitaxial growth of 3C-SiC on Si is a difficult technological challenge because of larger lattice ($\sim 20\%$) and thermal ($\sim 8\%$) mismatches. The breakthrough in material quality was achieved by the development of a buffer layer technology [147, 148]. The key process for the heteroepitaxial growth of SiC on Si is the carbonization of the Si substrate, providing a thin SiC buffer layer. After the demonstration of the buffer technique, numerous papers were published on 3C-SiC growth and characterization using Si substrates (for a review, see ref. [152] and references therein). 3C-SiC material grown on Si has been used for micromachining applications [100].

3C-SiC CVD on alpha substrates. It was suggested that crystal quality of 3C-SiC heteroepitaxial layers grown may be improved by using α-SiC substrates [142, 218]. Single-crystal 3C-SiC layers have been grown by CVD on 6H-SiC substrates [15, 90]. These 3C-SiC films have double positioning boundaries (DPBs) and often exhibit twining [90, 163, 219, 221]. Davis [29] described the growth of 3C-SiC on both the (0001)Si and (000$\bar{1}$)C faces of 6H-SiC Acheson substrates at 1400 to 1550°C, using SiH_4 and C_2H_4. The layers grown on the (0001)Si face had a smooth and reflective surface. The surface of the layer deposited on the (000$\bar{1}$)C face was, however, rough and unsuitable for device fabrication. High-resolution TEM displayed an abrupt and coherent 3C/6H interface, indicating that the 3C-SiC growth direction was $\langle 111 \rangle$. Plain-view TEM examination revealed the presence of DPBs in the 3C-SiC films. 3C-SiC films with a DPB-free 1-mm-square area were grown by Powell *et al.* [162, 164] on 6H-SiC substrates with a tilt angle from 0.1 to 0.6°. The substrates were either sawed or damaged in order to nucleate the 3C material at a single point so that step flow growth would occur and DPBs could be avoided. Prior to the growth, substrates were etched in gases HCl (5% in H_2) to eliminate unwanted nucleation sites. It was found that etching conditions are critical in order to control the polytype of the SiC layer. Low- and high-temperature etching are favorable for 3C-SiC formation, while etching at intermediate temperatures results in 6H-SiC deposition, even at a low tilt angle. It was suggested that surface disturbance (e.g., dislocations), not surface step density, controls the SiC polytype formation on low-tilt 6H-SiC wafers with an (0001) orientation.

SiC CVD on AlN. The close lattice match of SiC and AlN makes heterostructures of these materials very attractive for several applications. For example, AlN can provide a high-temperature gate dielectric for SiC metal-oxide semiconductor devices. Also, AlN can be used as a barrier layer in the fabrication of SiC high electron mobility transistor (HEMT) devices.

AlN has a 2H crystal structure that is closely lattice matched to hexagonal SiC. Single-crystal layers of AlN have been grown on sapphire, Si, and SiC substrate. Single-crystal SiC layers have been grown on AlN by CVD [34, 151, 193]. X-ray diffraction study, cathodoluminescence (CL), and photoluminescence measurements on some of these films showed the polytype to be 3C-SiC for SiC/AlN/Si structures [37]. SiC layers grown on the AlN/6H-SiC substrates at the same conditions produced x-ray diffraction patterns with peaks only from 6H-SiC and AlN. Based on the x-ray diffraction results, the authors suggested that the SiC layer had a hexagonal structure. Under ultraviolet excitation, these samples displayed blue photoluminescence (77 K). Crystal growth of SiC on AlN/sapphire substrates by CVD was reported by Nishino et al. [151] using Si_2H_6-C_2H_2-H_2 or HMDS/TMA/H_2 precursors. The samples displayed p-type conduction, and CL measurements made at 4 K revealed a main peak at 430 nm, indicating that the SiC film is hexagonal.

2. Liquid-Phase Epitaxy

SiC LPE growth takes place from a supersaturated solution of Si and C in a melt solvent. The main feature of LPE is that the growing films are in equilibrium with the liquid phase. The endpoints of the process are determined by the phase diagrams for Si, C, and the solvent material. LPE was

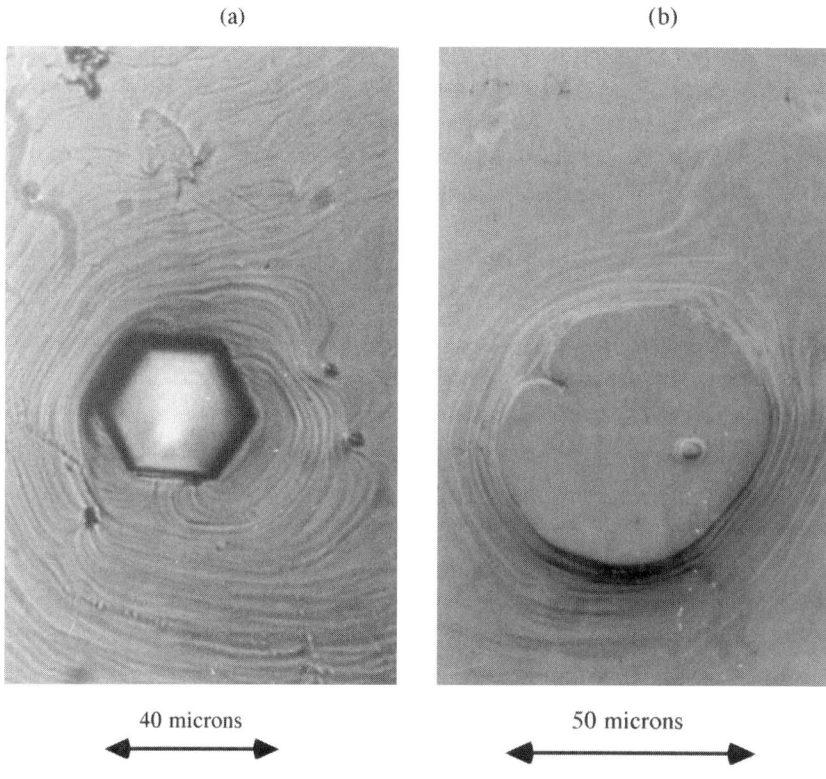

(a) (b)

40 microns   50 microns

Fig. 16. (a) Optical photograph of a micropipe defect on SiC substrate on the initial stage of LPE growth. (b) The same part of the sample after LPE growth; micropipe is closed [173].

used early in the development of SiC technology. Growth from Si melt as well as alloy melts was demonstrated. LPE has been performed in graphite boats by vertical dipping and by a novel levitation process called container-free epitaxy. Doping of SiC was accomplished in LPE over a large range of concentrations. Due to the difficulty in control of surface morphology, LPE techniques have lost ground in favor of CVD approaches. However, recently discovered unique properties of LPE, such as micropipe closing (Fig. 16) [173, 216] and the ability to produce very heavily doped p-type films ($>5 \times 10^{20}$ cm^{-3}) [173], may secure a future role for this technology. The usual melt for SiC LPE is silicon, but alternative materials, such as Sn, Ge, Ga, and their mixtures, are also used for SiC LPE (see [39] for a review).

a. LPE from an Si Melt

The phase diagram for the Si-C system has been studied by Scace and Slack [181]. The maximum amount of SiC grown from Si-C melt solution is determined by the Si-C phase diagram. Growth of SiC will take place either: (1) on the surface of SiC seed crystal, (2) in the Si melt, or (3) on the crucible walls. Usually, the SiC growth on SiC seed is epitaxial and the crystallographic orientations of the layer and of the crystal (substrate) are the same. Both isothermal and nonisothermal conditions have been used for SiC LPE growth.

LPE growth from a graphite boat. The first epitaxial SiC growth from carbon-saturated Si solution was developed more than 25 years ago by Brander and Sutton [16]. At a temperature of 1650°C, SiC single-crystal layers up to 100 μm thick were reproducibly grown. After the growth, the SiC wafer was recovered by opening the crucible and dissolving the silicon in an HF-HNO$_3$ solution. X-ray analysis showed that layers were single crystals of the same polytype as the substrate. With this technique, SiC p-n junctions were prepared using nitrogen, aluminum, and boron impurities. About 10 years later, LPE growth from an Si melt was investigated by Muench and Kurzinger [136]. A p-type SiC substrate was clamped to the bottom of a graphite crucible containing about 1 cm^3 of Si. The system was operated in 1 atm of high-purity argon, except during the nitrogen-doping period. Growth rates up to 1 μm/min were obtained at a melt temperature of 1800°C and a temperature gradient of 30°C/cm. The cool-down process after the growth was carefully controlled in order to avoid substrate cracking during the melt solidification phase. The background doping level of the epilayers was usually less than 10^{16} cm^{-3}. The n-type background electron concentration could be reduced to 10^{14} cm^{-3} by prolonged heating and degassing cycles. P-type layers were grown by the addition of aluminum to the melt.

b. *LPE Growth by Vertical Dipping*

A major problem encountered in LPE of SiC is damage, due to stresses induced from the Si melt solidification. To solve this problem, LPE growth of SiC by a vertical dipping was developed by Suzuki and co-workers [192]. A graphite crucible was employed; to avoid evaporation of silicon, a graphite lid was also used. The substrate was tied to the graphite holder with molybdenum or tantalum wire. After the silicon was melted, the substrate was dipped to a depth of 2 to 3 mm into the melt and maintained for 5 hours. The growth temperature was between 1500 and 1750°C. The temperature of the sample was kept 10 to 40°C lower than that at the crucible bottom, and the growth occurred in the temperature gradient. After the growth, the sample was pulled up from the crucible before the Si melt solidification. Using this technique, the crucible as well as the Si melt could be reused several times. Epitaxial films 20 to 40 μm thick were obtained. The dipping method has been employed for growth of SiC p-n junctions and multilayer structures. Ikeda *et al.* [60] used a growth system consisting of three crucibles, with the melts for the growth of p-layers, n-layers, and rinsing. The growth temperature was varied from 1600 to 1700°C. Undoped

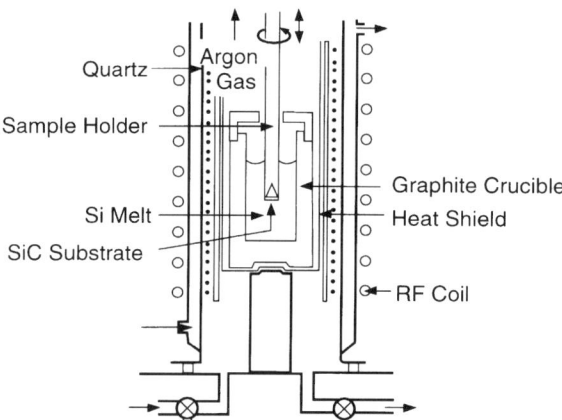

FIG. 17. Cross-sectional view of growth apparatus for SiC LPE. (After Nakata *et al.* [137a].)

6H-SiC layers grown using this technique were n-type at a carrier concentration of 5×10^{17} cm^{-3} and electron mobility of 174 cm^2/V s. Al, Ga, B, and N were investigated as dopants. The maximum electron mobility of 227 cm^2/V·sec was obtained for a Ga-doped n-layer with a carrier concentration of 1.2×10^{18} cm^{-3}. Hole mobility for p-type SiC was measured at 18 cm^2/V·sec at a carrier concentration of 2.4×10^{18} cm^{-3}. The arrangement for the growth in a vertical dipping configuration is shown in Fig. 17.

c. *Container-Free LPE Growth of SiC*

In addition to the melt solidification, another serious problem for LPE of SiC is the interaction of crucible materials with the solution. This is due to the aggressive behavior of liquid Si and its vapor. Muench and co-authors [134] designed a float-zone arrangement, without crucible, to grow SiC layers from Si melt. The growth of SiC was carried out at temperatures around 1800°C. P-type layers were grown by doping the melt with aluminum. Layers grown in an ambient of 1 torr nitrogen had an n-type carrier concentration of 5×10^{18} cm^{-3}. Dmitriev [38] developed container-free LPE (CFLPE) based on an electromagnetic crucible-free technique. In this method, liquid Si metal is suspended in a high-frequency electromagnetic field. The epitaxy was carried out in an He ambient. The pressure in the growth chamber was varied from 10^{-5} to 760 torr. To produce the electro-

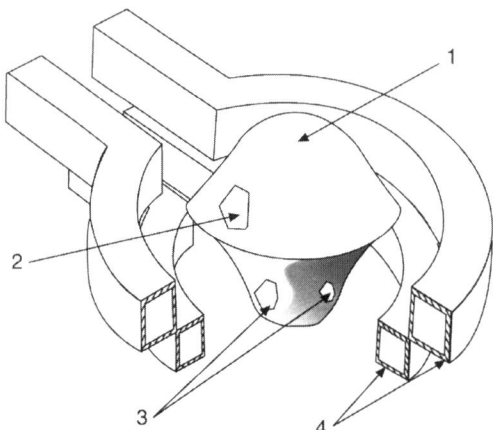

FIG. 18. Silicon melt (1) suspended in an electromagnetic field produced by the water-cooled inductor (4); SiC source crystals (3) and SiC substrate (2) are on the surface of the Si melt. (After Nikolaev *et al.* [141].)

magnetic field, a two-turn copper inductor was used (Fig. 18). The substrates were n-type 6H- and 4H-SiC crystals grown by the Lely method, usually with {0001} basal-plane orientation. A single run consisted of two to six substrates. The area of one substrate was 0.5 to 1.5 cm². Before growth, the substrates were etched in molten KOH to remove surface damage. Si with a resistivity of 100 to 200 Ω cm (300 K) was used as a solvent, and the melt volume was varied from 4 to 10 cm³. Aluminum was introduced into the melt in the form of metal pellets for p-type doping, while nitrogen gas provided n-type doping. Both nonisothermal and isothermal growth conditions were used. The isothermal growth was possible due to the temperature gradient that exists across the suspended Si drop. The top part of the melt is always cooler than the bottom part. Depending on growth conditions, the growth rate varied from 0.05 to 2.0 μm/min. Anisotropy of the SiC LPE growth rate was investigated by Nikolaev *et al.* [141] (Fig. 19). During these studies, it was found that dopants that are introduced in the melt affect the growth rate. Both aluminum and nitrogen appear to increase SiC growth rate. It was also found that at the same substrate temperature the CFLPE growth rate is higher than the growth rate of LPE performed with a crucible. The reason may be due to intensive melt mixing by the electromag-

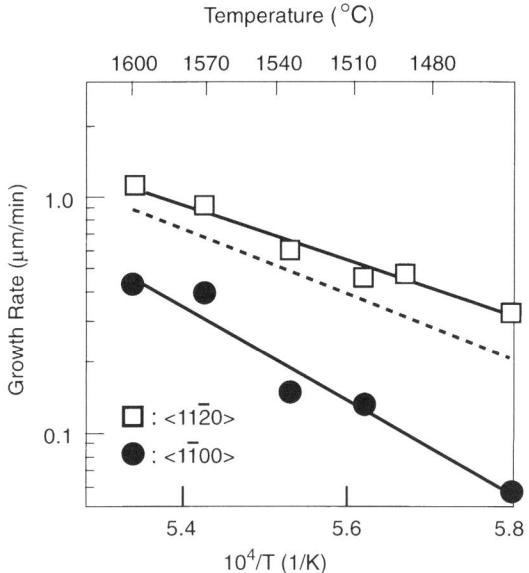

FIG. 19. Temperature dependence of SiC growth rate for different crystallographic directions; dashed line corresponds to the growth in the ⟨0001⟩ direction. (After Nikolaev *et al.* [141].)

netic field. Epitaxial layers grown by the CFLPE technique had single-crystal structure and usually replicated the substrate polytype. The main foreign polytype, if present, was cubic 3C-SiC. Dislocation density revealed by chemical etching was less than $10^3 \, \text{cm}^{-2}$ (for doping levels lower than $10^{19} \, \text{cm}^{-3}$). In the epitaxial material, regions larger than $100 \, \mu\text{m}$ were completely free of planar and linear defects. The FWHM of the x-ray rocking curve of the undoped 6H-SiC layers was from 10 to 25 arc sec. Undoped 6H-SiC layers with the uncompensated electrically active donor concentration, $N_d - N_a$, of $8 \times 10^{15} \, \text{cm}^{-3}$ was grown in vacuum 5×10^{-5} torr. The deep-level concentration in the undoped layers was less than $2 \times 10^{14} \, \text{cm}^{-3}$. Doping in the CFLPE technique is similar to that discussed for other LPE approaches.

d. Low-Temperature LPE

To grow SiC at temperatures below 1400°C, it is necessary to use alternative melts. SiC solubility in Sn, Ge, Ga, Pb, Al, and Si melts has been investigated. It was shown that the SiC solubility in an Si melt is equal to that obtained in alternative melts at lower temperatures. Tairov et al. [195a] were the first to propose Sn and Ga melts as solvents for SiC low-temperature LPE (LTLPE). These investigators used a graphite sliding boat to deposit SiC layers at temperatures between 1100 and 1400°C and at a temperature gradient of 10°C/cm. The resulting layers were single crystals with a thickness of 0.5 to 4.0 μm. Dmitriev et al. [35] reported 6H-SiC p-n junction growth and SiC selective growth at 1100 to 1200°C by LTLPE. This is the lowest reported growth temperature for SiC p-n junctions. The LPE growth was carried out under isothermal conditions with a temperature gradient of approximately 30°C/cm. A Ga melt and melt mixtures of Sn-Al-Si, Ga-Al-Si, and Ge-Si were used as solvents at a growth rate of about 0.1 μm/hr. To grow p-type layers, melts containing Al or Ga were used. P-n structures with p-layer grown from Ga melt on n-type 6H-SiC Lely crystals demonstrated uniform avalanche breakdowns. This was the first observation of the uniform breakdown of SiC p-n junctions grown by LPE. Using LTLPE, it may be possible to grow SiC p-n junctions at 1000°C or lower. The main limitations at temperatures less than 1000°C are inclusions of cubic polytype and low growth rate.

e. LPE Growth of 3C-SiC on 6H-SiC

Epitaxial layers of (111) 3C-SiC have been grown on the (0001)Si face of 6H-SiC Lely substrates at temperatures of 1500°C and lower by CFLPE [36]. The thickness of the films ranged from 3 to 30 μm. The minimum

FWHM of x-ray rocking curve for an undoped 3C-SiC layer was 11.5 arcsec, which is the best value ever reported for heteroepitaxial 3C-SiC. The presence of DPB in the layers was not investigated. If the substrate was misoriented by 1 to 3° from the (0001) plane, a 6H-SiC layer was formed while the 3C-SiC polytype grew on the singular surface. Highly nitrogen-doped 3C-SiC layers were also grown on 6H-SiC substrates using the CFLPE method [140].

3. Sublimation Epitaxy

The mechanism and principals of SiC sublimation epitaxy are similar to those for bulk SiC sublimation growth (see Section II.1). However, sublimation epitaxy is usually performed at lower temperatures with smaller growth rates and for shorter time periods than is bulk SiC sublimation growth. Growth of epitaxial SiC using the sublimation process had been the subject of many early investigations. (For early work, the reader is referred to the Proceedings of First, Second, and Third International Conferences on Silicon Carbide. The breakthrough in sublimation epitaxial technology was achieved with the development of "sublimation sandwich method" by Vodakov and Mokhnov [205]. They employed a nearly flat source positioned close to the substrate and performed the growth under near-equilibrium conditions. This method allowed for the vapor equilibrium to be constant over the substrate. The sublimation sandwich method made it possible to grow high-quality SiC layers in the temperature range of 1600 to 2100°C [7, 126, 195, 206]. An excellent review on SiC sublimation epitaxy was written by Konstantinov [93].

a. Sublimation Sandwich Method

The growth cell contains a vapor source (plates of mono- or polycrystalline SiC or the mixture of Si and C) and a substrate with 0.02 to 3.0 mm of clearance between them (Fig. 20). Supersaturation in the growth zone is varied by changing the temperature difference between the source and the substrate. To prevent leakage of Si from the growth cell, a closely spaced growth system made by dense graphite was developed. The distance between the vapor source and the substrate is small to prevent vapor loss from the growth zone. 6H-SiC, 15R-SiC, and 4H-SiC Lely crystals have been used as substrates. Before the growth, the substrates were etched in KOH at 500°C for a few minutes to remove surface layer damaged by mechanical treatment. The temperature gradient in the growth cell ranged from 5 to 30°/cm. The

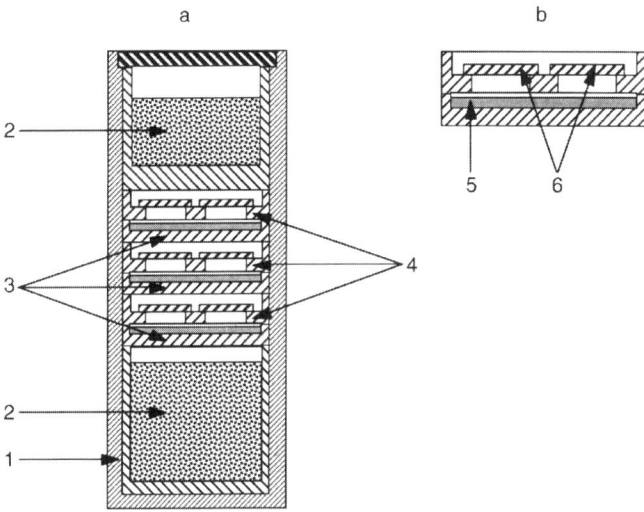

FIG. 20. (a) Crucible assembly and (b) growth cell for SiC sublimation epitaxy by sandwich method: 1, graphite crucible; 2, SiC powder (additional vapor source); 3, graphite holders for SiC vapor source; 4, graphite substrate holder; 5, SiC vapor source; 6, SiC substrates.

temperature difference between the source and the substrate was 0.5 to 10.0°C.

SiC layers grown by sublimation epitaxy usually range in thickness from 10 to 100 μm. Structural defects of SiC layers grown by the sublimation sandwich method are described by Mokhov *et al.* [123, 124, 126, 127]. It was shown [124] that SiC layers with homogeneous polytype structure and dislocation density $\leqslant 10^2$ cm^{-2} can be grown on Lely substrates with any crystallographic orientation if the growth conditions are close to equilibrium. On the other hand, the presence of impurities and deficiency of Si vapor usually lead to surface and crystal structure degradation. The purest undoped layers grown by sublimation epitaxy had background concentrations $N_d - N_a$ of about $(0.5-1.0) \times 10^{16}$ cm^{-3}. Doping control was obtained by the addition of impurities in the source material or by the use of an additional source [125, 209] (Fig. 21). A strong dependence of the doping concentration on growth rate and substrate orientation was found. Further, at high Si pressure ($P_{Si} \sim P_{Si_{sat}}$), the nitrogen concentration in SiC significantly decreased and boron concentration increased. Nitrogen concentration was also found to decrease in the presence of Ge vapor, whereas high Sn content in the vapor phase led to an increase in nitrogen concentration. Microsegregation of the impurities and second-phase formation were inves-

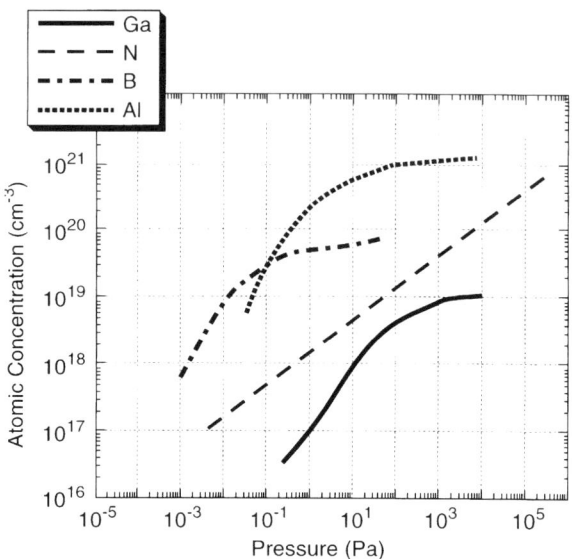

Fig. 21. Dependence of doping concentration in SiC epitaxial layer on impurity vapor pressure in the growth zone for Al, B, N, and Ga. (After Vodsakov et al. [209].)

tigated for Al, Ga, In, Ti, P, Cr, Mn, Cu, Y, Ho, Ta, W, Au, Sn, and Ge [223]. SiC lattice deformation due to doping with Al and B during sublimation growth was studied by Kutt et al. [101].

Polytype control was investigated in sublimation epitaxy. It was found that the polytype reproducibility was higher for (0001)Si substrate surface than for (000$\bar{1}$)C surface. Si vapor and the presence of the isovalent impurities were also found to play an important role in determining the resulting polytype. Excess Si vapor was found to lead to 3C-SiC formation, while isovalent impurities Ge, Sn, and Pb help to form 4H polytype, making it possible to grow 4H-SiC layer on 6H-SiC and 15R-SiC substrates. Mokhov and co-workers [128] grew 4H-SiC epitaxial layers without inclusions of other polytypes on the 6H-, 15R-, and 21R-SiC substrates, using Sn vapor to stabilize 4H-SiC formation.

4. Molecular Beam Epitaxy

Molecular beam epitaxy (MBE) of SiC has been demonstrated by several research groups [43, 44, 46, 47, 71, 72, 80, 121, 131–133, 174, 175, 199, 200, 220, 221, 225]. 3C-SiC layers were grown by gas-source MBE using periodic introduction of Si_2H_6 and C_2H_2 gases [121] at growth temperatures from 700 to 1160°C. At 1000°C, a growth rate of two to six monolayers per cycle was obtained. When 6H-SiC substrates were used [121], epitaxial growth of 3C-SiC on 6H-SiC was observed at 850°C, which is about 500°C lower than for conventional SiC CVD. Layers grown on 6H-SiC (0001) substrates contained DPBs, while the layers grown on 6H-SiC (01$\bar{1}$4) were free of DPBs [221]. Material properties were not reported in detail.

3C-SiC and 6H-SiC layers were grown by gas-source MBE using a simulations supply of Si_2H_6 and C_2H_4. The layers were grown on singular and misoriented 6H-SiC substrates [80] in the temperature range of 1100 to 1300°C. To our knowledge, this was the first report of 6H-SiC growth by MBE. Aluminum doping was achieved in the SiC MBE films, which resulted in p-type material with hole concentrations ranging from 5×10^{16} to 1×10^{18} cm^{-3} (300 K). At the highest doping level, Al inclusions in the films were observed [200].

Growth of SiC layers on AlN/SiC substrates by plasma-assisted, gas-source MBE was reported by Rowland et al. [175]. The growth rate was 6.2 nm/hr. High-resolution TEM showed that the interface between the AlN and top SiC layers was abrupt and that the SiC layer had a 3C polytype structure. Later, Tanaka and co-workers [200] reported the controlled growth of the rate 2H-SiC polytype on AlN.

Solid-source MBE has been developed for SiC growth [43, 44]. Epitaxial SiC films were grown on Si(111) and 6H-SiC(0001) in the temperature range of 800 to 1000°C. The source materials were polycrystalline Si and pyrolitic carbon. Silicon and carbon were evaporated separately by electron-beam guns. The SiC substrates were 3 to 4° off SiC(0001) wafers grown by the modified Lely method and 2 to 5° off platelets obtained by the Acheson method. Stoichiometric epitaxial SiC films were grown at 850°C at a growth rate of 2 nm/min. The crystal structure of the grown layers corresponded to the superposition of twinned 3C- and 2H-SiC. The authors emphasized that the polytype structure of the grown layer possibly may be controlled by variation of thickness of the Si adlayer existing on the growing surface. No other material properties were reported.

Epitaxial growth process are key for the production of high-quality devices. Step-controlled growth of SiC by CVD appears to be the workhorse technology. Using this technique, high-quality films can be produced over a large range of doping. Commercial CVD reactors are beginning to produce high-purity epitaxial films over large areas, although additional efforts are needed to extend the growth rate for thick power device layers and to improve doping-thickness uniformity. Investigation of defects in epitaxial films is beginning to provide answers to the nature of defects in SiC, but an understanding of the nature of epitaxial defects is essential to the fabrication of SiC devices that will compete favorably with Si technology.

Liquid-phase techniques are difficult to implement due to problems in handling the molten Si. However, these techniques have several unique features that may be useful in the future. Sublimation epitaxial techniques appear to have limited advantages over high-temperature CVD techniques and have been demonstrated only for small-area wafers. MBE technology for SiC is in its infancy, and material with competitive performance parameters is yet to be produced.

IV. SiC Doping

Wide bandgap materials are difficult to dope due to the large ionization energies and self-compensation effects of most substitutional impurities. Fortunately, for SiC, an excellent donor atom (nitrogen) exists and a workable acceptor atom also exists (aluminum). Progress in SiC technology has resulted in a wide doping range for both n- and p-type conductivity. The three techniques employed for doping are doping during growth, diffusion, and ion implantation.

1. Impurities in SiC

Impurities for SiC doping have been found for: (1) donors (mainly nitrogen), (2) acceptors (mainly aluminum), and (3) deep-level impurities to form semi-insulating material (practically only vanadium). Recently, Er was introduced in SiC and preliminary results for IR emission [27] look promising. There have been few attempts to study isoelectronic traps in SiC, but the subject requires further investigation. In general, information on impurity properties in SiC is limited even for impurities such as nitrogen and aluminum. For a review of the literature on dopants in SiC see refs. [61, 130].

The doping development for SiC is complicated because the doping process depends not only on surface orientation, doping technology, and doping concentration, but also on polytype structure. In SiC, dopants may take cubic-like positions or hexagonal-like positions, which will effect dopant characteristics, particularly activation energy. The number of cubic-like and hexagonal-like positions in an elemental crystal cell depends on the polytype. For example, 6H-SiC is 33% hexagonal, while 4H-SiC is 50% hexagonal.

2. Doping During Crystal Growth

SiC doping during bulk and epitaxial growth is described in Sections II and III of this chapter.

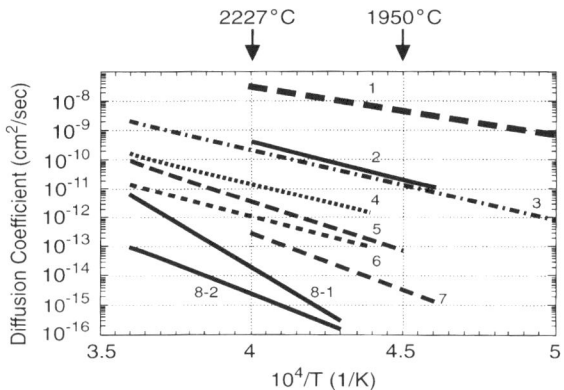

FIG. 22. Diffusion coefficients for different impurities in 6H-SiC: 1, Be (bulk diffusion); 2, Be (diffusion in the surface region); 3, B (bulk diffusion); 4, B (diffusion in the surface region); 5, Al; 6, Ga; 7, oxygen, 8-1 and 8-2, nitrogen. (After Mokhov et al. [124].)

3. Diffusion of Impurities in SiC

Impurity diffusion is one of the principal methds of doping. Although the effective diffusion coefficients for most of the major impurities in SiC are too small for practical applications, it is nevertheless important to understand diffusion processes in SiC, as these processes may take place during bulk and epitaxial growth, as well as during heat treatments. Diffusion coefficients for different impurities in 6H-SiC are shown in Fig. 22 [124]. Boron and aluminum diffusion has been used for SiC device fabrication. Boron diffusion has produced yellow [204] and green [9a] light-emitting diodes, while field effect transistors were fabricated by aluminum diffusion [23].

4. Ion Implantation

Ion implantation is the only alternative to doping during growth and is widely used in SiC device fabrication technology. For a review of ion implantation in SiC, see the refs. [30, 65, 214]. Ion implantation has been used for: (1) p-n junction formation [6, 151, 170, 184, 211, 215], (2) light-emitting diode fabrication [53], (3) highly doped contact layers [188, 187], (4) field effect transistor channels [5, 187], and (5) device isolation and termination [137, 178].

Since the sixties, many elements have been implanted into SiC (Al, B, Ga, In, Tl, N, P, Sb, Be, Bi, Kr, Ar, Er, Si, C) [40, 78, 87, 106], although the most commonly used are Al, B, and N. Ion-implanted dopant activation is achieved by thermal annealing using resistively or rf-heated furnaces [48]. Excimer laser activation has also been reported [2]. Simulation of implantation profiles in SiC and their comparison with experimental results have been performed [3, 158].

N-type doping by ion implantation has been developed into a production process. Annealing of nitrogen implantation has, in general, resulted in low residual damage due to the small size of the nitrogen atom. Nitrogen ion implantation is usually performed in 6H-SiC at elevated temperatures. Annealing at 1500 and 1600°C for 15 min, performed in SiC crucibles, results in Retherford back-scattering yields at the virgin crystal level, indicating a good recovery of the crystalline quality. Recently, activation processes of high-dose (3.8×10^{15} and 7.1×10^{15} cm^{-2}) nitrogen implants into 6H-SiC have been investigated [11]. It was shown that low resistivity of the implanted material can be obtained after long-time (2000 min) anneals at 900°C.

One of the current problems in ion implantation technology is activation of the p-type dopants. Because Al is a large atom, higher annealing temperatures and times are required to produce device-quality p-type layers [87]. Amorphization and recrystallization of Al-implanted 6H-SiC were investigated [78]. Ion implantation was performed in n-type 6H-SiC in the temperature range of room temperature to 1000°C. Aluminum ions were implanted with a dose range of 5×10^{13} to 5×10^{16} cm^{-2} and implant energies of 180 and 360 keV into 6H-SiC epitaxial layers having doping concentrations of 1×10^{16} cm^{-3}. After implantation, the samples were annealed between 800 and 1600°C for 30 min in an Ar flow. It was shown that density of defects induced by implantation decreased exponentially with implantation temperature. However, residual defects were detected even after annealing at 1600°C. At a 1600°C anneal temperature sublimation of SiC was observed. Carrier concentration and mobility were found to be independent of the implantation temperature, and the carrier concentrations at room temperature were measured to be about 5% of implanted dopant. The measured hole mobility was less than 1 cm^2/V s (300 K). Annealing p-type implants at lower temperatures has proven ineffective. For example, annealing at 1400°C resulted in a hole concentration of only 5×10^{17} cm^{-3} [172]. Also, C or Si coimplantation also did not improve Al activation efficiency [171]. Experimental data on ion implantation in other than the 6H polytype are limited.

V. Conclusions

The technology of SiC epitaxy and doping has shown dramatic progress in the last few years. It appears that sublimation is the current choice of technology for commercial substrates. Significant progress has been made in improving the quality of 6H and 4H material, and dramatic strides have been made in the reduction of micropipes. Still, for SiC to continue to advance and capture a larger percentage of the semiconductor market, the substrate size needs to be increased quickly to 4 in. in diameter, and subsequently to 6 in. in diameter. It remains to be seen whether a vapor-based technology (e.g., physical vapor transport) can meet the challenges posed by market demands for large-area wafers. It will also be interesting to see if there will be continued improvement in the quality of 3C substrates.

Epitaxial SiC is dominated by CVD processes and will continue to be so in the foreseeable future. New large-area commercial production tools are now available to handle multiple or large-size SiC wafers (as soon as they become available). New techniques of substrate preparation and etching have been developed, and the surface morphology of epitaxial films is vastly improved. However, defect density in SiC epitaxial structures is too high to demonstrate large-area (larger than a few sq mm) semiconductor devices. A major issue is the growth rate in CVD and to what extent that can be increased while maintaining material quality. Recent progress in SiC LPE, including the development of low defect density material, makes this technique promising for high power device structures. MBE is an interesting technique for the growth of SiC and exhibits a promising future for assisting in the fundamental understanding of this material; however, the technology has much to demonstrate before it can leave the laboratory.

A wide range of doping concentrations ($10^{14}-10^{20}$ cm^{-3}) have been demonstrated in the epitaxial technologies as well as by ion implantation. Still, the development of a p-type low-resistivity doping process is critical to a number of potential SiC devices.

In summary, SiC materials research is a healthy and growing field and is providing the basic tools for producing high-quality materials and advancing the state of the art in SiC devices.

REFERENCES

For the sake of brevity, the following abbreviations will be used:

1993 Washington ICSCRM—*Silicon Carbide and Related Materials*: Proceedings of the Fifth International Conference, Nov. 1–3, 1993, Washington, D.C. Eds. M. G. Spencer, R. P. Devaty, J. A. Edmond, M. Asif Kahn, R. Kaplan, and M. Rahman, Institute of Physics Conf. Series 137, Philadelphia and Bristol.

1995 Kyoto ICSCRM—*Silicon Carbide and Related Materials 1995*: Proceedings of the Sixth International Conference, Kyoto, Japan, September 18–21, 1995, Eds. S. Nakashima, H. Matsunami, S. Yoshida, and H. Harma, Institute of Physics Conf. Series 142, Philadelphia and Bristol.

1. Ahmed, S., Barbero, C. J., Sigmon, T. W., and Ericson, J. W. "Boron and aluminum implantation in SiC." *Appl. Phys. Lett.* **65** (1994) 67.

2. Ahmed, S., Barbero, C. J., and Sigmon, T. W. "Activation of ion implanted dopants in a-SiC." *Appl. Phys. Lett.* **66** (1995) 712.
3. Ahmed, S., Barbero, C. J., Sigmonh, T. W., and Ericson, J. W. "Empirical depth profile simulator for ion implantation in 6H-SiC." *J. Appl. Phys.* **77** (1995) 6194.
4. Allendorf, M. D. "Equilibrium predictions of the role of organosilicon compounds in the chemical vapor deposition of silicon carbide." *J. Electrochem. Soc.* **140** (1993) 747–753.
5. Alok, D., and Baliga, B. J. "High voltage (450 V) 6H-SiC substrate gate JFET (SG-JFET)" (1995) Kyoto ICSCRM, pp. 749–752.
6. Anikin, M. M., Lebedev, A. A., Popov, I. V., Sevastjanov, V. N., Syrkin, A. L., Suvorov, A. V., Chelnokov, V. E., and Shpinev, G. P. "SiC based rectifying diode." *Sov. Phys. Techn. Lett.* **10** (1984) 1053.
7. Anikin, M. M., Ivanov, P. A., Lebedev, A. A., Pyatko, S. N., Strel'chuk, A. M., and Syrkin, A. L., "High-temperature discrete devices in 6H-SiC: sublimation epitaxial growth, device technology and electrical performance," in *Semiconductor Interfaces and Microstructures* (World Scientific Publishing Co. Ltd., Singapore, 1993) pp. 280–311.
8. Acheson, A. G., Engl. Pat. 17911 (1892); the method has been described in Knippenberg, W. F. (1963). "Growth Phenomena in Silicon Carbide." Philips Research Reports 18, pp. 161–274, Edited by the Research Laboratory of N. V. Philips Gloeilampenfabriken, Eindhoven, Netherlands.
9. Bakin, A. S., Hallin, C., Kordina, O., and Janzen, E. "High resolution XRD study of silicon carbide CVD growth." (1995) Koyto ICSCRM, pp. 433–436.
9a. Barash, A. S., Vodakov, Yu. A., Kol'tsova, E. N., Mal'tsev, A. A., Mokhov, E. N., and Roenkov, A. D., "Green light emitting diodes fabricated on heteroepitaxial layers of 4H silicon carbide", *Sov. Phys. Techn. Lett.* **14** (1988) 2222.
10. Barlett, R. W., and Mueller, R. A. "Epitaxial growth b-SiC." *Mat. Res. Bull.* **4** (1969) S 341.
11. Brandt, C. D., Agarwal, A. K., Augustine, G., Barron, R. R., Burk, A. A., Jr., Clarke, R. C., Glass, R. C., Hobgood, H. M., Morse, A. W., Rowland, L. B., Seshadri, S., Siergiej, R. R., Smith, T. J., Sriram, S., Driver, M. C., and Hopkins, R. H., "Advances in SiC materials and devices for high frequency applications" (1995) Kyoto ICSCRM, pp. 659–664.
12. Barrett, D. L., Seidensticker, R. G., Gaida, W., Hopkins, R. H., and Choyke, W. J. "Sublimation Vapor Transport Growth of Silicon Carbide." Springer Proc. in Physics, v. 56, Eds. G. L. Harris, M. G. Spencer, and C. Y. Yang (Springer, Berlin, 1992) pp. 33–39.
13. Barrett, D. L., Mchugh, J. P., Hobgood, H. M., Hopkins, R. H., McMillin, P. G., Clark, R. C., and Choyke, W. J. "Growth of large SiC single crystal." *J. Crystal Growth* **128** (1993) 356–362.
14. Behrens, R. G., and Rinehard, G. H. "Vaporization Thermodynamics and Kinetics of Hexagonal Silicon Carbide." NBS Specil Publication N 561 (1979).
15. Berman, I., and Comer, J. "Heteroepitaxy of beta silicon carbides employing liquid metals." *Mater. Res. Bull.* **4** (1969) 107.
16. Brander, R. W., and Sutton, R. P. "Solution grown SiC p-n junctions." *Br. J. Appl. Phys.* **2** (1969) 309.
17. Buchan, N. I., Choi, S., Henshall, D., Woo Sik Yoo, Tischler, M. A., Roth, M. D., and Mitchel, W. C. "Characterization of impurity incorporation in 6H-SiC grown by physical vapor transport," Transactions of the Second Int. High Temperature Electronics Conf. June 5–10, 1994, Charlotte, NC, v. 1, pp. X-11–X-16.
18. Burk, A. A., Jr., Barrett, D. L., Hobgood, H. M., Siergiej, R. R., Braggins, T. T., Clark, R. C., Eldridge, G. W., Brandt, C. D., Larkin, D. J., Powell, J. A., and Choyke, W. J. "SiC epitaxial growth on a-axis SiC substrates." (1993) Washington ICSCRM, pp. 29–32.

19. Burk, A. A. Jr., Rowland, L. B., Agarwal, A. K., Sriram, S., Glass, R. C., and Brandt, C. D. "Vapor phase homoepitaxial growth of 6H and 4H silicon carbide." (1995) Kyoto ICSCRM, pp. 201–204.
20. Burk, A. A., Jr., and Rowland, L. B. "Reduction of unintentional aluminum spikes at SiC vapor phase epitaxial layer/substrate interfaces." *Appl. Phys. Lett.* **68** (1996) 382.
21. Burk, A. A., Jr., Rowland, L. B., Augustine, G., Hobgood, H. M., and Hopkins, R. H. "The impact of pre-growth conditions and substrate polytype on SiC epitaxial layer morphology." *Mat. Res. Soc. Symp. Proc.* **423** (1996) 275–280.
22. Campbell, R. B., and Chu, T. L. "Epitaxial growth of silicon carbide by the thermal reduction technique." *J. Electrochem. Soc.* **113** (1966) 825.
23. Campbell, R. B., and Berman, H. S. "Electrical properties of SiC devices," *Mat. Res. Bull* **4** (1969) S211–S222.
24. Chein, F. R., Nutt, S. R., Yoo, W. S., Kimoto, T., and Matsunami, H. "Terrace growth and polytype development in epitaxial β-SiC films on α-SiC (6H and 15R) substrates." *J. Mater. Res.* **9** (1994) 940.
25. Rai-Choudhury, P., and Formigoni, N. P. "β-Silicon carbide film." *J. Electrochem. Soc.* **116** (1969) 1440.
26. Chelnokov, V. E., Syrkin, A. L., and Dmitriev, V. A. (1996) "Overview of silicon carbide power electronics," Reported at the 1st European Conference on Silicon Carbide and Related Materials, Greece, October 6–9, 1996.
27. Choyke, W. J., Devaty, R. P., Clemen, L. L., Yoganathan, M., Pensl, G., and Hassler, C. "Intense erbium-1.54-μm photoluminescence from 2 to 525 K in ion-implanted 4H, 6H, 15R, and 3C-SiC." *Appl. Phys. Lett.* **65** (1994) 1668–1670.
28. Clemen, L. L., Devaty, R. P., Choyke, W. J., Burk, A. A., Jr., Larkin, D. J., and Powell, J. A. "Hydrogen in CVD films of 6H, 4H, and 15R SiC" (1993) Washington ICSCRM, pp. 227–230.
29. Davis, R. F. "Epitaxial growth and doping of and device development in monocrystalline β-SiC semiconductor thin films." *Thin Solid Films* **181** (1989).
30. Davis, R. F., Kelner, G., Shur, M., Palmour, J. A., and Edmond, J. A. "Thin film deposition and microelectronic and optoelectronic device fabrication and characterization in monocdrystalline alpha and beta silicon carbide." *Proc. IEEE* **79** (1991) 677–701.
31. Drowart, J., De Maria, G., and Inghram, M. G. "Thermodynamic study of SiC utilizing a mass spectrometer." *J. Chem. Phys.* **41**, 5 (1958) 1015.
32. Drowart, J., De Maria, G. "Thermodynamic study of the binary system carbon-silicon using a mass spectrometer," in *Silicon Carbide: A High Temperature Semiconductor*, Eds. J. R. O'Connor and J. Smiltens (Pergamon, New York, 1960) p. 16.
33. Dmitriev, V. A., Ivanov, P. A., Morozenko, Ya. V., Chelnokov, V. E., and Cherenkov, A. E. "Silicon carbide devices for active electronics and optoelectronics," in *Applications of Diamond Films and Related Materials*, Eds. Y. Tang, M. Yoshikawa, M. Murakawa, and A. Feldman (Elsevier Science, North Holland, 1991) pp. 769–774.
34. Dmitriev, V. A., Irvine, K. G., and Spencer, M. G. Presented at Office of Naval Research Workshop on SiC Materials and Devices, Charlottesville, VA, September, 1992.
35. Dmitriev, V. A., Elfimov, L. B., Il'inskaya, N. D., and Rendakova, S. V. "Liquid-phase epitaxy of silicon carbides at temperatures of 1100–1200°C," in Springer Proceedings in Physics, Vol. 56, *Amorphous and Crystalline Silicon Carbide III*, Eds. G. L. Harris, M. G. Spencer, and C. Y. Yang (Springer, Berlin) pp. 307–311.
36. Dmitriev, V. A., and Cherenkov, A. "Growth of SiC and SiC-AlN solid solution by container-free liquid phase epitaxy." *J. Cryst. Growth* **128** (1993) 343–348.

37. Dmitriev, V. A., Irvine, K. G., Spencer, M. G., and Nikitina, I. P. "Heteroepitaxial Growth of SiC on AlN by Chemical Vapor Deposition" (1993) Washington ICSCRM, p. 67.
38. Dmitriev, V. A. "Silicon carbide and SiC-AlN solid solution p-n structures grown by liquid-phase epitaxy. *Physica B* **185** (1993) 440.
39. Dmitriev, V. A. "LPE of SiC and SiC-AlN," in *Properties of Silicon Carbide*, Ed. G. L. Harris, EMIS data reviews series N13 (INSPEC, IEE, London, 1995) pp. 214–227.
40. Dunlap, H. L., and Marsh, O. J. "Diodes in silicon carbide by ion implantation." *Appl. Phys. Lett.* **15** (1969) 311–313.
41. Fazi, C., Dudley, M., Wang, S., and Ghezz, M. "Issues associated with large-area SiC diodes with avalanche breakdown" (1993) Washington ICSCRM, pp. 487–490.
42. Feng, Z. C., Rohatgi, A., Tin, C. C., Hu, R., Wee, A. T. S., and Se, K. P. "Structural, optical, and surface science studies of 4H-SiC epilayers grown by low pressure chemical vapor deposition." *J. Electron. Mater.* **25** (1996) 917.
43. Fissel, A., Schroter, B., and Richter, W. "Low-temperature growth of SiC thin films on Si and 6H-SiC by solid-source molecular beam epitaxy." *Appl. Phys. Lett.* **66** (1995) 3182–3184.
44. Fissel, A., Kaiser, U., Pfennighaus, K., Ducke, E., Schroter, B., and Richter, W. "Adlayer-determined epitaxial growth of SiC on Si-stabilized a-SiC(0001) by solid-source MBE," (1995) Kyoto ICSCRM, pp. 121–124.
45. Frank, F. C. "Capillary equilibria of dislocated crystals." *ActaCryst* **4** (1951).
45a. Furukawa, K., Tajima, Y., Saito, H., Fuji, Y., Suzuki, A., and Nakajima, S. "Bulk growth of single crystal cubic silicon carbide by vacuum sublimation method." *Jpn. J. Appl. Phys.* **32**, Pt. 2, No. 5A (1993) L645.
46. Fuyuki, T., Nakayama, M., Yoshinobu, T., Shiomi, H., and Matsunami, H. "Atomic layer epitaxy of cubic SiC by gas source MBE using surface superstructure." *J.Cryst. Growth* **95** (1989) 461.
47. Fuyuki, T., Yoshinobu, T., and Matsunami, H. "Atomic layer epitaxy controlled by surface superstructure in SiC." *Thin Solid Films* **225** (1993) 225.
48. Gardner, J. A., Rao, M. V., Tian, Y. L., Holland, O. W., Roth, E. G., Chi, P. H., and Ahmad, I. "Rapid thermal anneling of ion implanted 6H-SiC by microwave processing." *J. Electron. Mater.* **26** (1997) 144.
48a. Glasow, P. A. "6H-SiC studies and developments at the Corporate Research Laboratory of Siemens AG and the Institute for Applied Physics of the University in Erlangen (FRG)," in Springer Proceedings in Physics, Vol. 34, Eds. G. L. Harris and C. Y.-W. Yang (Springer, Berlin, 1989) pp. 13–34.
49. Glass, R. C., Henshall, D., Tsvetkov, V. F., and Carter, C. H., Jr. "SiC-seeded crystal growth." MRS Bull. March (1997) 30–35.
50. Glass, R. C., Kjellberg, L. O., Tsvetkov, V. F., Sundgren, J. E., and Janzen, E. "Structural macro-defects in 6H-SiC wafers," *J. Crystal Growth* **132** (1993) 504.
51. Glagovski, A. A., Grankovski, E. V., Drozdov, A. K., Efimov, V. M., Skrinnikova, G. Ya., Smirnov, B. V., Travadgan, M. G., Chetkov, M. P., and Shevchenko, V. A. "Some aspects of obtaining silicon carbide and epitaxial SiC-based structures," in *Problems of Physics and Technoology of Wide Band Gap Semiconductors* (LIYaF, Leningrad, 1980) pp. 226–240 [in Russian].
52. Gorin, S. N. and Pleutushkin, A. A. "Crystal structure peculiarities of cubic silicon carbide crystals obtained from a gas phase." Izvestiya Academii Nauk USSR, Ser. Fizicheskaya, Vol. 28, N 8 (1964) 1310–1315.
53. Gusev, V. M., Demakov, K. D., and Stolyarova, V. G. "Radiative recombination of Al^+-implanted alpha-SiC pn structures." *Radiation Effects* **69** (1983) 307–310.

54. Hallin, J. C., Konstantinov, A. O., Kordina, O., and Janzen, E. "The mechanism of cubic SiC nucleation on off-axis substrates" (1995) Kyoto ICSCRM, p. 85.
55. Hobgood, H. M., McHugh, J. P., Greggi, J., Hopkins, R. H., and Skowronski, M. "Large diameter 6H-SiC crystal growth for microwave device applications" (1993) Washington ICSCRM, pp. 7–12.
56. Hobgood, H. M., Glass, R. C., Augustine, G., Hopkins, R. H., Jenny, J., Skowronski, M., Mitchel W. C., and Roth, M. "Semi-insulating 6H-SiC grown by physical vapor transport." *Appl. Phys. Lett.* **66** (1995) 1364–1366.
57. Hofmann, D., Eckstein, R., Makarov, Y., Muller, M., Muller, St. G., Schmitt, E., and Winnacker, A. (1996) "Innovative aspects of SiC bulk growth technology," presented at the 1st European Conference on Silicon Carbide and Related Materials, Greece, October 6–9, 1996.
58. Hong, L., Misawa, S., Okumura, H., and Yoshida, S. "Gas-phase reaction in epitaxial growth of SiC films by chemical vapor deposition from SiH_4 and C_3H_8" (1993) Washington ICSCRM, pp. 239–242.
59. Hong, L., and Wu, C. "Film growth mechanism in synthesis of β-SiC films by using $SiH_2Cl_2/C_2H_2/H_2$-CVD reaction system." (1995) Kyoto ICSCRM, pp. 93–96.
60. Ikeda, M., Hayakawa, T., Yamagiwa, S., Matsunami, H., and Tanaka, T. "Fabrication of 6H-SiC light-emitting diodes by a rotation dipping technique: Electroluminescence mechanisms." *J. Appl. Phys.* (1979) 8215.
61. Ikeda, M., Matsunami, H., and Tanaka, T. "Site effect on the impurity levels in 4H, 6H, and 15R SiC." *Phys. Rev. B* **22**, 6 (1980) 2842–2854.
62. Ikoma, K., Yamanaka, M., Yamaguchi, H., and Shichi, Y. "Heteroepitaxial growth of β-SiC on Si(111) by CVD using a CH_3Cl-SiH_4-H_2 gas system" *J. Electrochem. Soc.* **138**, 10 (1991) 3031–3028.
63. Itoh, A., Akita, H., Kimoto, T., and Matsunami, H. "Step-controlled epitaxy of 4H-SiC and its physical properties" (1993) Washington ICSCRM, pp. 59–62.
64. Itoh, A., Akita, H., Kimoto, T., and Matsunami, H. "High-quality 4H-SiC homoepitaxial layer grown by step-controlled epitaxy." *Appl. Phys. Lett.* **65** (1994) 1400.
65. Ivanov, P. A., and Chelnokov, V. E. "Recent developments in SiC single-crystal electronics." *Semicond. Sci. Technol.* **7** (1992) 863–880.
66. Jang, S., Kimoto, T., and Matsunami, H. "Deep level in 6H-SiC wafers and step-controlled epitaxial layers." *Appl. Phys. Lett.* **65** (1994) 581.
67. Jayatirtha, H. N., and Spencer, M. G. "The effect of source powder height on the growth rate of 3C-SiC grown by the sublimation technique" (1995) Kyoto ICSCRM, pp. 61–64.
68. Janzen, E., and Kordina, O. "Recent progress in epitaxial gowth of SiC for power device applications" (1995) Kyoto ICSCRM, pp. 653–658.
69. Jennings, V. L., Sommer, A., and Chang, H. C. "The epitaxial growth of silicon carbide." *J. Electrochem. Soc.* **113** (1966) 728.
70. Kanaya, M., Takahashi, J., Fujiwara, Y., and Moritani, A. "Controlled sublimation growth of single crystalline 4H-SiC and 6H-SiC and identification of polytypes by x-ray diffraction." *Appl. Phys. Lett.* (1991) 58–60.
71. Kaneda, S., Sakamoto, Y., Nishi, C., Kanaya, M., and Hannai, S. I. "The growth of single crystal of 3C-SiC on Si substrate by the MBE method of using multi electron beam heating." *Jpn. J. Appl. Phys.* **25** (1986) 1307.
72. Kaneda, S., Sakamoto, Y., Mihara, T., and Tanaka, T. "MBE growth of 3C-SiC/6H-SiC and electrical properties of its pn junctions." *J. Cryst. Growth* **81** (1987) 536–542.
73. Karmann, S., Suttrop, W., Schoner, A., Schadt, M., Haberstroh, C., Engeibrecht, F., Helbig, R., Penal, G., Stein, R. A., and Leibenzeder, S. "Chemical vapor deposition and

characterization of undoped and nitrogen-doped crystalline 6H-SiC." *J. Appl. Phys.* **72** (1992) 5437.
74. Karmann, S., Haberstroh, C., Engelbrecht, F., Suttrop, W., Schoner, A., Schadt, M., Helbig, R., Pensl, G., G., Stein, R. A., and Leibenzeder, S. "CVD growth and characterization of single-crystalline 6H silicon carbide." *Physica B* **185** (1993) 75–78.
75. Karmann, S., Di Cioccio, L., Blanchard, B., Ouisse, T., Muyard, D., and Jaussaud, C. "Unintentional incorporation of contaminants during chemical vapor deposition of silicon carbide." *Mater. Sci. Eng. B* **29** (1995) 134–137.
76. Karpov, S. Y., Makarov, Y., N., Ramm, M. S., and Talalaev, R. A. "Excess phase formation during sublimation growth of silicon carbide" (1995) Kyoto ICSCRM, pp. 69–72.
77. Karpov, S. Y., Makarov, Y. N., and Ramm, M. S. "Theoretical consideration of Si-droplets and graphite inclusions formation during chemical vapor deposition of SiC epitaxial layers" (1995) Kyoto ICSCRM, pp. 177–180.
78. Kawase, D., Ohno, T., Iwasaki, T., and Yatsuo, T. "Amorphization and re-crystallization of Al-implanted 6H-SiC" (1995) Kyoto ICSCRM, pp. 513–516.
79. Kendall, J. T. "Electronic conduction in silicon carbide." *J. Chem. Phys.* **21** (1953) 821.
80. Kern, R. S., Tanaka, S., and Davis, R. F. "Growth and characterization of thin, epitaxial silicon carbide films by gas-source molecular beam epitaxy." Transactions of the Second International High Temperature Electronic Conf., Charlotte, NC, June 5–10, 1994, v. 2, pp. P-141–P-146.
81. Kimoto, T., Nishino, H., Yoo, W. S., and Matsunami, H. "Growth mechanism of 6H-SiC in step controlled epitaxy." *J. Appl. Phys.* **73** (1993) 726.
82. Kimoto, T., and Matsunami, H. "Growth model for step-controlled epitaxy of SiC: Surface kinetics of adatoms on vicinal 6H-SiC{0001} faces" (1993) Washington ICSCRM, pp. 95–98.
83. Kimoto, T., and Matsunami, H. "Two-dimensional nucleation and step dynamics in crystal growth of SiC" (1993) Washington ICSCRM, pp. 55–58.
84. Kimoto, T., Itoh, A., and Matsunami, H. "Step bunching in 6H and 4H-SiC growth by step-controlled epitaxy" (1993d) Washington ICSCRM, p. 241.
85. Kimoto, T., and Matsunami, H. "Surface kinetics of adatoms in vapor phase epitaxial growth of SiC on 6H-SiC{0001} vicinal surfaces." *J. Appl. Phys.* **75** (1994) 850–859.
86. Kimoto, T., Itoh, A., Matsunami, H., Ridhara, S. S., Clemen, L. L., Devaty, R. P., Choyke, W. J., Dalibor, T., Peppermuller, C., and Pensl, G. "Characterization of high-quality 4H-SiC epitaxial layers" (1995) Kyoto ICSCRM, pp. 393–396.
87. Kimoto, T., Itoh, A., Matsunami, H., Nakata, T., and Watanabe, M. "Aluminum and boron ion implantations into 6H-SiC epilayers." *J. Electron. Mater.* **25** (1996) 879.
88. Knippenberg, W. F. "Growth Phenomena in Silicon Carbide," Philips Research Reports 18, p. 161–274, Edited by the Research Laboratory of N. V. Philips Gloeilampenfabriken, Eindehoven, Netherlands. Phillips Research Reports 18 (1963) pp. 161–274.
89. Koga, K., Fujikawa, Y., Ueda, Y., and Yamaguchi, T. (1992) "Growth and Characterization of 6H-SiC Bulk Crystals by the Sublimation Method," in Springer Proc. in Physics, v. 71, pp. 96–100.
90. Kong, H. S., Jiang, B. L., Glass, J. T., Rozogonyi, G. A., and More, K. L. "An examination of double positioning boundaries and interface misfit in beta-SiC films on alpha-SiC substrates." *J. Appl. Phys.* **63** (1988) 2645–2650.
91. Kong, H. S., Glass, J. T., and Davis, R. F. "Chemical vapor deposition and characterization 6H-SiC thin films on off-axis 6H-SiC substrates." *J. Appl. Phys.* **64**, 5 (1988) 2672–2679.
92. Konstantinov, A. O., and Ivanov, P. A. "Sublimation growth of silicon carbide in the growth system free of carbon materials" (1993) Washington ICSCRM, p. 37.

93. Konstantinov, A. O. "Sublimation growth of SiC," in *Properties of SiC*, Ed. G. L. Harris, published by INSPEC, the Institution of Electrical Engineers, London, United Kingdom (1996) pp. 170–203.
94. Konstantinov, A. O., Hallin, C., Kordina, O., and Janzen, E., "Effect of vapor composition on polytype homogeneity of epitaxial silicon carbide." *J. Appl. Phys.* **80** (1996) 5704–5712.
95. Kordina, O., Hallin, C., Glass, R. C., Henry, A., and Janzen, E. "A novel hot-wall CVD reactor for SiC epitaxy" (1993) Washington ICSCRM, pp. 41–44.
96. Kordina, O., Henry, A., Hallin, C., Glass, R. C., Konstantinov, A. O., Hemmingsson, C., Son, N. T., and Janzen, E. "CVD-growth of low-doped 6H SiC epitaxial films." *Mat. Res. Soc. Symp. Proc.* **339** (1994) 405–410.
97. Kordina, O., Bergman, J. P., Henry, A., and Jansen, E. "Long minority carrier lifetimes in 6H-SiC grown by chemical vapor deposition." *Appl. Phys. Lett.* **66** (1995a) 189.
98. Kordina, O., Hallin, C., Ellison, A., Bakin, A. S., Ivanov, I. G., Henry, A., Yakimova, R., Touminen, M., Vehanen, A., and Janzen, E. "High temperature chemical vapor deposition of SiC." *Appl. Phys. Lett.* **69**, 10 (1996) 1456–1458.
99. Krishna, P., Jiang, S. S., and Lang, A. R. "An optical and x-ray study of giant screw dislocations in silicon carbide." *J. Crystal Growth* **71** (1985) 41.
100. Krotz, G., Wanger, C., Legner, W., Sonntag, H., Moller, H., and Muller, G. "Micromachining applications of heteroepitaxially grown β-SiC layers on silicon" (1995) Kyoto ICSCRM, pp. 829–832.
101. Kutt, R. N., Mokhov, E. N., and Tregubova, A. S. "Crystal lattice deformation and crystal quality of silicon carbide epitaxial layers doped with aluminum and boron." *Fizika Tverdogo Tela* **23** (1981) 3196.
102. Larkin, D. J. "An overview of SiC epitaxial growth." *Mater. Res. Soc. Bull.* **22** (1997) 36–40.
103. Larkin, D. J., Neudeck, P. G., Powell, J. A., and Matus, L. G. "Site-competition epitaxy for controlled doping of CVD silicon carbide" (1993) Washington ICSCRM, pp. 51–54.
104. Larkin, D. J., Neudeck, P. G., Powell, J. A., and Matus, L. G. "Site-competition epitaxy for superior silicon carbide electronics." *Appl. Phys. Lett.* **65** (1994) 1659–1661.
105. Larkin, D. J. "Site-competition epitaxy for n-type and p-type dopnt control in CVD SiC epilayers" (1995) Kyoto ICSCRM, pp. 23–28.
106. Leith, F. A., King, W. J., and McNaly, P. US Air Force Report, AF CRL-67-0123, June (1967).
107. Lely, J. A. "Silizium carbide von Art und mendge der eingebeunten verunrininungen." *Ber. Dtsch. Kerm. Ges.* **32** (1955) 229–231.
108. Lever, R. F. "Transport of solid through the gas phase using a single heterogeneous equilibrium." *J. Chem. Phys.* **37**, 6 (1962) 1174.
109. Lever, R. F. "Multiple reaction vapor transport of solids." *J. Chem. Phys.* **37**, 5 (1962) 1078.
110. Lever, R. F. and Mandel, G. "Diffusion and the vapor transport of solids." *J. Phys. Chem. Solids* **23** (1962) 599.
111. Lever, R. F. "Gaseous equilibria in the germanium iodine system." *J. Electrochem. Soc.* **110**, 7 (1963) 775.
112. Liaw, P., and Davis, R. F. "Epitaxial growth and characterization β-SiC thin films." *J. Electrochem. Soc.* **132** (1985) 642.
113. Lilov, D. S. K., Tairov, Y. M., Tsvetkov, V. F., and Chernov, M. V. "Structural and morphological peculiarities of the epitaxial layers and monocrystals of silicon carbide highly doped by nitrogen." *Phys. Stat. Solidi* (a) 37 (1976) 143.
114. Lilov, S. K., Tairov, Y. M., and Tsvetkov, V. F. "Study of silicon carbide epitaxial growth kinetics in the SiC-C system." *J. Crystal Growth* **46** (1979) 269.

115. Mandel, G. "Vapor transport of solids by vapor phase reactions." *J. Phys. Chem. Solids* **23** (1962) 587.
116. Mandel, G. "Multiple reactions and vapor transport of solids." *J. Chem. Phys.* **37**, 6 (1962) 1177.
117. Mandel, G. "Surface Limited vapor solvent growth of crystals." *J. Chem. Phys.* **40**, 3 (1964) 683.
118. Marshall, R. C. "Growth of silicon carbide from solution." *Mater. Res. Bull.* **4** (1969) S73–S84.
119. Matsunami, H., Nishino, S., Odaka, M., Tanaka, T. "Epitaxial growth of α-SiC layers by chemical vapor deposition technique." *J. Cryst. Growth* **31** (1975) 72.
120. Matsunami, H. "Recent progress in epitaxial growth of SiC," in *Amorphous and Crystalline Silicon Carbide IV*. Springer Proceedings in Physics, Vol. 71. Eds. C. Y. Yang, M. M. Rahman, and G. L. Harris (Springer, Berlin, 1992) pp. 3–12.
121. Matsunami, H. "Progress in epitaxial growth of SiC." *Physica B* **185** (1993) 65.
122. Minagawa, S., and Gatos, H. C. "Epitaxial growth of α-SiC from the vapor phase." *Jpn. J. Appl. Phys.* **10** (1971) 1680.
123. Mokhov, E. N., Shulpina, I. L., Tregubova, A. S., and Vodakov, Y. A. "Epitaxial growth of silicon carbide layers by sublimation 'Sandwich Method' (II)." *Crystal Res. Technol.* **16** (1981) 879.
124. Mokhov, E. N., Vodakov, Y. A., and Lomakina, G. A. "Issues in controllable fabrication of SiC-based doped structures," in *Problems of Physics and Technology of Wide Band Gap Semiconductors* (LIYaF, Leningrad, 1980) pp. 136–150 [in Russian].
125. Mokhov, E. N., Ramm, M. G., Roenkov, A. D., Vodakov, Y. A., Verenchikova, R. G., Zabrodski, G. A., Kol'tsova, E. N., Lomakina, G. A., Maltsev, A. A., and Oding, V. G., in *Properties of Doped Semiconductor Materials*, Ed. V. S. Zemskov (Nauka, Moscow, 1990) p. 51 [in Russian].
126. Mokhov, E. N., Radovanova, E. I., and Sitnikova, A. A. "Investigation of structural defects in SiC crystals by transmission electron microscopy," in *Amorphous and Crystalline Silicon Carbide III*, Springer Proceedings in Physics, Vol. 56, Eds. G. L. Harris, M. G. Spencer, and C. Y. Yang (Springer, Berlin, 1992) p. 231.
127. Mokhov, E. N., Radovanova, E. I., and Sitnikova, A. A. *Amorphous and Crystalline Silicon Carbide III*, Springer Proceedings in Physics, Vol. 56, Eds. G. L. Harris, M. G. Spencer, and C. Y. Yang (Springer, Berlin, 1992) p. 207.
128. Mokhov, E. N., Roenkov, A. D., Vodakov, Y. A., Saparin, G. V., and Obyden, S. K. "4H-SiC growth by sandwich method" (1995) Kyoto ICSCRM, pp. 245–248.
129. Moore, W. J. "Identification and activation energies of shallow donors in cubic SiC." *J. Appl. Phys.* **74**, 3 (1995) 1805–1809.
130. Moore, W. J., Freitas, J. A., Jr., Altaiskii, Y. M., Zuev, V. L., and Ivanova, L. M. "Donor excitation spectra in 3C-SiC" (1993) Washington ICSCRM, pp. 181–184.
131. Motoyama, S., and Kaneda, S. "Low temperature growth of 3C-SiC by the gas source molecular beam epitaxial technique." *Appl. Phys. Lett.* **54** (1989) 242.
132. Motoyama, S., Morikawa, N., and Kaneda, S. "Low temperature growth and its growth mechanisms of 3C-SiC crystal by gas source molecular beam epitaxial method." *J. Cryst. Growth* **100** (1990) 615.
133. Sugii, T., Aoyama, T., and Ito, T. "Low-temperature growth of β-SiC on Si by gas-source MBE." *J. Electrochem. Soc.* **137** (1990) 989.
134. Muench, W., Kurzinger, W., and Pfaffeneder, I. "Silicon carbide light-emitting diodes with epitaxial junctions." *Solid State Electron.* **19** (1975) 871–874.
135. Muench, W. V., and Pfaffeneder, I. "Epitaxial deposition of silicon carbide from silicon tetrachloride and hexane." *Thin Solid Films* **31** (1976) 39–51.

136. Muench, W. V. and Kurzinger, W. "Silicon carbide blue-emitting diodes produced by liquid phase epitaxy." *Solid State Electron.* **21** (1978) 1129.
137. Nadella, R. K., Capono, M. A. "High-resistance layers in n-type 4H-silicon carbide by hydrogen ion implantation." *Appl. Phys. Lett.* **70** (1997) 886.
137a. Nakata, T., Koga, K., Matsushita, Y., Ueda, Y., and Niina, T. "Single crystal growth of 6H-SiC by a vacuum sublimation method and blue LEDs," in Springer Proceedings in Physics, Vol. 43, Eds. M. M. Rahman, C. Y. Yang, and G. L. Harris (Springer, Berlin, 1989) pp. 26–34.
138. Nelson, W. E., Halden, F. A., and Rosengreen, A. "Growth and properties of β-SiC single crystals." *J. Appl. Phys.* **37** (1966) 333.
139. Neudeck, P. G. and Powell, J. A. "Performance limiting micropipe defects in silicon carbide wafers." *IEEE Electron Device Lett.* **15** (1994) 63–65.
140. Nikolaev, A. E., Nikitina, I. P., and Dmitriev, V. A. "Highly nitrogen doped 3C-SiC grown by liquid phase epitaxy" (1995) Kyoto ICSCRM, pp. 125–128.
141. Nikolaev, A. E., Ivantsov, V. A., Rendakova, S. V., Blashenkov, M. N., and Dmitriev, V. A. "SiC liquid-phase epitaxy on patterned substrates." *J. Cryst. Growth* **166** (1996) 607–611.
142. Nishino, K., Kimoto, T., and Matsunami, H. "Epitaxial growth of 3C-SiC on α-SiC substrates by chemical vapor deposition" (1993) Washington ICSCRM, p. 33.
143. Nishino, K., Kimoto, T., and Matsunami, H. "Homoepitaxial growth of 3C-SiC on 3C-SiC substrates grown by sublimation method" (1995) Kyoto ICSCRM, pp. 89–92.
144. Nishino, S., Matsunami, H., Tanaka, T. "Growth and morphology of 6H-SiC epitaxial layers by CVD." *J. Cryst. Growth* **45** (1978) 144–149.
145. Nishino, S., and Sorier, J. (Springer Proc.), Vol. 34, (1989) pp. 186–190.
146. Nishino, S., Ibaraki, H., Matsunami, H., Tanaka, T., *Jpn. J. Appl. Phys.* (Japan), Vol. 19, (1980) pp. 353–356.
147. Nishino, S., Powell, J. A., and Will, H. A. "Production of largest-area single crystal wafers of cubic SiC for semiconductor devices." *Appl. Phys. Lett.* **42** (1983) 460.
148. Nishino, S., Suhara, H., Ono, H., Matsunami, H. "Epitaxial growth and electric characteristics of cubic SiC on silicon." *J. Appl. Phys.* **61** (1987) 4889.
149. Nishino, S. and Saraie, J. "Heteroepitaxial growth of cubic SiC on a Si substrate using the Si_2H_6-C_2H_2-H_2 system," in *Amorphous and Crystalline Silicon Carbide II*, in Springer Proceedings in Physics, Vol. 43, Eds. M. M. Rahman, C. Y. Yang, and G. L. Harris (Springer, Berlin, 1989a) pp. 8–13.
150. Nishino, S., and Saraie, J. "Epitaxial growth of 3C-SiC on Si substrate using methyltrichlorosilane," in Springer Proceedings in Physics, Vol. 34, *Amorphous and Crystalline Silicon Carbide*, Eds. G. L. Harris and C. Y.-W. Yang (Springer, Berlin, 1989b) p. 45.
151. Nishino, S., Takahashi, K., Tanaka, H., and Saraie, J. "Crystal growth of SiC on AlN/sapphire by CVD method" (1993) Washington ICSCRM, p. 63.
152. Nishino, S. "Chemical vapor deposition of SiC," in *Properties of Silicon Carbide*, Ed. G. Harris (INSPEC, IEE London, 1995) pp. 204–213.
153. Nordell, N., Andersson, S. G., and Schoner, A. "A new reactor concept for epitaxial growth of SiC" (1995) Kyoto ICSCRM, pp. 81–84.
154. Nordell, N., Savage, S., and Sconer, A. "Aluminum doped 6H-SiC: CVD growth and formation of ohmic contacts" (1995) Kyoto ICSCRM, pp. 573–576.
155. Nordell, N., Schoner, A., Linnarsson, M. K. "Control of Al and B doping transients in 6H and 4H SiC grown by vapor phase epitacy." *J. Electron. Mater.* **26** (1997) 187.
156. Onoue, K., Nishikawa, T., Katsuno, M., Ohtani, N., Yashiro, H., and Kanaya, M. "Fabrication of low resistivity n-type 6H and 4H SiC substrates by the sublimation growth" (1995) Kyoto ICSCRM, pp. 65–68.

157. Palmour, J. W., Allen, S. T., Singh, R., Lipkin, L. A., and Waltz, D. G. "4H-silicon carbide power switching devices" (1995) Kyoto ICSCRM, pp. 813–816.
158. Pan, J. N., Cooper, J. A., Jr., and Melloch, M. R. "Activation of nitrogen implants in 6H-SiC." *J. Electron. Mater.* **26** (1997) 208.
159. Pazik, J. C., Kelner, G., Bottka, N., and Freitas, J. A., Jr., "Chemical vapor deposition of β-SiC on silicon-on-sapphire and silicon-on-insulator substrates." *Mater. Sci. Eng. B* **11** (1992).
160. Pensl, G., Afanas'ev, V. V., Bassler, M., Schadt, M., Troffer, T., Heindl, J., Strunk, H. P., Maier, M., and Choyke, W. J. "Electrical properties of silicon carbide polytypes" (1995) Kyoto ICSCRM, pp. 275–280.
161. Powell, J. A., Matus, L. G., and Kuczmarski, M. A. "Growth and characterization of cubic SiC single-crystal films on Si." *J. Electrochem. Soc.* **134** (1987) 1558–1564.
162. Powell, J. A., Petit, J. B., Edgar, J. H., Jenkins, I. G., Matus, L. G., Yang, J. W., Pirouz, P., Choyke, W. J., Clemen, L., and Yoganathan, M. "Controlled growth of 3C-SiC and 6H-SiC films on low-tilt-angle vicinal (0001) 6H-SiC wafers." *Appl. Phys. Lett.* **59** (1991) 333–335.
163. Powell, J. A., Petit, J. B., Matus, L. G., and Lempner, S. E. "Growth and characterization 3C-SiC and 6H-SiC films on 6H-SiC wafers," in Springer Proceedings in Physics, Vol. 56. *Amorphous and Crystalline Silicon Carbide III*, Eds. G. L. Harris, M. G. Spencer, and C. Y. Yang (Springer, Berlin, 1992) p. 313.
164. Powell, J. A., Larkin, D. J., Petit, J. B., and Edgar, J. H. "Investigation of the growth of 3C-SiC and 6H-SiC films on low-tilt-angle vicinal (0001) 6H-SiC wafers," in Springer Proceedings in Physics, Vol. 71, *Amorphous and Crystalline Silicon Carbide IV*, Eds. C. Y. Yang, M. M. Rahman, and G. L. Harris (Springer, Berlin, 1992) p. 23.
165. Powell, J. A., Neudeck, P. G., Larkin, D. J., Yang, J. W., and Pirouz, P. "Investigation of defects in epitaxial 3C-SiC, 4H-SiC and 6H-SiC films grown on SiC substrates" (1993) Washington ICSCRM, pp. 161–164.
166. Powell, J. A., Larkin, D. J., and Abel, P. B. "Surface morphology of silicon carbide epitaxial films." *J. Electron. Mater.* **24** (1995) 295–301.
167. Powell, J. A., Larkin, D. J., and Abel, P. B. "Effect of tilt angle on the morphology of SiC epitaxial films grown on vicinal (0001)SiC substrate" (1995b) Kyoto ICSCRM, pp. 77–80.
168. Powell, J. A., Larkin, D. J., Zhou, L., and Pirouz, P. "Sources of morphological defects in SiC epilayers." Transactions of the Third Int. High Temperature Electronics Conference, Albuquerque, NM, June 9–14, 1996, II-3.
169. Powell, J. A., and Neudeck, P. (1997) "Chemical vapor deposition (CVD) of 4H-SiC epitaxial films with reduced defects and improved morphology," reported at DARPA Silicon Carbide (SiC) High-Power Electronic Materials Program Review, Arlington, VA, February 10, 1997.
170. Ramungul, N., Khemkia, V., Tyagi, R., Chow, T. P., Ghezzo, M., Neudeck, P. G., Kretchmer, J., Hennessy, W., and Brown, D. M. "Comparison of aluminum- and boron-implanted vertical 6H-SiC p^+n junction diodes" (1995) Kyoto ICSCRM, pp. 713–716.
171. Rao, M. V., Griffiths, P., Holland, O. W., Kelner, G., Freitas, J. A., Jr., Simons, D. S., Chi, P. H., and Ghezzo, M. "Al and B ion-implantations in 6H- and 3C-SiC." *J. Appl. Phys.* **77**, 6 (1995) 2479–2485.
172. Rao, M. V., Gardner, J., Holland, O. W., Kelner, G., Ghezzo, M., Simons, D. S., and Chi, P. H. "Al and N ion implantations in 6H-SiC" (1995) Kyoto ICSCRM, pp. 521–524.
173. Rendakova, S. V., Ivantsov, V. A., Nikitina, I. P., Tregubova, A. S., and Dmitriev, V. A. "Defects in 6H-SiC and 4H-SiC layers grown by liquid phase epitaxy," reported at the Workshop on "Wide band semiconductors: Defects and Fundamental properties," NC, January 1997.

174. Rowland, L. B., Tanaka, S., Kern, R. S., and Davis, R. F. "Growth and characterization of β-SiC films grown on Si by gas-source molecular beam epitaxy," in Springer Proceedings in Physics, Vol. 71, Eds. C. Y. Yang, M. M. Rahman, and G. L. Harris (Springer, Berlin, 1992) pp. 84–89.
175. Rowland, L. B., Tanaka, S., Kern, R. S., and Davis, R. F. "Aluminum nitride/silicon carbide multilayer heterostructure produced by plasma-assisted, gas-source molecular beam epitaxy." *Appl. Phys. Lett.* **62** (1993) 3333–3335.
176. Rupp, R., Lanig, P., Schorner, R., Dohnke, K. O., Volkl, J., and Stephani, D. "Improvement of the SiC CVD epitaxial process in a vertical reactor configuration" (1995) Kyoto ICSCRM, pp. 185–188.
177. Rupp, R., Lanig, P., Volkl, J., and Stephani, D. "First results on silicon carbide vapor phase epitaxy growth in new type of vertical low pressure chemical vapor deposition reactor." *J. Cryst. Growth* **146** (1995b) 37.
178. Ryu, S., and Kornegay, K. T. "Design and fabrication of depletion load NMOS integrated circuits in 6H-SiC" (1995) Kyoto ICSCRM, pp. 789–792.
179. Sacuma, E., Yoshida, S., Okumura, H., Wisawa, S., and Endo, K. "Heteroeptiaxial growth of SiC polytypes." *Bull Electrotechnical Lab.* **51**, 12 (1987) 26.
180. Saddow, S. E., Tipton, C. W., Mazzola, M. S., Neudeck, P. G., and Larkin, D. J. "Capacitance spectroscopy on 6H-SiC pn junctions grown by site-competition epitaxy" (1995) Kyoto ICSCRM, pp. 289–292.
181. Scace, R. I., and Slack, G. A. "Solubility of carbon in silicon and germanium." *J. Chem. Phys.* **30** (1959) 1551–1555.
182. Schoner, A., Nordell, N., Rottner, K., Helbig, R., and Pensel, G. "Dependence of the aluminum ionization energy on doping concentration and compensation in 6H-SiC" (1995) Kyoto ICSCRM, pp. 493–496.
183. Sculte, F., Strauch, G., Jurgensen, H., Niemann, E., and Leidich, D. "Horizontal laminar flow CVD reactor for 4″ high temperature SiC growth." Transactions of the Second International High Temperature Electronics Conference, Charlotte, NC, June 5–11, 1994, v. 1, p. X-3.
184. Shenoy, P. M., and Baliga, B. J. "Planar, high voltage, boron implanted 6H-SiC P-N junction diodes" (1995) Kyoto ICSCRM, pp. 717–720.
185. Shields, V. B., Fekade, K., and Spencer, M. G. "Near-equilibrium growth of thick, high quality beta-SiC by sublimation." *Appl. Phys. Lett.* **62** (1993) 1919–1921.
186. Si, W., Dudley, M., Kong, H., Sumakeris, J., and Carter, C., Jr. "Investigations of 3C-SiC inclusions in 4H-SiC epilayers on 4H-SiC single crystal substrate." *J. Electron. Mater.* **26** (1997) 151.
187. Slater, D. B., Jr., Lipkin, L. A., Johnson, G. M., Suvorov, A. V., and Palmour, J. W. "NMOS and PMOS high temperature enhancement-mode devices and circuits in 6H-SiC" (1995) Kyoto ICSCRM, pp. 805–808.
188. Spiep, L., Nennewitz, O., and Pezoldt, J. "Improved ohmic contacts to p-type 6H-SiC" (1995) Kyoto ICSCRM, pp. 585–588.
189. Stein, R. A., Lanig, P., and Leibenzeder, S. "Influence of surface energy on the growth of 6H- and 4H-SiC polytypes by sublimation." *Mater. Sci. Eng. B* **11** (1992) 69–71.
190. Stein, R. A. "Formation of macrodefects in SiC." *Physica B* **185** (1993) 211–216.
191. Stinespring, C. D., and Wormhoudt, J. C. "Gas phase kinetics analysis and implications for silicon carbide chemical vapor depositions." *J. Cryst. Growth* **87** (1988) 481–493.
192. Suzuki, A., Ikeda, M., Matsunami, H., and Tanaka, T. "Liquid-phase epitaxial growth of 6H-SiC by vertical dipping technique." *J. Electrochem. Soc.* **122** (1975) 1741–1742.
193. Sywe, B. S., Yu, Z. J., Burchhard, S., and Edgar, J. H. "Epitaxial growth of SiC on sapphire substrates with an AlN buffer layer." *J. Electrochem. Soc.* **141** (1994) 510–513.

194. Takahashi, J., Kanaya, M., and Fujiwara, Y. "Sublimation growth of SiC single crystalline ingots on faces perpendicular to the (0001) basal plane." *J. Cryst. Growth* **135** (1994) 61–70.
195. Tairov, Y. M., Tsvetkov, V. F., Lilov, S. K., and Safaraliev, G. K. "Studies of growth kinetics and polytypism of silicon carbide epitaxial layers grown from the vapor phase." *J. Cryst. Growth* **36** (1976) 147–151.
195a. Tairov, Yu. M., Raihel, F. I., and Tsvetkov, V. F. "Silicon carbide solubility in tan and gallium." *Neorganicheskie Materiali* (1982) 1390–1391.
196. Tairov, Y. M., and Tsvetkov, V. F. "Investigation of growth processes of ingots of silicon carbide single crystals." *J. Cryst. Growth* **43** (1978) 209.
197. Tairov, Yu. M., and Tsvetkov, V. F. "General principals of growing large-size single crystals of various silicon carbide polytypes." *J. Cryst. Growth* **52** (1981) 146.
198. Tairov, Yu. M., and Tsvetkov, V. F. "Progress in controlling the growth of polytypic crystals," in *Progress in Crystal Growth and Characterization*, Vol. 7, Ed. P. Krishna (Pergamon, Oxford, 1983) pp. 111–161.
199. Tanaka, S., Kern, R. S., and Davis, R. F. "Effects of gas flow ratio on silicon carbide thin film growth mode and polytype formation during gas source molecular beam epitaxy." *Appl. Phys. Lett.* **65** (1994) 2851–2853.
200. Tanaka, S., King, S. W., Kern, R. S., and Davis, R. F. "Control of the polytypes (3C, 2H) of silicon carbide thin films deposited on pseudomorphic aluminum nitride (0001) surfaces" (1995) Kyoto ICSCRM, pp. 107–112.
201. Tsvetkov, V. F., Allen, S. T., Kong, H. S., and Carter, C. H., Jr. "Recent progress in SiC crystals growth" (1995) Kyoto ICSCRM, pp. 17–22.
202. Tuominen, M., Yakimova, R., Glass, R. C., Tuomi, T., and Janzen, E. "Investigation of structural defects in 4H SiC wafers." *Mater. Res. Soc. Symp. Proc.* **339** (1994) 729–734.
203. Tuominen, M., Yakimova, R., Bakin, R. S., Ivanov, I. G., Henry, A., Vehnanen, A., and Janzen, E. "Control of the rate-determining step of the silicon carbide sublimation growth" (1995) Kyoto ICSCRM, pp. 45–48.
204. Violin, E. E. "SiC-based light emitting devices," in *Problems of Physics and Technology of Wide Band Gap Semiconductors* (LIYaF, Leningrad, 1980) pp. 185–1197 [in Russian].
205. Vodakov, Y. A. and Mokhov, E. N., Patent USSR No. 403275 (1970), Patent France No. 7409089.
206. Vodakov, Y. A., Mokhov, E. N., Ramm, M. G., and Roenkov, A. D. "Epitaxial growth of silicon carbide layers by sublimation sandwich method (I)." *Kristall Technik* **14** (1979) 729–740.
207. Vodakov, Y. A., Mokhov, E. N., Roenkov, A. D., and Anikin, M. M. Effect of impurities on polytypism of silicon carbide." *Pisma J. Technich. Phisiki* **5** (1979) 367–369.
208. Vodakov, Y. A., Mokhov, E. N., Roenkov, A. D., and Saidbekov, D. T. "Effect of crystallographic orientation on the polytype stabilization and transformation of silicon carbide." *Phys. Stat. Solidi* (a) **51** (1979) 209–215.
209. Vodakov, Y. A., Mokhov, E. N., Ramm, M. G., and Roenkov, A. D., in Springer Proceedings in Physics, Vol. 56, *Amorphous and Crystalline Silicon Carbide III*, Eds. G. L. Harris, M. G. Spencer, and C. Y. Yang (Springer, Berlin, 1992) p. 329.
210. Wang, Y. A. and Davis, R. F. "Growth rate and surface microstructure in (6H)-SiC thin films grown by chemical vapor deposition." *J. Electron. Mater.* **20** (1991) 869.
211. Wang, Y., Xie, W., Cooper, J. A., Jr., Melloch, M. R., and Palmour, J. W. "Mechanisms limiting current gain in SiC bipolar junction transistors" (1995) Kyoto ICSCRM, pp. 809–812.

212. Wessels, B., Gatos, H. C., and Witt, A. F. "Epitaxial growth of silicon carbide by chemical vapor deposition," in *Silicon Carbide 1973*, Eds. R. C. Marshall, J. W. Faust, Jr., and C. E. Ryan (University of South Carolina Press, Columbia, 1974) p. 25.
213. Wolff, G. A., Das, B. N., Lamport, C. B., Mlavsky, A. I., and Trickett, E. A. "Principles of solution and traveling solvent growth of silicon carbide." *Mater. Res. Bull.* **4** (1969) S67–S72.
214. Wongchotigul, K. "Ion implantation and anneal characteristics of SiC," in *Properties of Silicon Carbide*, Ed. G. L. Harris (INSPEC, IEE, London, 1995) pp. 157–161.
215. Xie, W., Wang, Y., Melloch, M. R., Cooper, J. A., Jr., Johnson, G. A., Lipkin, L. A., Palmour, J. W., and Carter, C. H., Jr. "Development of nonvolatile random access memories in 6H-SiC" (1995) Kyoto ICSCRM, pp. 785–788.
216. Yakimova, R., Tuominen, M., Bakin, A. S., Fornall, J. O., Vehanen, A., and Janzen, E. "Silicon carbide liquid phase epitaxy in the Si-Sc-C system" (1995) Kyoto ICSCRM, pp. 101–104.
217. Yoo, W., Nishino, S., and Matsunami, H. "Epitaxial growth of thick single crystalline cubic silicon carbide by sublimation method." Memoirs of the Faculty of Engineering, Kyoto Univ., Vol. 49, Part 1 (1987) pp. 21–31.
218. Yoo, W. S., Hamaguchi, N., Carulli, J. N., Jr., Buchan, N. I., Tischler, M. A., Fen-Ren Chien, and Nutt, S. R. "Characterization of β-SiC CVD films on α-SiC substrates" (1993) Washington ICSCRM, pp. 259–262.
219. Yoshida, S., Sakuma, E., Okumura, H., Misawaand, S., and Kondo, K. "Heteroepitaxial growth of SiC polytypes." *J. Appl. Phys.* **62** (1987) 303.
220. Yoshinobu, T., Nakayama, M., Shiomi, H., Fuyuki, T., and Matsunami, H. "Atomic level control in gas source MBE growth of cubic SiC." *J. Cryst. Growth* **99** (1990) 520.
221. Yoshinobu, T., Mitsui, H., Izumlkawa, I., Fuyuki, T., and Matsunami, H. "Lattice-matched epitaxial growth of single crystalline 3C-SiC on 6H-SiC substrates by gas source molecular beam epitaxy." *Appl. Phys. Lett.* **60** (1992) 824–826.
222. Yoshikawa, T., Nishino, S., and Saraie, J. "Sublimation growth of cubic SiC bulk" (1995) Kyoto ICSCRM, pp. 57–60.
223. Yuldashev, G. F., Usmanova, M. M., Rachmatullaeva, H., Mokhov, E. N., Roenkov, A. D., and Ramm, M. G. "Macrosegregation of doping impurities in silicon carbide single crystals and epitaxial layers," in *Properties of Doped Semiconductor Materials*, Ed. V. S. Zemskov (Nauka, Moscow, 1990) pp. 152–156 [in Russian].
224. Zeigler, G., Lanig, P., Theis, D., and Weirich, C. "Single crystal growth of SiC substrate material for blue light emitting diodes." *IEEE Trans. Electron Devices*, **ED-30** (1983) 277–281.
225. Zhou, G. L., Ma, Z., Shen, T. C., Allen, L. H., and Markoc, H. "Low temperature growth of single crystalline cubic SiC on SiC by solid source molecular beam epitaxy." *J. Cryst. Growth* **134** (1993) 167.

CHAPTER 3

Building Blocks for SiC Devices: Ohmic Contacts, Schottky Contacts, and p-n Junctions

V. Saxena and A. J. Steckl

DEPARTMENT OF ELECTRICAL AND COMPUTER ENGINEERING AND COMPUTER SCIENCE
UNIVERSITY OF CINCINNATI
CINCINNATI, OH

I. INTRODUCTION	77
II. ELECTRICAL PROPERTIES OF METAL–SiC SYSTEMS	81
1. The Schottky-Mott and Bardeen Limits for Barrier Height	81
2. Metal Contacts for High-Power Applications	84
III. SURFACE PREPARATION TECHNIQUES	86
IV. METAL CONTACTS TO 6H-SiC	87
1. Ohmic Contacts to 6H-SiC	87
2. Schottky Contacts to 6H-SiC	96
3. High-Voltage Schottky Diodes on 6H-SiC	109
4. Edge Termination in 6H-SiC Schottky Diodes	113
5. Excess Leakage Current in SiC Schottky Diodes	114
V. METAL CONTACTS TO 4H-SiC	118
1. Ohmic Contacts to 4H-SiC	118
2. Schottky Contacts and High Voltage Schottky Diodes on 4H-SiC	118
VI. METAL CONTACTS TO 3C-SiC	125
1. Ohmic Contacts to 3C-SiC	125
2. Schottky Contacts and Diodes on 3C-SiC	126
VII. SiC p-n JUNCTIONS DIODE RECTIFIERS	131
1. 3C-SiC p-n Junctions	133
2. 6H-SiC p-n Junctions	138
3. 4H-SiC p-n Junctions	147
VIII. SUMMARY AND CONCLUSIONS	149
References	151

I. Introduction

In this chapter we provide a review of the essential building blocks necessary for operating SiC structures as electronic devices: ohmic contacts, Schottky contacts, and p-n junctions. Many of the advantages of SiC devices are due to its wide forbidden energy bandgap, enabling, for example, operation at high voltage or current and/or high temperature. However, this wide bandgap also creates difficulties in the fabrication of well-controlled metal–SiC contact properties. At the same time, operation under the extreme conditions of which SiC is capable introduces additional requirements on the physical and electronic properties of the metal contacts.

As discussed in detail in Chapter 1, SiC takes the form of many polytypes. Only one polytype takes the cubic form. This is the so-called β-SiC, with a three bilayer (Si/C) stacking period (or 3C). The noncubic polytypes, referred to as α-SiC, take on hexagonal, rhombohedral, and other structures. The most heavily investigated forms of SiC for electronic applications are the 3C, 4H, and 6H polytypes, of which only the hexagonal types are commercially available at this time. Some of the materials properties relevant to contact formation on each SiC polytype are summarized in Table I. They include the energy bandgap, electric field at which dielectric breakdown occurs; the saturated electron drift velocity; the carrier mobility; the electron work function; and affinity. While the breakdown field and saturated drift velocity appear to be close in value for all three polytypes, the mobility varies significantly. The cubic polytype has the highest reported electron mobility but unfortunately has not been grown in bulk crystal sizes larger than a few millimeters. Nelson et al. [Nelson 66] have reported a Hall electron mobility at 300 K ranging from ~ 700 to $\sim 1000\,\text{cm}^2/\text{V-sec}$ for n-type 3C-SiC with carrier concentrations of $\sim 10^{16}$ to $10^{17}/\text{cm}^3$. These crystals were grown from a Si melt solution in a graphite crucible. Aivazova et al. [Aivazova 77] have measured a Hall electron mobility of $1020\,\text{cm}^2/\text{V-sec}$ in a very lightly doped sample (with a carrier concentration of $7 \times 10^{13}/\text{cm}^3$ and a total impurity concentration of $2 \times 10^{15}/\text{cm}^3$). These crystals were grown by decomposition from trichlorosilane [Gorin 65] and have been shown [Steckl 96] to exhibit excellent structural and optical properties. 3C-SiC material can also be obtained by growth on α-SiC substrates or on Si. Kelner et al. [Kelner 89] have reported very good electron mobilities for n-type ($\sim 3 \times 10^{16}/\text{cm}^3$) 3C films grown on p-type Acheson 6H-SiC crystals: room-temperature Hall effect and field effect mobilities of 470 and $565\,\text{cm}^2/\text{V-sec}$, respectively.

The commercial hexagonal SiC substrates are typically cut perpendicular (or offset a few degrees) to the c-axis of the crystal. In these cases, one needs to be concerned with the carrier mobility either perpendicular to the c-axis (i.e., carrier motion in the basal plane of the lattice) or parallel to it, depending on the device structure and operation. In the α-SiC polytypes, the carrier transport properties exhibit an anisotropic behavior with regard to crystallographic orientation in each polytype. Schadt et al. [Schadt 94] have reported the electron mobility anisotropy present in heavily N-doped 4H-,

TABLE I

MATERIALS PROPERTIES OF THE THREE MAIN SiC POLYTYPES RELEVANT TO OHMIC AND RECTIFYING CONTACTS

Materials Properties at 300 K	4H-SiC	6H-SiC	3C-SiC
Energy bandgap (eV)	$\sim 3.24^a$ – excitonic	$\sim 2.86^b$ – optical	$\sim 2.2^a$ – optical
Breakdown electric field (MV/cm) at 10^{17}/cm^3 doping			1.5^d, 3.0^e
$\perp$ c-axis	—	—	
$\parallel$ c-axis	3.0^c	3.2^d	
Saturated electron drift velocity ($\times 10^7$ cm/s)	2.0^f	2.0^f	$\sim 2.0^g$; 2.5^h
$\perp$ c-axis	—	—	
$\parallel$ c-axis	—	—	
Electron (Hall) mobility (cm^2/V-sec) in n-type SiC ($\sim 10^{16}$/cm^3)			980^i
$\perp$ c-axis	790^i	370^i	
$\parallel$ c-axis	950^i	75^i	
Hole (Hall) mobility (cm^2/V-sec) in p-type SiC ($\sim 10^{16}$/cm^3)			40^f
$\perp$ c-axis	115^i	90^i	
$\parallel$ c-axis	—	—	
Work function (eV)	—	4.7^k for n-SiC	5.2^k
	—	4.85^k for p-SiC	—
Electron affinity (eV)	—	3.3^k	4.0^k

[a] Choyke et al. (64)
[b] Philip and Taft (60)
[c] Palmour et al. (94)
[d] Neudeck et al. (94a)
[e] Bhatnagar and Baliga (93): calculated based on 6H-SiC value
[f] Cree Research, 97 Product Data Sheets
[g] Kelner et al. (89)
[h] Ferry (75)
[i] Schaffer et al. (94)
[j] Nelson et al. (66)
[k] Porter and Davis (95a)

6H-, and 15-RSiC bulk crystals grown by the modified Lely method. Schaffer et al. [Schaffer 94] have investigated this effect using epitaxial SiC layers with total (donors plus acceptors) impurity levels varying from $\sim 10^{15}$/cm^3 to 10^{20}/cm^3. The room-temperature mobilities reported by Schaffer et al. [Schaffer 94] for lightly doped ($\sim 10^{16}$/cm^3) 4H- and 6H-SiC epitaxial layers are shown in Table I. In the case of 6H-SiC, the electron mobility in the direction of the c-axis is much lower than the mobility in the basal plane. This leads to a significant electron mobility anisotropy ratio for 6H-SiC:

$$A_e(6H) = \frac{\mu_\perp}{\mu_\parallel} \approx 4.9.$$

The electron mobility anisotropy of epitaxial layers doped to levels of 10^{17} and $10^{18}/cm^3$, [Schaffer 94] is essentially identical at room temperature and changes only slightly (from 4.7 to 5.1) over a wide temperature range (300–600 K). A similar anisotropy ratio of 4.8 is reported by Schadt et al. [Schadt 94] for bulk crystals with a total impurity level of $\sim 6 \times 10^{15}/cm^3$.

The basal plane electron mobility in 4H-SiC, which is the condition routinely measured using wafers cut perpendicular to the c-axis, is reported by Schaffer et al. [Schaffer 94] to be approximately twice the basal plane mobility of 6H-SiC, for total impurity concentrations up to $\sim 10^{17}/cm^3$. Furthermore, in 4H-SiC, the electron mobility parallel to the c-axis is reported by Schaffer et al. to be actually slightly higher than the mobility in the basal plane, leading to an anisotropy ratio smaller than unity:

$$A_e(4H) = \frac{\mu_\lambda}{\mu_\parallel} \approx 0.83.$$

Very similar values were also reported by Schadt et al. [Schadt 94]. The significantly larger electron mobility in 4H-SiC, along with the reduced anisotropy, indicate that this polytype should be superior to 6H-SiC in many applications. This is particularly true for power devices, where the current flow is frequently perpendicular to the substrate plane. However, one must also bear in mind that the larger energy bandgap of 4H SiC can also lead to larger Schottky barriers and more resistive contacts.

The work function and electron affinity of the SiC polytype chosen are important in determining the nature of the metal–SiC contact (ohmic vs. rectifying). The resulting "internal" characteristics (barrier height, resistivity) are also a function of additional basic materials properties such as mobility, carrier concentration, and so on. These in turn affect the important "external" characteristics, including forward DC bias current and associated voltage drop, reverse bias DC stand-off voltage and leakage current, DC and AC rectifying efficiency, and high-temperature operating capability.

In Section II, we discuss the basic properties and measurement issues of metal contacts to SiC. Surface preparation prior to metallization is briefly considered in Section III. The properties of ohmic and Schottky contacts to 6H-SiC, 4H-SiC, and 3C-SiC are reviewed in Sections IV, V, and VI, respectively. The properties and characteristics of SiC p-n junction diodes are reviewed in Section VII. Finally, in Section VIII, a brief summary and conclusions are given.

II. Electrical Properties of Metal–SiC Systems

Metal–semiconductor contacts are indispensable for operation of all electronic circuits. High-quality ohmic contacts are necessary to transfer signals to and from the semiconductor and the external circuitry. Stable rectifying (Schottky) contacts may also be required, either by themselves, to perform switching or to provide rectification, or for the efficient operation of other devices utilizing them. In this section, the theory behind both types of contacts is briefly discussed. The parameters determining the quality of these contacts are discussed next, followed by the techniques commonly utilized to measure these parameters.

1. THE SCHOTTKY-MOTT AND BARDEEN LIMITS FOR BARRIER HEIGHT

In the Schottky model for an ideal intimate metal–semiconductor contact, the respective work functions φ_M and φ_S of the metal and semiconductor determine the ohmic or rectifying nature of the contact. Figures 1 and 2 show the energy-level diagrams of the metal contacts to n-type and p-type semiconductors, respectively [van der Ziel 68]. As seen from Fig. 1b, metal contact to an n-type semiconductor with $\varphi_M > \varphi_S$ results in a rectifying contact with a barrier to the flow of electrons in either direction. The barrier to the movement of electrons from the n-type semiconductor into the metal is $\varphi_M - \varphi_S$. The barrier to the reverse flow of electrons from the metal to the semiconductor is $(\varphi_M - \chi_S)$, where χ_S is the electron affinity of the semiconductor. Under forward bias conditions (n-type semiconductor-negative with respect to the metal), the forward barrier reduces to $q(V_D - V_F)$, where V_D is the built-in potential and V_F is the applied voltage. However, the barrier to reverse flow, which is not affected by the applied bias, is defined to be the barrier height for the metal–semiconductor pair

$$\varphi_{Bn} = \varphi_M - \chi_S. \tag{1a}$$

For a metal to n-type semiconductor contact with $\varphi_M < \varphi_S$, it is seen from Fig. 1d that virtually no barrier exists in the conduction band and that the nature of the contact is ohmic.

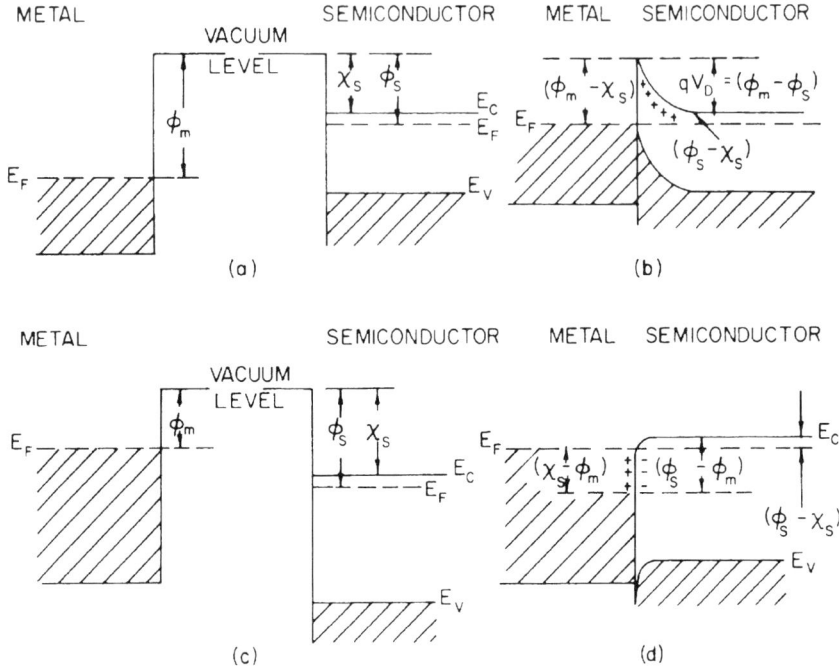

FIG. 1. Energy-level diagram of a metal contact to an n-type semiconductor [van der Ziel 68].

In Fig. 2, the corresponding band diagrams for a metal contact to a p-type semiconductor are shown. For $\varphi_M < \varphi_S$, the contact is rectifying because a barrier ($\varphi_S - \varphi_M$) exists to the flow of holes from semiconductor to metal. This barrier decreases under forward bias condition (p-type semiconductor–positive with respect to the metal). The barrier to reverse flow, however, remain unaffected and is defined as the barrier height

$$\varphi_{Bp} = E_G - (\varphi_M - \chi_S). \tag{1b}$$

For a metal to p-type semiconductor contact with $\varphi_M > \varphi_S$, there is virtually no barrier for current flow from either side, and the contact is ohmic.

The barrier height φ_B ($= \varphi_{Bn}$ or φ_{Bp}) for a metal–semiconductor pair is also known as the Schottky-Mott limit [Mott 38] and is independent of the applied bias and doping level of the semiconductor. If the contact is rectifying, φ_B is positive and is called the Schottky barrier height (SBH). In the case of an ohmic contact, the barrier has a small negative value and is referred to as the dipole surface charge barrier (DSCB). As seen from Eq.

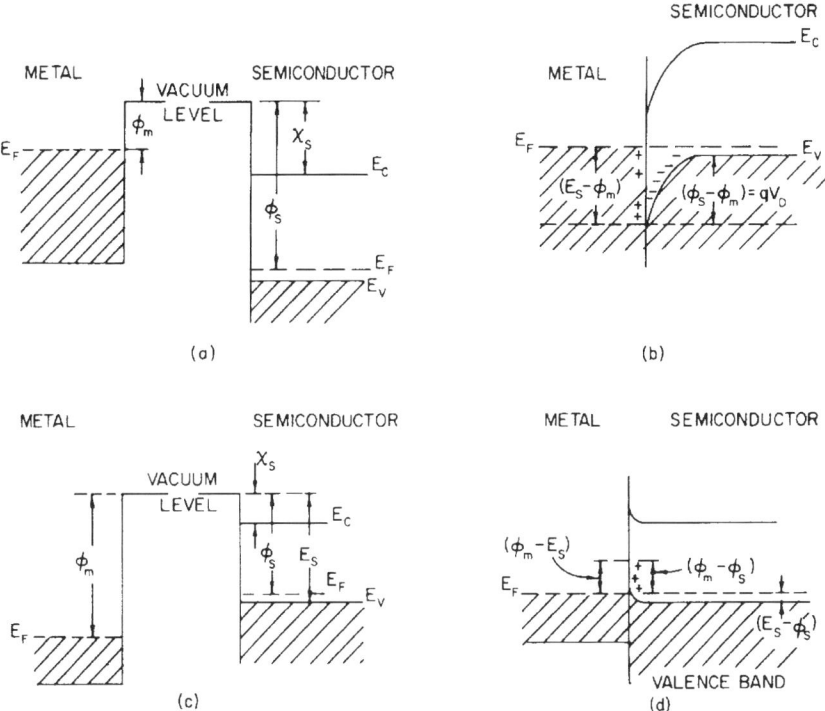

FIG. 2. Energy-level diagram of a metal contact to a p-type semiconductor [van der Ziel 68].

(1), this theory predicts that for any given semiconductor, the *sum* of the barrier heights in n-type and p-type materials for the same metal should be equal to the bandgap of the semiconductor.

Schottky-Mott theory for ideal metal–semiconductor therefore predicts that the SBH should be linearly dependent on the metal work function φ_M. Frequently, actual Schottky contacts fabricated on various semiconductors exhibit a much weaker dependence than that predicted by Eq. (1). Indeed, under certain conditions for many semiconductors, the SBH is *independent* of the choice of metal. A possible explanation suggested by Bardeen [Bardeen 47] is the presence of a continuous distribution of surface states at the interface between the semiconductor and the thin interfacial oxide layer. These surface states are characterized by a neutral level, φ_0, measured from the top of the valence band. For a large density of surface states, φ_0 approaches the surface Fermi level (E_F), and the barrier height becomes

$$\varphi_{Bn} = E_G - \varphi_0, \quad \text{for n-type} \quad (2a)$$

and

$$\varphi_{Bp} = \varphi_0, \qquad \text{for p-type.} \tag{2b}$$

This is the so-called Bardeen limit. In this situation, the barrier height is said to be "pinned" by the high density of surface states. In most practical metal–SiC Schottky contacts, the SBH takes on a value between the limits given by Eqs. 1 and 2.

2. Metal Contacts for High-Power Applications

Many important applications for SiC devices place a requirement of high blocking voltages and/or high current-handling capability. In this section, we discuss the important parameters that characterize the metal-SiC contacts in light of these requirements for high-power applications.

a. Specific Contact Resistivity and Ohmic Contacts

The intrinsic electrical property of an ohmic contact is characterized by a parameter termed *specific contact resistivity*, which is given as

$$\rho_C = R_C A_E \qquad (\Omega\text{-cm}^2), \tag{3}$$

where R_C is the contact resistance and A_E is the effective contact area of the device. The frequency and noise characteristics, and the power capability of a device are significantly affected by the quality of the ohmic contact. A condition for good ohmic contact is $R_C \ll R_{ON}$, where R_{ON} is the total on-resistance of the device. Satisfying this condition results in negligible voltage drop across the contact, which is necessary for high device efficiency. A lower value of R_C is particularly important in high-power devices because the power dissipated across R_C can cause substantial heating of the devices.

The specific contact resistivity of ohmic contacts can be measured by using a variety of different techniques. The size of the contacts and the resistivity of the metal may affect the accuracy of the measurements [Marlow 82]. Some techniques of measuring the contact resistivity include the linear transmission line method (TLM) [Shockley 64; Berger 72], the circular TLM method [Chen 95], the Kelvin technique [Proctor 82, 83], and the Cox and Strack method [Cox 67]. The experimental details for these and several other techniques have also been reviewed by Schroder [Schroder 90].

b. Schottky Barrier Height and Schottky Contacts

For power devices based on Schottky contacts, the most important parameter is the SBH. Optimization of SBH plays a key role in the determination of the on-state voltage drop and the leakage current in such devices. A larger barrier height of the metal–SiC junction leads to a higher on-state voltage drop, but also results in a much smaller leakage current density. For high-temperature device operation, it is safer to choose metals that result in a higher barrier height to SiC because the leakage current increases exponentially with temperature.

Rhoderick and Williams [Rhoderick 88] have reviewed the different techniques for measuring the SBH. The current–voltage characteristics [Norde 79; Cheung 86] are extensively utilized to obtain the SBH. The SBH can also be obtained from capacitance–voltage measurements [Goodman 63], photoelectric measurements [Anderson 75], and photoelectron emission spectroscopy [Rhoderick 88].

c. On-Resistance and Power Losses in Power Rectifiers

For efficient operation of high-power devices, it is also essential to minimize their overall power losses. The current-handling capability of power rectifiers is often limited by their on-resistance. The theoretical on-resistance for these (Schottky and p-n junction) rectifiers can be written as follows:

$$R_{on} = R_c + R_{on}(epi) + R_{on}(sub) \tag{4a}$$

$$= R_c + \rho_{epi} W_{epi} + \rho_{sub} W_{sub}, \tag{4b}$$

where R_c is (are) the contact resistance (s), R_{on} (epi) and R_{on} (sub) are the resistances per unit area of the epilayer and substrates, ρ_{epi} and ρ_{sub} are the epilayer and substrate resistivities, and W_{epi} and W_{sub} are the thickness of the epilayer and substrate, respectively. For p-n rectifiers, R_c would include contacts to both p- and n-type layers, and R_{on} (epi) would include the resistance from all the epilayers.

The average power loss dissipated per unit area of the diode during the on-state and off-state for 50% duty cycle is given by [Itoh 96a]

$$P_L = \frac{1}{2}(J_F V_F + J_R V_R), \tag{5}$$

where V_F and J_F are the forward voltage drop and the corresponding forward current density, while V_R and J_R are the reverse blocking voltage and corresponding leakage current density. Since V_F depends on R_{ON} and φ_B, and J_R depends on φ_B, the decrease in R_{ON} and the optimization of φ_B leads to the reduction of P_L.

III. Surface Preparation Techniques

The properties of an ohmic contact can strongly depend on the surface cleaning procedure used prior to forming the contact. The electrical characteristics of Schottky barrier diodes are also very sensitive to surface preparation techniques. Presence of contamination or oxide at the metal–semiconductor interface can reduce metal adhesion, affect barrier heights, increase the specific contact resistance, and create electrically active defects that may alter the conduction of carriers through the contact. Thus, it is of utmost importance to provide a reproducibly clean semiconductor surface prior to performing the metallization to form the contact. This involves removing surface contamination and any interfacial oxide layer. From a purely materials standpoint, it may be possible to obtain a very clean and oxide-free SiC surface by doing a wet chemical cleaning followed by a heat treatment. Studies utilizing procedures involving such combination of chemical cleaning processes and subsequent thermal processing have been reviewed by Porter and Davis [Porter 95a]. However, most studies on the electrical behavior of metal contacts to SiC have reported using only the wet chemical cleaning procedure prior to doing metallization. These procedures may involve the use of solvents such as trichloroethylene, trichloroethane, acetone, methanol, and propanol to degrease the SiC wafer. Buffered hydrofluoric acid ($HF + NH_4F + H_2O$) is widely used for oxide removal. Varying mixtures of acids and bases such as HF, H_2SO_4, HCl, NH_4OH, H_2O_2, HNO_3, and KOH are commonly utilized for surface cleaning.

Some specific procedures utilized for cleaning SiC surfaces prior to forming metal contacts are discussed next. The standard RCA clean procedure has routinely been used by several researchers, sometimes with certain modifications. A Huang clean procedure (10-min dip in NH_4OH: H_2O_2:H_2O::1:1:5 followed by a 10-min dip in HCl:H_2O_2:H_2O::1:1:5 solution at 70°C) has been used by the NCSU group [Raghunathan 95] before depositing the Schottky contact. Researchers at Kyoto clean the as-grown epitaxial layers in organic solvents, heated K_2CO_3, aqua regia, and HF, and then rinse in deionized water. Yoshida and co-workers at the Electrotechnical Laboratory [Yoshida 85] cleaned the epilayer surface in four steps: etching in 5% HF for 5 min, followed by a 10-min hot bath in 20% K_2CO_3, a 5-min dip in HCl, and a 5-min etch in 5% HF once again prior to doing

Schottky metallization. Waldrop and co-workers at Rockwell [Waldrop 90, 92, 93] cleaned the wafers serially in solutions of detergent, 5:1:1 $NH_4OH:H_2O_2:H_2O$, 5:1:1 $HCl:H_2O_2:H_2O$, and then followed the four-step procedure used by Yoshida et al. [Yoshida 85]. At Cincinnati, initially an RCA cleaning is performed on the SiC wafers. The successive steps for this procedure are solvent removal, organic cleaning, oxide removal, ionic cleaning, and nitrogen dry. The solvents used for cleaning are warm TCE, acetone, and methanol. Organic cleaning is performed in 5:1:1 NH_4OH: $H_2O_2:H_2O$ for 5 min, and the ionic cleaning in 5:1:1 $HCl:H_2O_2:H_2O$ for 5 min. BHF is used for oxide removal, and DI water for the rinsing performed between all the previously mentioned steps. An oxide layer is then grown thermally on the epilayer, and it serves as a sacrificial layer for regions where the contacts are formed. Prior to Schottky metal deposition, this sacrificial oxide layer is etched away using BHF from the regions where contacts are to be formed. At regions outside the Schottky contact, the oxide on the epilayer serves as a passivation layer.

IV. Metal Contacts to 6H-SiC

1. Ohmic Contacts to 6H-SiC

Ohmic contacts to both n- and p-type 6H-SiC for various metallization schemes studied are summarized in Table II. Early schemes for forming ohmic contacts on either n- or p-type SiC were reported by Hall, who utilized the alloying of SiC with tungsten at a temperature of about 1900°C [Hall 58]. High-temperature heating with other metal alloys, such as Si-Al or Si-B for p-type and Si-P for n-type, also resulted [Hall 58] in good ohmic characteristics of the contacts.

The first deposited ohmic contacts on 6H-SiC [Palmour 91] were fabricated using Ni and subsequent high-temperature annealing. Glass et al. [Glass 91, 92] obtained as-deposited ohmic contacts on fairly highly doped (1.6×10^{18} cm^{-3}) SiC by ion-assisted reactive evaporation in ultra high vacuum of TiN, with the substrates held at 350°C. In this case, the ohmic contact is due to the formation of a thin (<10 Å) layer of amorphous silicon nitride (a-Si-N), resulting in the metal–insulator–semiconductor (MIS) structure shown in Fig. 3. The a-Si-N layer appears to free the Fermi level at the SiC surface and promotes ohmic behavior. The contact resistance was found to be 4×10^{-2} Ω-cm^2.

Petit and Zeller [Petit 92] performed studies of ohmic contacts using metals (Ni and Ni/Mo) and silicides ($MoSi_2$, $TaSi_2$, $TiSi_2$). As expected, the silicide films were stable and showed little reaction on either the Si-

TABLE II
OHMIC CONTACTS TO 6H-SiC

Metallization	SiC Type	Annealing Conditions	SiC Carrier Conc. (cm^{-3})	ρ_c at RT (Ω-cm^2)	Method of ρ_c Measurement	Ref.
Al	p	700°C, 10 min	1.8×10^{18}	1.7×10^{-3}	TLM	[Crofton 92]
Al	p	800°C, 10 min	8×10^{18}	10^{-2}–10^{-3}	TLM	[Porter 95a]
Al-Si	p	1700°C	NR	NR	—	[Hall 58]
Al-Si	p	900–1000°C	NR	NR	—	[Shier 70]
Al-Ti	p	950°C, 5 min	NR	NR	—	[Nakata 89]
Al-Ti	p	1000°C, 5 min	5×10^{15}–2×10^{19}	2.9×10^{-2}–1.5×10^{-5}	Circular TLM	[Crofton 93]
Cr	n	Melting (≥ 2130°C)	NR	NR	—	[Addamiano 70]
Cu-Ti	p	>880°C	NR	NR	—	[Hall 58]
Mo	n	as-deposited	$>1 \times 10^{19}$	$\sim 1 \times 10^{-4}$	TLM, 4 pt. probe	[Petit 94]
Mo	p	as-deposited	$>1 \times 10^{19}$	2×10^{-4}	TLM, 4 pt. probe	[Petit 94]
Ni	n	1000°C, 20 s	4.5×10^{17}	1.7×10^{-4}	TLM	[Kelner 91]
Ni	n	950°C, 5 min	4.7×10^{18}	mid 10^{-2}	4 pt. probe	[Crofton 92]
Ni	n	950°C, 2 min	7.9×10^{18}	$<5 \times 10^{-6}$	TLM	[Crofton 94]
Ni	n	1050°C, 5 min	9.8×10^{17}	10^{-3}–10^{-4}	TLM	[Porter 95a]
Ni	n	1000°C, 5 min	4.5×10^{20}	1×10^{-6}	Contact area	[Uemoto 95]
Ni-Cr	n	950°C, 5 min	4.7×10^{18}	1.8×10^{-3}	Circular TLM	[Crofton 92]
Ni/3C-SiC	n	1000°C, 30 s	1.2×10^{18}	1.7×10^{-5} (Si-face); 6×10^{-5} (C-face)	Cox and Strack	[Dmitriev 94]

Metal	Type	Conditions	Doping (cm⁻³)	Resistivity (Ω·cm²)	Method	Reference
Pt	p	450–850°C, 20 min	$>1 \times 10^{18}$	NR	—	[Glass 94]
Si-B	p	1700–2000°C	NR	NR	—	[Hall 58]
Ta	n	as-deposited	$>1 \times 10^{19}$	$\sim 1 \times 10^{-4}$	TLM	[Petit 94]
					4 pt. probe	[Pelletier]
Ta	p	as-deposited	$>1 \times 10^{19}$	7×10^{-4}	TLM, 4 pt. probe	[Petit 94]
Ti	n	as-deposited	$4.5 \times 10^{17} - 1 \times 10^{20}$	$1 \times 10^{-2} - <2 \times 10^{-5}$	Circular TLM	[Alok 93]
Ti	p	as-deposited	$>1 \times 10^{19}$	3×10^{-4}	TLM, 4 pt. probe	[Petit 94]
Ti/Al	p	1000°C, 5 min	NR	NR	—	[Uemoto 95]
Ti/Al/3C-SiC	p	950°C, 2 min	NR	$2-3 \times 10^{-5}$	—	[Dmitriev 94]
TiN	n	600°C, 30 min	$\sim 1 \times 10^{18}$	4×10^{-2}	TLM	[Glass 91, 92; Crofton 92]
TiW	n	O$_2$ plasma + 600°C, 5 min	4.7×10^{18}	7.8×10^{-4}	Circular TLM	[Crofton 92]
TiW	n	750°C, 5 min	$7-8 \times 10^{18}$	$\sim 8 \times 10^{-4}$	circular TLM	[Crofton 93]
W	n	1200–1600°C	$3 \times 10^{18} - 1 \times 10^{19}$	$5 \times 10^{-3} - 1 \times 10^{-4}$ (Si-face); $1 \times 10^{-2} - 5 \times 10^{-4}$ (C-face)	4 pt. probe	[Anikin 92]
W	p	1900°C	NR	NR	—	[Hall 58]
W/AuW/W/Al	p	1900°C, 120 s	NR	$(2-5) \times 10^{-4}$	4 pt. probe	[Anikin 92]
W/Pt/Al	p	800–850°C, 10 min	8×10^{18}	$10^{-2} - 10^{-3}$	TLM	[Porter 95a]
W/Ti/Ni	n	1050°C, 5 min	9.8×10^{17}	$10^{-3} - 10^{-4}$	TLM	[Porter 95a]

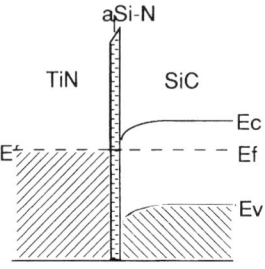

FIG. 3. Schematic representation of band diagram expected for MIS structure [Glass 92].

or C-face. Ni was unstable on both the faces and reacted to yield free carbon at the interface. The Ni/Mo bilayers films were rectifying, with SBH of 1.8 eV on the C-face and 0.9 eV on the Si-face. High-temperature anneal made these contacts ohmic, with specific contact resistance values between $(7-20) \times 10^{-4}\,\Omega\text{-cm}^2$. Later investigations [Petit 94] on refractory metals found that Mo, Ta, Ti, and Zr contacts were ohmic as-deposited on highly doped ($>10^{19}\,\text{cm}^{-3}$) 6H-SiC. These contacts resulted in contact resistivity in the $10^{-4}\,\Omega\text{-cm}^2$ range. Mo contacts resulted in the lowest as-deposited ρ_c for n-type SiC, and Ti contacts had the lowest ρ_c for p-type SiC.

Anikin et al. [Anikin 92] investigated W/Au-based ohmic contacts on n-type and W/Au/Al-based ohmic contacts on p-type 6H-SiC. For W/Au contacts to n-type SiC, ρ_c varied from $1 \times 10^{-2}\,\Omega\text{-cm}^2$ to $5 \times 10^{-4}\,\Omega\text{-cm}^2$ for the C-face, and from $1 \times 10^{-4}\,\Omega\text{-cm}^2$ to $5 \times 10^{-3}\,\Omega\text{-cm}^2$ for the Si-face. For W/Au/Al contacts to p-type SiC, ρ_c varied from $(2-5) \times 10^{-4}\,\Omega\text{-cm}^2$.

Crofton et al. [Crofton 92] found that TiW, Ni, and NiCr form ohmic contacts on n-type SiC and that Al and AlSi form ohmic contacts on p-type 6H-SiC after high-temperature annealing. For contacts to n-type SiC, TiW resulted in the lowest contact resistivity of $7.8 \times 10^{-4}\,\Omega\text{-cm}^2$. For contacts to p-type, Al metallization resulted in a contact resistivity of $1.7 \times 10^{-3}\,\Omega\text{-cm}^2$. The authors hypothesized that for TiW contacts annealing is necessary only to remove the interfacial oxide layer. In contrast to the Ni and NiCr contacts, where the ohmic nature is due to a reaction between Ni and SiC at the metal–semiconductor interface, using TiW it should be possible to form as-deposited or nonalloyed contacts with ohmic characteristics. The ohmic nature in Al and AlSi contacts is due to a heavy doping of the p-type SiC under the contact that occurs upon high-temperature annealing. The same group [Crofton 93] reported the specific contact resistance as a function of p-type doping of the epitaxial 6H-SiC for Al-Ti contacts. As shown in Fig. 4, ρ_c varied from 2.9×10^{-2} to $1.5 \times 10^{-5}\,\Omega\text{-cm}^2$ for doping concentrations ranging from 5.5×10^{15} to $2 \times 10^{19}\,\text{cm}^{-3}$. A good theoreti-

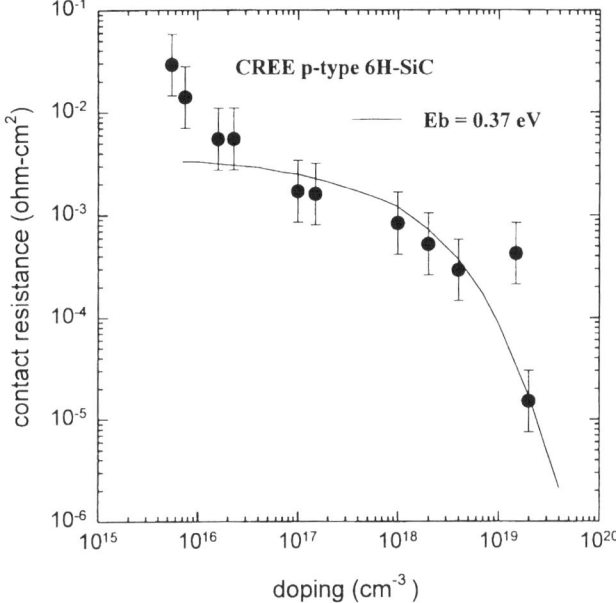

FIG. 4. Specific contact resistance versus doping for Al-Ti ohmic contacts on p-type 6H-SiC [Crofton 93].

cal fit to the contact resistance data was obtained by assuming the barrier height for the contact to equal 0.37 eV. This low value of ϕ_B has been attributed to the image force lowering of the Schottky barrier. In another study, Crofton et al. [Crofton 95] reported Ni/6H-SiC ohmic contacts capable of operating with low contact resistance at high temperatures. ρ_c less than $5 \times 10^{-6}\,\Omega\text{-cm}^2$ was obtained at room temperature, and it decreased with increasing temperature up to 500°C.

The dependence of the specific contact resistance on doping of n-type 6H-SiC was also studied. Crofton et al. [Crofton 95] reported that a good Ni/6H-SiC ohmic contact is possible on moderately to heavily doped SiC (i.e., for material with carrier concentration higher than the mid-10^{17} range). As shown in Fig. 5, the specific contact resistance decreases from 2×10^{-4} $\Omega\text{-cm}^2$ for a doping concentration of $3.2 \times 10^{17}\,\text{cm}^{-3}$ to $5 \times 10^{-6}\,\Omega\text{-cm}^2$ for $7.8 \times 10^{18}\,\text{cm}^{-3}$. The specific contact resistance for this doping range was found to further decrease to the lower $10^{-6}\,\Omega\text{-cm}^2$ range after the contacts were annealed at 1200°C. For doping concentrations below $10^{17}\,\text{cm}^{-3}$, annealing temperatures in excess of 1400°C were necessary. Reliable measurements could not be made after subjecting the contacts to such high

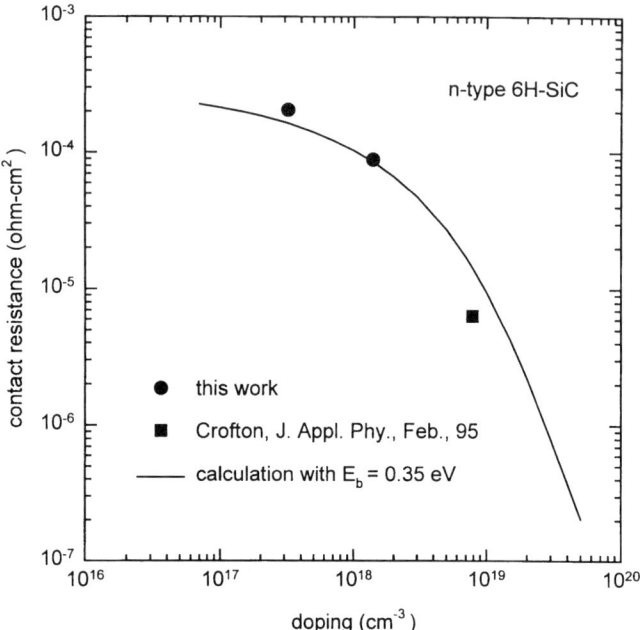

FIG. 5. Specific contact resistance as a function of doping for annealed Ni on n-type 6H-SiC [Crofton 94].

temperature, possibly due to graphitization occurring at the unprotected SiC surface, which in turn resulted in excessive leakage.

Dmitriev and co-workers [Dmitriev 94] reported low resistivity $(1.7 - 3 \times 10^{-5}\,\Omega\text{-cm}^2)$ ohmic contacts to n and p-type 6H-SiC by depositing the metal on a thin 3C-SiC layer grown by CVD on 6H-SiC. Ni was used for forming n-type contacts, and Al/Ti for the p-type contacts. This procedure for forming ohmic contacts requires single crystal heteroepitaxy of 3C-SiC films on 6H-SiC.

Uemoto [Uemoto 95] reported the dependence of specific contact resistance of Ni contacts on the doping level of n-type 6H-SiC. As shown in Fig. 6, ρ_c values as low as $10^{-6}\,\Omega\text{-cm}^2$ were obtained at a doping level of $(2-5) \times 10^{20}\,\text{cm}^{-3}$, when the contacts were annealed at 1000°C for 5 min. He also reported that it is possible to form ohmic contacts to n-SiC by using Ti/Al, which formed good ohmic contacts to p-SiC. Therefore, the same metallization scheme could now be used to form good contacts to both n- and p-type regions in a single device.

Goesmann and Schmid-Fetzer [Goesmann 95] showed that as-deposited

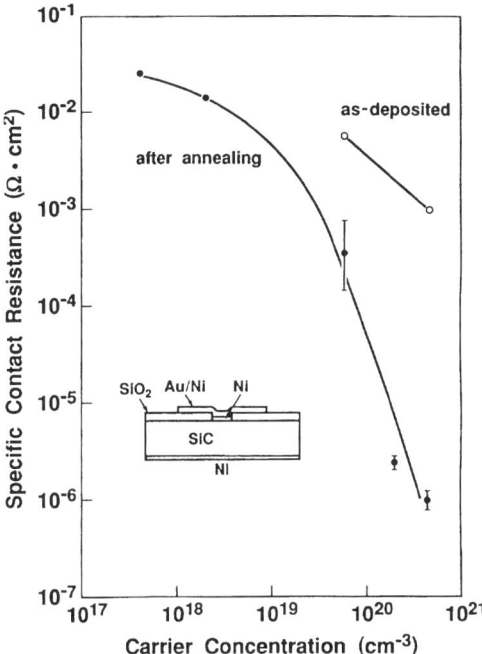

FIG. 6. Specific contact resistance before and after annealing of Ni contacts on 6H-SiC as a function of carrier concentration [Uemoto 95].

Ti/6H-SiC Schottky contacts became ohmic after a 5-min anneal at 900°C. TiC ohmic contacts to n-type 6H-SiC with a doping concentration of 4×10^{19} cm^{-3} were reported by Chaddha et al. [Chaddha 95]. The resulting contact resistivity, sheet resistance (R_s), contact resistance (R_c), and transfer length (L_T) were calculated from total resistance (R_T) versus contact spacing (d) measurements on TLM structures. The average values of these parameters were found to be $R_c = 1.3 \times 10^{-5}$ Ω-cm^2, $R_s = 14.4$ Ω/square, $R_T = 1.6$ Ω, for a spacing of d = 9.5 μm.

Hallin et al. [Hallin 96] compared Ni and Ni-Al metallizations for ohmic contacts on n-type SiC. As-deposited Ni contacts were found to be rectifying, whereas the Ni/Al contacts had an ohmic behavior with contact resistivity of 4.4×10^{-4} Ω-cm^2. The specific contact resistivity for the Ni and Ni-Al contacts after a 5-min anneal at 1000°C were 2.1×10^{-4} and 1.2×10^{-4} Ω-cm^2, respectively. Annealing for longer duration resulted in an increase in the contact resistivity, which according to the authors is due to the formation of a relatively large content of graphite in the subsurface

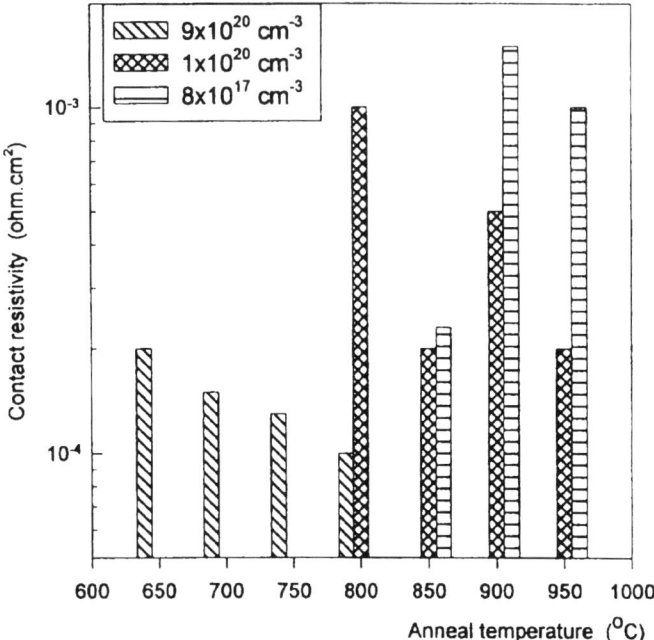

FIG. 7. Contact resistivity of Al/Ti/Al ohmic contacts as a function of anneal temperature on wafers with different carrier concentration in the active region [Nordell 96].

region. Other interesting conclusions drawn from the XPS studies by this group include that: (1) Al forms Al_2O_3 by reducing the residual SiO_2 when it comes in contact with SiC; (2) Ni or Al may have a catalytic effect in dissociating SiC into Si and C at higher temperatures ($>400°C$); (3) in Ni/SiC contacts, C remains mainly in the graphite state; on the other hand, in Ni/Al/SiC contacts, C reacts with Al to form Al_4C_3, which is a stable compound; and (4) ohmic contact is formed at high temperatures through the formation of Al and/or Ni silicide at the interface.

Nordell et al. [Nordell 96] reported that the contact resistivity of Al/Ti/Al contacts on 6H-SiC samples is dependent on the doping concentration of the samples and on the postmetallization annealing conditions for the contact. Figure 7 indicates the values of contact resistivity as a function of anneal temperature on SiC wafers with different active doping concentrations. It is seen that for highly doped samples ($\sim 9 \times 10^{20}\,cm^{-3}$), the contact resistivity decreases to values below $10^{-4}\,\Omega\text{-}cm^2$ as the anneal temperature is increased from 650 to 800°C. However, much higher anneal temperatures are required for forming comparable ohmic contacts to

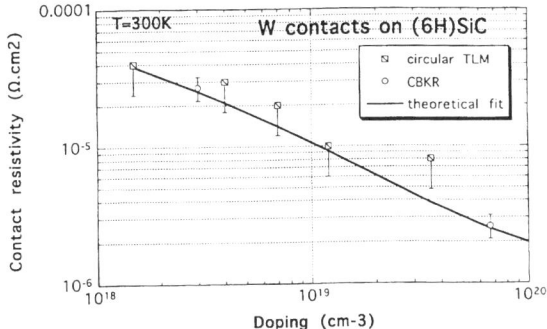

FIG. 8. Specific contact resistance as a function of doping concentration of n-type 6H-SiC [Baud 96].

samples with lower doping levels. These higher temperatures produce a resistive film over the metal surface, thus making measurements difficult. Hence, good ohmic contacts could not be realized for samples with doping levels $< 1 \times 10^{20}$ cm^{-3}.

The presence of Ni$_2$Si and of carbon-containing regions near the interface of Ni/6H-SiC contacts was also reported by Rastegaeva [Rastegaeva 96]. After annealing these contacts at 1000°C for 2 min, the specific contact resistivity was measured as $(8-9) \times 10^{-5}$ Ω-cm^2.

Teraji et al. [Teraji 96] reported ohmic behavior in as-deposited Ti and Al contacts on n-type 6H-SiC. To prevent Fermi-level pinning, they removed the surface layer (which is likely to have imperfections) by thermal oxidation followed by HF etching of the oxide. The dangling bonds left behind were passivated by dipping in pH-modified BHF or in boiling water (BW) for 10 min. ρ_c values in the range of 6×10^{-3} to 1.5×10^{-2} Ω-cm^2 were obtained on samples with doping concentrations of 2×10^{17} cm^{-3}. A similar procedure was utilized to produce Ti ohmic contacts on 6H-SiC requiring no annealing or excessive doping of the contacting layer Teraji et al. [Teraji 97]. These contacts were realized by sacrificial oxidation followed by HF and BW dip prior to Ti deposition. The resulting contact resistivity was found to be $(6 \pm 1) \times 10^{-3}$ Ω-cm^2).

Baud et al. [Baud 96] measured ρ_c of W contacts on n-type 6H-SiC using the circular TLM pattern and a cross bridge Kelvin resistor (CBKR). As shown in Fig. 8, the contact resistivity obtained by the two methods for different doping levels provides a good linear fit of decreasing ρ_c as the doping increases.

The high work function combined with the wide bandgap of SiC makes it difficult to form ohmic contacts with low contact resistivity, especially for lightly doped substrates or epilayers. Al forms a good ohmic contact to p-SiC only after diffusing into SiC during annealing. This creates a highly doped surface and reduces the width of the barrier through which carriers can then tunnel. Unfortunately, the resulting contact resistivity is not low enough. Furthermore, Al oxidizes rapidly, thereby degrading the contact. The high contact resistivity for Al/p-SiC contacts could be due to several factors

1. Interfacial oxide layer
2. Oxidation of the contact metal surface
3. Depletion of carriers in the SiC surface below the contact due to the formation of SiO_2 upon annealing; this depletion would result in a higher contact resistivity because current transport across the contact occurs by carrier tunneling and is thus dependent on the carrier concentration in SiC at the surface.

Porter and Davis [Porter 96] deposited metal contacts by electron-beam evaporation in UHV. Ni and Au contacts were rectifying and resulted in high SBHs ($\sim 1.3\,eV$). As-deposited Ni/Ni-Al/6H-SiC contacts were also rectifying. After annealing at 1000°C for 80 s, the Ni/Ni-Al contacts showed ohmic behavior with somewhat large contact resistivity ranging from 1.9×10^{-2} to $3.1 \times 10^{-2}\,\Omega\text{-cm}^2$.

Spiess et al. [Spiess 96] reported Al-Ti ohmic contacts to p-type 6H-SiC, with room-temperature values of ρ_c varying between 5×10^{-4} and $5 \times 10^{-3}\,\Omega\text{-cm}^2$. These contacts were deposited on two types of samples. The first type of samples as unimplanted but had different epitaxial doping levels. The second type of samples was implanted with different doses of Al at 50 keV, and subsequently annealed at 1650°C for 30 min. Table III shows the summary of their results. For contacts on unimplanted samples, the lowest ρ_c achieved was $1.2 \times 10^{-2}\,\Omega\text{-cm}^2$ after annealing at 1050°C. The implanted contacts with an implant dose of $1 \times 10^{15}\,cm^{-2}$ show the best specific contact resistance of $5.6 \times 10^{-4}\,\Omega\text{-cm}^2$ after annealing at 550°C.

2. Schottky Contacts to 6H-SiC

Schottky diodes based on both n- and p-type 6H-SiC reported to date are summarized in Table IV. The first rectifying junctions on 6H-SiC were prepared by fusing small pellets of Si-Al and Si-B alloys to SiC single crystals at 1700°C and >2000°C, respectively [Hall 58]. These rectifiers showed a sharp breakdown at voltages around $\sim 50\,V$, and several of them had reverse currents less than 1 mA before breakdown occurred. Typically these junctions had a reverse current of 1 mA at a reverse bias of 14 V, and a forward drop of 4.5 V at a forward current of 0.5 A.

The first barrier height studies on Schottky contacts to 6H-SiC were reported by Mead and Spitzer [Mead 64] and by Mead [Mead 66] using and Al and Au on n-type SiC. These studies concluded that the SBH (determined using differential capacitance–voltage techniques) is almost independent of the work function of the metals utilized for the contact, and has a value of 1.95–2.0 V. This was strong evidence for Fermi-level pinning due to the presence of surface states. Similar conclusions were also drawn by Hagen [Hagen 68] from studies on the SBHs of Au, Ag, and Al on both p- and n-type samples of cleaved and etched 6H- and 15R-SiC polytypes. The ideality factor determined from the I–V characteristics of these diodes was found to be between 1.5 and 2.0.

TABLE III

Specific Contact Resistance for the Nonimplanted and Implanted Al/Ti Contacts on p-Type 6H-SiC [Spiess 96]

	Nonimplanted Samples			
N_A [cm^{-3}]	Annealing Temperature [K]	ρ [Ω-cm^2]	Sheet Resistance [Ω]	Effective N_A [cm^{-3}] (calculated)
1.2×10^{15}	500–1320	Not ohmic	6.0×10^7	9.5×10^{14}
2.6×10^{17}	500–1320	Not ohmic	5.0×10^6	2.9×10^{16}
1.4×10^{18}	Unannealed	Not ohmic	7.3×10^5	7×10^{16}
	500	7.1×10^{-1}		
	620	5.4×10^{-1}		
	820	1.9×10^{-1}		
	1320	8.2×10^{-2}		

	Implanted Samples			
Implant Dose [Al$^+$/cm^2]	Annealing Temperature [K]	ρ [Ω-cm^2]	Sheet Resistance [Ω]	Effective N_A [cm^{-3}] (Calculated)
3.3×10^{13}	Unannealed	4.5×10^{-1}	1.71×10^5	1.5×10^{17}
	500	7.9×10^{-2}		
	770	2.5×10^{-2}		
1.0×10^{14}	Unannealed	1.7×10^{-1}	1.14×10^5	2.1×10^{17}
	500	3.1×10^{-2}		
	770	1.8×10^{-2}		
3.3×10^{14}	Unannealed	3.3×10^{-2}	8.41×10^4	4.8×10^{17}
	500	3.6×10^{-2}		
	770	2.0×10^{-2}		
1.0×10^{15}	Unannealed	9.7×10^{-3}	2.33×10^4	8.7×10^{17}
	500	5.6×10^{-3}		
	770	5.6×10^{-4}		

TABLE IV
Schottky Contacts to 6H-SiC

Metallization	SiC Type	SBH φ_B (eV)	Method of φ_B Measurement	Comments	Ref.
Ag	n	1.45	C-V, PR	Samples cleaved in vacuum	[Hagen 68]
Ag	n	1.10–1.21 (C-face), 0.83–0.97 (Si-face)	I-V, C-V, XPS		[Waldrop 92]
Ag	n	0.699 (I-V) 1.50 (C-V)	I-V, C-V	Interface layer suspected	[Zhang 96]
Ag	p	1.15	PR		[Hagen 68]
Al	n	2	C-V		[Mead and Spitzer 64]
Al	n	1.45	C-V, PR		[Hagen 68]
Al	n	1.7	C-V	Anneal: 900°C, 3 min	[Yasuda 87]
Al	n	0.84–0.96 (C-face) 0.26–0.30 (Si-face)	I-V, C-V, XPS		[Waldrop 92]
Al	n	1.36–1.68 (C-face) 0.82–1.21 (Si-face)	I-V, C-V, XPS	Anneal: 600°C, 30 sec	[Waldrop 93]
Al	n	0.678 (I-V) 3.96 (C-V)	I-V, C-V	Oxide-like interface layer suspected	[Zhang 96]
Al	p	1.36, 1.45	PR, C-V		[Hagen 68]
Al	p	1.71	I-V		[Lang 96]
Au	n	1.95	C-V		[Mead and Spitzer 64]
Au	n	1.45	C-V, PR		[Hagen 68]
Au	n	1.40	I-V, C-V, PR	AlTi backside ohmic contact	[Wu 74]
Au	n	1.4–1.63	I-V	Vacuum deposition: 500°C Properties improved with etching damage layer	[Anikin 91]
Au	n	2	C-V		[Dmitriev 92]
Au	n	1.14–1.19 (C-face) 1.37–1.40 (Si-face)	I-V, C-V, XPS		[Waldrop 92]
Au	n	1.4	I-V		[Urushidani 94]
Au	n	0.693 (I-V) 1.54 (C-V)	I-V, C-V	Oxide-like interface layer suspected	[Zhang 96]
Au	p	1.07, 1.37	PR, C-V		[Hagen 68]
Au	p	1.27	I-V		[Porter 94]
Co	n	0.79	I-V, C-V		[Lundberg 93]
Co	n	1.06–1.15	I-V, C-V, XPS		[Porter 95b]
CoSi$_2$	n	1.05 (C-V) 0.96 (I-V)	C-V, I-V	Anneal: 500–900°C	[Lundberg 94]

Metal	Type	Barrier height (eV)	Method	Notes	Reference
CoSi$_2$	p	1.41 (I-V) 1.9 (C-V)	I-V, C-V	Anneal: 500–900 °C	[Lundberg 94]
Cr	n	0.681 (I-V) 1.32 (C-V)	I-V, C-V	Oxide-like interface layer suspected	[Zhang 96]
Cs	n	0.57 ± 0.05	MIGS	Fermi-level pinning	[van Elsbergen 96]
Cs	p	2.28 ± 0.1	MIGS	Fermi-level pinning	[van Elsbergen 96]
Hf	n	0.97	I-V, C-V		[Porter 93]
Hf	n	1.01–0.86	I-V, C-V		[Porter 93]
Mg	n	0.33 (C-face) 0.30–0.34 (Si-face)	I-V, C-V, XPS	Anneal: 700 °C, 20–60 min	[Waldrop 92]
Mn	n	0.79–0.81	I-V, C-V, XPS		[Waldrop 92]
Ni	n	1.59–1.76 (C-face) 1.24–1.29 (Si-face)	I-V, C-V, XPS		[Waldrop 93]
Ni	n	1.51–1.66 (C-face) 1.23–1.25 (Si-face)	I-V, C-V, XPS		[Waldrop 93]
Ni	n	1.16–1.41 (Si-face) 1.17	I-V, C-V, XPS I-V	Anneal: 400 °C	[Waldrop 93] [Lang 96]
Ni	n	0.696 (I-V) 1.02 (C-V)	I-V, C-V	Anneal: 600 °C Sacrificial oxidation, Ni backside ohmic contact Oxide-like interface layer suspected	[Zhang 96]
Ni	p	0.96	I-V	AlTi backside ohmic contact	[Lang 96]
Ni	p	1.36	I-V		[Porter 96]
NiAl	p	1.37	I-V		[Porter 96]
Pd	n	1.58–1.62 (C-face) 1.11–1.26 (Si-face)	I-V, C-V, XPS	Unannealed	[Waldrop 92]
Pt	n	1.04–1.10	I-V, C-V	Films deposited at 140 °C	[Bhatnagar 92]
Pt	n	1.06–1.33	I-V, XPS		[Porter 93, 95d]
Pt	n	1.15–1.26	I-V	Anneal: 450–750 °C, 20 min	[Porter 93, 95d]
TaSi$_2$		0.71 (before anneal) 0.62 (after anneal)	I-V	Au-TaSiN overlayer; anneal: 600 °C for 30 min	[Shalish 96]
Ti	n	0.84–0.88	I-V, C-V, XPS	Unannealed	[Spellman 92; Porter 93, 95c]
Ti	n	0.86–1.04	I-V, C-V	Anneal: 700 °C, 40–60 min	[Spellman 92; Porter 93, 95a]
Ti	n	1.00–1.09 (C-face) 0.73–0.75 (Si-face)	I-V, C-V, XPS		[Waldrop 93]
Ti	n	0.98–1.05 (C-face) 0.93–0.97 (Si-face)	I-V, C-V, XPS	Anneal: 400 °C	[Waldrop 93]
Ti	p	0.51	I-V	AlTi backside ohmic contact	[Lang 96]
W	n	0.79–0.87	I-V	Anneal: 800 °C, 2 hrs	[Lundberg 96a, 96b]
W	p	1.57–1.8	I-V	Anneal: 800 °C, 2 hrs	[Lundberg 96a, 96b]

Wu and Campbell [Wu 74] performed a study of the I–V characteristics of Au/6H-SiC Schottky barrier diodes. They found the ideality factor to be 1.07 ± 0.02 for voltages between 0.35 and 0.85 V. The forward I–V characteristics were found to agree quantitatively with the theory based on thermionic emission with the barrier height modified by image-force lowering. The barrier height for contacts on n-type samples was deduced from photoresponse, differential capacitance–voltage, and forward I–V characteristics, and was found to be 1.40 ± 0.05 V.

Yasuda et al. [Yasuda 87] studied the annealing effects on Al/n-type 6H-SiC Schottky contacts deposited on both the Si- and the C-face. Their study showed significant improvement in the electrical characteristics of the diodes when they were annealed at 900°C. The ideality factor decreased from a wide range of 3.2 to 10.0 to the fairly narrow range of 2.0 to 2.3 as the diodes were annealed at 660 to 900°C. The built-in voltage (V_{BI}) for diodes on the Si-face was found to be stable at 1.7 V and did not change as the anneal temperature was increased from 660 to 900°C. However, V_{BI} for diodes on the C-face decreased from 3.5 to 1.7 V as the temperature was increased progressively in this range. The capacitance characteristics of the diodes on the C-face also showed dependence on the anneal temperature, which is believed to be due to the possible formation of Al_2O_3 or SiO_2 at the Al/SiC interface.

Anikin et al. [Anikin 91] reported Au/6H-SiC diodes capable of operating at temperatures up to 300°C. A forward current density exceeding 300 A/cm² was observed at a forward drop of ~ 4 V. The barrier height of Au on n-type 6H-SiC was found to be 1.4 to 1.63 V at room temperature, and an abrupt breakdown was observed under revese bias at 100 to 170 V. The capacitance was found to be independent of frequency in the 1 to 100-kHz range.

Schottky barriers for Pd, Au, Ag, Tb, Er, Mn, Al, and Mg to n-type SiC were investigated by Waldrop et al. [Waldrop 92]. They reported on the interface chemistry and also presented values of SBH for these metals on both Si- and C-faces of 6H-SiC obtained by different methods, including I-V and C-V characteristics, and x-ray photoemission spectroscopy (XPS). Figure 9 shows a plot of these values of SBH for both Si- and C-face versus the respective metal work function value. ϕ_{Bn} for different metals is seen to take a wide range of values, extending from 0.3 eV for Mg (Si-face) to 1.6 eV for Pd (C-face), thus providing strong evidence against Fermi-level pinning. In later studies, Waldrop and Grant [Waldrop 93] reported the crystal-face dependence of barrier height and interface chemistry for Ni, Ti, and Al on n-type 6H-SiC. As the contacts were annealed at higher tempertures (400°C), it was observed that Al had a limited reactivity and no crystal-face dependence, whereas Ni and Ti were found to be significantly more reactive

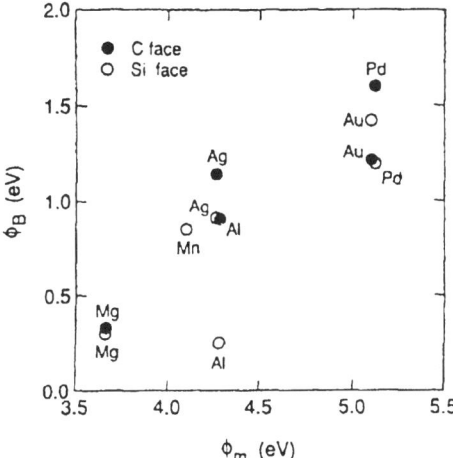

FIG. 9. Plot of the SBH of various metal contacts to n-type 6H-SiC as a function of ϕ_m [Waldrop 92].

to the C-face than the Si-face. Figure 10 shows the representative I-V data for the Ti and Ni contacts that were characterized by XPS studies. The solid curve is for unannealed contacts, whereas the dashed curves are for contacts annealed at 400°C for 30 s. SBHs of Pd, Ni, Au, Ag, Mg, Ti, and Al to p-type 6H-SiC were also reported by Waldrop [Waldrop 94]. Figure 11 summarizes these results in a form similar to that of Fig. 9. Values of ϕ_{Bp} are shown to be in the range of 1.17 to 2.56 eV and are dependent on the metal work function and the crystal face. A comparison with the previously reported values of ϕ_{Bn} for similar metals showed that, as predicted by theory, the sum of ϕ_{Bp} and ϕ_{Bn} is close to the value of the bandgap for 6H-SiC.

Dmitriev et al. [Dmitriev 92] reported that the SBH of Au/6H-SiC contacts is sensitive to damage on the surface due to substrate polishing or to high-temperature annealing during final stages of epitaxial growth. These damaged regions can evidently be removed by etching a thin layer off the top surface. As seen in Fig. 12, this damage layer is 1 to 2 µm thick for epitaxial films grown by the sublimation technique at high temperature (~2000°C) and ~5 µm thick for substrates grown by the Lely technique. ϕ_B was consistently determined to be 2 eV when the surface was etched beyond this damage layer. The I-V characteristics also showed an improvement when the damaged layer was etched off.

Spellman et al. [Spellman 91, 92], Porter et al. [Porter 93, 94], and Porter and Davis [95c] reported the effect of annealing conditions on the electrical characteristics of Ti/6H-SiC Schottky diodes. The SBH of unannealed

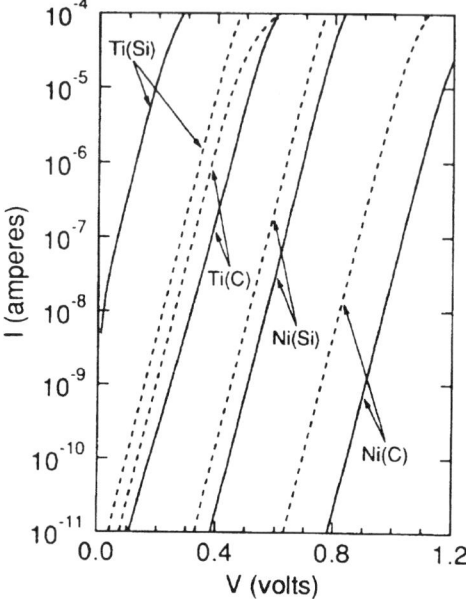

FIG. 10. Forward current-voltage characteristics of Ni and Ti Schottky diodes on Si-face and C-face 6H-SiC. Solid curves are for unannealed contact and dashed curves are for contact annealed at 400°C for 30 s. [Waldrop 93].

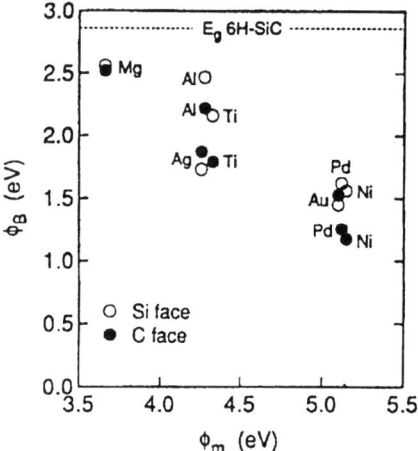

FIG. 11. SBH versus ϕ_m for different metals on p-type 6H-SiC [Waldrop 94].

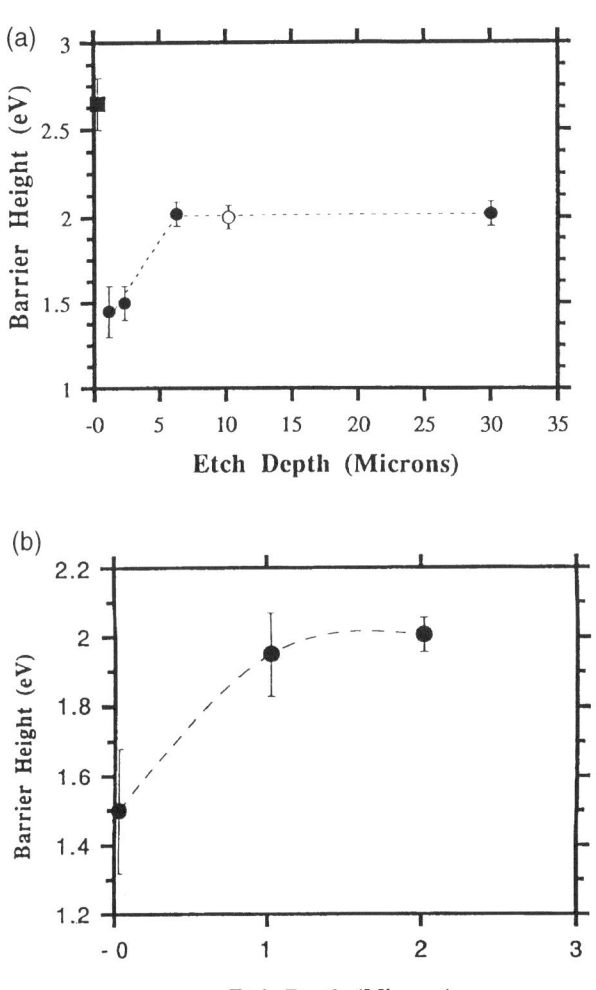

FIG. 12. Variation of SBH on the etch depth of: (a) 6H-SiC substrate; (b) 6H-SiC epitaxial film [Dmitriev 92].

contacts was found to be $\phi_{Bn} = 0.79 \pm 0.1$ eV. Low ideality factors (1.02–1.09) and reverse leakage currents (9×10^{-8} A/cm^2 at -10 V) were reported for these epitaxially grown rectifying contacts. Figure 13 shows the effect of annealing conditions on the forward current–voltage characteristics of these diodes. A degradation of the characteristics is evidenced from the more restricted linear regime after a 20-min anneal in UHV at 700°C.

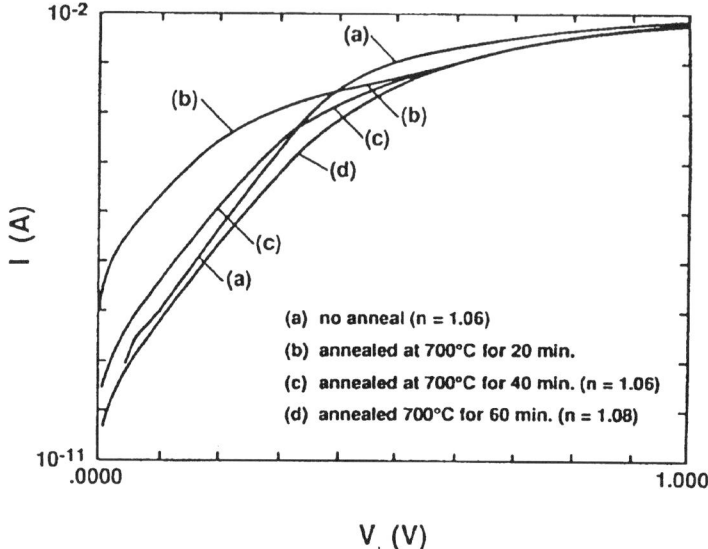

FIG. 13. Effect of annealing conditions on forward current–voltage characteristics of Ti/6H-SiC Schottky diodes [Spellman 92].

However, the characteristics are seen to improve again after each subsequent 20-min anneal. The SBH of these contacts determined from differential capacitance–voltage measurements was found to increase from 0.88 eV for as-deposited contacts to 1.04 eV after a 60-min anneal at 700°C.

Porter et al. [93, 94] and Porter and Davis [95b, 95c] the interfacial chemistry, microstructure, and the effect of high-temperature annealing on the electrical behavior of Pt, Co, and Hf contacts on n-type 6H-SiC. The metals were deposited in UHV by electron beam evaporation and formed Schottky contacts in the as-deposited state. The SBH for the unannealed Pt and Co contacts as determined from XPS analysis was found to be 1.33 ± 0.1 and 1.06 ± 0.1 eV, respectively. The SBH for the unannealed Hf contact was reported to be 0.97 eV. Ideality factors below 1.1 and leakage currents in the 10^{-8} A/cm^2 range at a reverse voltage of 10 V were reported for Schottky diodes formed using these metals. As shown in Fig. 14, Pt contacts remained rectifying even after annealing at 700 to 750°C, and their SBH was found to increase with anneal temperature. Hf contacts also maintained rectifying behavior upon high-temperature annealing at 700°C. The Hf-SiC SBH first increased after a 20-min anneal at 700°C but then successively decreased after the 40-min and 60-min anneals. Co contacts became ohmic after heating to 1000°C, but their resistivity increased when the anneal time was increased from 2 to 3 min.

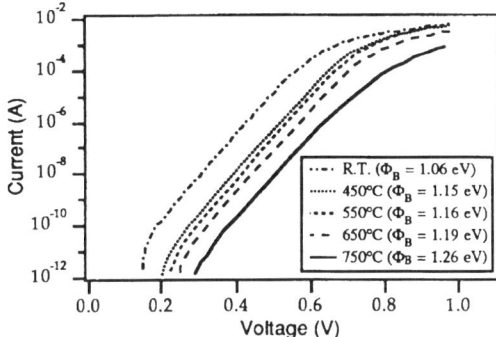

Figure 4. Log I vs. V of Pt/SiC (contact area = 2.0 x 10^{-3} A/cm^2).

FIG. 14. Effect of annealing on forward characteristics of Pt/6H-SiC diodes [Porter 93].

Cobalt disilicide Schottky contacts on both n- and p-type 6H-SiC were investigated by Lundberg and Östling [94] and Lundberg [95]. The contacts showed good rectifying characteristics after annealing at 700°C, and the SBH was found to be 1.05 ± 0.05 and 1.90 ± 0.05 eV to n- and p-type SiC, respectively. Lundberg *et al.* [96a and Lundberg and Östling 96b] also studied tungsten contacts deposited by chemical vapor deposition (CVD) on n- and p-type 6H-SiC. With n-type SiC these contacts had a low room-temperature barrier height of 0.87 eV and ideality factors close to unity. Figure 15a shows the I-V characteristics of these diodes as a function of temperature. It can be seen that there is only a slight change in the value of the SBH and ideality factor when the temperature was increased up to 160°C. Upon annealing these contacts at 800°C for 2 hr, ϕ_{Bn} at room temperature decreased to 0.79 eV. For contacts to p-type SiC, the room-temperature barrier height decreased from 1.8 to 1.57 eV, and the ideality factor decreased from 1.51 to 1.2 when the contacts were annealed under similar conditions. Figure 15b shows the variation of I–V characteristics of these contacts as a function of temperature. It can be seen that for these diodes, the current grows nearly exponentially over more than six decades. The SBH increases and the ideality factor becomes closer to unity as the temperature is increased to 200°C.

The effect of post deposition annealing of Al Schottky contacts to n-type 6H-SiC has also been studied by Reddy *et al.* [Reddy 96]. Figure 16 shows the systematic improvement in the forward current–voltage characteristics when these contacts are annealed at different temperatures. Low–temperature electrical behavior of these Schottky barriers was also studied, and it

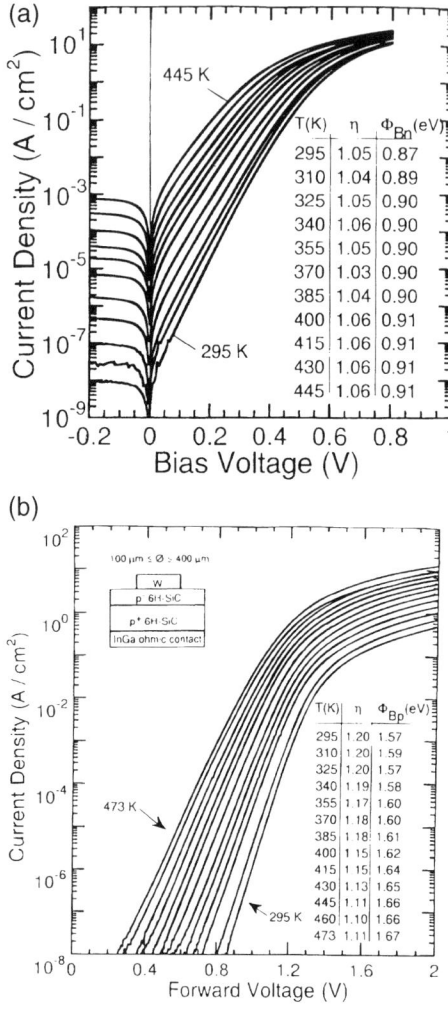

FIG. 15. Forward current–voltage characteristics as a function of temperature for: (a) n-type W/6H-SiC contacts; (b) p-type W/6H-SiC contacts [Lundberg 96b].

was observed that the ideality factor increased, while the SBH decreased significantly as the temperature was reduced to 77 K.

Andreev et al. [Andreev 95] found that the barrier height of Au, Mo, Cr, and Al on n-type 6H-SiC was in the range of 1.22 to 1.40 eV and was virtually independent of the metal work function. This was attributed to the large surface state density resulting from their surface preparation techniques. They also reported that ϕ_{Bn} decreases with increasing doping concentration of the epilayer.

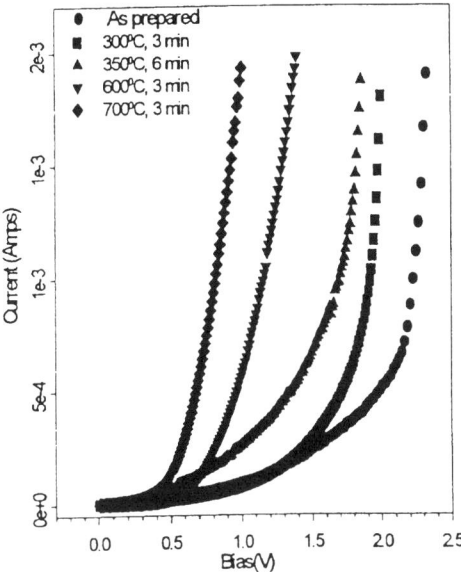

FIG. 16. Improvement of I–V characteristics of Al/6H-SiC diodes as a function of anneal temperature [Reddy 96].

Smith *et al.* [Smith 95] observed that the barrier height of metal Schottky contacts to 6H-SiC changes dramatically with temperature. This change is 10 to 20 times the change in the bandgap over the same temperature range. The change in the diffusion potential for contacts to n-type SiC, and the SBH for contacts to p-type SiC, are shown in Figs 17a and 17b, respectively. It is seen that for n-type material, the diffusion potential, and hence ϕ_{Bn}, decreases with temperature, whereas for p-type material, ϕ_{Bp} increases with temperature. For the same metal, the rate of change in barrier heights to n- and p-type material with temperature is not the same, and therefore their sum deviates from the bandgap of SiC at any given temperature. Fröjdh and Petersson [Fröjdh 1996] argued that for the barrier height extracted from the CV measurements to be reliable, no frequency dependence should exist at the measurement frequencies, and the slope of the $1/C^2$ vs voltage curve should not be temperature dependent. They stated that the existence of this frequency and temperature dependence indicates the presence of deep levels in the SiC samples studied. However, in their reply to the argument of Fröjdh and Peterson, Smith et al. [Smith 1996] reported the deep level transient spectroscopy (DLTS) spectrum which showed the absence of deep levels in their samples. They concluded that the argument of Fröjdh and Petersson is valid only at relatively lower temperatures (up to 320 K), and

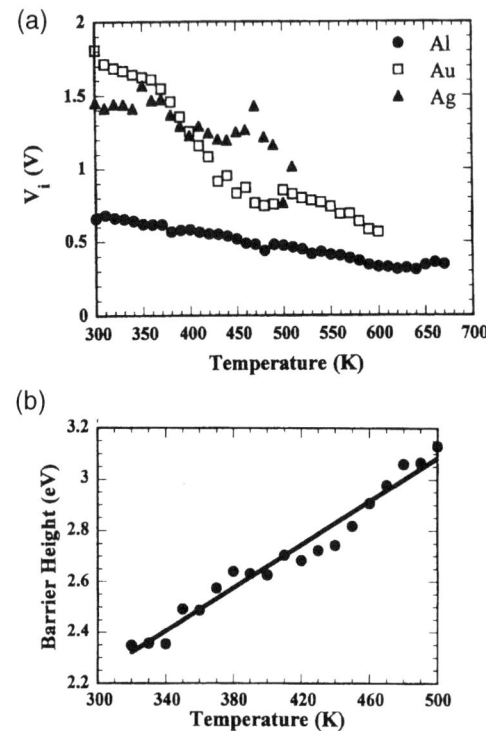

FIG. 17. Contacts to 6H-SiC: (a) n-type diffusion potential versus temperature for Al, Au, and Ag contacts; (b) p-type SBH versus temperature for Al contacts [Smith 95].

that at higher temperatures (300–673 K), this temperature dependence of the barrier height does exist.

Cs/6H-SiC Schottky contacts were studied by van Elsbergen et al. [van Elsbergen 96]. The surface preparation technique followed by them resulted in Fermi-level pinning at 1.2 eV above the valence-band maximum, irrespective of doping. The barrier height of the contact was found to be 0.57 ± 0.05 eV with n-type and 2.28 ± 0.1 eV with p-type doped samples.

The effect of postdeposition annealing conditions on the SBH and ideality factors for Au-Ta$_{0.2}$Si$_{0.4}$N$_{0.4}$-TaSi$_2$/6H-SiC and Re/6H-SiC Schottky diodes on n-type SiC have been reported by Shalish et al. [Shalish 96]. The SBH and the ideality factor of contacts with Au overlayers decrease from 0.71 eV and 1.55 for the as-deposited case to 0.62 eV and 1.18 when annealed at 600°C for 30 min. ϕ_{Bn} and η for Re contacts changed from 0.71 eV and 1.6 to 1.04 eV and 1.16 after annealing at 700°C and to 0.76 eV and 1.36 after annealing at 900°C.

3. HIGH-VOLTAGE SCHOTTKY DIODES ON 6H-SiC

SiC-based Schottky diodes that are capable of operating at high (>100 V) reverse voltages are summarized in Table V. A predictive model of SiC power Schottky diode has been reported by Kneifel et al. [Kneifel 1996]. The great majority of Schottky diodes to date have been fabricated on n-type SiC. The first high-voltage Schottky barrier rectifiers were reported by the North Carolina group [Bhatnagar 92]. As shown in Figs. 18a and 18b, these Pt/6H-SiC diodes had a blocking voltage in excess of

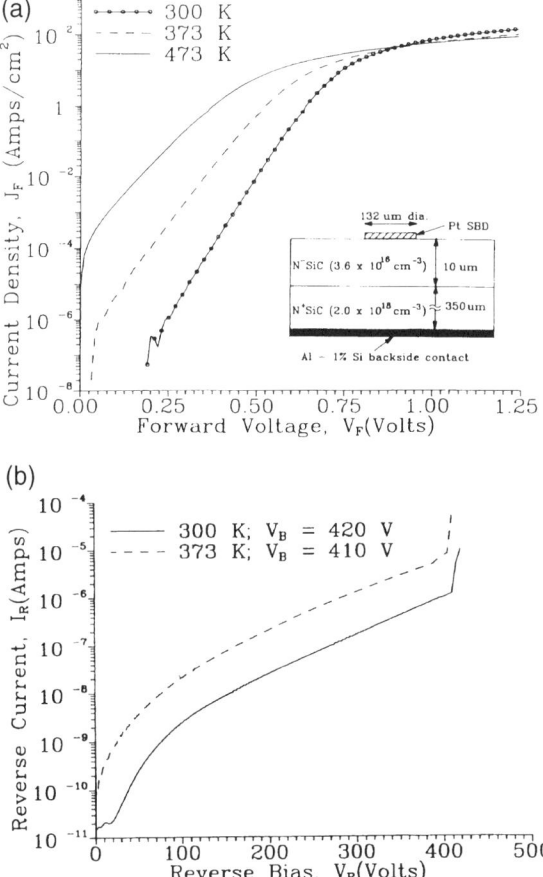

FIG. 18. Current–voltage characteristics of Pt/6H-SiC diode: (a) forward bias; (b) reverse bias [Bhatnagar 92].

TABLE V
HIGH VOLTAGE (> 100 V) SCHOTTKY BARRIER DIODES ON 4H- AND 6H-SiC

Scheme: M_1/M_2 M_1: Schottky Metal M_2: Ohmic Metal	Doping Conc. of Drift Layer (n-type (cm^{-3}))	Thickness of Drift Layer (μm)	Diode Ideality Factor	Forward Current Density at X Volts (A/cm^2)
Au/Ni/4H	5×10^{15}	10	1.08	100 at 1.67
Ni/Ni/4H	5×10^{15}	10	1.01	100 at 1.50
Ti/Ni/4H	5×10^{15}	10	1.03	100 at 1.12
Ti-Al (1:1)/Ti-Al (1:1)/4H	1×10^{16}	10	1.2	100 at 1.06
Ti/Ni/4H edge termination: B$^+$ impl.	$7-20 \times 10^{15}$	10	1.03	100 at 1.3 (200 at 2.5)
Ti-Al (1:3)/Ni/6H	9×10^{15}	2.4	1.07	~ 1 at 1
Ti-Al/Ni/6H with guard ring	1.3×10^{16}	4.7	NR	NR
Ni/Ni/6H	1.3×10^{16}	10	1.08–1.12	55 at 2
Ti-Al/Ti-Al/6H edge termination: Ar$^+$ impl.	2×10^{16}	10	NR	NR
Au/Ni/6H	5.8×10^{15}	9.6	1.1–1.3	42 at 2
Ni/Ni/6H	7.2×10^{15}	10	1.19	63.37 at 2
Ni/Ni/4H	6.1×10^{15}	10	1.29	100 at 1.86
Pt/Ni/4H	6.1×10^{15}	10	1.11	100 at 1.76
Au/NR/6H	$5-10 \times 10^{16}$	NR	1.05–1.07	333 at 4
Pt/Al-Si/6H	4×10^{16}	10	1.1	100 at 1.1
4H edge termination: Ar$^+$ impl.	7.5×10^{15}	10	—	100 at 1.1
Ni/4H edge termination: B$^+$ impl.	3.5×10^{15}	13	—	100 at 2
Ti/4H edge termination: B$^+$ impl.	3.5×10^{15}	13	—	70 at 1

400 V and a low forward drop of ~ 1.1 V at an on-state current density of 100 A/cm^2. The reverse current density was $\sim 7.3 \times 10^{-3}$ A/cm^2 at a reverse bias of 400 V. This translates into an I_{ON}/I_{OFF} ratio of $\sim 1.4 \times 10^4$. The breakdown voltage obtained for these diodes was only 40 to 70% of the ideal parallel plane breakdown voltage. This was explained to be due to the formation of a surface depletion layer on SiC caused by the presence of a high density of defect states at the interface. The fabrication of these diodes was a significant achievement, and it convinced researchers of the potential success of Schottky contact-based devices on SiC in the coming years.

Subsequently, high-voltage Au/6H-SiC Schottky diodes were reported by the Kyoto group [Urushidani 94, Kimoto 93]. As seen in Fig. 19, these diodes had breakdown voltages exceeding 1100 V and were shown to be capable of operating at temperatures up to 400°C. The ideality factor and the SBH were determined [Kimoto 93] as 1.2 and 1.4 eV, respectively, and a forward current density of 42 A/cm^2 was obtained at a voltage drop of 2 V. Figure 20 shows the temperature dependence of the specific on-resistance for these SiC Schottky diodes (experimental data) and for a *Si* Schottky rectifier with the same blocking voltage (theoretical data). The R_{on} for a Si

Breakdown Voltage (V)	% Theoretical Breakdown Voltage	Reverse Leakage Current Density (A/cm^2) at X Volts	Specific On-Resistance Ω-cm^2	Ref.
790	31.2	2.9×10^{-4} at 700	1.4×10^{-3}	[Itoh 95a, 95c]
~790	31.2	$~2.9 \times 10^{-4}$ at 700	$~2 \times 10^{-3}$	[Itoh 95a, 95c]
~790	31.2	$~2.9 \times 10^{-4}$ at 700	$~2 \times 10^{-3}$	[Itoh 95a, 95c]
1000	48.4	6×10^{-5} at 100	2.1×10^{-3}	[Raghunathan 95]
1100–1750	46.9–74.6	$~1 \times 10^{-4}$ at 500	$1-3 \times 10^{-3}$	[Itoh 96]
250	34.7	$~3 \times 10^{-5}$ at 200	NR	[Ueno 95, 96]
550	44.5	$~0.1$ at 500	NR	[Ueno 95, 96]
1100	55.4	1.4×10^{-3} at 900	NR	[Su 96]
~1000	75.2	$~0.1$ at 600	NR	[Alok 94]
>1100	42.8	2.1×10^{-3} at 1100	8.5×10^{-3}	[Kimoto 93]
800–1000	31.6–40.0	4.8×10^{-3} at 800	3.5×10^{-2}	[Saxena 96]
1000	41.2	3.6×10^{-4} at 600	8×10^{-3}	[Saxena 98]
1000	41.2	1.14×10^{-6}	8.39×10^{-3}	[Saxena 98]
100–170	—	$~3.3 \times 10^{-6}$ at 150	$6-9 \times 10^{-3}$	[Anikin 91]
400	58.5	2.19×10^{-5} at 100	3×10^{-3}	[Bhatnagar 92]
1400	—	0.8 at 1000	1.5×10^{-3}	[Weitzel 96]
1720	—	10^{-4} at 1000	5.6×10^{-3}	[Cooper 98]
1500	—	0.1 at 1000	—	[Cooper 98]

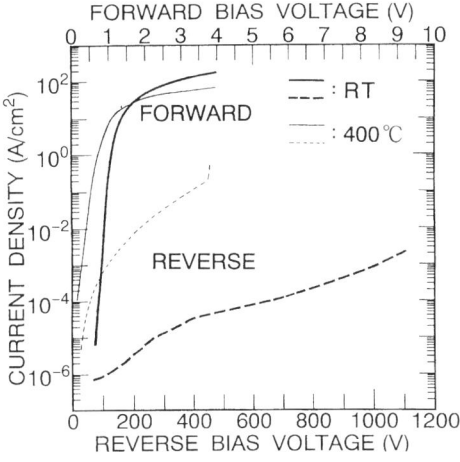

FIG. 19. Current–voltage characteristics of 1100-V 6H-SiC SBD at room temperature and 400°C [Kimoto 93].

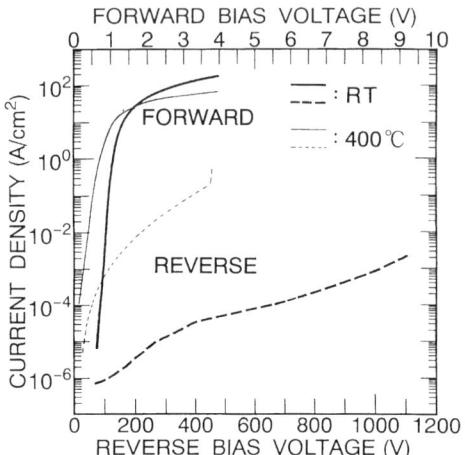

Fig. 20. Temperature dependence of specific on-resistance for SiC (experimental results) and Si (theoretical results) [Kimoto 93].

device is calculated to be ~ 20 times larger than the value measured for the SiC device. In addition, R_{on} for SiC is somewhat less sensitive to temperature as compared to that for Si.

High-voltage Schottky diodes utilizing Ni for both ohmic and Schottky contacts were reported by the Cincinnati group [Su 95, 96; Saxena 96, 98]. These Ni/6H-SiC SBDs showed a high breakdown voltage (> 1000 V) at both 25 and 300°C. The diodes utilized grown oxide for surface passivation and device isolation. Figure 21a shows the cross-sectional view of the Ni/6H-SiC diode structure. As seen in Fig. 21b, these diodes had breakdown voltages exceeding 1100 V and operated at elevated temperatures up to 300°C. The current mechanisms for these diodes in the temperature range of 100 to 573 K have also been reported [Saxena 96]. Under forward bias, the diodes showed exponential behavior over a large voltage range at all temperatures. Figure 21c shows the forward voltage drop (V_F) as a function of temperature for different current densities. At low current-density levels (0.1–10.0 A/cm^2), V_F was found to decrease with increasing temperature, from 1.5 to 2.0 V at 100 K to 0.6 to 1.3 V at 300°C. At higher current-density levels (50–80 A/cm^2), V_F exhibits a minimum at 200 to 250 K, after which it increases slightly, with temperature reaching values of 2.5 to 3.5 V at 300°C. The rectifier efficiency of these diodes was essentially independent of temperature, with near ideal values (37–39%) up to 600 kHz. The AC power dissipation was also fairly constant over the entire high-temperature range for current values up to 0.1 A, but showed a monotonic increase with temperature for higher current levels.

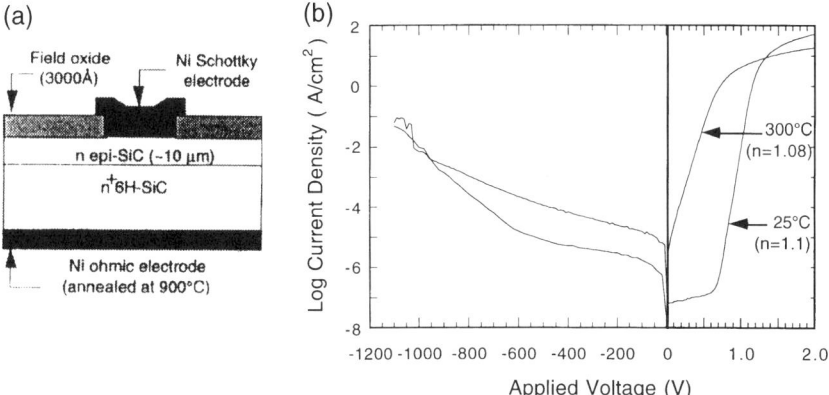

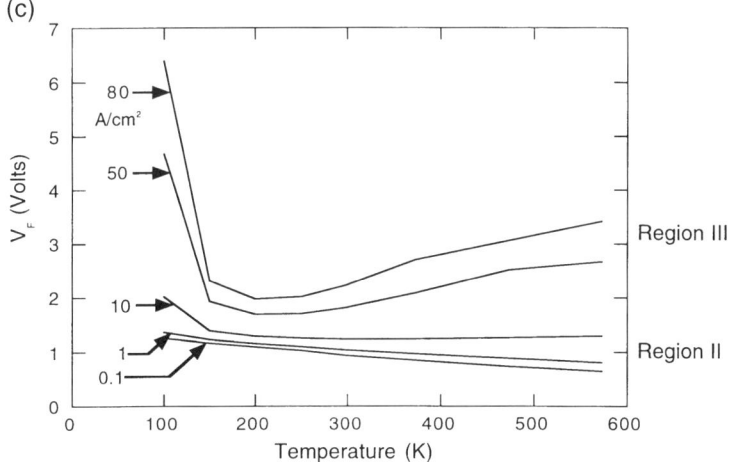

FIG. 21. Ni/6H-SiC Schottky diode with field plate structure: (a) cross-sectional view; (b) I–V characteristics of Ni/6H-SiC SBD at 25 and 300°C; (c) forward voltage drop as a function of temperature for different levels of J_F [Saxena 96]. Region II: current dominated by thermionic emission. Region III: current limited by series resistance.

4. Edge Termination in 6H-SiC Schottky Diodes

For high-voltage Schottky diodes, edge termination plays an important role in determining the breakdown voltage. The experimentally obtained values of breakdown voltages for Schottky diodes without edge termination are usually considerably less than the theoretically calculated values for a parallel plane structure. This is due to the electric field crowding at the diode edges. For high-voltage SiC SBDs, several techniques have been shown to reduce field crowding at the edges, thus resulting in higher breakdown voltages. The simplest of these is the SiO_2 field plate structure, in which the metal overlaps the oxide so that the maximum electric field (E_{max}) is at the SiO_2-metal interface. In this case, the breakdown voltage should ideally not be affected by electric field crowding. However, since the field plate length in SiC is shorter than in Si because of the smaller space charge region (due to higher E_{max}), a more precise alignment is necessary between the Schottky metal pattern and the contact–hole pattern [Bhatnagar 93a]. Thus, other techniques for edge termination have been investigated. These include: (1) the use of floating metal field rings (FMRs) and resistive Schottky barrier field plates (RESPs) as reported by Bhatnagar and Baliga [Bhatnagar 93b]; (2) the use of implantation of neutral species at the periphery of the diode to form an amorphous layer, reported by Alok et al. [Alok 94]; and (3) the guard ring termination, reported by Ueno et al. [95, 96].

In their study performed on SiC Schottky barrier diodes with a theoretical breakdown voltage of 1125 V, Bhatnagar et al. [96] calculated the breakdown voltage for the unterminated diode and the diode with FMR termination to be 225 and 600 V, respectively. The experimentally determined breakdown voltages for these two diodes were 220 and 400 V, respectively.

Alok et al. [94] reported a further improvement in blocking voltages of diodes, with edge termination achieved by Ar implantation. As shown in Fig. 22, nearly ideal breakdown voltage was obtained when a dose of $10^{15}/cm^2$ was utilized to implant the periphery of Ti/6H-SiC diodes.

Figure 23a shows the cross-sectional view of the Al-Ti/6H-SiC Schottky barrier diode with the guard ring termination reported by Ueno et al. [95, 96]. Al-Ti forms a Schottky contact with the n-type drift region, and an ohmic contact with p-type guard rings. Hence, the guard ring is electrically at the same potential as the Schottky contact. E_{max} therefore appears at the edge of the guard ring and not at the Schottky contact. The breakdown voltage is thus not affected by electrical field crowding. Figure 23b shows the improvement in the reverse characteristics of the Schottky diodes with the guard ring structure. These diodes show a breakdown voltage in excess of 550 V, which is about 70% of the ideal breakdown voltage for the epilayer.

5. EXCESS LEAKAGE CURRENT IN SiC SCHOTTKY DIODES

Experimental tests on SiC Schottky diodes have shown that they generate larger leakage currents than theoretically predicted. Figure 24 shows the reverse leakage current observed in Ti and Pt/6H-SiC Schottky barrier

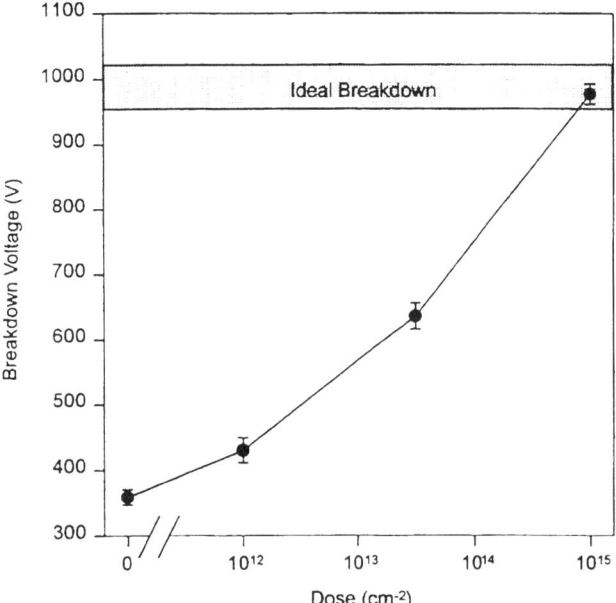

FIG. 22. Breakdown voltage as a function of implant dose on the edge of SBDs [Alok 94].

diodes, and the calculated theoretical values [Bhatnagar 96]. The theoretical values have been calculated assuming thermionic emission and by using the barrier height extracted from C-V measurements [Bhatnagar 92, Urushidani 94]. It can be seen that the observed leakage current is about two orders of magnitude larger than the theoretical value. An attempt has been made to explain the excess leakage currents using a model based on localized defects in the epitaxial layer at the metal–SiC interface. These defects lower the barrier height of the contact at the defect site [Bhatnagar 96]. Analytical and numerical analysis have been performed to study the effect of these barrier height inhomogenities on the I-V of 6H-SiC SBDs, and good agreement has been achieved with the measured characteristics.

A related explanation for the high reverse currents in SiC Schottky diodes has been offered along similar lines by Schroder et al. [Schroder 96]. This explanation is based on the formation of a very thin inhomogeneous interfacial layer between metal and semiconductor. The SBH therefore becomes voltage-dependent. The formation of this layer is postulated to cause the voltage dependence of the reverse current in SiC Schottky contacts, due to pinch-off effects of low barrier regions that surround high barrier regions on the inhomogeneous interfacial layer.

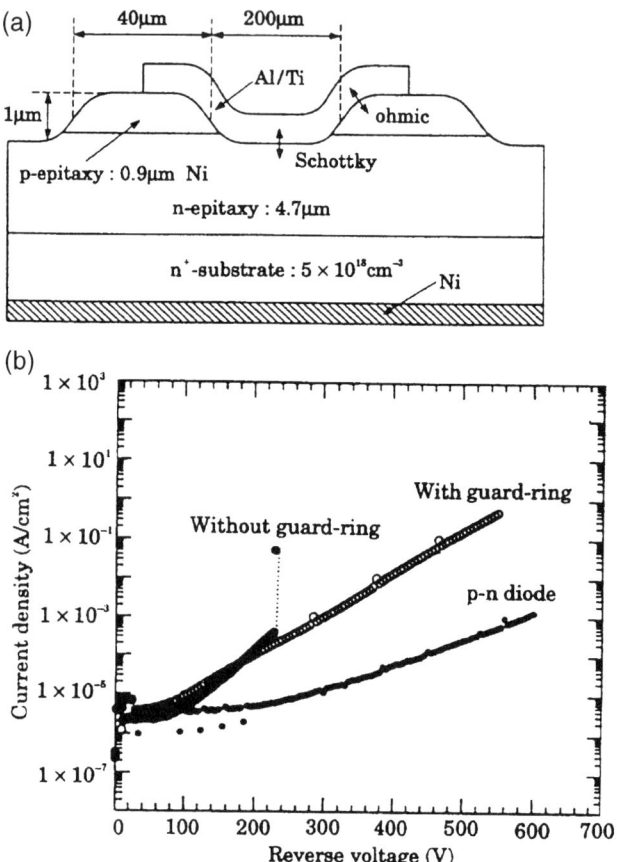

FIG. 23. Al-Ti Schottky barrier diode with guard ring: (a) cross-sectional view; (b) reverse current–voltage characteristics of SBDs with and without guard ring, and the p-n junction [Ueno 95].

Yet another possible cause for larger reverse currents in SiC Schottky diodes could be quantum mechanical tunneling through the Schottky barrier. Crofton and Sriram [Crofton 96b] reported calculations of the reverse current density, which were performed for 6H-SiC using a WKB evaluation of the tunneling probability through a reverse biased Schottky barrier. The image force lowering of the SBH was taken into account in these calculations. Figure 25 shows the comparison of the calculated and experimental reverse leakage currents for Pt-SiC Schottky diodes on n-type

3 BUILDING BLOCKS FOR SiC DEVICES

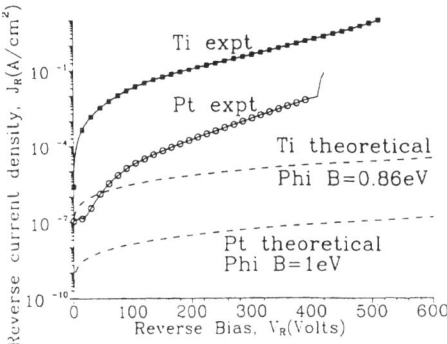

FIG. 24. Experimental and theoretical reverse leakage currents at 300 K for Ti/6H-SiC and Pt/6H-SiC Schottky diodes [Bhatnagar 96].

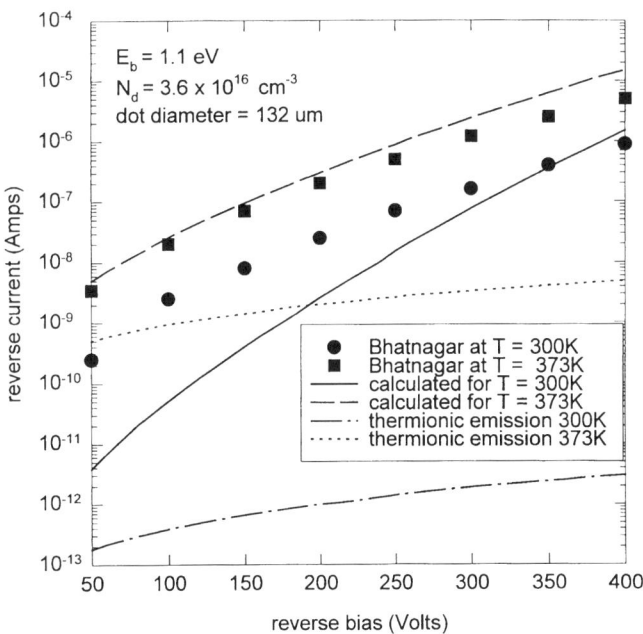

FIG. 25. Comparison of calculated and experimental reverse leakage currents for Pt/6H-SiC Schottky diodes [Crofton 96b].

6H-SiC. It can be seen from this figure that the calculated values of the reverse currents based on the quantum mechanical tunneling theory show better agreement with the previously reported experimental values.

V. Metal Contacts to 4H-SiC

1. OHMIC CONTACTS TO 4H-SiC

Relatively few studies have been performed on ohmic contacts to the 4H polytype of SiC. These studies are summarized in Table VI. Arnodo et al. [Arnodo 96] studied ohmic contacts using Ni and Mo. Ni contacts were annealed in the temperature range of 950 to 1000°C, whereas higher temperatures in the 1000 to 1600°C range were utilized for Mo contacts. At doping levels no greater than $5 \times 10^{18}\,cm^{-3}$, both metals offered similar contact resistance (R_c), with best R_c values in the low $10^{-5}\,\Omega\text{-}cm^2$ range. At higher doping levels, however, R_c for Ni ohmic contacts continues to decrease to values in the low $10^{-6}\,\Omega\text{-}cm^2$ range, whereas for Mo contacts, no improvement is observed.

Low-resistivity alloyed Al/Ni/Al contacts on 4H-SiC with an R_c value of $1.8 \times 10^{-5}\,\Omega\text{-}cm^2$ have been reported by Hallin et al. [Hallin 97]. This is lower than that obtained for single-layer Ni contacts. Their studies indicate that the first layer of Al prevents void formation at the interface and reduces the interface oxide layer. The Al is also believed to react with the excess carbon left behind due to silicide formation as a result of strong Ni-Si chemical affinity.

2. SCHOTTKY CONTACTS AND HIGH-VOLTAGE SCHOTTKY DIODES ON 4H-SiC

Table VII gives a summary of the studies performed on Schottky barriers for different metals to 4H-SiC. The first high-voltage Schottky barrier diodes on the 4H-SiC polytype were reported by Itoh et al. [Itoh 95a, 95c]. Au, Ni, and Ti were employed for forming the Schottky contact. The SBH for these metals was determined by I-V and C-V analysis to be 1.73 to 1.8, 1.6 to 1.7, and 1.1 to 1.15 eV, respectively. Figure 26 shows the SBH for these metals as a function of the respective metal work functions. It can be seen that for metal–4H-SiC structures, the pinning of the Fermi level at the surface does not occur, and the barrier height depends on the metal work function with

TABLE VI
Ohmic Contacts to 4H-SiC

Metallization	Annealing Condition	SiC Carrier Conc. (cm^{-3})	SiC Type	ρ_c at RT (Ω-cm^2)	Method of ρ_c Measurement	Ref.
Ni	1000–1200°C, 1 min	3.2×10^{17} and 1.4×10^{18}	n	1.3×10^{-5}– 3.6×10^{-6}	TLM	[Crofton 96a]
Ni	950–1000°C	2×10^{18}– 2×10^{19}	n	4×10^{4}– $\sim 10^{-6}$	TLM	[Arnodo 96]
Mo	950–1000°C	2×10^{18}– 2×10^{19}	n	4×10^{-4}– $\sim 10^{-5}$	TLM	[Arnodo 96]
Ni, Ni/W, Ni/Ti/W, Ni/Cr/W	1000–1050°C, 5–10 min	10^{17}–10^{18}	n	10^{-3}–10^{-6}	TLM	[Liu 96]
Cr/W, Cr/Mo/W	1000–1050°C, 5–10 min	10^{17}–10^{18}	n	10^{-2}–10^{-4}	TLM	[Liu 96]
Al/Ni/Al	1000°C, 5 min	1×10^{19}	n	1.8×10^{-5}	XPS	[Hallin 97]

TABLE VII
SCHOTTKY CONTACTS TO 4H-SiC

Metallization	Carrier conc. of active region (cm^{-3})	SiC Type	SBH φ_B(eV)	Method of φ_B Measurement	Ref.
Au	5×10^{15}	n	1.73–1.8	I-V, C-V	[Itoh 95b]
Ni	5×10^{15}	n	1.6–1.7	I-V, C-V	[Itoh 95b]
Ni	6×10^{15}	n	1.67	I-V	[Saxena 98]
Pt	6×10^{15}	n	1.31	I-V	[Saxena 98]
Ti	5×10^{15}	n	1.1–1.15	I-V, C-V	[Itoh 95b]
Ti	1×10^{16}	n	0.99	I-V	[Raghunathan 95]

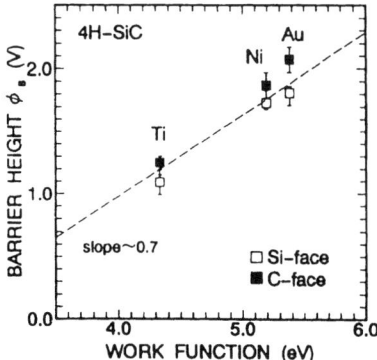

FIG. 26. Barrier height as a function of metal work function for 4H-SiC [Itoh 96b].

a slope of ~0.7. The room-temperature breakdown voltage for these diodes was reported to be 800 V. A current density of 100 A/cm^2 was obtained at a forward drop of 1.67 V. The functional device yield was ~70%. Interestingly, the ideal breakdown voltage calculated using the breakdown field and the thickness of the drift region for this device structure should be ~2000 V. The variation of R_{on} as a function of breakdown voltage for Si, 6H-, and 4H-SiC SBDs is plotted in Fig. 27 [Kimoto 93]. The specific on-resistance for these Au/4H-SiC diodes is 1.4×10^{-3} Ω-cm^2, which is two orders of magnitude lower than silicon SBDs with comparable breakdown voltage, and less than one fifth of that observed for similar Schottky diodes on 6H-SiC [Kimoto 93], as discussed before. More recently, the same group [Itoh 96] improved the breakdown of Ti/4H-SiC Schottky diodes by using

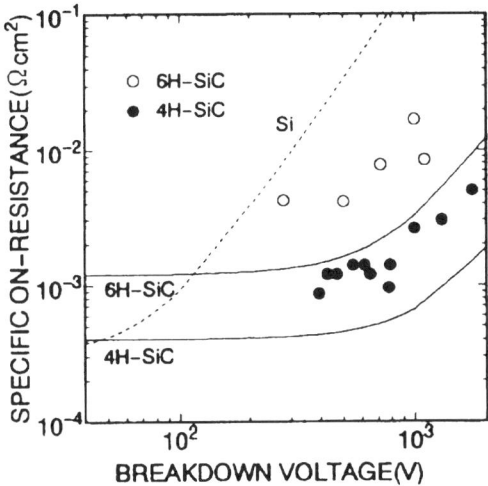

FIG. 27. Dependence of specific on-resistance on the breakdown voltage in Si, 6H-, and 4H-SiC rectifiers [Itoh 95c].

a highly resistive layer at the periphery of the contact for edge termination. The layer was formed by B^+ implantation, followed by a heat treatment to reduce damage to the implanted areas. These diodes showed a higher breakdown voltage of 1100 V compared to 760 V for diodes that did not have any edge termination. Figure 28 shows the current–voltage characteristics of diodes with and without edge termination. Though there is no significant difference in the forward bias characteristics, the leakage current under reverse bias is reduced for diodes with edge termination, thereby allowing higher breakdown voltage exceeding 1100 V.

Raghunathan et al. [Raghunathan 95] also reported high breakdown (1000 V) Ti/4H-SiC SBDs with a forward drop of only 1.06 V at a forward current density of 100 A/cm². The SBH for Ti was determined to be 0.99 V by I-V analysis. The specific on-resistance was found to be 2×10^{-3} Ω-cm².

Weitzel et al. [Weitzel 1996] reported 1400 V 4H-SiC Schottky diodes with an on-resistance of 1.5 mΩ-cm². These diodes were fabricated on 10 μm thick epitaxial layers with a doping level of 7.5×10^{15} cm^{-3} and utilized Ar implant damage termination. The diodes had a high forward current density of 732 A/cm² at 2 V, but exhibited higher leakage currents (~ 0.8 A/cm² at -1000 V).

Recently, Saxena and Steckl [Saxena 97, 98] reported SBDs on 4H-SiC using Ni and Pt for Schottky contacts. The current–voltage characteristics

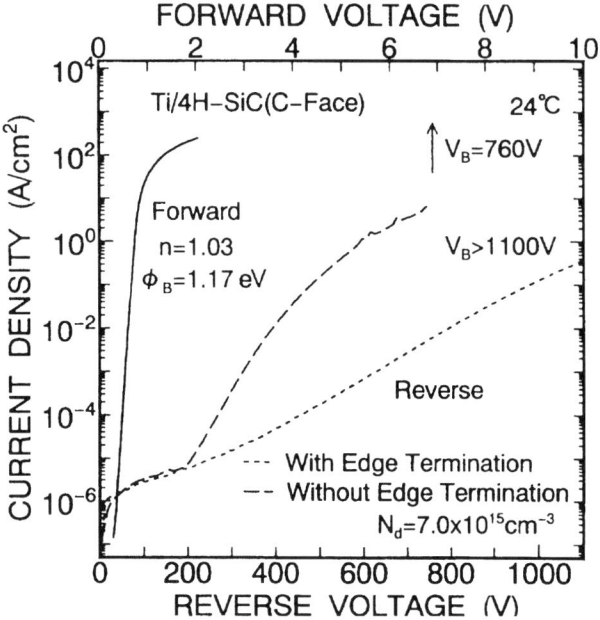

FIG. 28. I–V characteristics of Ti/4H-SiC Schottky rectifiers with and without edge termination [Itoh 96].

of typical Pt/4H SiC and Ni/4H SiC SBDs at room temperature are shown in Fig. 29. A forward current density (J_F) of 100 A/cm^2 was achieved at a forward voltage drop of 1.76 and 1.86 V, respectively. The ideality factors for the Pt and Ni diodes were calculated from the forward J-V plots as 1.11 and 1.29, respectively. The Schottky barrier height (ϕ_B) and the specific on-resistance (R_{on}) for the Ni/4H-SiC diodes were found to be 1.31 eV and 8 mΩ-cm^2, respectively. The saturation current density was found to be 4.7×10^{-15} A/cm^2. Both Ni and Pt diodes were able to withstand reverse voltages in excess of 1000 V. Some diodes had a breakdown voltage as high as 1200 V. Under reverse bias of 600 V, leakage current densities of 1.14×10^{-6} and 3.6×10^{-4} A/cm^2 were measured for the Pt and Ni diodes. The current "on-off" ratio (corresponding to J_F at 2 V divided by J_R at -600 V) was measured at 25°C to be 1.623×10^8 and 2.85×10^6 for the Pt and Ni diodes. Figure 30 shows the forward drop for Ni/4H-SiC diodes versus temperature for current densities 1, 10, and 80 A/cm^2. It can be seen that the forward voltage drop increases with temperature when the diode is operating at higher current densities, such as

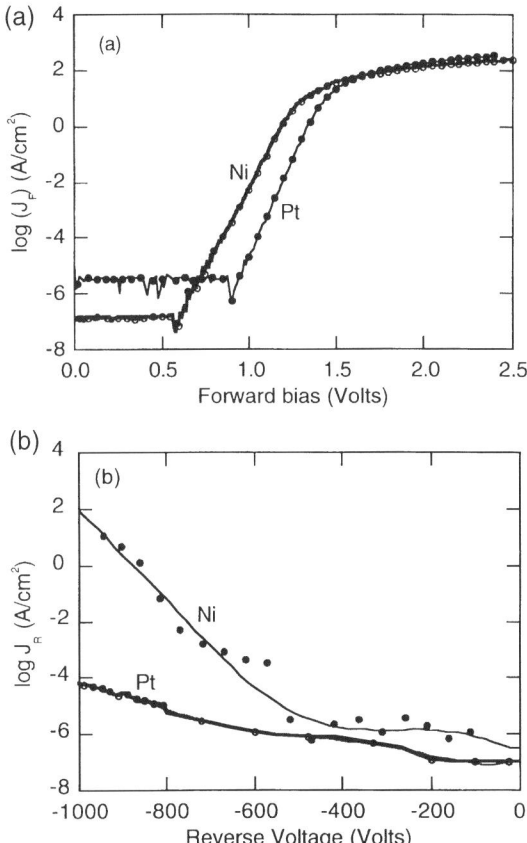

FIG. 29. Current–voltage characteristics of Pt/4H-SiC and Ni/4H-SiC SBD at 25°C: (a) forward bias; (b) reverse bias [Saxena 98].

80 A/cm^2. This allows for parallel operation [Shenoy 94] of several such diodes for power-switching applications, without current filamentation. The current on-off ratio for the Ni/4H-SiC diodes does not show a significant reduction with temperature up to more than 300°C, where this ratio is still in excess of 10^6.

Schottky diodes with B-implanted edge termination were reported by Cooper [Cooper 98] on 13 μm 4H-SiC epitaxial layers with doping concentration 3.5×10^{15} cm^{-3}. The Ni/4H-SiC diodes had a high blocking voltage of 1720 V, and a specific on-resistance of 5.6 mΩ-cm^2. The barrier heights for Ti and Ni on 4H-SiC were determined to be 0.8 and 1.3 eV, respectively.

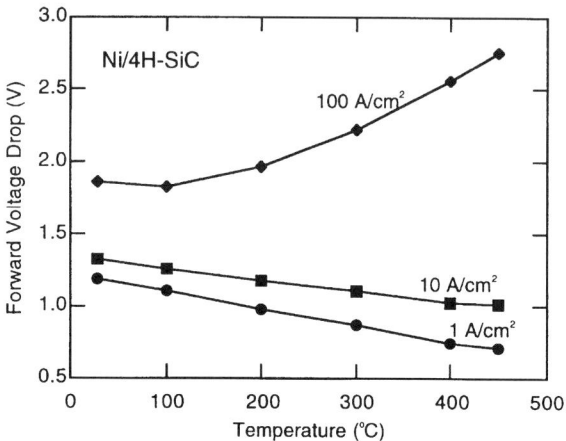

FIG. 30. Forward voltage drop versus temperature for different current densities in Ni SBDs on 4H-SiC [Saxena 98].

Kimoto et al. [Kimoto 97, Wahab 98] at Linkoping have also recently reported high-voltage (2.2–3.0 kV) Schottky rectifiers fabricated on thick low-doped epilayers grown by high-temperature CVD [Kordina 95a]. The thickness of the epilayer was 42 μm, and the doping level was $\sim 2 \times 10^{15}$ cm^{-3}. Figure 31 compares the I-V characteristics of the Schottky diodes fabricated on substrates grown at Cree Research and at Linkoping. They utilized the "all-Ni" approach for metallization developed at Cincin-

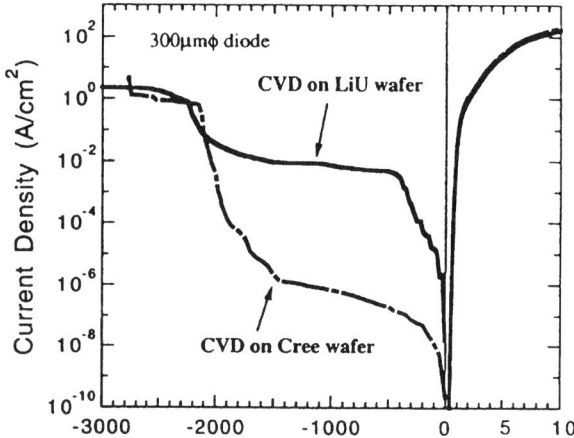

FIG. 31. Comparison of the I–V characteristics of Schottky diodes fabricated on 4H-SiC substrates grown at Cree Research and Linkoping [Kimoto 97].

nati [Steckl 93, 94; Su 95, 96; Saxena 96, 98], wherein Ni was used for forming both ohmic and Schottky contacts to the diode. The barrier height was estimated to be 1.69 eV, and an on-resistance of 34 mΩ-cm^2 was measured, possibly due to the high resitivity of the substrate and the ohmic contacts. A low reverse leakage current of 7×10^{-7} A/cm^2 was achieved at a reverse bias of 1 kV, and the forward voltage drop at 100 A/cm^2 was 7.4 V. Successful operation of these diodes demonstrates that the high-temperature CVD process has great potential to provide thick epitaxial films useful for future high-power devices.

VI. Metal Contacts to 3C-SiC

Among the SiC polytypes investigated for device fabrication, cubic SiC (3C-SiC) has the highest reported electron mobility (see Table 1). However, the lack of monocrystalline 3C-SiC has limited the development of electronic devices on this polytype material and has led researchers to focus their efforts on hexagonal polytypes (4H- and 6H-SiC) for which the substrates are readily available. 3C-SiC has been grown on Si wafers by several research groups world-wide but the quality of the epitaxial layers requires improvement and does not compare with that of epitaxial layers grown on monocrystalline hexagonal polytype substrates. For the sake of completeness, this section briefly reviews metal contacts to the 3C-SiC polytype. A more detailed discussion has been presented by Porter and Davis [Porter 95].

1. Ohmic Contacts to 3C-SiC

Edmond *et al.* [Edmond 88a] studied ohmic contacts to n-type 3C-SiC using Ni, Au-Ta, Cr, TaSi$_2$, and Al, and to p-type 3C-SiC using Al-TaSi$_2$ and Al. TaSi$_2$ contacts annealed for 5 min at 850°C had the lowest contact resistivity (2×10^{-2} Ω-cm^2) for n-type, and Al contacts annealed for 3 min at 875°C had the lowest contact resistivity (3.1×10^{-2} Ω-cm^2) for p-type. These contacts were reported to be stable during electrical operation for 8 h in air at temperatures up to 400°C. At 400°C, the contact resistivity of TaSi$_2$ and Al on 3C-SiC decreased by a factor of 2 and 10, respectively.

Geib *et al.* [Geib 89] applied Auger electron spectroscopy to study the metal–semiconductor interface of contacts formed by W deposition on 3C-SiC. They reported that although some W-Si and W-C bonding did occur at the interface as a result of the deposition process, the extent of this bonding did not change as the contacts were annealed at higher temperatures, up to 850°C. This was consistent with results on the measured contact resistance, which also did not change considerably with high-temperature annealing. ρ_c was found to be 0.24 Ω-cm^2 at 23°C and dropped to 8×10^{-2} Ω-cm^2 at 900°C. Baud et al. [Baud 95] also investigated the effect of anneal temperature (in the range from 600°C to 1100°C) on the resulting W/3C-SiC interface. They found the contact resistivity to be about 10^{-3} Ω-cm^2.

Chaudhry et al. reported Ni, NiCr, W, and Ti, WSi$_2$ and TiSi$_2$ ohmic contacts on 3C-SiC [Chaudhry 90, 91]. The minimum contact resistance for Ti and W contacts was 7.6×10^{-3} and 6.1×10^{-3} Ω-cm^2, respectively. The silicides of these metals yielded a lower contact resistance of 1.1×10^{-4} and 3×10^{-4} Ω-cm^2 for TiSi$_2$ and WSi$_2$, respectively.

Shor and co-workers [Shor 92, 94] examined the high-temperature operation of W/Pt/Au and Ti/TiN/Pt/Au ohmic contacts on n-type 3C-SiC up to 700°C. Figure 32 shows the variation of the specific contact resistance for these W and Ti contacts, as a function of anneal time. In Fig. 32a, the anneal temperature is 650°C, while in Fig. 32b, each curve represents a different Ti contact annealed at 650°C and 750°C. For W contacts, little change in contact resistivity was observed even after 8 h of annealing. For Ti contacts, annealing at 650°C reduces ρ_c for up to 2 h of anneal time, beyond which time ρ_c increases slowly with anneal time, eventually becoming rectifying after 31 h of annealing.

Low specific resistance ($< 6 \times 10^{-6}$ Ω-cm^2) TiC ohmic contacts to n-type 3C-SiC (doping concentration of 2×10^{19} cm^{-3}) were reported by Parsons and co-workers [Parsons 94]. The 1500 Å TiC layer was epitaxially grown by CVD using gas precursors at 1260°C.

The effect of post-deposition anneal temperature on the electrical characteristics of Ni-Mo and Ni-Au ohmic contacts to 3C-SiC was studied by Arugu and co-workers [Arugu 95]. The temperature range investigated was from 400 to 1200°C. Lowest contact resistance of 1.41×10^{-4} and 3.05×10^{-4} Ω-cm^2 was observed for Ni-Mo and Ni-Au, respectively, corresponding to anneal temperatures of 1000°C and 700°C.

Moki et al. [Moki 95] studied the effect of increasing the surface doping concentration achieved by ion implantation on the contact resistivity of Al and Ti ohmic contacts to 3C-SiC. Figure 33 shows the decrease in ρ_c for Al and Ti contacts as a function of the surface doping concentration. The minimum contact resistivity of 1.4×10^{-5} and 1.5×10^{-5} Ω-cm^2 were measured for Al and Ti contacts, respectively, at the highest surface doping concentration of 3×10^{20} cm^{-3}.

2. SCHOTTKY CONTACTS AND DIODES ON 3C-SiC

The first Schottky diodes on 3C-SiC were reported by Yoshida et al. [Yoshida 85] on epitaxially grown n-type layers. Au was used for forming

(a)

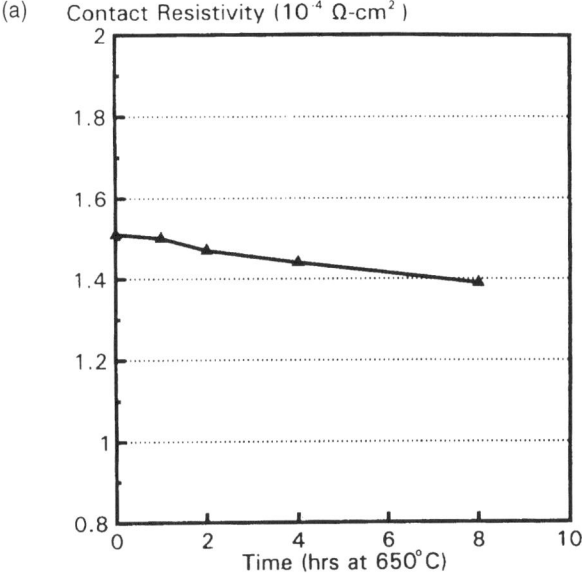

(b)

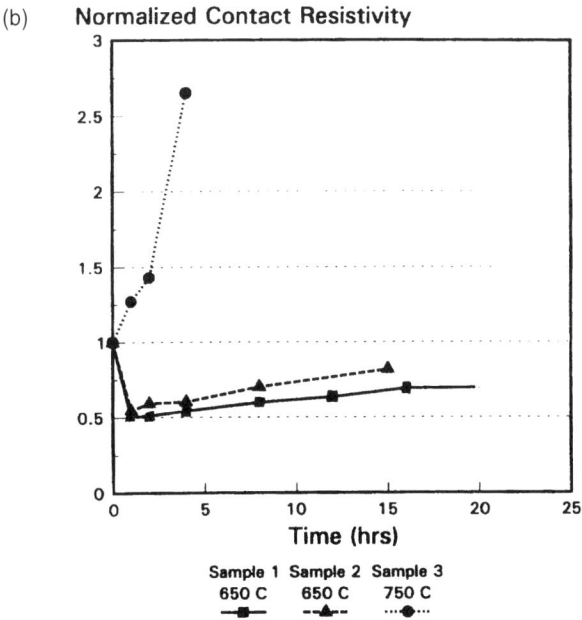

FIG. 32. Specific contact resistivity to 3C-SiC as a function of anneal time for: (a) W contacts; (b) Ti contacts [Shor 94].

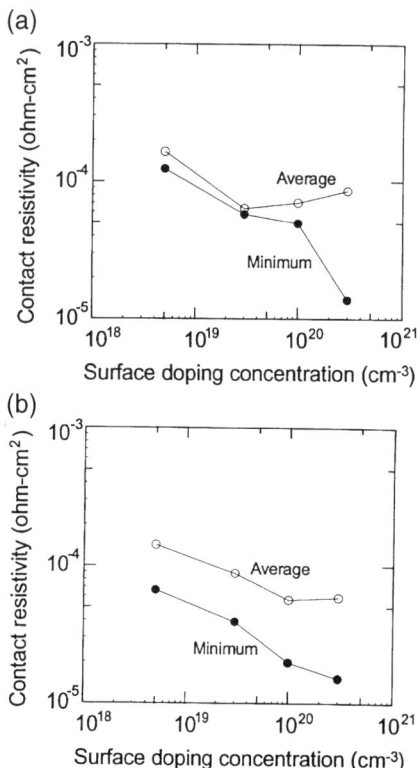

FIG. 33. Specific contact resistivity to 3C-SiC as a function of doping concentration for: (a) Al; (b) Ti [Moki 95].

the Schottky contact and the barrier height was determined to be in the range from 1.11–1.15 eV by C-V and photoresponse measurements.

Daimon et al. [Daimon 86] investigated Al and Al-Si metals contacts on n- and p-type 3C-SiC. They observed that as-deposited Al and Al-Si contacts were ohmic on n-type SiC, and their nature changed to rectifying upon annealing at 900°C. On the contrary, Al contacts on p-type Al showed good ohmic behavior only when annealed at 900°C.

Ioannou et al. [Ioannou 87] reported Au/3C-SiC Schottky diodes with an ideality factor of 1.5 ± 0.2, and an SBH of 1.2 eV. The diodes had a turn-on voltage of 0.6 to 1.0 V and breakdown voltages in the range of 8 to 10 V. The contacts remained stable during a 1-hr anneal at 300°C in Ar.

Bermudez [Bermudez 88] reported a barrier height of 1.4 eV for as-deposited Al contacts on 3C-SiC grown on (001) Si. He also studied the physical and electronic structures of the Al-SiC interface using XPS, LEED, and ELS.

Fujii et al. [Fujii 88] reported on the dependence of the electrical characteristics of 3C-SiC Schottky diodes on the crystal orientation of the Si substrate [($n11$), $n = 6, 5, 4, 3, 1$ and (100)], upon which the SiC films were grown by CVD. The I-V characteristics of diodes that were fabricated on SiC films on Si(611), Si(411), and Si(111) were found to be excellent compared to the conventional diodes on Si (100). They had a lower reverse leakage current and ideality factors closer to unity. The barrier height of the Pt-Schottky diodes was found to be 1.3 to 1.8 eV and that of Au-Schottky diodes was 1.0 to 1.6 eV, depending on the substrate orientation. ϕ_{Bn} for SiC films on Si(111) and Si(611) were found to be larger than for the other orientations.

Papanicolaou et al. [Papanicolaou 89] examined the electrical properties of Pt-Schottky contacts on 3C-SiC as a function of annealing temperature. They used Auger analysis to study the metallurgical reactions at the Pt/SiC surface and found that short annealing cycles in the 350 to 800°C temperature range led to the formation of a mixed structure of PtSi$_x$ and PtC at the interface. The interfacial reaction is dominated by the diffusion of Pt into the SiC layer. Figure 34 shows the variation of the Pt/3C-SiC structure SBH ϕ_B, the ideality factor η, and the reverse saturation current I_R as a function of anneal temperature. Annealing at each temperature is performed for 20 min. The barrier height was found to increase steadily as the anneal temperature was increased, from 0.95 eV for the as-deposited contacts to 1.35 eV for contacts annealed at 800°C. Minimum leakage current is obtained after the 450°C anneal. This anneal temperature also corresponds to the operation of these diodes, with an ideality factor close to unity.

Waldrop and Grant [Waldrop 90] investigated the Schottky barriers for several metals, including Pd, Au, Co, Ti, Ag, Tb, and Al, on 3C-SiC. The interface chemistry and the SBH were studied using XPS, and the electrical behavior was studied by measuring the current–voltage and capacitance–voltage characteristics. I-V characteristics showed that Pd, Au, and Co formed rectifying contacts, whereas Ti, Ag, Tb, and Al contacts showed ohmic behavior. Figure 35 shows the SBH obtained for various metals from the XPS studies plotted as a function of the work function of the respective metals. Similar to 6H and 4H polytypes, we see a strong correlation between ϕ_B and ϕ_M, once again ruling out the possibility of Fermi-level pinning on the surface. A wide range of ϕ_B (0.95–0.16 eV) is observed for different metals.

Wahab et al. [Wahab 91] reported Au Schottky diodes on sputtered 3C-SiC films deposited on Si substrates. The ideality factor of the diodes was 1.27 and a barrier height of 1.04 eV was determined from both I-V and C-V characteristics.

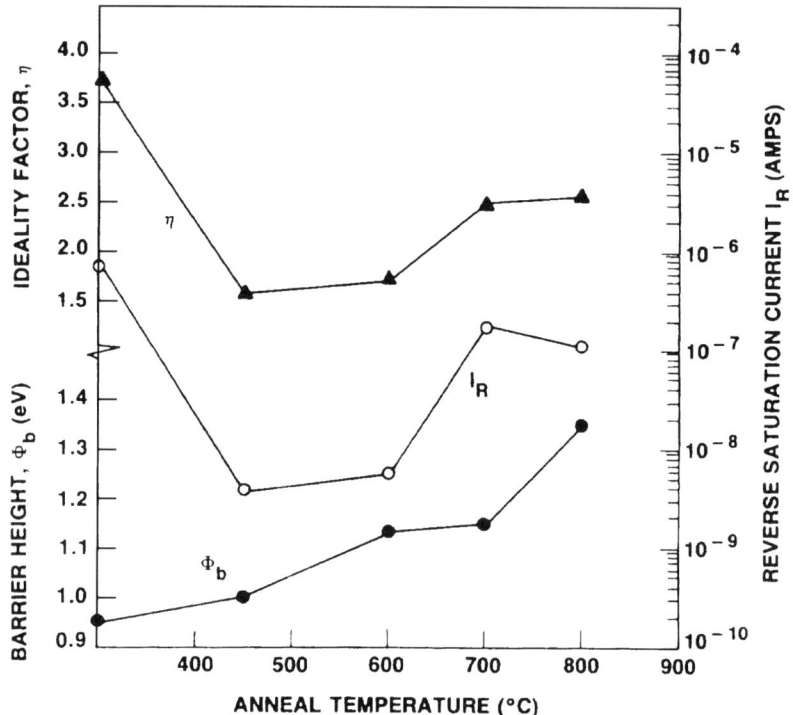

FIG. 34. SBH, ideality factor, and reverse saturation current for Pt/3C-SiC contacts as a function of anneal temperature [Papanicolau 89].

Steckl and Su [Steckl 93, 94] reported the "all-Ni technology" for the first time in 1993 utilizing Ni for forming both the ohmic and Schottky contacts. The resulting Schottky diodes exhibited high breakdown voltages in excess on 150 V.

Shenoy and co-workers [Shenoy 94] reported Pt-Schottky barrier diodes on n^-/n^+ 3C-SiC grown on n^+-Si substrates. These diodes had a low specific on-resistance of $\sim 6.1 \times 10^{-4}$ Ω-cm^2 and a relatively large breakdown voltage of ~ 85 V for a 4-μm-thick drift layer. Figure 36 shows the forward and reverse I-V characteristics for these diodes as a function of postmetallization anneal temperature. The log (I) versus V plot for the forward characteristics showed linearity over four decades with an ideality factor of 1.25. The SBH for these contacts was found to be ~ 0.85 V.

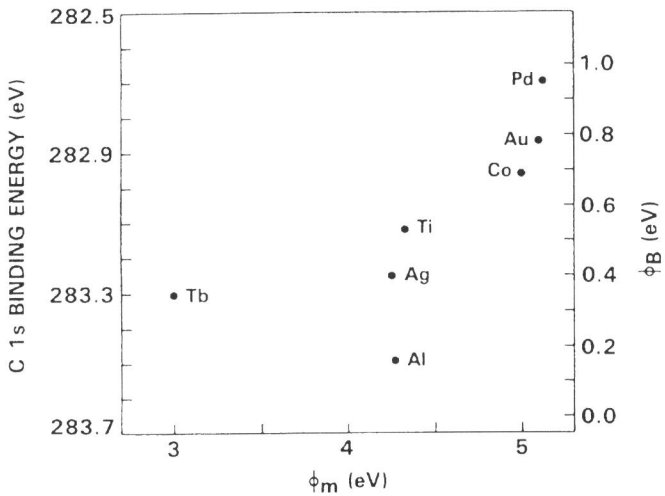

FIG. 35. SBH as a function of metal work function for 3C-SiC [Waldrop 90].

VII. SiC p-n Junctions Diode Rectifiers

Early techniques for fabricating *grown* p-n junctions in SiC used the traveling solvent (TS) method and the solution growth (SG) technique. Both of these techniques require very high temperatures, and the junctions obtained are often contaminated. In the TS technique [Campbell 71, 81], SiC crystals were grown together, and p-n junctions were formed by maintaining a heat zone across two SiC crystals of opposite conductivity type separated by a solvent metal (Pt, Cr). The temperature gradient across the thin solvent zone causes dissolution at both solvent-solid interfaces. However, the solubility of SiC in the solvent at the hotter interface is higher than that at the cooler interface. Thus, SiC diffuses through the solute and precipitates on the cooler crystal. In the SG technique [Campbell 71, 81], a small amount of SiC is dissolved in a molten metal (Fe, Cr, or Si). As the melt is cooled slowly, SiC crystals nucleate and grow in crucibles fabricated on graphite substrates.

More recently, p-n junction fabrication techniques have relied on the incorporation of impurities during either growth of the p- or n-type layers (so-called *in situ* doping) or incorporation after growth by impurity diffusion or by ion implantation. Commonly used techniques to grow epitaxial layers include sublimation epitaxy (SE), container-free liquid-phase epitaxy

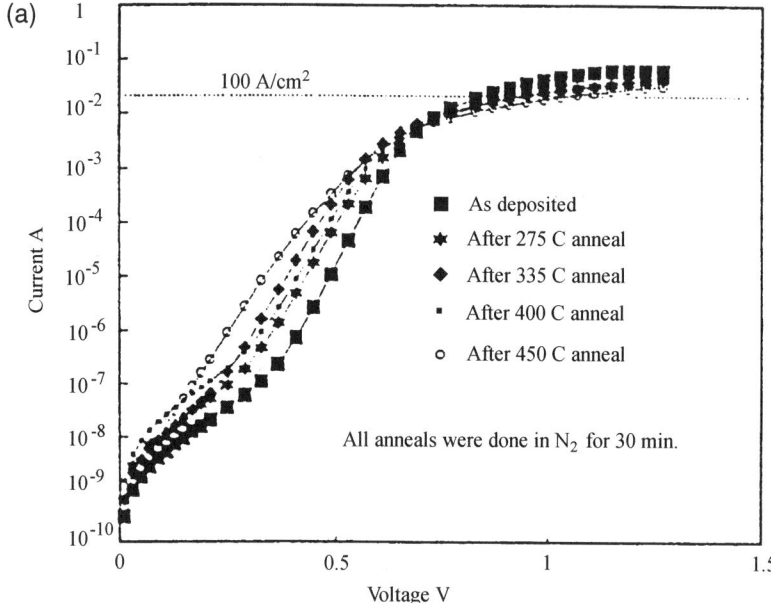

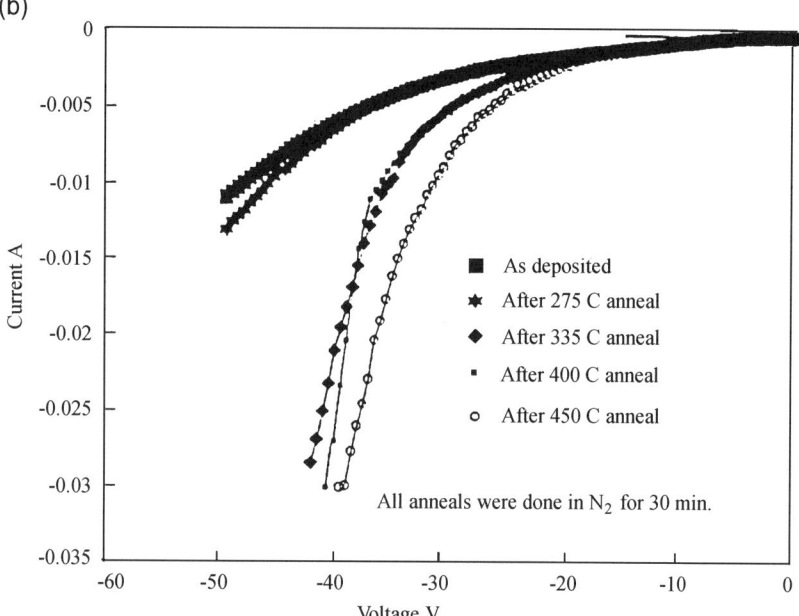

FIG. 36. Variation of current–voltage characteristics of Pt/3C-SiC SBDs with post metallization annealing: (a) forward bias; (b) reverse bias [Shenoy 94].

(CFLPE), and CVD [Lebedev 96]. *In situ* doping is performed by adding proper dopants to the ambient during the growth of these epitaxial layers. This technique, however, does not lend itself to the fabrication of planar structures. Therefore, etching of a mesa structure is required to produce individual p-n junction diodes. Postgrowth diffusion processes require high temperatures (in excess of 2000°C) and relatively long diffusion times to accomplish the mass transport required for device fabrication [Davis 91]. Prolonged exposure to such high temperatures results in the decomposition of SiC. Furthermore, oxide masks cannot be used for selective doping, as they vaporize at these temperatures. Thus, diffused junctions are very difficult to form in SiC.

The majority of the SiC planar p-n junctions diodes are fabricated using the ion implantation technique. Implantation can be performed using a thin film mask and therefore eliminates the need for subsequent mesa formation. For p-type impurities, group IIIA elements (B, Al, Ga, In, Tl) have been implanted in 6H-SiC, while group VA elements (N, P, Sb, Bi) have been implanted for n-type doping [Davis 91]. A quantitative discussion on ion implantation for SiC doping is presented in Chapter 2.

The following discussion summarizes the electrical properties and characteristics of the p-n junctions on different polytypes of SiC fabricated using the aforementioned techniques. Properties of selected high-voltage (>100 V) SiC p-n junction diodes are summarized in Table VIII.

1. 3C-SiC p-n JUNCTIONS

Suzuki *et al* [Suzuki 86] fabricated mesa structure p-n junction diodes on 3C-SiC and studied their high-temperature operation up to 500°C. At room temperature, the diode has a forward turn-on voltage of 1.2 V and a reverse current of $5 \mu A$ at 5 V. The diodes showed different current behaviors in three distinct voltage regimes under forward bias operation. At voltages lower than about 0.6 V, the diodes showed excess current. The current increased exponentially in the voltage range of 0.6 to 1.1 V. Beyond 1.1 V, there was a gradual increase in the current.

Furukawa *et al.* [Furukawa 86] studied the I-V and C-V characteristics of mesa structure diodes obtained by reactive ion etching (RIE) of CVD grown 3C-SiC on Si substrates. These diodes had a cut-in voltage of ~ 1.4 V, an ideality factor of 3.3, and a reverse leakage current smaller than $10 \mu A$ at -5 V.

Avila and co-workers [Avila 87] reported p-n junction diodes formed by ion implantation of B or Al in 3C-SiC/Si and annealing at 1365°C. The Al-implanted diodes did not show good rectification. The B-implanted

TABLE VIII

HIGH VOLTAGE (> 100 V) p-n JUNCTION DIODES ON 3C, 4H, AND 6H-SiC

Polytype/Structure	Growth Technique	Background Doping Concentration (cm^{-3})	Junction Depth or Epilayer Thickness (μm)
6H epitaxially grown layers	Growth using "double charge" method	n layer: 10^{17}–10^{19} p layer: 5×10^{19}	0.6
6H p-n mesas on grown layers on n-**6H** substrates; oxide passivation	Epilayers: CVD; substrates: Acheson	n-epi: 2×10^{16}	n-epi: 2 p-epi: 2
6H p-n mesas on grown layers on n-6H substrate; oxide passivation	Epilayers: CVD; substrates: Acheson and Lely	7×10^{16} 1.5×10^{17}	n-epi: 3 p-epi: 1.5
6H p-n mesas on grown layers on n-**6H** substrate; oxide passivation	Epilayers: CVD—site competition epitaxy; substrates: Cree Res.	n-epi: 2×10^{16}	n-epi: 4 p-epi: 0.75
6H p-n mesas, B^+ into epitaxial n/n^+ **6H** wafers; oxide passivation	B ion-implantation; substrates: Cree Res.	n-epi: 3.5×10^{15}	n-epi: 3.5
6H p-n mesas on grown layers on n-**6H** substrate; oxide passivation	Epilayers: CVD; substrates: Cree Res.	n-epi: 2×10^{16}	NR
3C p^+-n-n^+ mesas grown on **6H** n/n^+ epitaxial wafer: (i) with oxide passivation (ii) with no oxide	Epilayers: CVD—site competition epitaxy; substrates: Cree Res.	(i) n-epi: 5–20×10^{15} (ii) n-epi: 3–5×10^{15}	n-epi: 5 n-epi: 24
6H p^+-n-n^+ mesas grown on **6H** n^+ wafer	Epilayers: site competition epitaxy; substrates: Cree Res.	n-epi: 2–5×10^{15}	p^+-epi: 1 n-epi: 24 n^+-epi: 8
3C heterojunction (SiC/Si) mesa diodes; with oxide passivation	n-type 3C-SiC layer grown by RTCVD on p-Si substrate.	NR (unintentional)	n-epi: 1.2
6H planar, B^+ into epitaxial **6H** n/n^+ wafers; oxide passivation	Conversion of n-layer to p-type by B^+; substrates: Cree Res.	n-epi: 6–7×10^{15}	n-epi: 10 junction depth: 0.62
6H p^+-n-n^+ mesa on **6H** substrates	Hot-wall CVD	p^+-epi: 1×10^{18} n-epi: 1×10^{15}	p^+-epi: 2 n epi: 45

3 Building Blocks for SiC Devices

Diode Ideality Factor	Saturation Current Density (A/cm^2)	Forward Current Density at X Volts (A/cm^2)	Breakdown Voltage and Nature of Breakdown	Reverse Leakage Current Density (A/cm^2) at X Volts	Ref.
NR	NR	NR	167,	NR	[van Opdorp 1969]
NR	NR	NR	100, soft	NR	[Shibahara 87, Nishino 87]
2.5	1.2×10^{-32}	0.158 at 3.4 V	200, soft	3×10^{-4} at -100 V	[Wang 91]
NR	NR	~ 0.8 at 3	1000, hard	Leakage current: 0.4 mA at -300 V	[Matus 91]
1.77	NR	$\sim 1 \times 10^{-2}$ at 1.6	650, hard	10^{-10} at -10	[Ghezzo 93]
1.5–2	NR	$\sim 5 \times 10^{-6}$	710, soft	$< 5 \times 10^{-4}$ at -710, $\sim 10^{-9}$ at -200	[Edmond 93]
3.07	3.8×10^{-11}	$\sim 4.5 \times 10^{-2}$ at 2	200 V, soft	~ 0.3 at -200	[Neudeck 93, 94a]
NR	NR	~ 0.15 at 2	300 V, soft	~ 0.15 at -300	
2.05	5×10^{-22}	$\sim 1 \times 10^{-5}$ at 2	2000	0.1 at -2000	[Neudeck 94a, 94c]
1.05	NR	0.16 at 0.4	150	3×10^{-4} at -10	[Yih 94]
3.2–3.5	NR	100 at 7–8	800	$\sim 5 \times 10^{-5}$ at -800	[Shenoy 95]
—	—	100 at ~ 6		$< 5 \times 10^{-3}$ at -1100 ~ 10 at 4500	[Kordina 95b]

diodes showed rectification with ideality factors of 2.2, which increased with temperature up to 270°C. The diodes had breakdown voltages between 5 and 10 V and series resistance of the order of 20 kΩ.

Edmond and co-workers [Edmond 88b] prepared mesa structure junction diodes by ion implantation of Al in n-type and N in p-type 3C-SiC films. Rectification was observed in both types of diodes in the temperature range studied (300–673 K), with ideality factors in excess of 2. The forward characteristics also showed that the space-charge limited current in the presence of traps was the dominating mechanism for current transport. The temperature dependence of the forward and reverse bias characteristics for N- and Al-implanted diodes are shown in Figs. 37 and 38, respectively.

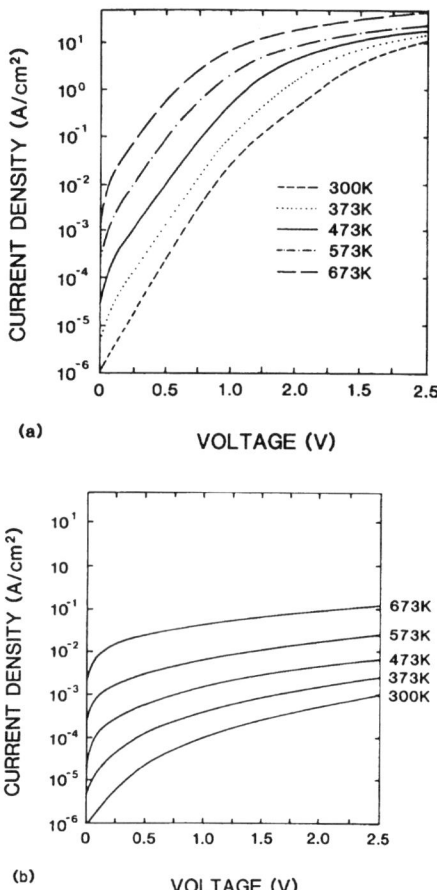

FIG. 37. Temperature dependence of current density versus voltage for N-implanted 3C-SiC p-n diode: (a) forward-bias; (b) reverse-bias [Edmond 88a].

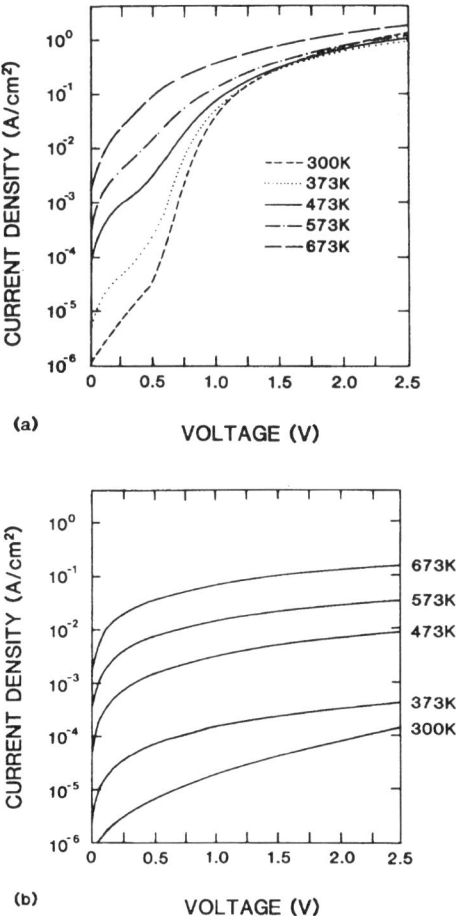

FIG. 38. Temperature dependence of current density versus voltage for Al-implanted 3C-SiC p-n diode: (a) forward-bias; (b) reverse-bias [Edmond 88a].

Neudeck and co-workers at NASA Lewis [Neudeck 93, 94a] reported improved electrical characteristics of p-n junctions with mesa structures on 3C-SiC layer grown by CVD on commercially available 6H-SiC substrates. The resulting 3C-SiC layers had increased crystal purity, no double positioning boundary (DPB) defects, and greatly reduced stacking fault densities. Best results were observed from diodes on 3C material with low doping densities $(3-5 \times 10^{15} \, \text{cm}^{-3})$ grown by utilizing the dopant control process developed by Larkin et al.: "site competition epitaxy" [Larkin 94]. These diodes showed a breakdown at $-300 \, \text{V}$. The soft breakdown characteristics was attributed to current leakage along the mesa perimeter. Diodes with higher doping densities ($>8.5 \times 10^{16}$ cm^{-3}) showed avalanche characteristics indicating bulk junction breakdown. The diodes on 3C-SiC layers with doping in the range of 0.5 to 2×10^{16} cm^{-3} were tested at temperatures up to 300°C. The forward characteristics of these diodes (ideality factor, current density at a given voltage drop) improved at higher temperatures, but the blocking voltage decreased slightly (from -200 V at 24°C to -170 V at 300°C). The saturation current density was 3.8×10^{-11} A/cm^2 at room temperature and increased to 1.1×10^{-6} A/cm^2 at 300°C.

The SiC group at Cincinnati [Yih 94] reported SiC/Si heterojunction p-n diodes fabricated by two different techniques. The first technique utilized self-selective growth of n-type SiC on p-type Si wafers patterned with an oxide layer to produce planar diodes. The second technique utilized mesa etching after a blanket deposition of an n-type SiC layer on p-type Si. The SiC layers were deposited using the rapid thermal chemical vapor deposition (RTCVD) process developed by Steckl and Li [Steckl 92]. The planar diodes had a breakdown voltage of 50 V, an ideality factor of 1.36, and a diode rectification ratio (defined at ± 1 V bias) of 1.6×10^4 at room temperature. The corresponding numbers for the mesa structure diode were 150 V, 1.05, and 1.6×10^4.

2. 6H-SiC p-n Junctions

Early p-n junctions on SiC were grown on the 6H polytype at high temperatures, such as 1740°C (traveling solvent method) and 1650°C (epitaxial growth from saturated silicon melts) [Campbell 71, 81]. The first p-n junctions fabricated by room-temperature implantation of 6H-SiC were reported by Dunlop and Marsh [Dunlop 69]. These junctions were obtained by the formation of n-type layers with ion-implanted N or Sb into p-type 6H-SiC. With this technique, SiC p-n junctions were produced without having to maintain high temperatures for hours for the impurity diffusion to take place during growth. Instead, only a few minutes of postimplant annealing was needed at 1400 to 1600°C for the activation of impurities. The resulting p-n junction diodes obtained from N and Sb implants had similar electrical characteristics. In the temperature range from 23 to 400°C, the diodes showed an ideality factor of 1.45. The dominating current mechanism at lower forward bias was suspected to be recombination in the depletion region. At higher forward bias, the diffusion current component dominated. The reverse characteristics showed avalanche breakdown at ~ 46 V.

van Opdorp and Vrakking fabricated 1×1 mm^2 p-n junctions on 6H-SiC by epitaxial growth procedure using the "double charge" method [van Opdorp 1969]. They used grinding and sawing to produce diodes with junction planes perpendicular to the c-axis. Alloyed ohmic contacts were used to make electrical connections to the n and p type layers. Breakdown voltage up to 167 V were obtained and the maximum breakdown electric field was found to be in the range 2.8–5 MV/cm.

Muench and Pfaffeneder [Muench 77] studied the electrical breakdown of p-n junctions in 6H-SiC obtained by vapor growth and mesa etching. Breakdown fields in the range of 2.0 to 3.7×10^6 V/cm were observed for these p-n junctions. The diodes exhibited a linear dependence between breakdown voltage and background p-type doping concentration in the range of 6×10^{17} to 2×10^{18} cm^{-3}. At lower doping levels, the breakdown voltage did not increase as expected. This could be attributed to the leakage around the mesa periphery and to local crystal imperfections [Edmond 93].

6H-SiC p-n junctions and devices based on them were reported by Dmitriev *et al.* [Dmitriev 85a, 85b, 87] and Anikin *et al.* [Anikin 84, 88, 89]. These devices were mostly fabricated on bulk crystals or on epitaxial films. The earlier reports were for devices on n$^+$-n-p$^+$ and n$^+$-n-p-p$^+$ structures on n$^+$ substrates [Dmitriev 85]. Cr and Al were utilized in these structures for ohmic contacts to n- and p-type SiC, respectively. The effective doping concentration of the n-type region was $N_d - N_a = 5 \times 10^{17}$ to 10^{18} cm^{-3}, and that of the p-type region was $N_a - N_d = 2 \times 10^{18}$ to 10^{19} cm^{-3}. The diode ideality factors were in the range of 1.76 to 1.96. The diode cut-in voltage was 2.7 V, and the saturation current density varied between 10^{-14} and 10^{-19} A/cm^2. The critical avalanche breakdown field was determined to be approximately 1.5×10^6 V/cm. Later investigations [Dmitriev 87] resulted in better quality diodes, with ideality factors somewhat closer to unity and with higher breakdown fields ($2-3 \times 10^6$ V/cm) [Davis 91].

Anikin *et al.* reported p$^+$-n mesa structures grown on 6H-SiC by the sublimation sandwich method on the Si (0001) face of single-crystal 6H-SiC wafers. The n-n$^+$ structures were fabricated [Anikin 84, 88] by implantation of Al ions followed by annealing. Final mesa structures were obtained by reactive ion etching. In the initial study [Anikin 84], the doping density for the n-type base region in the two samples studied was 6×10^{16} cm^{-3} and 2×10^{17} cm^{-3}. A study of the I-V characteristics at high temperatures (293–780 K) indicated a positive temperature coefficient of 5×10^{-3} V/degree for the current–voltage dependence under forward bias. Two distinct linear regimes were observed: one for voltages up to 2 V, and the other for voltages from 2 to 2.3 V. The ideality factors for different temperatures for these regimes were found to be 1.7 to 1.98 and 1.15 to 1.4, respectively. The I-V characteristics at voltages higher than 2.3 V were limited by the series resistance of the diodes. Later studies [Anikin 88] indicated similar behavior in the I-V characteristics with two distinct linear regimes. The ideality factor for the regime corresponding to lower voltages was found to be in the range of 1.65 to 2.00 for the temperature range from 370 to 720 K. The ideality factor for higher voltages was independent of temperature and was found to be ~1.5. The reverse leakage currents in these diodes were attributed to leakage along the periphery of the mesa structures.

Mesa-type 6H-SiC diodes fabricated by *in situ* doping and capable of high blocking voltages have been reported by Edmond and co-workers [Edmond 92, 93; Davis 91]. Rectifiers with breakdown voltages as high as 1400 V were reported in their studies. Low-current forward and reverse bias testing was performed on these diodes for the temperature range from 300 to 623 K. Figure 39 shows the I-V characteristics of a 710-V SiC diode as a function of temperature. As seen in Fig. 39a, the forward I-V characteristics show two distinct linear regimes: the generation-recombination current regime with ideality factor close to 2.0, followed by the regime with both diffusion and generation-recombination currents, with somewhat lower ideality factors. This behavior is similar to that observed by Anikin *et al.* [Anikin 84] in diodes fabricated by implantation. At 300 K, these regimes occur in the voltage range of 1.5 to 2.2 V and 2.2 to 2.55 V, respectively. As shown in Fig. 39a, the respective ideality factors are 2.0 and 1.5. At 623 K, the ideality factors corresponding to these two regimes were found to be 1.9 and 1.6. As expected, the reverse leakage current density (Fig. 39b) increases with temperature. However, the leakage at 623 K and 710 V (prior to the onset of avalanche) was still very low ($\sim 5 \times 10^{-4}$ A/cm^2). The diodes reported by this group also displayed a fast switching speed. As shown in Fig. 40, the reverse recovery time (t_{rr}) for a typical high-voltage (1200 V) diode was found to be less than 14 ns at room temperature, and increased by only 20% at 625 K.

Kaneda *et al.* [Kaneda 87] reported p-n junctions formed by growing p-type 3C-SiC by Molecular Beam Epitaxy (MBE) on n-type 6H-SiC substrates. A high breakdown electric field 6.7×10^5 V/cm was determined from the avalanche breakdown characteristics of the diodes.

High-voltage mesa-style p-n diodes were fabricated on CVD-grown n- and p-type 6H-SiC layers by Matus and co-workers [Matus 91]. These diodes showed excellent rectification up to the highest temperature tested, 600°C, and high breakdown voltage of 1000 V. The maximum electric field at breakdown was found to be 2.7×10^6 V/cm, and the reverse bias leakage current measured at 300 V increased from 0.4 μA at room temperature to 5 μA at 600°C. The built-in potential obtained from C-V characteristics was -2.7 V.

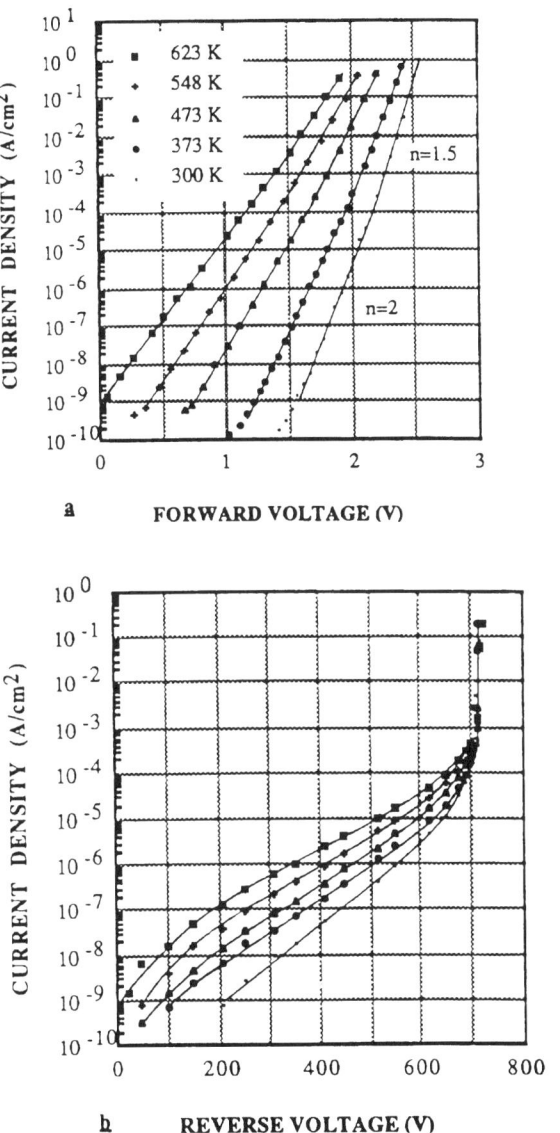

FIG. 39. Current–voltage characteristic of 6H-SiC p-n junction diode as a function of temperature [Edmond 93].

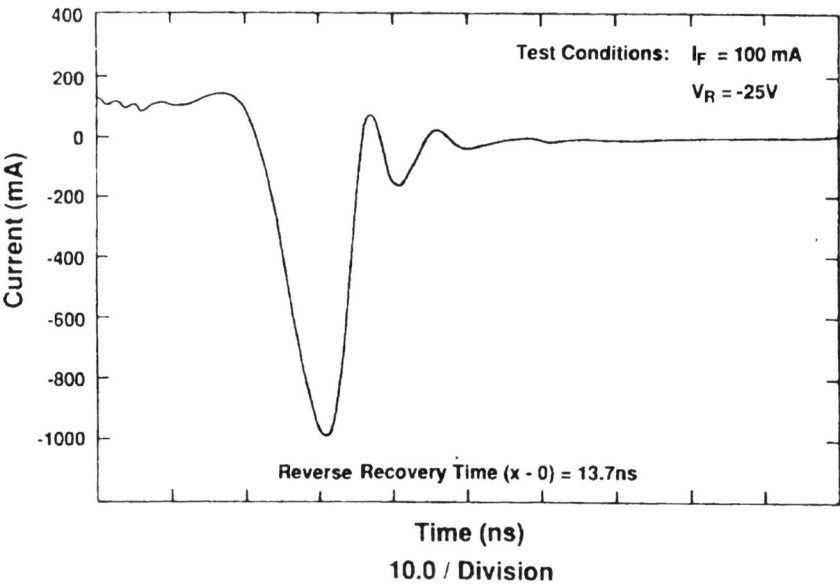

FIG. 40. Switching speed of high-voltage (1200 V) 6H-SiC p-n diode at 293 K [Davis 91].

Ghezzo et al. [Ghezzo 93] applied N implantation at high temperatures (up to 1000°C) to fabricate low leakage current diodes on 6H-SiC. These diodes also had a mesa structure and were patterned using reactive ion etching. The diodes had an ideality factor of 1.95 and a leakage current of 5×10^{-11} A/cm^2 at a reverse bias of 10 V. In a later study, Ghezzo and co-workers [93] reported p-n junction diodes on 6H-SiC fabricated by boron implantation at room temperature and at 1000°C, followed by postimplant anneal at 1300°C. As shown in Fig. 41, diodes fabricated by high-temperature implantation had better forward and reverse characteristics. Figure 41a shows the forward characteristics of these diodes at room and elevated temperatures (170 or 181°C), for both room-temperature implantation (left-hand side) and 1000°C implantation (right-hand side). The diode obtained by high-temperature implantation clearly has lower forward drops for any given current density. Figure 41b shows a histogram of reverse diode current densities measured at 21°C and -10 V for 6H-SiC diodes with the same implantation conditions. Once again, the diodes with high-temperature implant often had superior characteristics: a tighter distribution and a much smaller median value for the revese current density. The best diodes had an ideality factor of 1.77, a reverse bias leakage of 10^{-10} A/cm^2 at -10 V, and a breakdown voltage of -650 V.

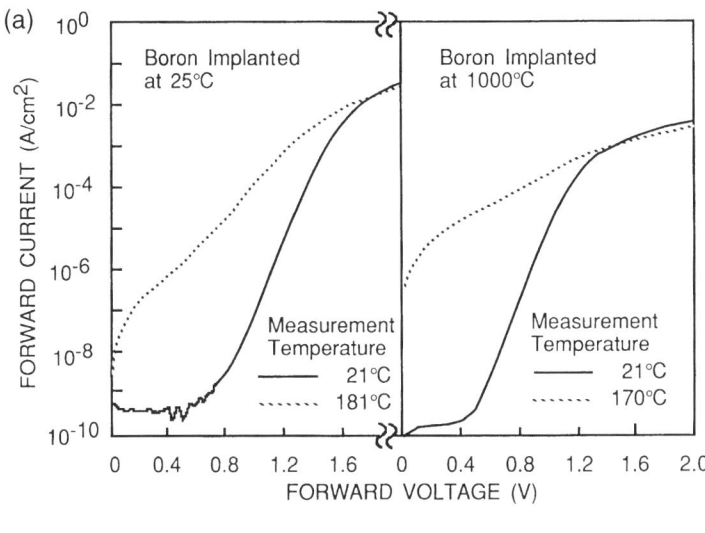

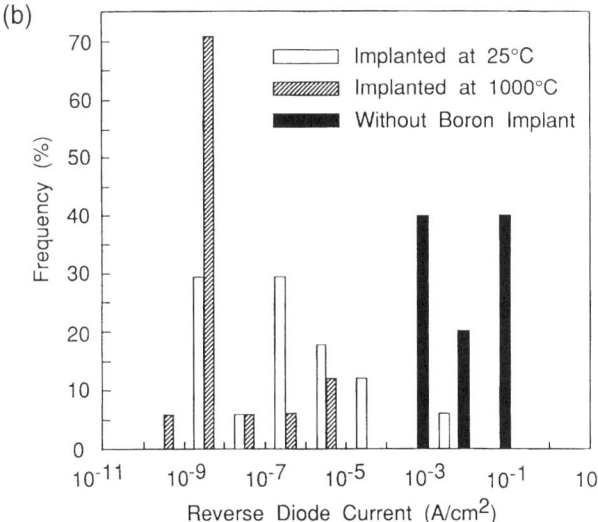

Fig. 41. Electrical characteristics of boron-implanted 6H-SiC diodes: (a) forward bias characteristics as a function of implantation and measurement temperature; (b) histogram of room-temperature reverse bias current density at $-10\,\text{V}$ as a function of implantation conditions [Ghezzo 93].

Vassilevski et al. [Vassilevski 93, 94] reported the pulsed power characteristics of 6H-SiC p-n diodes grown by liquid-phase epitaxy (LPE). These diodes operated at a high current density of 60 kA/cm² (input power density of 9 MW/cm²) for a current pulse duration of 60 ns, prior to experiencing avalanche breakdown at ~80 V. The temperature coefficient of avalanche breakdown ($\beta = V^{-1} dV/dT$) was found to first decrease with temperature over a broad range (300–750 K) and then increase with temperature from 750 to 900 K.

Neudeck and co-workers at NASA Lewis Research Center [Neudeck 94a, 94b, 94c] first reported high-voltage rectification exceeding 2000 V using SiC p-n junction diodes. The mesa structure 6H-SiC p^+-n junction diodes were fabricated on 6H-SiC epilayers grown by atmospheric CVD utilizing site competition epitaxy [Larkin 94] on commercially available 6H-SiC wafers. Figure 42 shows the current–voltage characteristics for these diodes. The forward characteristics of the diodes are very well behaved, exhibiting saturation current densities below 10^{-20} A/cm², and ideality factors close to 2. The peak junction electric field was calculated to be in the range of 1.4 to 1.8 MV/cm.

High-voltage p-n diodes fabricated on 6H-SiC grown by CVD in a hot-wall reactor [Kordina 95a] were reported by Kordina et al. [Kordina 95b, Janzen 96]. The mesa structure p-n junction was fabricated by growing a 2-μm-thick p-type layer over a 45-μm-thick n-type epitaxial layer on an n^+ substrate. As shown in Fig. 43, these diodes [Kordina 95b] had a

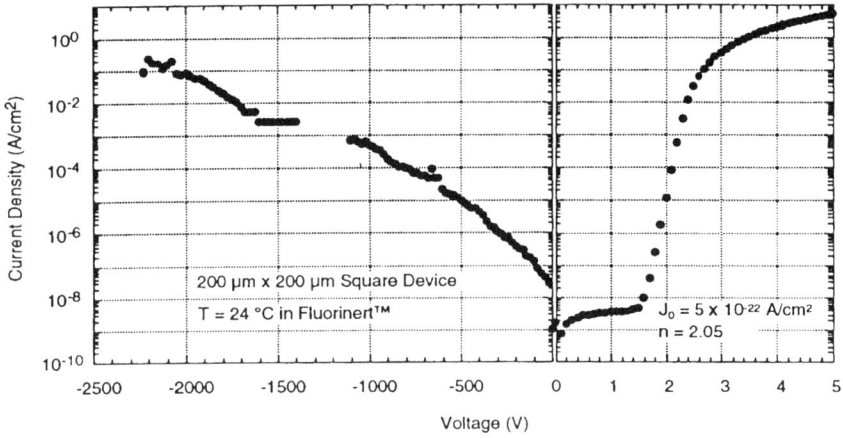

FIG. 42. 6H-SiC p-n junction current–voltage characteristics at room temperature. Note different voltage scales for forward and reverse bias [Neudeck 94c].

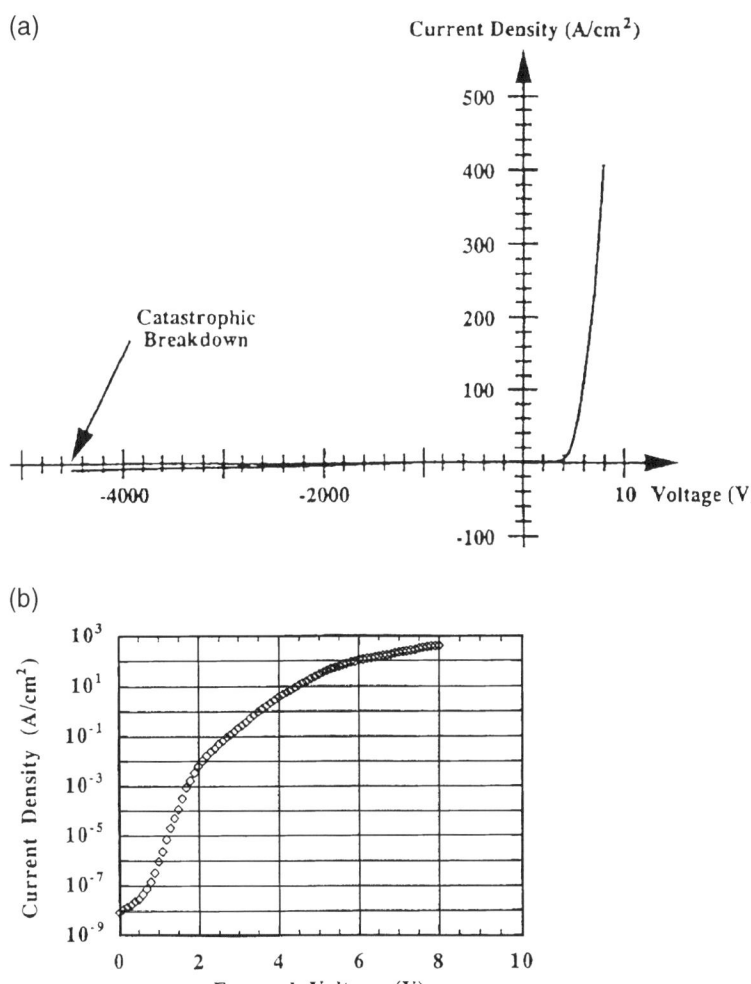

FIG. 43. Characteristics of a 4.5-kV Si rectifier: (a) I–V characteristics; (b) measured forward voltage drop versus current density [Kordina 95b].

breakdown voltage of 4.5 V before a catastrophic failure occurred. The forward voltage drop at a current density of 100 A/cm^2 was ~6 V.

The first *planar* p-n junction diodes on 6H-SiC were fabricated by Shenoy and Baliga [Shenoy 95, 96] by room-temperature boron implantation through a pad oxide. These diodes had a breakdown voltage of 800 V and had a low leakage current of less than 5×10^{-5} A/cm^2 until breakdown occurred. Figure 44 shows the current–voltage characteristics of these diodes for two junction depths (x_j), 0.2 and 0.62 μm. The diodes exhibit forward bias linearity over three decades of the log (I) versus V plots, with an ideality factor of 3.2 to 3.5. Series resistance effects were observed at current densities above $1\,A/cm^2$. This resulted in a high forward voltage drop of 7 to 8 V at $100\,A/cm^2$. In reverse bias, the breakdown voltage increases significantly with the increase in junction depth. For $x_j = 0.2\,\mu m$, the highest breakdown was 480 V, whereas for $x_j = 0.62\,\mu m$, it was up to 810 V.

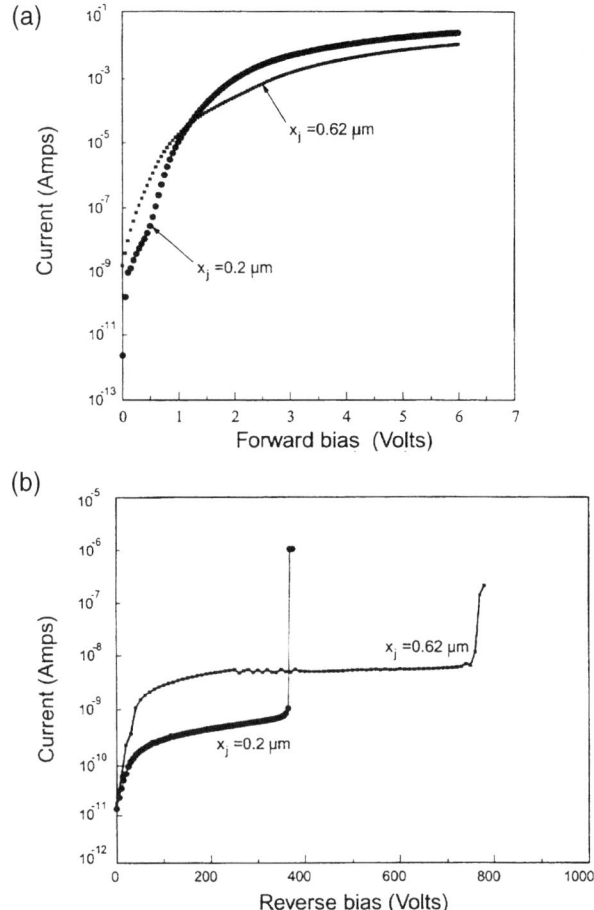

FIG. 44. Current–voltage characteristics of planar 6H-SiC p-n diodes for junction depths of 0.2 and 0.62 μm: (a) forward characteristics; (b) reverse characteristics [Shenoy 95].

3. 4H-SiC p-n JUNCTIONS

Palmour and co-workers first reported p-n junction diodes on 4H-SiC [Palmour 94]. These diodes had a mesa structure and were fabricated on n-type 4H substrates with doping concentration in the range of (4 to 10) $\times 10^{18}\,\mathrm{cm}^{-3}$. The epitaxial layers were doped in the range of 8×10^{15} to $2.2 \times 10^{18}\,\mathrm{cm}^{-3}$. Both p^+-n and n^+-p diode structures were studied on Si- and C-face SiC wafers. The thickness of the low-doped epitaxial layer was in the range of 0.5 to 15.0 μm. Figure 45 shows the breakdown voltage and the corresponding electric field as a function of background doping. All diodes shown in the illustration, except for the lowest doped sample, showed avalanche breakdown characteristics. The I-V characteristics of the diode showing highest breakdown are shown in Fig. 46. The breakdown voltage of the diode was 1130 V, and a forward current density of 68 A/cm² was obtained at 3.6 V.

Mesa structure p^+-n diodes on 4H-SiC were reported by Neudeck and Fazi [Neudeck 97]. The homoepitaxial 4H-SiC layers were grown on commercially available 4H-SiC substrates. The n-type doping was found to be in the range of 0.25 to $1.5 \times 10^{18}\,\mathrm{cm}^{-3}$. The majority of smaller size diodes tested were reported to have a stable positive temperature coefficient behavior indicating stable breakdown characteristics. However, the larger size diodes ($>1 \times 10^{-4}\,\mathrm{cm}^2$) exhibited negative temperature breakdown

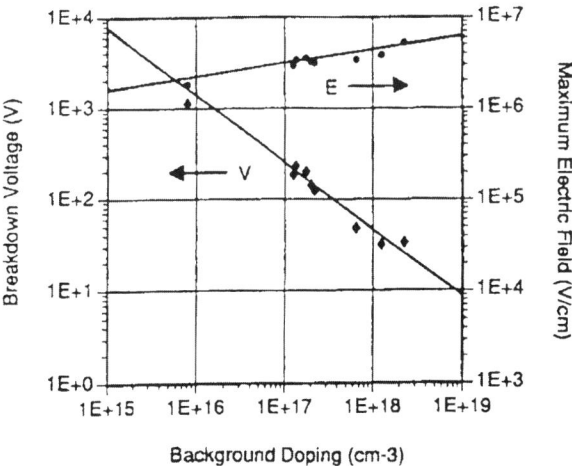

FIG. 45. Breakdown voltage and electric breakdown field as a function of background doping for 4H-SiC p-n diodes [Palmour 94].

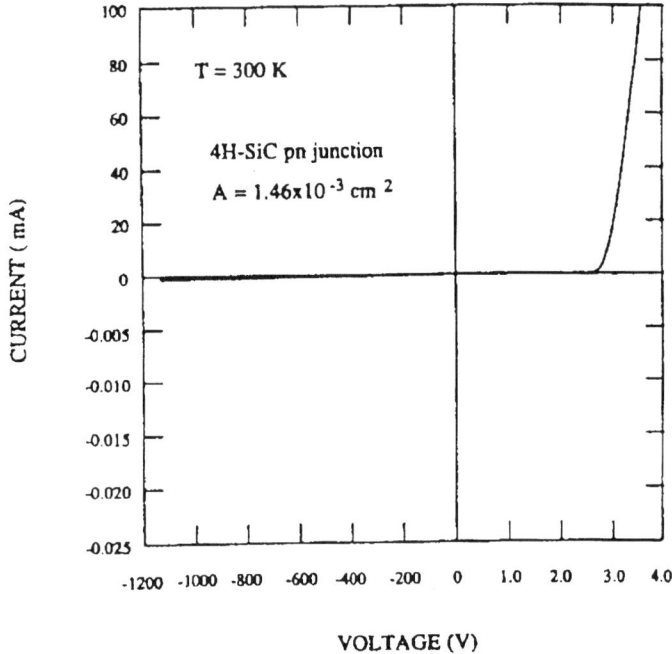

FIG. 46. I–V characteristics of a 4H-SiC mesa p$^+$-n diode [Palmour 94].

coefficient behavior, thus leading to the conclusion that the latter is related to the defects (dislocations, micropipes, and deep levels) that negatively impact the breakdown properties of the SiC p-n junctions.

4H-SiC p-n junctions grown by LPE on 30- and 35-mm wafers have also been reported [Rendakova 97]. The breakdown across the bulk region occurred at fields in the range of 2 to 4 MV/cm. No degradation in device characteristics occurred upon passing current densities as high as 100 A/cm^2 during reverse bias.

Recent results from the Cree Research group [Irvine 98] indicate the enhanced blocking voltage obtainable with thick, high quality epi-layers on 4H-SiC. Irvine et al. [Irvine 98] report a blocking voltage of 5.5 kV for a p$^+$ emitter layer of 83 μm.

Kuznetsov and Dmitriev [Kuznetsov 96] reported linearly graded p-n junctions on 4H-SiC substrates. The diodes had a p$^+$-p-n$^+$ mesa structure patterned by reactive ion etching. The three epitaxial layers were grown sequentially on 4H-SiC wafers. The n$^+$ layer grown first was 1 μm thick and had a net concentration of $\sim 2 \times 10^{18}$ cm^{-3}. Next a 3-μm-thick p-layer was grown, followed by a 0.5-mm-thick p$^+$ contact layer. The N_a-N_d profile in the p-layer determined by C-V measurements was found to increase linearly with distance from the junction. The impurity gradient was determined to be $\sim 5 \times 10^{21}\,\text{cm}^{-4}$. The I-V characteristics of these diodes were studied as a function of temperature in the range from 200 to 600 K. The ideality factor was found to be independent of temperature, with a measured value of 1.9. The current conduction in the temperature range investigated was attributed to the recombination of carriers in the space-charge region, through a deep level located near the middle of the bandgap. The saturation current density was determined to be 10 to $25\,\text{A/cm}^2$ at 300 K. The diodes showed an abrupt breakdown at 300 V at all temperatures, and the breakdown electric field was calculated as $3 \times 10^6\,\text{V/cm}$.

Neudeck and Fazi [Neudeck 96a, 96b] reported that p-n diodes fabricated on both 6H- and 4H-SiC failed catastrophically when subjected to single-shot reverse bias pulses. This failure occurred at pulse voltages much lower (60–80%) than the DC reverse bias voltages which these diodes were capable of blocking. Microscopic examination of the diodes subjected to these pulses showed highly localized damage to the device mesa and contact. It is believed that a current-filamentation type of failure occurs in the bulk of the device. When such a filament occurs, the current density in a localized spot drastically increases, greatly stressing the junction material, often to the point of failure.

VIII. Summary and Conclusions

SiC SBDs and p-n junctions have been shown to exceed the performance of Si-based counterparts for applications requiring high-power and high-temperature operation. SiC SBDs could rapidly become commercially viable as an attractive alternative to the slower Si-based p-i-n diodes currently being used in several power electronics applications. A useful figure of merit for rectifiers is the on:off current ratio, which combines the current-carrying capability in forward bias with the leakage current in reverse bias. A summary of published ON:OFF ratios of 4H- and 6H-SiC SBDs is shown in Fig. 47 as a function of reported breakdown voltage. This summary was only able to include results where all the necessary characteristics (forward current, leakage current, and breakdown voltage) were available in the published literature. As can be seen, ON:OFF ratios of one million or more have been obtained by several groups for diodes with breakdown voltages of ~ 1000 to 1100 V. SiC SBDs are also likely to allow operation beyond 300°C, which is the upper limit of Si SOI-based devices. SiC p-n junctions also have the potential to serve as rectifiers with very low off-state leakage

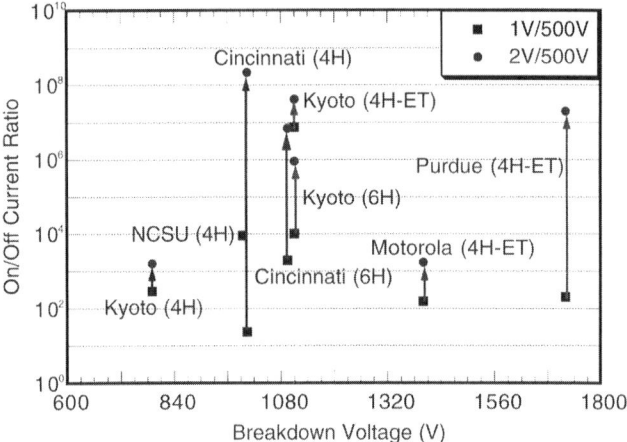

FIG. 47. Summary of published results on Schottky diode ON/OFF ratio versus breakdown voltage for various groups. The ON current is taken at either 1- or 2-V forward bias, while the OFF current is taken at 500-V reverse bias. ET: edge termination.

currents and as switches that are capable of blocking high voltages during reverse bias state, while allowing high currents to flow during the conduction or forward bias state.

The material quality of SiC commercially available today imposes restrictions on the device yield achievable from SiC wafers. The best available wafers typically still have 10 to 100 micropipes/cm^2. These micropipes severely limit the maximum practical device size. Another economic limitation is the size of the currently available SiC wafers (2in. or less in diameter). While these limitations do exist, our general view of the future of SiC rectifiers is very positive, as rapid progress has been made in the device technology. At the same time, significant progress has been made in reducing micropipe density and similar progress is expected in expanding the diameter of SiC crystals and substrates.

There is a growing need for high-temperature electronic components for a wide range of military and commercial uses, including many aerospace and automotive applications. SiC-based electronic devices are expected to play a critical role in such applications, which may require stable operation at high temperatures in the range of 200 to 600°C. Under conditions of severe mechanical stress or excessive heating caused by prolonged exposures to high temperature, failure of the SiC devices are likely to occur due to the degradation of the contacts (Schottky or ohmic) rather than the breakdown limits imposed by the device dimensions. This degradation of the metal contacts may occur due to oxidation, metal–semiconductor reaction at the interface, interdiffusion, or a combination of these possibilities. Studies on the reliability, stability, and lifetime of these contacts are thus necessary for satisfactory device performance.

ACKNOWLEDGMENTS

The authors are grateful to P. G. Neudeck of the NASA Lewis Research Center for taking the time to review this chapter and providing many useful suggestions for its improvement. The authors also wish to thank J. W. Palmour at Cree Research and J. Scofield of USAF Wright Laboratory for many useful discussion on the properties of SiC diodes. The research at Cincinnati has been supported by grants from the Ohio Aerospace Institute and Wright-Patterson Air Force Base. Finally, the authors gratefully acknowledge J. Muenchen for assistance with manuscript preparation.

REFERENCES

Addamiano, A. *"Semiconductive Crystals of Silicon Carbide with Improved Chromium-Containing Electrical Contacts".* US Patent No. 3510733, 1970.

Aivazova, L. S., Gorin, S. N., Sidyakin, V. G., and Shvarts, I. M. (1977). "Electrical properties of cubic silicon carbide," *Sov. Phys. Semicond.* **11** (Sep), pp. 1069–1070.

Alok, D., Baliga, B. J., and McLarty, P. K. (1994). "A simple edge termination for silicon carbide devices with nearly ideal breakdown voltage," *IEEE Electron. Dev. Lett.* **15** (Oct), pp. 394–395.

Anderson, C. L., Crowell, C. R., and Kao, T. W. (1975). "Effects of thermal excitation and quantum-mechanical transmission on photothreshold determination of Schottky barrier height," *Solid-State Electron.* **18**, pp. 705–713.

Andreev, A. N., Lebedev, A. A., Rastegaeva, M. G., Snegov, F. M., Syrkin, A. L., Chelnokov, V. E., and Shestopalova, L. N. (1995). "Barrier height in n-SiC-6H based Schottky diodes," *Semiconductors* **29** (Oct), pp. 957–962.

Anikin, M. M., Lebedev, A. A., Popov, I. V., Sevast'yanov, V. E., Syrkin, A. L., Suvorov, A. L., Chelnokon, V. E., and Shpynev, G. P. (1984). "Silicon carbide rectifying diode," *Sov. Tech. Phys. Lett.* **10** (Sep), pp. 444–445.

Anikin, M. M., Lebedev, A. A., Popov, I. V., Rastegaev, V. P., Strel'chuk, A. M., Syrkin, A. L., Tairov, Y. M., Tsvetkov, V. F., and Chelnokov, V. E. (1988). "Electrical characteristics of epitaxial p^+-n-n^+ structures made of the 6H polytype of silicon carbide," *Sov. Phys. Semicond.* **22** (Feb), pp. 181–183.

Anikin, M. M., Ivanov, P. A., Syrkin, A. L., Tsarenkov, B. V., and Chelnokov, V. E. (1989). "High-temperature SiC-6H field effect transistor with record high transconductance for silicon carbide transistors," *Sov. Tech. Phys. Lett.* **15** (Aug), pp. 636–638.

Anikin, M. M., Andreev, A. N., Lebedev, A. A., Pyatko, S. N., Rastegaeva, M. G., Savkina, N. S., Strel'chuk, A. M., Syrkin, A. L., and Chelnokov, V. E. (1991). "High-temperature Au-SiC-6H Schottky diodes," *Sov. Phys. Semicod.* **25** (2), pp. 198–201.

Anikin, M. M., Rastegaeva, M. G., Syrkin, A. L., and Chuiko, I. V. (1992). "Ohmic contacts to silicon carbide devices," in: *Amorphous and Crystalline Silicon Carbide III – Washington 1990*, Springer, *Proc. Phys.* **56**, pp. 183–189.

Arnodo, C., Tyc, S., Wyczisk, F., and Brylinski, C. (1996). "Nickel and molybdenum ohmic cotacts on silicon carbide," in: *Silicon Carbide and Related Materials VI—Kyoto 1995, Inst. Phys. Conf. Ser.* **142**, pp. 577–580.

Arugu, D. O., Harris, G. L., and Taylor, C. (1995). "Effect of anneal temperature on the electrical characteristics of nickel-based ohmic contracts to β-SiC," *Electronics Lett.* **31** (Apr), pp. 678–680.

Avila, R. E., Kopanski, J. J., and Fung, C. D. (1987). "Behavior of ion-implanted junction diodes in 3C-SiC," *J. Appl. Phys.* **62** (October), pp. 3469–3471.

Bardeen, J. (1947). "Surface states and rectification at a metal semiconductor contact," *Phys. Rev.* **71**, 717–727.

Baud, L., Jaussaud, C., Madar, R., Bernard, C., Chen, J. S., and Nicolet, M. A. (1995). "Interfacial reactions of W thin film on single-crystal (001) β-SiC," *Mater. Sci. Eng.* **29**, pp. 126–130.

Baud, L., Billon, T., Lassagne, P., Jaussaud, C., and Madar, R. (1996). "Low contact resistivity W ohmic contacts to n-type 6H-SiC," in: *Silicon Carbide and Related Materials VI—Kyoto 1995, Inst. Phys. Conf. Ser.* **142**, pp. 597–600.

Berger, H. H. (1972). "Contact resistance and contact resistivity," *J. Electrochem. Soc.* **119** (Apr.), pp. 507–514.

Bermudez, V. M. (1988). "Growth and structure of aluminum films on (001) silicon carbide," *J. Appl. Phys.* **63** (May), pp. 4951–4959.

Bhatnagar, M., McLarty, P. K., and Baliga, B. J. (1992). "Silicon carbide high-voltage (400 V) Schottky barrier diodes," *IEEE Electron Dev. Lett.* **13** (Oct), pp. 501–503.

Bhatnagar, M., Nakanishi, H., Bothra, S., McLarty, P. K., and Baliga, B. J. (1993a). "Edge Terminations for SiC High Voltage Schottky Rectifiers," in: *International Symposium on Power Semiconductor Devices & ICs, IEEE Cat. No.* **93CH3314-2**, pp. 89–94.

Bhatnagar, M. and Baliga, B. J. (1993b). "Comparison of 6H-SiC, 3C-SiC, and Si for Power Devices," *IEEE Trans. Electr. Dev.* **40** (3), pp. 645–655.

Bhatnagar, M., Baliga, B. J., Kirk, H. R., and Rozgonyi, G. A. (1996). "Effect of surface inhomogenities on the electrical characteristics of SiC Schottky contacts," *IEEE Trans. on Electron Dev.* **43** (Jan), pp. 150–156.

Campbell, R. B. (1971). "Silicon Carbide Junction Devices," in *Semiconductors and Semimetals, Part B*, P. K. Willardson and A. C. Beer. Academic Press: NY, **7**, pp. 625–683.

Campbell, R. B. (1981). "What happened to silicon carbide?" in: *Conference on High Temperature Electronics—Tucson 1981*, The Institute of Electrical and Electronics Engineers, Piscataway, NJ (IEEE Catalog No. 81CH1658-4), pp. 71–74.

Chaddha, A. K., Parsons, J. D., and Kruaval, G. B. (1995). "Thermally stable, low specific resistance ($1.30 \times 10^{-5} \Omega \, cm^2$) TiC ohmic contacts to n-type 6Hα-SiC," *Appl. Phys. Lett.* **66**, pp. 760–762.

Chaudhry, M. I., Berry, W. B., and Zeller, M. V. (1990). "Ohmic contacts on β-SiC," *Mat. Res. Soc. Symp. Proc.* **162**, pp. 507–512.

Chaudhry, M. I., Berry, W. B., and Zeller, M. V. (1991). "A study of ohmic contacts on β-SiC," *Int. J. Electronics* **71**, pp. 439–444.

Chen, J. S., Bachli, A., Nicholet, M.-A., Baud, L., Jaussaud, C., and Madar, R. (1995). "Contact resistivity of Re, Pt and Ta films on n-type β-SiC: preliminary results," *Mater. Sci. Eng.* **29**, pp. 185–189.

Cheung, S. K. and Cheung, N. W. (1986). "Extraction of Schottky diode parameters from forward current-voltage characteristics," *Appl. Phys. Lett.* **49** (Jul), pp. 85–87.

Choyke, W. J., Patrick, L., and Hamilton, D. R. (1964). In *Int. Conf. Semicond.—Paris 1964*, p. 751.

Cooper, J. A., "Recent advances in SiC Power devices" in *Silicon Carbide, III-Nitrides and Related Materials—VII Stockholm 1997*, edited by G. Pensl, H. Morkoc, B. Monemar and E. Janzén, Materials Science Forum **264–268**, pp. 895–900 (1998).

Cox, R. H. and Strack, H. (1967). "Ohmic contacts for GaAs devices," *Solid-State Electron.* **10** (Dec), pp. 1213–1218.

Crofton, J., Ferrero, J. M., Barnes, P. A., Williams, J. R., Bozack, M. J., Tin, C. C., Ellis, C. D., Spitznagel, J. A., and McMullin, P. G. (1992). "Metallization studies on epitaxial 6H-SiC," in: *Amorphous and Crystalline Silicon Carbide IV—Santa Clara 1991*, Springer, *Proc. Phys.* **71**, pp. 176–182.

Crofton, J., Barnes, P. A., and Williams, J. R. (1993). "Contact resistance measurements on p-type 6H-SiC," *Appl. Phys. Lett.* **62** (Jan), pp. 384–386.

Crofton, J., McMullin, P. G., Williams, J. R., and Bozack, M. J. (1995). "High-temperature ohmic contact to n-type 6H-SiC using nickel," *J. Appl. Phys.* **77** (Feb), pp. 1317–1319.

Crofton, J., Luckowski, E. D., Williams, J. R., Isaacs-Smith, T., Bozack, M. J., and Siergiej, R. (1996a). "Specific contact resistance as a function of doping for n-type 4H and 6H-SiC," in: *Silicon Carbide and Related Materials VI–Kyoto 1995, Inst. Phys. Conf. Ser.* **142**, pp. 569–572.

Crofton, J. and Sriram, S. (1996b). "Reverse leakage current calculations for SiC Schottky contacts," *IEEE Trans. on Electron. Dev.* **43** (Dec), pp. 2305–2307.

Daimon, H., Yamanaka, M., Sakuma, E., Misawa, S., and Yoshida, S. (1986). "Annealing effects on Al and Al-Si contacts with 3C-SiC," *Jpn. J. Appl. Phy.* **25** (Jul), pp. 592–594.

Davis, R. F., Kelner, G., Shur, M., Palmour, J. W., and Edmond, J. A. (1991). "Thin film deposition and microelectric and optoelectric device fabrication and characterization in monocrystalline alpha beta silicon carbide," *Proc. of IEEE* **79** (May), pp. 677–701.

Dmitriev, V. A., Ivanov, P. A., Strelchuk, A. M., Syrkin, A. L., Popov, I. V., and Chelnokov, V. E. (1985a). "SiC tunnel diodes," *Sov. Tech. Phys. Lett.* **11** (Aug), pp. 403–404.

Dmitriev, V. A., Ivanov, P. A., Korin, I. V., Monozenko, Y. V., Popov, I. V., Sidorva, T. A., Strelchuk, A. M., and Chelnokov, V. E. (1985b). "Silicon carbide p-n structures synthesized by liquid-phase epitaxy," *Sov. Tech. Phys. Lett.* **11** (Feb), pp. 98–99.

Dmitriev, V. A., Ivanov, P. A., Levin, V. I., Popov, I. V., Strelchuk, A. M., Tarisov, Y. M., Tsvetkov, V. F., and Chelnokov, V. E. (1987). "Fabrication of epitaxial silicon carbide structures obtained from bulk SiC crystals," *Sov. Tech. Phys. Lett.* **13** (Nov), pp. 489–490.

Dmitriev, V. A., Fekade, K., and Spencer, M. G. (1992). "Dependence of the Au-SiC(6H) Schottky barriers height on the SiC surface treatment," in *Amorphous and Crystalline Silicon Carbide IV—Santa Clara 1991, Springer Proc. Phys.* **71**, pp. 352–355.

Dmitriev, V. A., Irvine, K., and Spencer, M. (1994). "Low resistivity ($\sim 10^{-5}\,\Omega\,cm^2$) ohmic contacts to 6H silicon carbide fabricated using cubic silicon carbide contact layer," *Appl. Phys. Lett.* **64**, (Jan), pp. 318–320.

Dunlop, H. L. and Marsh, O. J. (1969). "Diodes in silicon carbide by ion implantation," *Appl. Phys. Lett.* **15** (Nov), pp. 311–313.

Edmond, J. A., Ryu, J., Glass, J. T., and Davis, R. F. (1988a). "Electrical contacts to beta silicon carbide thin films," *J. Electrochem. Soc.* **135** (Feb), pp. 359–362.

Edmond, J. A., Das, K., and Davis, R. F. (1988b). "Electrical properties of ion-implanted p-n junction diodes in β-SiC," *J. Appl. Phys.* **63** (Feb), pp. 922–929.

Edmond, J. A., Kong, H.-S., and Carter, Jr., C. H. (1992). "High-temperature rectifiers, UV photodiodes, and blue LEDs in 6H-SiC," in: *Amorphous and Crystalline Silicon Carbide IV—Santa Clara 1991, Springer Proc. Phys.* **71**, pp. 344–351.

Edmond, J. A., Kong, H.-S., and Carter, Jr., C. H. (1993). "Blue LEDs, UV photodiodes and high-temperature rectifiers in 6H-SiC," *Physica B* **185**, pp. 453–460.

Ferry, D. K. (1975). "High-field transport in wide-band-gap semiconductors", *Phys. Rev. B* **12**, p. 2361–2369.

Fröjdh, C., and Patterson, C. S. (1996). "Comment on Temperature dependence of the barrier height of metal-semiconductor contacts on 6H-SiC", *J. Appl. Phys.* **80**, pp. 6570–6571.

Fujii, Y., Shigeta, M., Furukawa, K., and Suzuki, A. (1988). "Dependence on the Schottky metal and crystal orientation of the Schottky diode characteristics of β-SiC single crystals grown by chemical vapor deposition," *J. Appl. Phys.* **64** (Nov), pp. 5020–5025.

Furukawa, K., Uemoto, A., Shigeta, M., Suzuki, A. and Nakajima, S. "3C-SiC p-n junction diodes", *Appl. Phys. Lett.* **48**, pp. 1536–1537.

Geib, K. M., Mahan, J. E., and Wilmsen, C. W. (1989). "W/SiC contact resistance at elevated temperatures," in: *Amorphous and Crystalline Silicon Carbide and Related Materials II—Santa Clara 1989*, **43**, pp. 224–229.

Ghezzo, M., Brown, D. M., Downey, E., Kretchmer, J., Hennessy, W., Polla, D. L., and Bakhru, H. (1992). "Nitrogen-Implanted SiC Diodes Using High-Temperature Implantation", *IEEE Electron Device Letters*, **13**, pp. 639–641.

Ghezzo, M., Brown, D. M., Downey, E., and Kretchmer, J. (1993). "Boron-implanted 6H-SiC diodes," *Appl. Phys. Lett.* **63** (Aug), pp. 1206–1208.

Glass, R. C., Spellman, L. M., and Davis, R. F. (1991). "Low energy ion-assisted deposition of titanium nitride ohmic contacts on alpha (6H)-silicon carbide," *Appl. Phys. Lett.* **59** (22), pp. 2868–2870.

Glass, R. C., Spellman, L. M., Tanaka, S., and Davis, R. F. (1992). "Chemical and structural analyses of the titanium nitride/alpha (6H)-silicon carbide interface," *J. Vac. Sci. Technol. A* **10** (4), pp. 1625–1631.

Glass, R. C., Palmour, J. W. Davis, R. F., and Porter, L. M. (1994). "Method of forming ohmic contacts to p-type wide bandgap semiconductors and resulting ohmic contact structure." US Patent No. 5323022.

Goesmann, F. and Schmid-Fetzer, R. (1995). "Temperature-dependent interface reactions and electrical contact properties of titanium on 6H-SiC," *Semiconductor Sci. & Tech.* **10** (Dec), pp. 1652–1658.

Goodman, A. M. (1963). "Metal-semiconductor barrier height measurement by the differential capacitance method-one carrier system," *J. Appl. Phys.* **34** (Feb), pp. 329–338.

Gorin, S. N. and Pletyushkin, A. A. (1965). *Growth Crystal* **6** (USSR), p. 210.

Hagen, S. H. (1968). "Surface-barrier diodes on silicon carbide," *J. Appl. Phys.* **39** (3), pp. 1458–1461.

Hall, R. N. (1958). "Electrical contacts to silicon carbide," *J. Appl. Phys.* **29** (6), pp. 914–917.

Hallin, C., Yakimova, R., Krastev, V., Marinova, T. S., and Janzén, E. (1996). "Interface chemistry and electrical properties of annealed Ni and Ni/Al-6H SiC structures," in *Silicon Carbide and Related Materials VI—Kyoto 1995, Inst. Phys. Conf. Ser.* **142**, pp. 601–604.

Hallin, C., Yakimova, R., Pécz, B., Georgieva, A., Marinova, T. S., Kasamakova, L., Kakanakov, R., and Janzén, E. (1997). "Improved Ni Ohmic contact on n-type 4H-SiC," *J. Electron. Mat.* **26** (3), pp. 119–122.

Ioannou, D. E., Papanicolaou, N. A., and Nordquist, P. E. (1987). "The effect of heat treatments on Au Schottky contacts on β-SiC," *IEEE Trans. Electron. Dev.* **34** (Aug), pp. 1694–1699.

Itoh, A., Kimoto, T., and Matsunami, H. (1995a). "High performance of high-voltage 4H-SiC Schottky barrier diodes," *IEEE Electron Dev. Lett.* **16** (Jun), pp. 280–282.

Itoh, A. Takemura, O., Kimoto, T., and Matsunami, H. (1995b). "Barrier height analysis of metal/4H-SiC Schottky contacts," in *Silicon Carbide and Related Materials VI—Kyoto 1995, Inst. Phys. Conf. Ser.* **142**, pp. 685–688.

Itoh, A., Kimoto, T., and Matsunami, H. (1995c). "Low power-loss 4H-SiC Schottky rectifiers with high blocking voltage," in *Silicon Carbide and Related Materials VI—Kyoto 1995, Inst. Phys. Conf. Ser.* **142**, pp. 689–692.

Itoh, A., Kimoto, T., and Matsunami, H. (1996). "Excellent reverse blocking characteristics of high-voltage 4H-SiC Schottky rectifiers with boron-implanted edge termination," *IEEE Electron Dev. Lett.* **17** (3), pp. 139–141.

Janzén, E. and Kordina, O. (1996). "Recent progress in epitaxial growth of SiC for power device applications," in: *Silicon Carbide and Related Materials VI—Kyoto 1995, Inst. Phys. Conf. Ser.* **142**, pp. 653–658.

Kaneda, S., Sakamoto, Y., Mihara, T., and Tanaka, T. (1987). "MBE growth of 3C SiC/6H SiC and the electric properties of its p-n junction," *J. Crystal Growth* **81**, pp. 536–542.

Kelner, G., Shur, M. S., Binari, S., Sleger, K. J. and Kong, H. S. (1989). "High-transconductance β-SiC Burried-Gate JFETs," *IEEE Trans. Electron. Dev.* **36**, pp. 1045–1048.

Kelner, G., Binari, S., Shur, M. and Palmour, J. W. (1991). "High temperature operation of α-silicon carbide buried-gate junction field-effect transistors," *Electron. Lett.* **27**, pp. 1038–1040.

Kimoto, T., Urushidani, T., Kobayashi, S., and Matsunami, H. (1993). "High-voltage ($>1\,\mathrm{kV}$) SiC Schottky barrier diodes with low on-resistances," *IEEE Electron Dev. Lett.* **14** (Dec), pp. 548–550.

Kimoto, T., Wahab, Q., Ellison, A., Forsberg, U., Tuominen, M., Yakimova, R., Henry, A., and Janzén, E. (1998). "High-voltage ($>2.5\,\mathrm{kV}$) 4H-SiC Schottky rectifiers processed on hot-wall CVD and high-temperature CVD layers," in: *Silicon Carbide, III-Nitrides and Related Materials VII — Stockholm 1997*, edited by G. Pensl, H. Morkoc, B. Monemar and E. Janzén, *Materials Science Forum*, **264**–268, pp. 921–924.

Kneifel, M., Silber, D., and Held, R. (1996). "Predictive modeling of SiC power Schottky diode for investigations in power electronics," in: *Proc. IEEE Appl. Power Elec. Conf. and Expo., IEEE Cat. No.* **95CH35748**, pp. 239–245.

Kordina, O., Henry, A., Bergman, J. P., Son, N. T., Chen, W. M., Hallin, C., and Janzén, E. (1995a). "High quality 4H-SiC epitaxial layers grown by chemical vapor deposition," *Appl. Phys. Lett.* **66** (Mar), pp. 1373–1375.

Kordina, O., Bergman, J. P., Henry, A., and Janzén, E. (1995b). "A 4.5 kV 6H silicon carbide rectifier," *Appl. Phys. Lett.* **67** (Sept), pp. 1561–1563.

Kuznetsov, N. I. and Dmitriev, V. A. (1996). "Electrical characteristics of 4H-SiC pn structures," in: *Third International High Temperature Electronic Conference — Albuquerque 1996*, **2**, pp. P/77–P82.

Lang, M., Isaac-Smith, T., Tin, C. C., and Williams, J. R. (1996). "Ni, Al, and Ti Schottky diodes and their electrical characterization on 6H-SiC," in: *Silicon Carbide and Related Materials VI — Kyoto 1995, Inst. Phys. Conf. Ser.* **142**, pp. 681–684.

Larkin, D. J., Neudeck, P. G., Powell, A. P., and Matus, L. G. (1994). "Site-competition epitaxy for controlled doping of CVD silicon carbide," in: *Silicon Carbide and Related Materials V — Washington 1993, Inst. Phys. Conf. Ser.* **137**, pp. 51–54.

Lebedev, A. A., Strel'chuk, A. M., Ortolland, S., Raynaud, C., Locatelli, M. L., Planson, D., and Chante, J. P. (1996a). "The negative temperature coefficient of the breakdown voltage of SiC p-n structures and deep centers in SiC," in: *Silicon Carbide and Related Materials VI — Kyoto 1995, Inst. Phys. Conf. Ser.* **142**, pp. 701–704.

Lebedev, A. A., Rastegaeva, M. G., Savkina, N. S., Tregubova, A. S., Chelnokov, V. E., and Scheglov, M. P. (1996b). "New results and prospects in the development of high power and high temperature SiC based diodes," in: *Third International High Temperature Electronic Conference — Albuquerque 1996*, **2**, pp. P/161–166.

Liu, S., Reinhardt, K., Severt, C., and Scofield, J. (1996). "Thermally stable ohmic contacts on n-type 6H- and 4H-SiC based on silicide and carbide," in: *Silicon Carbide and Related Materials VI — Kyoto 1995, Inst. Phys. Conf. Ser.* **142**, pp. 589–592.

Lundberg, N., Zettering, C. M., and Östling, M. (1993). "Temperature stability of cobalt Schottky contacts on n- and p-type 6H silicon carbide," *Applied Surface Science* **73** (Nov), pp. 316–321.

Lundberg, N. and Östling, M. (1994). "Cobalt disilicide ($CoSi_2$) Schottky contacts to 6H-SiC," *Physica Scripta* **134**, pp. 273–277.

Lundberg, N. (1995). "$CoSi_2$ ohmic contacts to n-type 6H-SiC," *Solid-State Electronics* **38** (Dec), pp. 2023–2028.

Lundberg, N., Tägtström, P., Jansson, U., and Östling, M. (1996a). "CVD of tungsten Schottky diodes to 6H-SiC," in: *Silicon Carbide and Related Materials VI—Kyoto 1995, Inst. Phys. Conf. Ser.* **142**, pp. 677–680.

Lundberg, N. and Östling, M. (1996b). "Chemical vapor deposition of tungsten Schottky diodes to 6H-SiC," *J. Electrochem. Soc.* **143** (May), pp. 1662–1667.

Lundberg, N. and Östling, M. (1996c). "Thermally stable low ohmic contacts to p-type 6H-SiC using cobalt silicides," *Solid-State Electronics* **39** (Nov), pp. 1559–1565.

Marlow, G. S. and Das, M. B. (1982). "The effects of contact size and non-zero metal resistance on the determination of specific contact resistance," *Solid-State Electronics* **25** (2), pp. 91–94.

Mead, C. A. and Spitzer, W. G. (1964). "Fermi level position at metal-semiconductor interfaces," *Physical Review* **134** (3A), pp. A713–A716.

Mead, C. A. (1966). "Metal-semiconductor surface barriers," *Solid-State Electronics* **9**, pp. 1023–1033.

Moki, A., Shenoy, P., Alok, D., Baliga, B. J., Wongchotigul, K., and Spencer, M. G. (1995). "Low resistivity as-deposited ohmic contacts to 3C-SiC," *J. Electron. Mat.* **24** (Apr), pp. 315–318.

Mott, N. F. (1938). "Note on the contact between a metal and an insulator or a semiconductor," *Proc. Camb. Phil. Soc.* **34**, p. 568.

Muench, W. v. and Pfaffeneder, I. (1977). "Breakdown field in vapor-grown silicon carbide p-n junctions," *J. Appl. Phys.* **48** (Nov), pp. 4831–4833.

Nelson, W. E., Halden, F. A., and Rosengreen, A. (1966). "Growth and properties of β-SiC single crystals," *J. Appl. Phys.* **37** (Jan), pp. 333–336.

Neudeck, P. G., Larkin, D. J., Starr, J. E., Powell, J. A., Salupo, C. S., and Matus, L. G. (1993). "Greatly improved 3C-SiC p-n junction diodes grown by chemical vapor deposition," *IEEE Electron Dev. Lett.* **14** (Mar), pp. 136–139.

Neudeck, P. G., Larkin, D. J., Salupo, C. S., Powell, J. A., and Matus, L. G. (1994a). "2000 V 6H-SiC on junction diodes," in: *Silicon Carbide and Related Materials V—Washington 1993, Inst. Phys. Conf. Ser.* **137**, pp. 475–479.

Neudeck, P. G., Larkin, D. J., Powell, J. A., and Matus, L. G. (1994b). "2000 V 6H-SiC p-n junction diodes grown by chemical vapor deposition," *Appl. Phys. Lett.* **64** (Mar), pp. 1386–1388.

Neudeck, P. G., Larkin, D. J., Starr, J. E., Powell, J. A., Salupo, C. S., and Matus, L. G. (1994c). "Electrical properties of epitaxial 3C- and 6H-SiC p-n junction diodes produced side-by-side on 6H-SiC substrates," *IEEE Trans. Electron Dev.* **41** (May), pp. 826–835.

Neudeck, P. G., Fazi, C., and Parsons, J. D. (1996). "Fast risetime reverse bias pulse failures in SiC PN junction diodes," in: *Third International High Temperature Electronic Conference—Albuquerque 1996*, **2**, pp. XVI/15–20.

Neudeck, P. G. and Fazi, C. (1997). "Positive temperature coefficient of breakdown voltage in 4H-SiC pn junction rectifiers," *IEEE Electron Dev. Lett.* **18** (Mar), pp. 96–98.

Norde, H. (1979). "A modified forward I-V plot for Schottky diodes with high series resistance," *J. Appl. Phys.* **50** (Jul), pp. 5052–5053.

Nordell, N., Savage, S., and Schöner, A. (1996). "Aluminum doped 6H-SiC: CVD growth and formation of ohmic contacts," in: *Silicon Carbide and Related Materials VI—Kyoto 1995, Inst. Phys. Conf. Ser.* **142**, pp. 573–576.

Palmour, J. W., Kong, H. S., Waltz, E. D., Edmond, J. A., and Carter, C. H. (1991). In: *First International High Temperature Electronics Conference, Albuquerque, NM*, edited by D. B. King and F. V. Thome.

Palmour, J. W., Edmond, J. A., Kong, H. S., and Carter, C. H. (1994). "Vertical power devices in silicon carbide," in: *Silicon Carbide and Related Materials V—Washington 1993, Inst. Phys. Conf. Ser.* **137**, pp. 499–502.

Papanicolaou, N. A., Christou, A., and Gipe, M. L. (1989). "Pt and PtSi$_x$ Schottky contacts on n-type β-SiC," *J. Appl. Phys.* **65** (May), pp. 3526–3530.

Parsons, J. D., Kruaval, G. B., and Chaddha, A. K. (1994). "Low specific resistance ($<6 \times 10^{-6}\,\Omega\,\text{cm}^2$) TiC ohmic contacts to n-type β-SiC," *Appl. Phys. Lett.* **65** (Oct), pp. 2075–2077.

Petit, J. B. and Zeller, M. V. (1992). "Electrical and chemical characterization of contacts to silicon carbide," *Mat. Res. Soc. Symp. Proc.* **242**, pp. 567–572.

Petit, J. B., Neudeck, P. G., Salupo, C. S., Larkin, D. J., and Powell, J. A. (1994). "Metal contacts to n- and p-type 6H-SiC: electrical characteristics and high-temperature stability," in: *Silicon Carbide and Related Materials V — Washington 1993, Inst. Phys. Conf. Ser.* **137**, pp. 679–682.

Philip, H. R. and Taft, E. A. (1960). "Intrinsic optical absorption in single crystal silicon carbide," in: *Silicon Carbide: A High Temperature Semiconductor,* Eds. J. R. O'Connor and J. Smitens. Pergamon Press: Oxford, pp. 366–370.

Porter, L. M., Glass, R. C., Davis, R. F., Bow, J. S., Kim, M. J., and Carpenter, R. W. (1993). "Chemical and electrical mechanism in titanium, platinum, and hafnium contacts to alpha (6H) silicon carbide," *Mat. Res. Soc. Symp. Proc.* **282**, pp. 471–477.

Porter, L. M., Davis, R. F., Bow, J. S., Kim, M. J., and Carpenter, R. W. (1994). "Deposition and characterization of Schottky and Ohmic contacts on n-type alpha (6H)-SiC (0001)," in: *Silicon Carbide and Related Materials V — Washington 1993, Inst. Phys. Conf. Ser.* **137**, pp. 581–584.

Porter, L. M. and Davis, R. F. (1995a). "A critical review of ohmic and rectifying contacts for silicon carbide," *Materials Science and Engineering B* **34**, pp. 83–105.

Porter, L. M. and Davis, R. F. (1995b). "Chemistry, microstructure, and electrical properties at interfaces between thin films of cobalt and alpha (6H) silicon carbide (0001)," *J. Mat. Res.* **10** (Jan), pp. 26–33.

Porter, L. M. and Davis, R. F. (1995c). "Chemistry, microstructure, and electrical properties at interfaces between thin films of titanium and alpha (6H) silicon carbide (0001)," *J. Mat. Res.* **10** (Mar), pp. 668–679.

Porter, L. M. and Davis, R. F. (1995d). "Chemistry, microstructure, and electrical properties at interfaces between thin films of platinum and alpha (6H) silicon carbide (0001)," *J. Mat. Res.* **10** (Sep), pp. 2336–2342.

Porter, L. M. and Davis, R. F. (1996). "Issues and status of ohmic contacts for p-type silicon carbide," in: *Third International High Temperature Electronic Conference — Albuquerque 1996,* **1**, pp. VII/3–8.

Proctor, S. J. and Linholm, L. W. (1982). "A direct measurement of interfacial contact resistance," *IEEE Electron Dev. Lett.* **EDL-3** (Oct), pp. 294–296.

Proctor, S. J. and Linholm, L. W. (1983). "Direct measurements of interfacial contact resistance, end resistance, and interfacial contact layer uniformity," *IEEE Electron Dev.* **ED-30** (Nov), pp. 1535–1542.

Raghunathan, R., Alok, D., and Baliga, B. J. (1995). "High voltage 4H-SiC Schottky barrier diodes," *IEEE Electron Dev. Lett.* **16** (Jun), pp. 226–227.

Ramungul, N., Khemka, V., Tyagi, R., Chow, T. P., Ghezzo, M., Nuedeck, P. G., Kretchmer, J., Henessy, W., and Brown, D. M. (1996). "Comparison of aluminum- and boron-implanted vertical 6H-SiC p$^+$n junction diodes," in: *Silicon Carbide and Related Materials VI — Kyoto 1995, Inst. Phys. Conf. Ser.* **142**, pp. 713–717.

Rastegaev, M. G., Andreev, A. N., Zelenin, V. V., Babanin, A. I., Nikitina, I. P., Chelnokov, V. E., and Rastegaev, V. P. (1996a). "Nickel-based metallization in processes of the 6H-SiC device fabrication: ohmic contacts, masking and packaging," in: *Silicon Carbide and Related Materials VI — Kyoto 1995, Inst. Phys. Conf. Ser.* **142**, pp. 581–584.

Raynaud, C., Ducroquet, F., Brounkow, P. N., Gullot, G., Porter, L. M., Davis, R. F., Jaussaud, C., and Billon, T. (1996). "Electrical characterization of epitaxial 6H-SiC admittance spectroscopy," *Materials Sci. and Tech.* **12** (Jan), pp. 94–97.

Reddy, C. V., Fung, S., Beling, C. D., and Brauer, G. (1996). "Current transport at low temperature in Al/6H-SiC Schottky barriers," in: *Silicon Carbide and Related Materials VI—Kyoto 1995, Inst. Phys. Conf. Ser.* **142**, pp. 669–673.

Rendakova, S., Ivantsov, V., and Dmitriev, V. (1998). "High quality 6H and 4H-SiC pn structures with stable electric breakdown grown by liquid phase epitaxy," in: *Silicon Carbides, III—Nitrides and Related Materials VII—Stockholm 1997*, edited by G. Pensl, H. Morkoc, B. Monemar and E. Janzén, Materials Science Forum, pp. 163–166.

Rhoderick, E. H. and Williams, R. H. (1988). "Metal-Semiconductor Contacts," Clarendon Press: Oxford.

Saxena, V., Steckl, A. J., Vichare, M., Ramalingam, M. L., and Reinhardt, K. (1996). "Temperature effects in the operation of high voltage Ni/6H-SiC Schottky rectifiers," in: *Third International High Temperature Electronic Conference—Albuquerque 1996*, **1**, pp. VII/15–20.

Saxena, V. and Steckl, A. J. (1997). "High Temperature Operation of High Breakdown Voltage Schottky Diodes on 4H and 6H-SiC" in International Semiconductor Device Research Symposium—1997, pp. 539–542.

Saxena, V. and Steckl, A. J. (1998). "High voltage Pt and Ni Schottky barrier diodes on 4H-SiC," in: *Silicon Carbide, III—Nitrides and Related Materials—VII Stockholm 1997*, edited by G. Pensl, H. Morkoc, B. Monemar and E. Janzén, Materials Science Forum, **264–268**, pp. 937–940.

Schadt, M., Pensl, G., Devaty, R. P., Choyke, W. J., Stein, R., and Stephani, D. (1994). "Anisotropy of the electron Hall mobility in 4H, 6H and 15R silicon carbide," *Appl. Phys. Lett.* **65** (Dec), pp. 3120–3122.

Schaffer, W. J., Negley, G. H., Irvine, K. G., and Palmour, J. W. (1994). "Conductivity anisotropy in epitaxial 6H and 4H SiC," *Mat. Res. Soc. Symp. Proc.* **337**, pp. 595–600.

Schroder, D. K. (1990). "Semiconductor Material and Device Characterization," John Wiley & Sons, Inc.: New York.

Schroder, C., Heiland, W., Held, R., and Loose, W. (1996). "Analysis of reverse current-voltage characteristics of Ti/6H-SiC Schottky diodes," *Appl. Phys. Lett.* **68** (Apr), pp. 1957–1959.

Shalish, I., Gasser, S., Kolawa, E., and Nicolet, M.-A. (1996). "Stability of Schottky contacts with Ta-Si-N amorphous diffusion barriers and Au overlayers on 6H-SiC," in: *Third International High Temperature Electronic Conference—Albuquerque 1996*, **1**, pp. VII/21–26.

Shenoy, P., Moki, A., Baliga, B. J., Alok, D., Wongchotigul, K., and Spencer, M. (1994). "Vertical Schottky barrier diodes on 3C-SiC grown on Si," in: *Tech. Digest Intl. Electron Dev. Meeting, IEEE Cat. No.* **94CH35706**, pp. 411–414.

Shenoy, P. M., and Baliga, B. J. (1995). "Planar, ion implanted, high voltage 6H-SiC p-n junction diodes," *IEEE Electron Dev. Lett.* **16** (Oct), pp. 454–456.

Shenoy, P. M. and Baliga, B. J. (1996). "Planar, high voltage, boron implanted 6H-SiC p-n junction diodes," in: *Silicon Carbide and Related Materials VI—Kyoto 1995, Inst. Phys. Conf. Ser.* **142**, pp. 717–721.

Shibahara, K., Kuroda, N., Nishino, S., and Matsunami, H. (1987). "Fabrication of p-n junction diodes using homoepitaxially grown 6H-SiC at low temperature by chemical vapor deposition," *Jpn. J. Appl. Phys.* **26** (Nov), pp. 1815–1817.

Shier, J. S. (1970). "Ohmic contacts to silicon carbide," *J. Appl. Phys.* **41**, pp. 771–773.

Schockley, W., Goetzberger, A., and Scarlet, R. M. (1964). "Research and investigation of inverse epitaxial UHF power transistors." U.S. Air Force Avionics Lab., Wright Patterson Air Force Base (Report No. AFAL-TDR-64-207): Dayton, OH.

Shor, J. S., Weber, R. A., Provost, L. G., Goldstein, D., and Kurtz, A. D. (1992). "High temperature ohmic contacts for n-type β-SiC sensors," *Mat. Res. Symp. Proc.* **242**, pp. 573–581.

Shor, J. S., Weber, R. A., Provost, L. G., Goldstein, D., and Kurtz, A. D. (1994). "High temperature ohmic contact metallizations for n-type 3C-SiC," *J. Electrochem. Soc.* **141** (Feb), pp. 579–581.

Smith, S. R., Evwaraye, A. O., and Mitchel, W. C. (1995). "Temperature dependence of the barrier height of metal-semiconductor contacts on 6H-SiC," *J. Appl. Phys.* **79** (Jan), pp. 301–304.

Smith, S. R., Evwaraye, A. O. and Mitchel, W. C. (1996). "Reply to 'Comment on 'Temperature dependence of the barrier height of metal-semiconductor contacts on 6H-SiC," *J. Appl. Phys.* **80**, pp. 6572–6573.

Spellman, L. M., Glass, R. C., Davis, R. F., Humphreys, T. P., Jeon, H., Nemanich, R. J., Chevacharoenkul, S., and Parikh, N. R. (1991). "Heteroepitaxial growth and characterization of titanium films on alpha 6H silicon carbide," *Mat. Res. Soc. Symp. Proc.* **221**, pp. 99–104.

Spellman, L. M., Glass, R. F., Davis, R. F., Humphreys, T. P., Nemanich, R. J., Das, K., and Chevacharoenkul, S. (1992). "Electrical characteristics of epitaxial titanium contacts to alpha (6H) silicon carbide," in: *Amorphous and Crystalline Silicon Carbide IV — Santa Clara 1991, Springer Proc. Phys.* **71**, pp. 417–422.

Spiess, L., Nennewitz, O., and Pezoldt, J. (1996). "Improved ohmic contacts to p-type 6H-SiC," in: *Silicon Carbide and Related Materials VI — Kyoto 1995, Inst. Phys. Conf. Ser.* **142**, pp. 585–588.

Steckl, A. J. and Li, J. P. (1992). "Epitaxial growth of β-SiC on Si by RTCVD with C_3H_8 and SiH_4," *IEEE Trans. Electron Dev.* **39** (Jan), pp. 64–74.

Steckl, A. J. and Su, J. N. (1993). "High voltage, temperature-hard 3C-SiC Schottky diodes using all-Ni metallization," *Tech. Digest Intl. Electron Dev. Meeting, IEEE Cat. No.* **93CH3361-3**, pp. 695–698.

Steckl, A. J., Su, J. N., Yih, P. H., Yuan, C., and Li, J. P. (1994). "Ohmic and rectifying contacts to 3C-SiC using all-Ni technology," in: *Silicon Carbide and Related Materials V — Washington 1993, Inst. Phys. Conf. Ser.* **137**, pp. 653–656.

Steckl, A. J., Devrajan, J., Tlali, S., Jackson, H. E., Tran, C., Gorin, S. N., and Ivanova, L. M. (1996). "Characterization of 3C-SiC grown by thermal decomposition of methyltrichlorosilane," *Appl. Phys. Lett.* **69** (Dec), pp. 3824–3826.

Su, J. N. and Steckl, A. J. (1995). "300°C operation of high voltage SiC Schottky rectifier," in: *Workshop on High Temperature Power Electronics — Ft. Monmouth, NJ*, pp. 50–53.

Su, J. N. and Steckl, A. J. (1996). "Fabrication of high voltage SiC Schottky barrier diodes by Ni metallization," in: *Silicon Carbide and Related Materials VI — Kyoto 1995, Inst. Phys. Conf. Ser.* **142**, pp. 697–700.

Suzuki, A., Uemoto, A., Shigota, M., Furukawa, K., and Makajima, S. (1986). "p-n junction diodes in beta-SiC," Extended Abstracts, *18th Int. Conf. Solid State Devices and Materials*, p. 101.

Teraji, T., Hara, S., Okushi, H., and Kajimura, K. (1996). "Ti ohmic contact without post-annealing process to n-type 6H-SiC," in: *Silicon Carbide and Related Materials VI — Kyoto 1995, Inst. Phys. Conf. Ser.* **142**, pp. 593–596.

Teraji, T., Hara, S., Okuslei, H., and Kajimura, K. (1997). "Ideal ohmic contact to n-type 6H-SiC by reduction of Schottky barrier height," *Appl. Phys. Lett.* **71** (Aug), p. 689.

Uemoto, T. (1995). "Reduction of ohmic contact resistance on n-type 6H-SiC by heavy doping," *Jpn. J. Appl. Phys.* **34** (Jan), pp. 7–9.

Ueno, K., Urushidani, T., Hashimoto, K., and Seki, Y. (1995). "The guard-ring termination for the high-voltage SiC Schottky barrier diodes," *IEEE Electron Dev. Lett.* **16** (Jul), pp. 331–332.

Ueno, K., Urushidani, T., Hashimoto, K., and Seki, Y. (1996). "The guard-ring termination for 6H-SiC Schottky barrier diodes," in: *Silicon Carbide and Related Materials VI — Kyoto 1995, Inst. Phys. Conf. Ser.* **142**, pp. 693–696.

Urushidani, T., Kobayashi, S., Kimoto, T., and Matsunami, H. (1994). "High-voltage Au/6H-SiC Schottky barrier diodes," in: *Silicon Carbide and Related Materials VI — Washington, Inst. Phys. Conf. Ser.* **137**, pp. 471–474.

van der Ziel, A. (1968). "Solid State Physical Electronics," Prentice Hall: Englewood Cliffs, NJ.

van Elsbergen, V., Kampen, T. U., and Monch, W. (1996). "Electronic properties of cesium on 6H-SiC surfaces," *J. Appl. Phys.* **79** (Jan), pp. 316–321.

van Opdorp, C. and Vrakking, J. (1969). "Avalanche breakdown in epitaxial SiC p-n junctions," *J. Appl. Phys.* **40**, pp. 2320–2322.

Vassilevski, K. V., Dmitriev, V. A., and Zorenko, A. V. (1993). "Silicon carbide diode operating at avalanche breakdown current density of $60\,kA/cm^2$," *J. Appl. Phys.* **74** (12), pp. 7612–7614.

Vassilevski, K. V., Zorenko, A. V., and Novozhilov, V. V. (1994). "Temperature dependence of avalanche breakdown voltage of pn-junctions in 6H-SiC at high current density," in: *Silicon Carbide and Related Materials V — Washington 1993, Inst. Phys. Conf. Ser.* **137**, pp. 659–661.

Wahab, Q., Karlsteen, M., Willander, M., and Sundgren, J. E. (1991). "Au Schottky barrier diodes on β-SiC thin films deposited on silicon substrates by reactive magnetron sputtering technique," *J. Electron. Mat.* **20** (Nov), pp. 899–901.

Wahab, Q., Kimoto, T., Ellison, A., Hallin, C., Tuominen, M., Yakimova, R., Henry, A., Bergman, J. P. and Janzén, E. (1998). "A 3kV Schottky barrier diode in 4H-SiC", *Appl. Phys. Lett.* **72** (Jan. 25), pp. 445–447.

Waldrop, J. R. and Grant, R. W. (1990). "Formation and Schottky barrier height of metal contacts to β-SiC," *Appl. Phys. Lett.* **56** (Feb), pp. 557–559.

Waldrop, J. R., Grant, R. W., Wang, Y. C., and Davis, R. F. (1992). "Metal Schottky barrier contacts to alpha 6H-SiC," *J. Appl. Phys.* **72** (Nov), pp. 4757–4760.

Waldrop, J. R. and Grant, R. W. (1993). "Schottky barrier height and interface chemistry of annealed metal contacts to alpha 6H-SiC: Crystal face dependence," *Appl. Phys. Lett.* **62** (21), pp. 2685–2687.

Waldrop, J. R. (1994). "Schottky barrier height of metal contacts to p-type alpha 6H-SiC," *J. Appl. Phys.* **75** (May), pp. 4548–4550.

Wang, Y. C. and Davis, R. F. (1991). "In-situ incorporation of Al and N and p-n junction diode fabrication in alpha (6H)-SiC thin films," *J. Electron. Mat.* **20**, pp. 289–294.

Weitzel, C. E., Palmour, J. W., Carter, C. H., Moore, K., Nordquist, K. J., Allen, S., Thero, S., and Bhatnagar, M. (1996). "Silicon carbide high-power devices," *IEEE Trans Electron Dev.* **43** (Oct), pp. 1732–1741.

Wu, S. Y. and Campbell, R. B. (1974). "Au-SiC Schottky barrier diodes," *Solid-State Electronics* **17**, pp. 683–687.

Yasuda, K., Hayakawa, T., and Saji, M. (1987). "Annealing effects of Al/n-type 6H-SiC rectifying contacts," *IEEE Trans. Electron Dev.* **34** (9), pp. 2002–2008.

Yih, P. H., Li, J. P., and Steckl, A. J. (1994). "SiC/Si heterojunction diodes fabricated by self-selective and by blanket rapid thermal chemical vapor deposition," *IEEE Trans. Electron Dev.* **41** (Mar), pp. 281–287.

Yoshida, S., Sasaki, K., Sakuma, E., Misawa, S., and Gonda, S. (1985). "Schottky barrier diodes on 3C-SiC," *Appl. Phys. Lett.* **46** (April), pp. 766–768.

Zhang, Y. G., Li, X. L., Li, A. Z., and Milnes, A. G. (1996). "Characterization of Schottky contacts on n type 6H- SiC," in: *Silicon Carbide and Related Materials VI — Kyoto 1995, Inst. Phys. Conf. Ser.* **142**, pp. 665–668.

CHAPTER 4

SiC Transistors

Michael S. Shur

CENTER FOR INTEGRATED ELECTRONICS AND ELECTRONICS MANUFACTURING
DEPARTMENT OF ELECTRICAL, COMPUTER, AND SYSTEMS ENGINEERING
RENSSELAER POLYTECHNIC INSTITUTE
TROY, NY

    I. INTRODUCTION 161
   II. SiC FIELD-EFFECT TRANSISTORS: MOSFETs, MESFETs, AND JFETs 162
      1. *Principle of Operation* 162
      2. *SiC MOSFETs* 168
      3. *SiC MESFETs and JFETs* 174
  III. SiC MICROWAVE FIELD-EFFECT TRANSISTORS 177
  IV. SiC DIGITAL INTEGRATED CIRCUITS 180
   V. SiC BIPOLAR TRANSISTORS AND THYRISTORS 182
  VI. TWO-DIMENSIONAL MODELING OF SiC TRANSISTORS 185
 VII. ANALYTICAL TRANSISTOR MODELS AND CIRCUIT SIMULATION (AIM-SPICE) . . 186
VIII. POTENTIAL PERFORMANCE AND APPLICATIONS OF SiC TRANSISTORS
      AND INTEGRATED CIRCUITS 188
      References 189

I. Introduction

The development of improved bulk crystal SiC growth and progress in modern epitaxial techniques of SiC growth have led to a rapid improvement in the material quality of several SiC polytypes (primarily 6H-SiC and 4H-SiC) and made practical SiC transistors a reality. These transistors, which include high-temperature, high-power transistors and switches, and microwave devices, are rapidly moving past the demonstration stage toward applications in defense and commercial systems.

In this chapter, we discuss different types of SiC field-effect transistors, including microwave and power devices; review emerging SiC digital integrated circuits; and describe SiC bipolar transistors and thyristors. We also describe simulation and modeling tools that have been used for the design of SiC transistors, discuss novel device ideas applied to SiC, and project possible improvements related to the expected progress in materials and contact technologies.

In Section II, we describe different types of SiC field-effect transistors (MOSFETs, and MESFETs, and JFETs), compare their designs, and discuss their relative advantages and disadvantages. Sections III deals with SiC microwave field-effect transistors and SiC power field-effect transistors, respectively. In Section IV, we review recent work on emerging SiC digital integrated circuits. Section V reviews SiC bipolar transistors and thyristors. Sections VI and VII are devoted to SiC transistor numerical and analytical modeling, respectively. Finally, in Section VIII, we discuss the potential performance and applications of SiC devices and integrated circuits.

II. SiC Field-Effect Transistors: MOSFETs, MESFETs, and JFETs

1. PRINCIPLE OF OPERATION

All field-effect transistors (FETs) have much in common. These devices rely on the concept first proposed by Lilienfeld [1] and developed by William Shockley in the early 1950s. Figure 1 shows the basic FET structure. In a FET, a gate contact and the device conducting channel form two plates of a parallel plate capacitor. The sheet charge density, Q_s, of the mobile carriers in point x along the channel is controlled by the voltage difference between the gate and the channel:

$$Q_S = c_i[V_{GS} - V(x) - V_T] \tag{1}$$

where V_T is the threshold voltage, V_{GS} is the voltage difference between the gate and the source, $V(x)$ is the channel potential (measured with respect to the source), and the gate capacitance per unit area, c_i, is approximately equal to

$$c_i = \frac{\varepsilon_i}{d_i} \tag{2}$$

Here, ε_i and d_i are the dielectric permittivity and thickness of the gate barrier layer separating the gate and the channel.

The current, I_D, between the drain and source increases when Q_s increases and depends on the drain-to-source voltage, V_{DS} as well. At small V_{DS}, I_D is

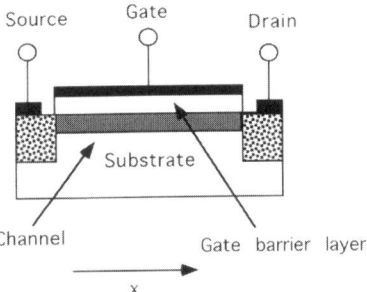

FIG. 1. Basic FET structure.

proportional to V_{DS}, but at larger drain-to-source voltages, the mobile carrier velocity saturation in the channel and the decreased values of Q_s at larger $V(x)$ [(see Eq. (1)] cause the drain current to saturate. As a result, at gate voltages above threshold, FET I–V characteristics look like those shown in Fig. 2 [2].

Typically, the gate-to-source voltage is an input voltage, and the drain-to-source voltage is an output voltage. Since the gate contact is isolated from the channel by the gate barrier layer, the input (gate-to-source) impedance is very high, a major advantage compared to bipolar junction transistors (BJTs).

One of the most important parameters of a FET is its threshold voltage, V_T, which separates the above-threshold regime (or on-state) where the charge induced in the device channel is proportional to the gate voltage

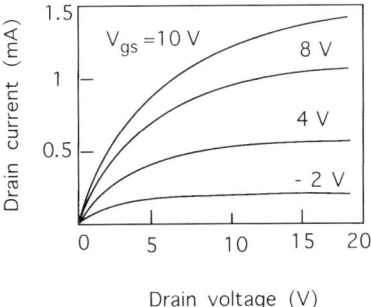

FIG. 2. Measured current–voltage characteristics of a self-aligned 6H-SiC MOSFET [2]. Gate length, 3 μm; gate width, 60 μm; gate oxide thickness, 80 nm. Threshold voltage is approximately −4 V.

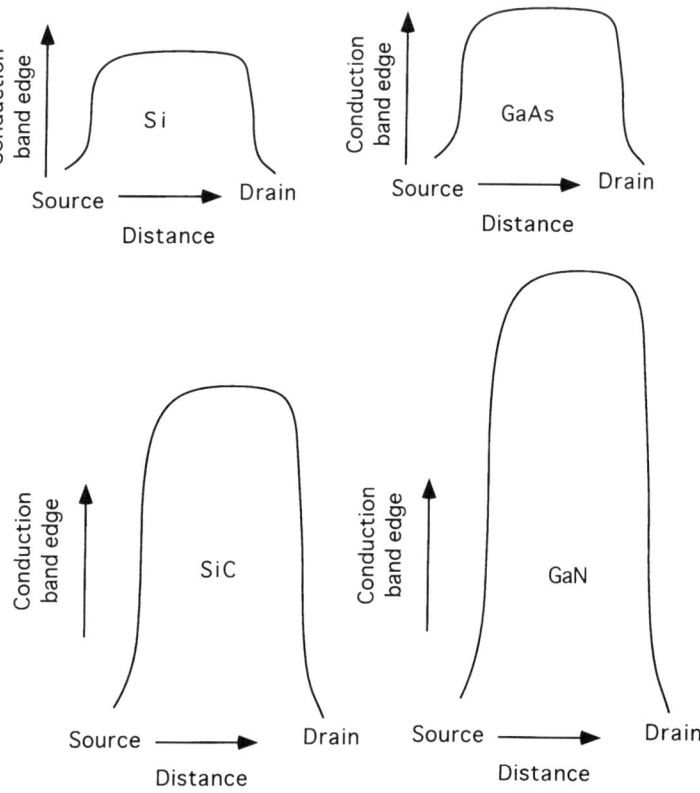

FIG. 3. Schematic conduction band profiles in the channel of a FET below threshold, illustrating the barrier between the source and drain at zero drain bias. The barrier heights shown are proportional to the energy gaps for Si, GaAs, SiC, and GaN, respectively.

swing [see Eq. (1)] and the below-threshold regime (or off-state). Below threshold, the source and drain contact are separated by a potential barrier, as shown in Fig. 3. This barrier effectively turns off most of the channel conduction current. The height of the barrier, which determines the transistor off-current, and hence its on-to-off current ratio, is largely determined by the semiconductor energy gap. (If the gate barrier layer can support a leakage current, this leakage current may be responsible for the before-threshold conduction.) In the below-threshold regime, the gate bias changes the height of the barrier separating the source and drain, and, therefore, the current depends exponentially on the gate bias (see Fig. 4 [2]).

The principle of operation of this type of FET allows us to understand the requirements to the semiconductor and gate barrier layer materials.

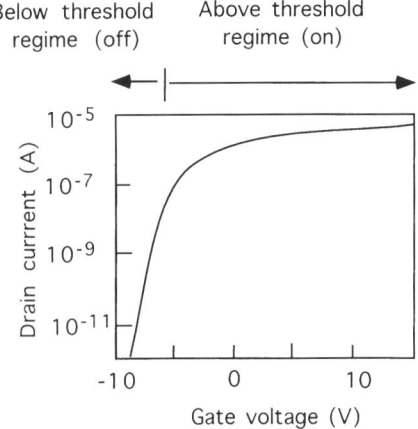

FIG. 4. Transfer characteristics (i.e., the dependence on the drain current on the gate voltage) of a self-aligned 6H-SiC MOSFET [2]. Gate length, 10 μm; gate width, 60 μm; gate oxide thickness, 80 nm; drain bias, 20 V.

These requirements include high electron mobility and saturation velocity (for higher speed operation), high thermal conductivity (for an ability to dissipate large power densities), and a large breakdown field for high power. The relevant parameters for 6H-SiC and 4H-SiC, which determine the transistor operation, are compared with those for Si, GaAs, GaN, AlN, and diamond in Table I (from [3]). (In Table I, we changed the breakdown field for GaN to 5 MV/cm and the electron mobility for GaN to 1500 cm^2/Vs, more in line with recent experimental data [81, 86], even though the largest breakdown electric field measured for GaN is on the order of 3 MV/cm [86]). As seen from the Table I, the thermal conductivity of SiC is superior to other semiconductors (with the exception of diamond). Its breakdown field, electron mobility, and saturation velocity are all quite high. SiC also has advantages compared to Si and GaAs in chemical inertness and its ability to operate at high temperatures. All of these factors may be combined to form different figures of merit characterizing a relative advantage of a given material for various device applications. Three different figures of merit are listed in Table I. Johnson's figure of merit (JFM) [4] estimates the potential of a material for high-frequency and high-power applications.

$$\mathrm{JFM} = \frac{E_B^2 v_s^2}{4\pi^2} \tag{3}$$

TABLE I

COMPARISON OF SEMICONDUCTOR MATERIALS[a]

Property	Si	GaAs	3C-SiC	4H-SiC	6H-SiC	GaN	AlN	Diamond
Lattice Constant (Å)	5.43	5.65	4.36	3.073 10.05 c	3.0806 15.117 c	4.51	3.11	3.567
Thermal expansion ($\times 10^{-6}$ °C)	2.6	5.73	4.7	4.2 4.68 c		5.6 4.68 c_0	4.5	0.8
Density (g cm^{-3})	2.329	5.32	3.210		3.211	6.095	3.255	3.515
Melting point (°C)	1420	1240	2830	2830				4000
Bandgap (eV)	1.1	1.43	2.4	3.26	3.101	3.45	6.2	5.45
Satur. electr. velocity ($\times 10^7$ cm/s)	1.0	1.0	2.2	2.0	2.0	2.2		2.7
Carrier mobility (cm^2/Vs)								
Electron	1500	8500	1000		1140	1500		2200
Hole	600	400	50	120	850			1600
Breakdown ($\times 10^5$ V/cm)	3	6	20	30		50	?	100
Dielectric constant	11.8	12.5	9.7	9.6/10		9	8.5	5.5
Resistivity (Ω-cm)	1000	10^8	150	$>10^{15}$		$>10^{10}$	$>10^{13}$	$>10^{13}$
Thermal conductivity (W/cm K)	1.5	0.46	3.2	3.7	4.9	1.3	3.0	22
Absorption edge (mm)	1.4	0.85	0.50	0.37	0.436	0.36	0.12	0.22
Refractive index	3.5	3.4	2.7	2.7	2.7		3.32	2.42
Hardness (kg/mm^2)	1000	600	3980	2130 c_0			1200	10,000
Johnson figure of merit ($\times 10^{23}$ WΩs^2)	9.0	62.5	2533	4410		15,670		
Keyes figure of merit ($\times 100$ W/cm s °C)	13.8	6.3	90.3	229		118		444
Baliga's figure of merit (relative to silicon)	1.0	15.7	4.4			24.6		
Temperature figure of merit	220	394	650	815		1,060		101

[a]Many parameters are taken from Yoder [3], Harris [83], and Levinshtein et al. [84, 85].

where E_B is the breakdown electric field and v_s is the electron saturation velocity. In terms of this figure of merit, SiC is 260 times better than Si and is inferior only to diamond.

Keyes' figure of merit (KFM) [5] is relevant to integrated circuits applications:

$$\text{KFM} = \chi \sqrt{\frac{cv_s}{4\pi\varepsilon_0}} \quad (4)$$

where c is the velocity of light, ε_0 is the relative static dielectric constant, and χ is the thermal conductivity.

Baliga's figure of merit (BFOM) [6] characterizes material properties for applications in high-power switches:

$$\text{BFOM} = \varepsilon_0 \mu E_B^3 \quad (5)$$

Here, μ is the electron mobility. Since all of these figures of merit give different assessment of material properties, we propose to introduce a combined dimensionless figure of merit (CFOM):

$$\text{CFOM} = \frac{\chi \varepsilon_0 \mu v_s E_B^2}{(\chi \varepsilon_0 \mu v_s E_B^2)_{\text{silicon}}} \quad (6)$$

When χ is in W/cm degree K, E_B is in kV/cm, μ is cm^2/V-s, and v_s is in m/s, the denominator in Eq. (6) is equal to 2.4×10^{14}. The values of CFOM for Si, GaAs, 6H-SiC, 4H-SiC, GaN, and diamond are given in Table II. However, we should notice that 6H-SiC has anisotropic properties, which make CFOM much worse for devices where the electron transport occurs along the crystallographic directions with a low electron mobility.

Of course, many other factors, such as the availability and cost of defect-free bulk substrates and of high-quality epitaxial layers, doping control, impurity and trap concentration, the availability of a good-quality native oxide, and the ease of processing, are all very important and may even be crucial. Therefore, CFOM may, for example, grossly overestimate the potential of diamond for device applications. Clearly, no matter what all of these figures of merit are, silicon now (and for the foreseeable future) is the most important material of semiconductor electronics. However, tremendous progress has been achieved in all areas of SiC technology, making practical applications of SiC transistors a reality. Among wide-bandgap semiconductors, 4H-SiC has come closer to practical applications than any other material, followed closely by GaN [7].

TABLE II

CFOM for Si, GaAs, 6H-SiC, 4H-SiC, GaN, and Diamond

Material	Combined Factor of Merit
Si	1
GaAs	7.36
6H-SiC (disregarding anisotropy of mobility)	393
4H-SiC	404
GaN	404
Diamond	30,080

2. SiC MOSFETs

FETs use different semiconductors and different ways of isolating the gate from the channel. In silicon metal-oxide semiconductors field-effect transistor (MOSFET), a native silicon oxide (SiO_2) is used as a gate dielectric. In SiC MOSFET (Fig. 5 [2]), SiO_2 is not a native oxide, and the SiO_2–SiC interface is not nearly as good as Si–SiO_2.

MOSFETs can have an n-type channel (NMOS) or a p-type channel (PMOS). In an NMOS, the channel is created by a free electron gas induced by the gate voltage in the semiconductor at the silicon–silicon dioxide interface. These electrons come from an n^+ source and flow into an n^+ drain. The drain and source contacts are implanted or diffused into a p-type substrate. A MOSFET is actually a four-terminal device with source, gate, drain, and bulk (substrate) contact. The n^+ source and drain contacts and the n-type channel are isolated from the p-type substrate by the depletion regions. The substrate bias can control the width of these depletion regions [8].

For a PMOS, conductivity types of all the NMOS layers should be changed from n-type to p-type and vice versa. Since holes (that carry current in a PMOS) have a much smaller mobility than electrons, the performance of PMOS is usually much worse. However, PMOS are widely used together with NMOS in complementary MOS (CMOS). In this technology, the basic element of electronic circuits—the inverter—is formed by connecting two MOSFETs (n-channel and p-channel) fabricated on the same wafer. To achieve such an integration, the n-channel and p-channel devices have to be isolated. Just as in a regular MOSFET, depletion regions between n-type and p-type doped regions can be used for the isolation.

CMOS transistors can be fabricated on the same chip with BJTs. This technology is called BiCMOS.

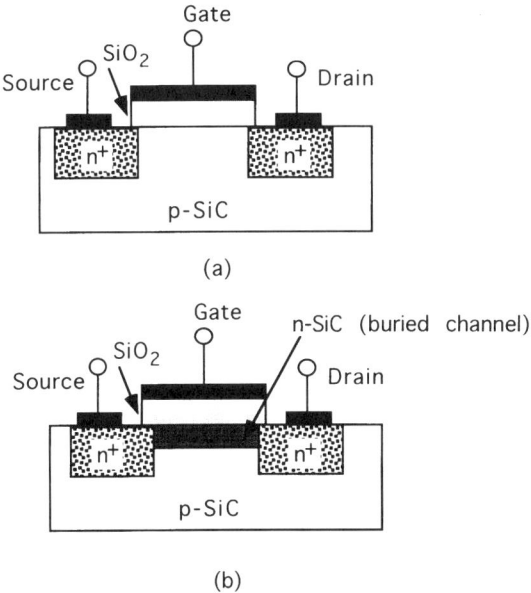

FIG. 5. Schematic structure of (a) inversion-channel and (b) buried-channel SiC MOSFETs.

The MOSFET design, especially the gate and channel geometry, determines the device characteristics. Such designs vary a great deal depending on the application and on the particular semiconductor material.

The earlier work on SiC FETs was reviewed by Kelner *et al.* [9]. The first SiC MOSFETs were fabricated by Suzuki *et al.* [10] in 3C-SiC (i.e., cubic SiC or β-SiC). SiC MOSFET research has been carried out by several groups, including Cree Research, Inc. [11–13], Northrop Grumman (formerly Westinghouse) [14], Purdue University [2, 15], and GE [16]. (See also a review paper from Mororola [17].)

A typical *p*-type 6H-SiC substrate (available from Cree Research, Inc.) is doped by approximately 4×10^{18} cm^{-3} acceptors. For MOSFETs described in [15], an epitaxial *p*-type layer grown of such a substrate was doped at 2.8×10^{16} cm^{-3} with Al. The n^+ contacts were formed by ion implantation. Such an implant was performed at high temperature through a 0.9-μm titanium–nickel mask and a 300-Å SiO$_2$ screen layer. Multiple implants were used to achieve a total dose of 10^{15} cm^{-2} with a rectangular profile 0.55 μm deep, with a concentration of 1×10^{20} cm^{-3} [15]. The fabrication technology is described in detail in Chapter 2. (See also [18].)

A native oxide for SiC is SiO_2 (i.e., the same oxide as for silicon). However, in inversion MOS structures (see Fig. 5a), the density of the surface states at the SiC–SiO_2 interface was approximately $4 \times 10^{12}\,\text{cm}^{-2}$ [15]—more than three orders of magnitude larger than for the Si–SiO_2 interface. Such a large density of the surface states led to a fairly small value of the field-effect mobility of $20\,\text{cm}^2/\text{V-s}$ at room temperature. However, further studies and optimization of surface cleaning of SiC resulted in the demonstration of the surface-state density at the SiO_2–SiC interface as low as $1.5 \times 10^{11}\,\text{cm}^{-2}\,\text{eV}^{-1}$ [19]. More recently, Lipkin and Palmour demonstrated that a reoxidation process can lower the density of states to $1 \times 10^{11}\,\text{cm}^{-2}\,\text{eV}^{-1}$, which resulted in the highest channel mobility reported to date ($72\,\text{cm}^2/\text{V-s}$) [89]. Palmour *et al.* [12] reported an operation of inversion 6H-SiC MOSFETs up to 923 K.

In the buried-channel MOSFET described in [15], the channel mobility is determined by the bulk mobility of SiC, which is higher than the effective electron mobility in an inversion MOSFET channel in SiC. In the buried-channel MOSFET, the n-type channel was formed by ion implantation. It was approximately $0.25\,\mu\text{m}$ deep, with a doping level of $\approx 4.9 \times 10^{17}\,\text{cm}^{-3}$. The effective field-effect mobility was approximately $180\,\text{cm}^2/\text{V-s}$. These values of the field-effect mobility are comparable to those achieved for SiC epitaxial junction field-effect transistors (JFETs) [20] described in Section II.3.

The ion-implantation process for fabrication of buried-channel SiC MOSFETs allows one to fabricate both enhancement and depletion mode transistors on the same chip, which can be used for monolithic integrated circuits. As seen from Tables I and II, SiC devices are especially attractive for high-power, high-voltage applications. Their operation is projected for voltages exceeding 1 kV [21]. For such applications, vertical SiC MOSFETs (Fig. 6) have the most promising potential.

The first SiC UMOSFETs were reported by the Cree Research, Inc., team [22]. A major advantage of SiC is its ability to support much higher electric fields, which allows one to increase the doping level of the drift layer and shrink the drift region to decrease the on-resistance of the device. The devices reported in [22] blocked 150 V and were rated at $100\,\text{A/cm}^2$ (67 mA). Figure 7 (from [17]) compares the analytical predictions and experimental data for a subsequent 260-V device (reported in [55]) with the corresponding data for silicon devices. Figure 8 (from [55]) shows the room-temperature I-V characteristic of a 2-A 4H-SiC UMOSFET. This power 4H-SiC UMOSFET had specific on-resistance of 18 ohms [55].

As pointed out by the Purdue group [23], one serious limitation in these devices is oxide breakdown at the bottom of the etched trench. The device may have reliability problems if the oxide field is higher than approximately $2.5\,\text{MV/cm}$. This limits the electric field in SiC to $2.5\,\text{MV/cm} \times \varepsilon_{SiO_2}/\varepsilon_{SiC} \approx$

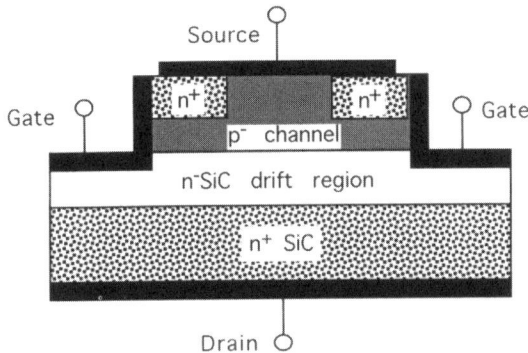

Fig. 6. Cross-section of SiC UMOSFET.

1 MV/cm, which is much smaller than the SiC breakdown field. The UMOS blocking voltage scales as the peak electric field squared. Hence, SiC in a UMOS does not perform at its full potential. The Purdue group announced that they have submitted two patent disclosures describing novel methods for eliminating this problem [23].

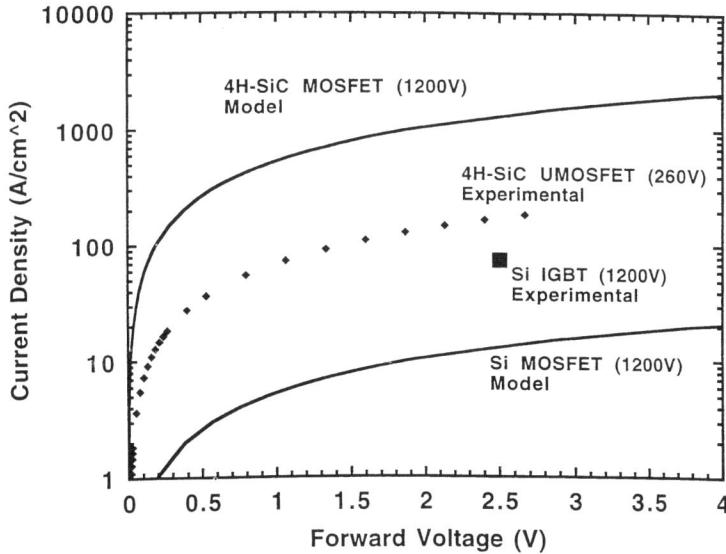

Fig. 7. Current density versus forward voltage for Si MOSFET, Si Insulated Gate Bipolar Transistor (IGBT), and 4H-SiC UMOSFET [17].

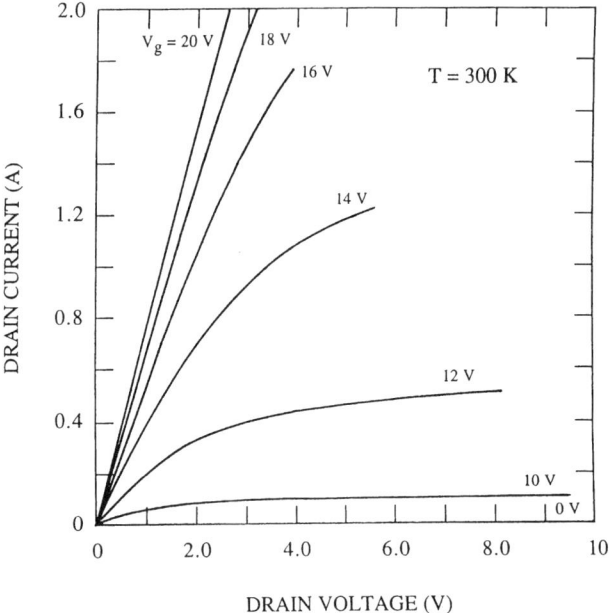

FIG. 8. Room temperature I–V characteristic of a 4H-SiC UMOSFET [55]. Device area is 0.01 cm^2. Specific on-resistance is 13.2 mΩ/cm^2.

The Purdue University group reported on a planar 6H-SiC high-voltage MOSFET fabricated using a double implant; they called this device DIMOS [24, 25]. The device cross-section is shown in Fig. 9. SiC DIMOS is similar to the silicon double-diffused MOS (Si DMOS) power transistor. However, in SiC the diffusion coefficients of impurities are very low, and, therefore, in SiC DIMOS, the p base and n^+ source regions are produced by ion implantation. Just like in Si DMOS, a positive bias applied to the polysilicon DIMOS gate produces a surface inversion layer at the SiO_2–p-layer interface, creating a conducting path from the n^+ source to the n-drift region. In the off-state, the lightly doped n^- drift region is depleted and can withstand a large drain voltage. Since the breakdown field in SiC is 10 times larger than that in Si, a 10 times thinner depletion region is required in order to maintain (block) the same voltage without a breakdown. This means that the doping level in the drift region can be increased by two orders of magnitude, with a commensurate drop in the on-resistance. These devices (with circular geometry) demonstrated blocking action up to 760 V. This corresponded to an estimated electric field of 1.35 MV/cm. The blocking mode and conduction mode characteristics are shown in Fig. 10. The

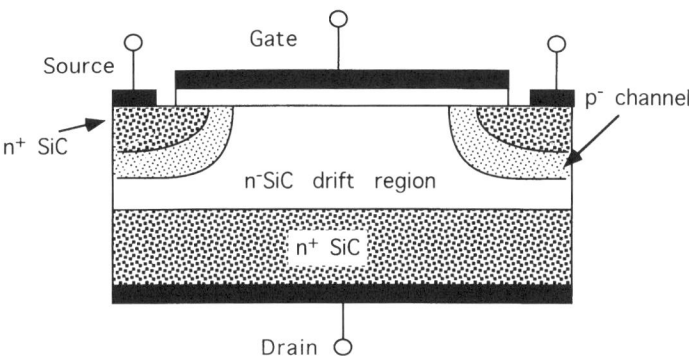

Fig. 9. DIMOS cross-section (after 24, 25).

Purdue group demonstrated specific on-resistances of 125 mΩ-cm^2 for the DIMOS blocking 760 V and 66 mΩ-cm^2 for a 2-μm channel length SiC DIMOS blocking 500 V. These devices were fabricated using 6H-SiC. Much lower on-resistances are expected for 4H-SiC DIMOS because of much higher bulk mobility of the 4H polytype (see Table I). Compared to a SiC UMOSFET, DIMOS requires a larger surface area. However, it has the advantage of less complicated processing.

Devices closely related to MOSFETs are charge-coupled devices (CCDs). These devices consist of a series of closely spaced MOS capacitors located

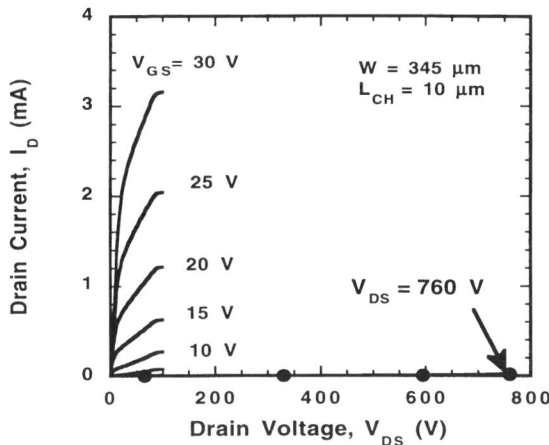

Fig. 10. DIMOS blocking mode and conduction mode characteristics [24].

between two ohmic contacts. The application of a positive gate bias to a MOS gate creates a potential well at the semiconductor surface under that gate where a charge packet may be localized. By changing the gate biases applied to these gates in a synchronous fashion, these localized charge packets can be moved from gate to gate. The first SiC CCDs were described in [26]. These devices may find applications in specialized image sensors, which are transparent to visible light, for applications in defense systems and UV astronomy.

3. SiC MESFETs and JFETs

As mentioned, a native oxide for SiC is SiO_2, the same oxide as for silicon. However, the achieved density of surface states at the $SiC-SiO_2$ interface is much greater than that for Si. Often, in SiC, higher quality transistors can be fabricated by using a Schottky contact. The depletion region between the Schottky metal and a SiC conducting layer isolates the gate from the FET channel. Such a FET with Schottky contact gate is called a metal semiconductor field-effect transistor (MESFET).

In a MESFET (Fig. 11), the gate voltage and the built-in voltage of the Schottky gate partially or totally deplete the channel. The change in the gate bias changes the cross-section of the conducting channel and, hence, the gate voltage controls the drain-to-source current.

SiC MESFETs, SiC JFETs, and SiC static induction transistors (described later) have emerged as leading contenders for microwave applications. SiC MESFETs [12, 17, 27–29] have exhibited higher field-effect mobilities than SiC MOSFETs with inversion channels. 6H-SiC MESFETs reported by Palmour *et al.* [12] operated up to 773 K.

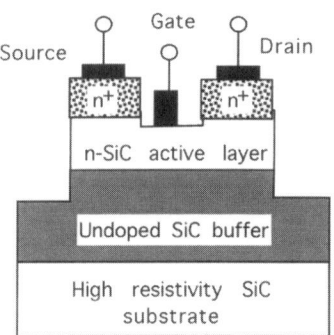

Fig. 11. Cross-section of SiC MESFET.

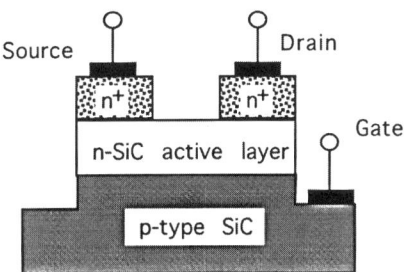

FIG. 12. Cross-section of n-channel buried-gate SiC JFET.

At IEDM-96, the Northrop-Grumman group reported on SiC MESFETs with breakdown voltages over 100 V and maximum drain currents over 500 mA/mm [30]. Figure 12 shows the cross section of an n-channel buried-gate SiC JFET [31]. In this structure, the depletion region between the buried p-type gate layer and the conducting n-type channel controls the channel cross-section and the drain-to-source current. An obvious disadvantage of this design is a large parasitic capacitance between the gate and the source and drain contact regions. Palmour et al. solved this problem in top-gated SiC JFETs, which operated at microwave frequencies (with maximum frequency of oscillations of 8 GHz) [90].

Kelner et al. fabricated and evaluated buried gate JFETs in 3C-SiC grown on an 6H-SiC substrate [31]. Devices with a 4-μm gate length had $g_m = 20$ mS/mm. The data obtained from this study were analyzed using a charge-control model. The analysis showed that the value of the field-effect mobility (≈ 560 cm^2/Vs) was close to the measured value of the Hall mobility (≈ 470 cm^2/Vs) and that the electron saturation velocity in the channel was consistent with the theoretical value of 2×10^7 cm/s. Kelner et al. [32], Anikin et al. [33], and Palmour et al. [12] reported on the high-temperature operation of 6H-SiC buried-gate JFETs. Devices fabricated with a 4-μm gate length had a maximum g_m in the saturation region of 17 mS/mm and a maximum drain saturation current of 450 mA/mm at zero gate voltage at room temperature. The device transconductance decreased with temperature because of the decreasing electron mobility.

More recently, Neudeck et al. [34] demonstrated the operation of 6H-SiC JFETs operating at voltages up to 100 V with peak currents over 1 A at room temperature. The devices operated up to 600°C with minimum degradation over 30 hr at this temperature.

SiC has a relatively low mobility but a relativley high electron saturation drift velocity (see Table I). As shown in [35-37], the performance of a FET made from such a material can be substantially improved when the electric

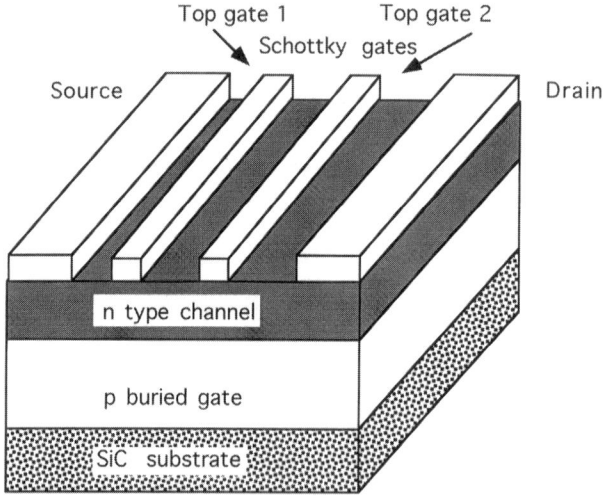

FIG. 13. Device structure of 6H-SiC JFET with multiple gates for improved field uniformity in the channel [37].

field distribution along the channel (in the direction from the source to the drain) is made more uniform by using a doping profile varying along the channel; or multiple (separately biased) gates, as shown in Fig. 13. In this device, the gate voltage swing, $V_g - V_T$, varies along the channel, with the highest swing near the drain. Our simulations showed that when the top gate (gate 2), which is closer to the drain, is biased approximately 2 V higher than the other top gate (i.e., $V_{\text{Topgate2}} = V_{\text{Topgate1}} + 2$ V), the maximum drain current in a SiC FET should more than double [37].

Figure 13 shows a buried-gate SiC JFET with two top gates. However, a similar approach can be also used for other SiC FETs, such as MOSFETs or MESFETs.

As mentioned one of the major advantages of SiC technology is its ability to operate at elevated temperatures. The Cree Research, Inc., team reported on SiC MOSFETs operating at up to 350°C, SiC PMOS circuits operating at up to 400°C, enhancement-mode SiC NMOS operating at up to 350°C, and buried-gate SiC JFETs and SiC MESFETs operating at up to 500°C [38].

A major drawback of buried-gate JFETs is large parasitic gate-to-source and gate-to-drain capacitances associated with the overlaps between the gate and the ohmic contact regions. Lateral SiC JFETs (Fig. 14), which are very similar to SiC MESFETs, where the Schottky metal is substituted by

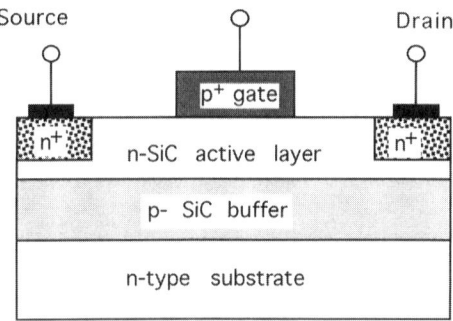

FIG. 14. Lateral SiC JFET [39].

a p^+ epitaxially grown gate, are more suitable for microwave applications [39]. The use of the semi-insulating substrate should lead to a considerable improvement in microwave performance.

III. SiC Microwave Field-Effect Transistors

SiC MESFETs, SiC JFETs, and SiC static induction transistors (SITs) (described next) have emerged as leading contenders for microwave applications. At IEDM-96, the Northrop-Grumman team reported on high-frequency 4H-SiC MESFETs with maximum frequency of oscillations of 42 GHz and gain over 10 dB at 10 GHz [30]. The improvement in device performance is linked to the use of high-resistivity substrates, which dramatically decreases parasitic capacitances [40]. These devices operated as class A amplifiers with power densities of 3 W/mm at 1.8 GHz (slightly higher than the previous record of the 4H-SiC MESFETs operation at 54 V of drain bias with a power density of 2.8 W/mm, with power-added efficiency of 12.7% at 1.8 GHz [41]). For class B amplifiers, power densities of 2 W/mm with power-added efficiencies of 66 and 50% at 850 MHz and 1.8 GHz, respectively, have been achieved [41]. These devices can be operated in the frequency range of 6 to 10 GHz.

Cree Reasearch, Inc., reported on SiC microwave FETs with a cutoff frequency of 32 GHz [91]. For 6H-SiC lateral JFETs, the power of 1.3 W/mm was obtained at 850 MHz for the drain bias of 40 V [42]. A potential advantage of SiC JFETs is lower gate leakage current compared to SiC MESFETs, since a p^+ gate has a larger barrier height than a metal

Schottky gate. This may allow one to operate SiC JFETs at higher temperatures than SiC MESFETs. As mentioned in [17], the performance of SiC JFETs may be dramatically improved by using 4H-SiC and by replacing the conductive substrate with a high-resistivity substrate.

A SIT was first proposed for Si by Nishizawa et al. [43] and is different from conventional FETs. In this device (Fig. 15), the conduction between the drain and source is controlled by a metal or a $p-n$ junction grid embedded into the semiconductor. This device can operate in two different modes. When the depletion regions between the metal grid fingers overlap, electrons traveling from the source to the drain encounter a potential barrier located between the grid fingers. In this case, the gate bias changes the barrier height, leading to a strong modulation of the drain current. At a more positive gate bias, the depletion regions around the grid fingers do not overlap. In this case, the gate bias changes the cross-section of the undepleted regions between the fingers. This regime of operation is very similar to that of a MESFET.

A SIT was demonstrated in SiC and exhibited an excellent performance [45–47]. SiC SITs operated in the frequency range of 3 to 6 GHz. At 600 MHz, SiC SITs with breakdown voltages of 300 V produced 450-W power under pulsed operation with low duty cycle. The Northrop-Grumman group reported on the 1-kW multitransistor SIT module, which was used for a high-definition television (HDTV) transmission at the National Association of Broadcasters convention in Las Vegas, Nevada, April 1996

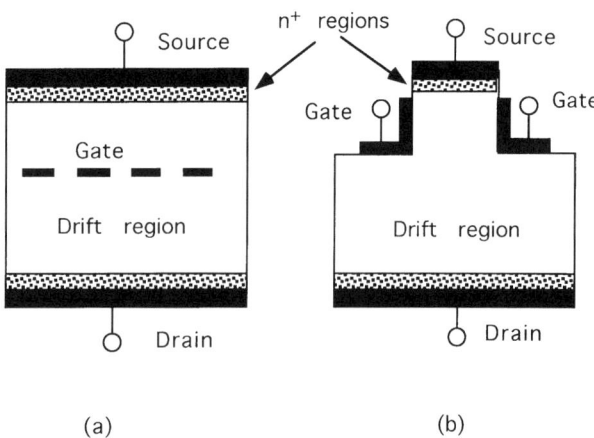

Fig. 15. Schematic diagram of a SIT with (a) embedded p-type grid (may be substituted by metal grid) and (b) the top gates.

TABLE III
SUMMARY OF MICROWAVE PERFORMANCE FOR SiC FETs AND SITs

Device	Minimum Feature Size (μm)	Power (W/mm)	Drain Bias (V)	f_T (GHz)	f_{max} (GHz)	Power-added Efficiency (%)	Ref.
4H-SiC MESFET	0.45	3 at 1.8 GHz	54		42	66% class B	38
6H-SiC MESFET		2.8 at 1.8 GHz		16.2	32	12.7	39
SiC SIT	0.5	1.36 at 600 MHz					38
6H-SiC lateral JFET	0.3	1.3 at 850 MHz	40	7.3	9.2		30

[45]. Eastman *et al.* [44] proposed a different version of such a device, in which the controlling gates are located at the top surface (see Fig. 15b). Table III provides a summary of microwave performance for SiC FETs and SITs.

IV. SiC Digital Integrated Circuits

The wide energy bandgap of SiC should result in very low leakage currents and in very large on-to-off current ratio, even at high temperatures. These factors (as well as the high thermal conductivity of SiC) make this material very attractive for high-temperature digital integrated circuits and nonvolatile solid-state memories. These circuits will find applications at temperatures at which silicon VLSI cannot operate reliably.

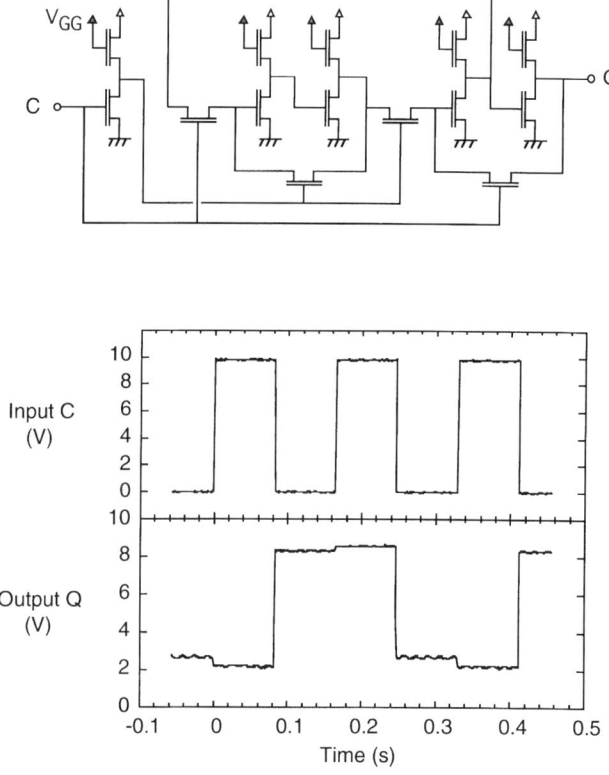

FIG. 16. SiC NMOS binary counter and operating waveforms at room temperature [49].

In 1992, Harris et al. [48] reported on a prototype 3C-SiC MESFET-type inverter circuit. This inverter was operational at room temperature with a maximum DC gain of 2.5 and a unity gain at a frequency of 0.143 MHz.

SiC digital integrated circuits, described in [49], were fabricated using enhancement-load NMOS, with the load transistor operated in the non-saturating regime using a separate gate-bias voltage supply. These circuits included inverters, NAND and NOR gates, XNOR gates, RS flip-flops, binary counters, and half adders. All of the circuits operated from room temperature to above 300°C. Figures 16 and 17 show the binary counter and half adder circuits along with operating waveforms at room temperature and at 304°C, respectively.

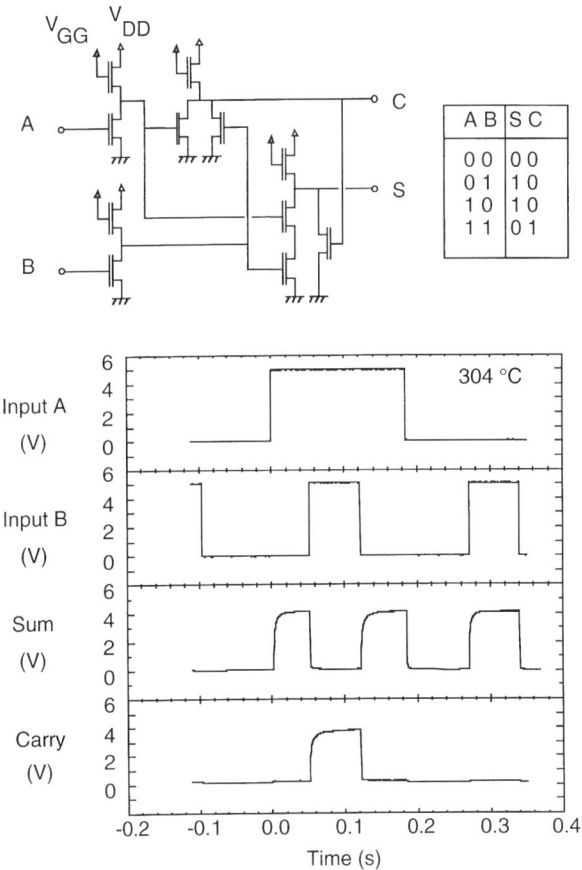

FIG. 17. SiC NMOS half adder and operating waveforms at 304°C [49].

Cree Research, Inc., reported on 6H-SiC CMOS ICs [50, 51]. More recently, SiC CMOS ICs with better p-channel transistors (including NAND, NOR, XOR, half adders, flip-flops and 11-stage ring oscillators were reported by the Purdue group [23]. Their CMOS inverters demonstrated stable operation at room temperature and 300°C with the power supplies of 10 V and 15 V.

Cree Research, Inc., demonstrated 17-stage SiC ring oscillators with gate delays of 15 ns at room temperature and SiC CMOS operational amplifiers with open-loop gain over 10,000 and input offset voltage less than 100 mV at room temperature [52].

V. SiC Bipolar Transistors and Thyristors

For applications ranging from DC transmission over long lines to power delivery for motors or actuators, SiC bipolar devices may have application-specific advantages compared to both Si bipolar transistors and thyristors and SiC power MOSFETs, since SiC devices may use higher doping and have a smaller on-resistance because of a much higher breakdown field compared to Si. Also, SiC BJTs may be more reliable than SiC power MOSFETs because they do not use an SiO_2 gate.

Most work on SiC bipolar devices is concentrated on SiC thyristors. Vainshtein et al. [53] and Dmitriev et al. [54] described the first 6H-SiC dynistor diode. This diode is an $n-p-n-p$ thyristor-like structure but without gate contacts. For this device, an $n-p^+-n^+$ structure was epitaxially grown on p^+ 6H-SiC substrates using container-free liquid-phase epitaxy in a continuous epitaxial process. The turn-on time constant varied from 1 to 10 ns. The recovery time varied from 150 to 200 ns. The static switching voltage varied between 10 and 50 V, with an inverse voltage of up to 90 V. Edmond et al. [80] fabricated the first 6H-SiC thyristor, which operated up to 350°C.

The SiC thyristor structure is shown in Fig. 18. Cree Research, Inc., has demonstrated high-power 4H-SiC npnp power thyristors with a blocking voltage of 900 V and on-current of 2.0 amps, as well as thyristors with maximum currents of 10 amps with a blocking voltage 700 V [52]. These 4H-SiC npnp thyristors operated up to 500°C. They had turn-off times ranging from 360 to 640 ns [55], which favorably compares with typical silicon thyristors. Xie et al. [93] reported the switching of 6H-SiC thyristors at a frequency of 640 kHz. SiC BJTs operated at up to 400°C but had fairly low gains because of a short minority carrier diffusion length [92].

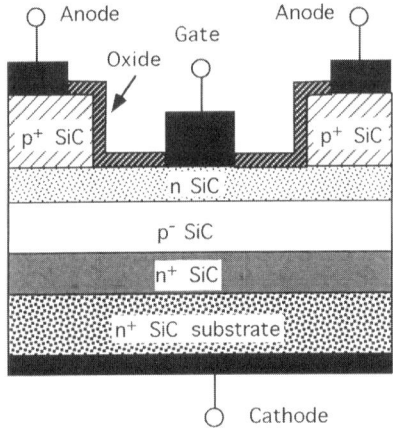

FIG. 18. Cross-section of SiC thyristor.

Since both SiC MOSFETs and SiC bipolar devices have been sucessfully demonstrated, the insulated-gate bipolar transistor (IGBT) [56] and MOS-controlled thyristor (MCT) can also be fabricated in SiC. An expected improvement in material quality, such as increasing minority carrier lifetimes in SiC (typically from 40–400 ns) and the improvement of the $SiC-SiO_2$ interface, will facilitate the development of these devices. Lifetimes of 2.1 µs have been obtained in very thick, lightly doped 4H-SiC epilayers [81].

As discussed by Chelnokov et al. [87], the most important limitation of SiC power devices is related to defects, which cause dependence of the breakdown voltage on the device area. A high breakdown voltage has been observed only for devices with a cross-section smaller than 10^{-3} cm^2. These defects are present even in SiC devices, which are free of micropipes.

Another important application of SiC bipolar technology is the development of nonvolatile random-access memories (NVRAMs) [57–60]. The basic cell, invented at Purdue and Cree Research, Inc., consists of a bipolar access transistor vertically merged with an $n-p-n$ storage capacitor, as shown in Fig. 19 [59]. Room-temperature capacitance transients of SiC $n-p-n$ capacitances are plotted as a function of temperature in Fig. 20 [60]. As seen, the data extrapolate out to a room-temperature storage time of over 100 years for 6H-SiC NVRAMs.

A 64-bit bipolar NVRAM cell array, complete with SiC NMOS sense amplifiers and control logic, has been fabricated at Purdue and at Cree

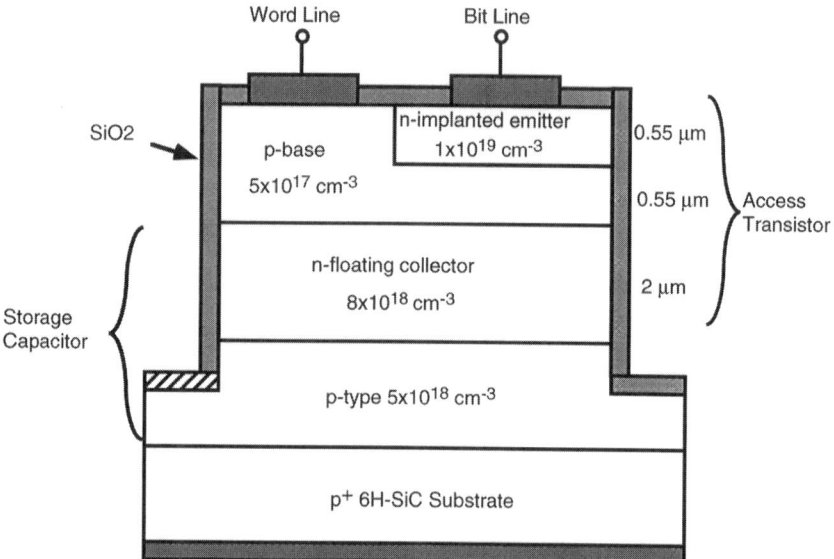

FIG. 19. Cross-section of the SiC vertically integrated bipolar NVRAM. The NVRAM cell consists of an n-p-n bipolar access transistor connected to a p-n junction storage capacitor [59].

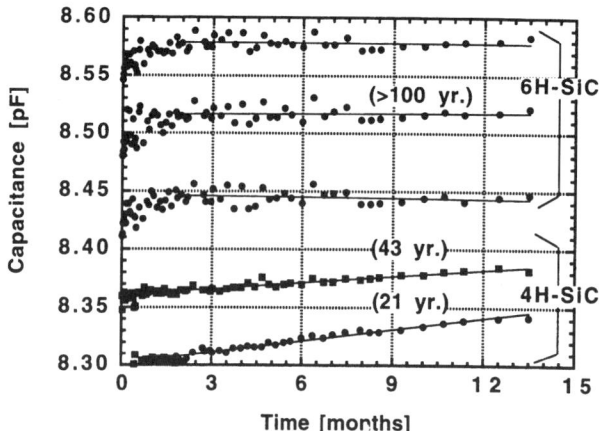

FIG. 20. Room-temperature capacitance transients of charged SiC NVRAM cells [60].

Research. The development of a full 1-kbit memory is now under way at Cree Research [91].

VI. Two-Dimensional Modeling of SiC Transistors

Two-dimensional device simulation provides insight into device operation and may help in device design and optimization. A general purpose two-dimensional device simulator BLAZE (more recently replaced by the ATLAS device simulator) designed to simulate heterojunction technologies based on various material systems has been applied to simulating SiC JFETs and MESFETs. BLAZE was used to simulate SiC p–n–n^+ diodes [61] and a buried-gate SiC heterojunction JFET [61, 62].

The simulated transistor had an unintentionally doped n-type 3C-SiC layer on top of a p-type 6H-SiC substrate. The current in the channel was modulated using the p-type 6H-SiC substrate as a gate. This transistor was fabricated and measured by Kelner et al. [63]. The bandgaps of 3C-SiC and 6H-SiC are 2.2 and 2.9 eV, respectively, so that the materials form a heterojunction at their interface. Structures with channel doping levels of $N_d = 2 \times 10^{16}$ cm^{-3} and $N_d = 3 \times 10^{16}$ cm^{-3} and a substrate doping of 2×10^{17} cm^{-3} were simulated. Doping and field-dependent mobilities were modeled using the Scharfetter-Gummel expressions [64], with lattice mobilities set to agree with measured data [64]. The simulated device was fully depleted for $V_G \sim 28$ V for $N_d = 2 \times 10^{16}$ cm^{-3}, and for $V_G \sim 17$ V for $N_d = 3 \times 10^{16}$ cm^{-3}. The simulated current–voltage characteristics for $N_d = 3 \times 10^{16}$ cm^{-3} were in reasonably good agreement with the measured data presented in [63]. The simulation showed that BLAZE (ATLAS) is capable of simulating devices that include SiC heterostructures. In this particular device, the heterostructure affected only the FET threshold voltage. However, this capability is indispensable for simulating more advanced and more practical SiC devices, such as an SiC heterostructure IMPATT or an SiC heterostructure bipolar transistor.

An 6H-SiC MESFET [65] was also simulated. The device had a gate length of 10 μm, a gate width of 1 mm, and a source-to-drain distance of 30 μm. The calculated current–voltage characteristics were in reasonably good agreement with measured data.

Ramungul et al. [56] used two-dimensional modeling for the design and optimization of a 6H-SiC UMOS IGBT. These simulations suggested ways to improve and alleviate field crowding and increase device breakdown voltage.

VII. Analytical Transistor Models and Circuit Simulation (AIM-Spice)

Analytical modeling of SiC FETs used in this chapter was done using FET models implemented in AIM-Spice [66, 67]. These models describe the charge control by the gate and drain biases using the following empirical equation—the unified charge-control model (UCCM) [66]:

$$V_{GT} - \alpha V_F = a(n_s - n_0) + \eta V_{th} \ln(n_s/n_0) \quad (7)$$

Here, n_s is the sheet carrier concentration in the channel; $V_{GT} = V_{GS} - V_T$; V_{GS} is the intrinsic gate-source voltage; V_F is the Fermi potential along the channel referred to the source; α (less than but close to unity) is the so-called body effect parameter that accounts for the variation in the depletion width along the channel in MOSFETs; $a = q/c_a$ where q is the electronic charge and c_a is the gate-channel capacitance per unit area; n_0 is the electron concentration per unit area in the channel at threshold; η is the subthreshold ideality factor, which accounts for the gate voltage division between the layer separating the gate and the channel (oxide or wide-gap semiconductor) and the semiconductor channel; and V_{th} is the thermal voltage (approximately 26 mV at room temperature).

Equation (7) reduces to the well-known parallel plate capacitor behavior well above threshold, [i.e., $qn_s = c_a(V_{GT} - \alpha V)$], where V_F is replaced by the channel potential V, and the characteristic exponential behavior determined by Boltzmann carrier statistics, $n_s = n_0 \exp[(V_{GT} - \alpha V_F)/\eta V_{th}]$, below threshold, where V_{GT} is negative [6]. An approximate version of the UCCM expression of Eq. (7) is given by [66]:

$$n_s \approx 2n_0 \ln\left[1 + \frac{1}{2}\exp\left(\frac{V_{GT} - \alpha V_F}{\eta V_{th}}\right)\right] \quad (8)$$

As shown in [68], UCCM provides an accurate description of n_s versus V_{GT}.

In AIM-Spice, we express the drain current in each regime of operation—subthreshold, above-threshold linear, and above-threshold saturation—and then proceed to join the various expressions using interpolation functions. The resulting model, which is applicable to all types of FETs, has the following universal form:

$$I_d = \frac{g_{ch} V_{ds}(1 + \lambda V_{ds})}{[1 + (g_{ch} V_{ds}/I_{sat})^m]^{1/m}} \quad (9)$$

Here, m is a shape parameter that determines the transition between the above-threshold linear and saturation regimes, V_{ds} is the extrinsic (indicated by lowercase subscripts) drain-source voltage, I_{sat} is the saturation current for a given extrinsic gate-source voltage V_{gs}, and g_{ch} is the extrinsic channel conductance at small V_{ds}. The factor $(1 + \lambda V_{ds})$ describes the finite output conductance in saturation.

To simulate a transient response, we also need to model capacitance voltage (C-V) characteristics. The simplest possible C-V model, proposed by Meyer [69], is based on calculating derivatives of the gate charge with respect to the terminal voltages. However, it is known that reciprocity in a MOSFET is inconsistent with charge conservation [70]. Although charge conservation is not strictly enforced by using the set of Meyer capacitances, the resulting error is usually small. Nonetheless, in transient analyses of certain demanding circuits (RAM cells, switched capacitors circuits, charge pumps), the Meyer model is known to give erroneous results [71–73].

A more accurate model of the intrinsic FET capacitances is based on "partitioning" the channel charges between the source and drain, so that a source charge $Q_S = K_p Q_I$ and a drain charge $Q_D = (1 - K_p)Q_I$, where K_p is a partitioning factor between 0 and 1. The substrate depletion charge Q_B below the channel is assigned to the substrate terminal. The total gate charge $Q_G = -(Q_I + Q_B)$, which assures charge conservation [67].

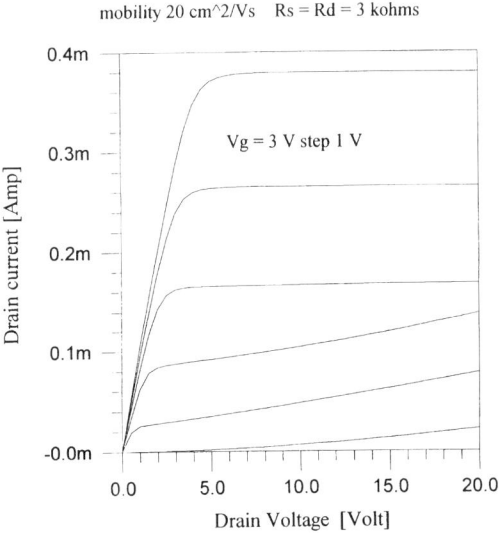

FIG. 21. Current–voltage characteristics of enhancement-mode SiC MOSFET [15], simulated using a level 7 AIM-Spice model.

Figure 21 shows the current–voltage characteristics of an enhancement-mode SiC MOSFET [15], simulated using a level 7 AIM-Spice model. The simulated curves agree fairly well with experimental data obtained in reference [15].

Kobayashi et al. [78] used a simplified model for a double-diffused MOSFET [79] to evaluate the performance potential of 6H-SiC and 4H-SiC power MOSFETs. They concluded that these devices can become practical, even at high drain voltages, only if the field-effect mobility exceeds $50 \text{ cm}^2/\text{V-s}$.

VIII. Potential Performance and Applications of SiC Transistors and Integrated Circuits

At the present time, two of the most promising potential applications of SiC transistors are for power switches and for the generation of high microwave power. However, material defects still present a serious limitation of overall SiC transistor performance. As stated in [17], even the record low defect density of 3.5 defects/cm^2 achieved for SiC [74] may be not low enough for applications of this material in power devices. Other, still unidentified defects, with densities on the order of 10^3 cm^2 seem to limit the breakdown voltage in SiC devices, even in the absence of micropipes [87]. Still, further improvement in SiC materials technology is expected to lead to the development of SiC power switches that are greatly superior to those implemented in Si or GaAs [88]. SiC SITs are being commercialized for applications in HDTV transmitters.

The projected power densities for microwave SiC MESFETs are close to 20 W/mm [17], more than an order of magnitude higher than for GaAs power microwave MESFETs. (See also [75–76].) SiC MESFETs may be very close to commercialization. The potential competition from GaN-based FETs, which reached higher frequencies of operation (see, for example, [77], [81]) is tempered by a low thermal conductivity of sapphire substrates used in the present generation of GaN-based microwave FETs, even though the use of SiC substrate for GaN-based FETs may help overcome the gap in power performance.

Acknowledgment

The author is grateful to Dr. Phil Neudeck, Professor Michael Melloch, Dr. Charles Weitzel, Dr. John Palmour, and Professor Mikhail Levinshtein for valuable comments and suggestions.

References

1. J. E. Lilinfeld, US Patent 1,745,175 (1930).
2. J. N. Pan, J. A. Cooper, and M. R. Melloch, Electronic Letters, vol. 31, no. 14, pp. 1200–1201 (1995).
3. M. Yoder, IEEE Trans. Electron Devices, vol. 43, no. 10, pp. 1633–1636 (1997).
4. E. O. Johnson, RCA Rev., vol. 26, 163 (1965).
5. R. W. Keyes, Proc. IEEE, vol. 60, p. 225 (1972).
6. B. J. Baliga, IEEE Electron Dev. Lett., vol. 10, p. 455 (1989).
7. M. Shur and A. Khan, GaN based field effect transistors, in "High Temperature Electronics," Eds. M. Willander and H. Hartnagel, Chapman, London (1996).
8. M. S. Shur, "Physics of Semiconductor Devices," Prentice Hall, New Jersey (1990).
9. G. Kelner, M. S. Shur, and G. L. Harris, SiC devices and ohmic contacts, in "Properties of Silicon Carbide," Ed. G. Harris, pp. 247–264, INSPEC, IEE, United Kingdom (1995).
10. A. Suzuki, H. Ashida, N. Furui, K., Mameno, and H. Matsunami, Jpn. J. Appl. Phys., vol. 21, p. 579 (1982).
11. J. W. Palmour, H. S. Kong, and R. F. Davis, J. Appl. Phys., vol. 64, pp. 2168–2177 (1988).
12. J. W. Palmour, H. S. Kong, D. G. Waltz, J. A. Edmond, C. H. Carter, Jr., Proc. of First Intern. High Temperature Electronics Conference, Albuquerque, NM, pp. 511–518 (1991).
13. J. W. Palmour and J. A. Edmond, Proc. 14th IEEE Biannual Cornell Conf., Ithaca, NY 1991.
14. D. L. Barrett, P. G. McMullin, R. G. Seidensticker, R. H. Hopkins, J. A. Spitznagel, Proc. of First International High Temperature Electronics Conference, Albuquerque, NM, pp. 180–185 (1991).
15. S. T. Sheppard, M. R. Melloch, and J. A. Cooper, IEEE Trans. Electron Devices, vol. 41, no. 7, pp. 1257–1264 (1994).
16. V. Krishnamurthy, D. M. Brown, M. Ghezzo, J. Kretchmer, W. Hennessy, E. Downey, and G. Michon, in Proc. 5th Int. Conf. on SiC and Related Compounds, Washington, DC, Nov. 1–3, 1993, Eds. M. G. Spencer, R. P. Devaty, J. A. Edmond, M. A. Khan, R. Kaplan, and M. Rahman, IOP Publishing, Bristol, UK, No. 137, pp. 483–486 (1994).
17. C. E. Weitzel, J. W. Palmour, C. H. Carter, Jr., K. Moore, K. J. Nordquist, S. Allen, C. Thero, and M. Bhatnagar, Silicon carbide high power devices, IEEE Trans. Electron Devices, vol. 43, no. 10, pp. 1754–1759 (1996).
18. M. R. Melloch and J. A. Cooper, Jr., MRS Bulletin, vol. 22, pp. 42–47 (1997).
19. M. R. Melloch and J. A. Cooper, Journ. Electonic Mat., vol. 24, no. 4, pp. 303–309 (1995).
20. G. Kelner, S. Binari, M. S. Shur, K. Sleger, J. Palmour, and H. Kong, 6H-SiC buried gate JFET, Materials Science and Engineering, B11, no. 1/4, pp. 121–124 (1992).
21. M. Bhatnagar and B. J. Baliga, IEEE Trans. Electron Devices, vol. 40, p. 645 (1993).
22. J. W. Palmour, J. A. Edmond, H. S. Kong, and C. H. Carter, in Proc. 5th Int. Conf. SiC and Related Compounds, Washington, DC, Nov. 1–3, 1993, Eds. M. G. Spencer, R. P. Devaty, J. A. Edmond, M. A. Khan, R. Kaplan, and M. Rahman, IOP Publishing, Bristol, UK, No. 137, pp. 499–502 (1994).
23. see http://www.ecm.purdue.edu/WBG/Index.html.
24. J. N. Shenoy, M. R. Melloch, and J. A. Cooper, IEEE Electron Dev. Lett., vol. 18, pp. 93–95 (1997).
25. J. N. Shenoy, M. R. Melloch, and J. A. Cooper, Jr., "High-Voltage Double-Implanted MOS Power Transistors in 6H-SiC," IEEE Device Research Conf., Santa Barbara, CA, June 24–26, 1996.
26. S. T. Sheppard, M. R. Melloch, and J. A. Cooper, Jr., Experimental demonstration of a buried-channel charge-coupled device in 6H silicon carbide, IEEE Electron Device Letters, vol. 17, no. 1, pp. 4–6 (1996).

27. D. L. Barrett, P. G. McMullin, R. G. Seidensticker, R. H. Hopkins, J. A. Spitznagel, Proc. of First International High Temperature Electronics Conference, Albuquerque, NM, pp. 180–185 (1991).
28. R. C. Clarke, T. W. O'Keefe, P. G. McMullin, T. J. Smith, S. Sriram, T. J. D. L. Barrett, 50th Device Research Conference, Cambridge, MA, June 22–24, 1992.
29. S. Yoshida, K. Endo, E. Sakuma, S. Misawa, H. Okumura, H. Daimon, E. Muneyama, and M. Yamanaka, Proc. of MRS, vol. 97, pp. 259–264 (1987).
30. A. K. Agarwal, G. Augustine, V. Balakrishna, C. D. Brandt, A. A. Burk, Li-Shu Chen, R. C. Clarke, P. M. Esker, H. M. Hobgood, R. H. Hopkins, A. W. Morse, L. B. Rowland, S. Seshadri, R. R. Siergiej, T. J. Smith, Jr., S. Sriram, SiC Electronics, in IEDM Technical Digest, paper 9.1.1, IEEE, San Francisco (1996).
31. G. Kelner, M. S. Shur, S. Binari, K. Sleger, and H. S. Kong. "A High Transconductance 3C-SiC Buried-Gate Junction Field Effect Transistor," in Proceedings of 2nd International Conference on Amorphous and Crystalline Silicon Carbide and Related Materials (ICACSC'88), Santa Clara, pp. 184–190, December (1988).
32. G. Kelner, S. Binari, M. Shur, K. Sleger, J. Palmour, and H. Kong, Electronics Letters, vol. 27, no. 12, pp. 1038–1040 (1991).
33. M. M. Anikin, P. A. Ivanov, A. L. Syrkin, B. V. Tsarenkov, V. E. Chelnokov, Sov. Tech. Phys. Lett., vol. 15, no. 8, pp. 636–637, Aug. (1989).
34. P. G. Neudeck, J. B. Petit, and C. S. Salupo, in Transactions of Second International High Temperature Electronics Conference, vol. 1, pp. 1–6, June 5–10 (1994).
35. M. Shur, Appl. Phys. Lett., vol. 54, no. 2, pp. 162–164 (1989).
36. M. Shur, Proceedings of Advanced Material Concepts Conference, Denver, pp. 432–440, Ed. F. W. Smith, Advanced Materials Institute, Denver (1989).
37. G. Kelner and M. S. Shur, Junction Field Effect Transistor with Lateral Gate Voltage Swing (GVS-JFET), United States Patent 5,309,007, May 1994.
38. J. W. Palmour, H. S. Kong, D. E. Waltz, J. A. Edmond, C. H. Carter, Jr., Proceedings of the First International High Temperature Conference, Eds. B. King and F. V. Thome, p. 511, Sandia National Laboratory, Albuquerque, NM (1991).
39. C. E. Weitzel, J. W. Palmour, C. H. Carter, Jr., K. J. Nordquist, K. Moore, and S. Allen, SiC Microwave Power MESFETs and JFETs, in "Compound Semiconductors 1994," Eds. H. Goronkin and U. Mishra, Inst. Phys. Pub., Bristol, no. 141, pp. 389–394 (1994).
40. S. Sriram, R. C. Clarke, A. A. Burk, Jr., H. M. Hogwood, P. G. MacMillan, P. A. Orphanos, R. R. Siergiej, T. J. Smith, C. D. Brandt, M. C. Driver, and R. H. Hopkins, RF performance of SiC MESFETs on high resistivity substrates, IEEE Electron Device Lett., vol. 15, no. 11, pp. 458–459 (1994).
41. C. E. Weitzel, J. W. Palmour, C. H. Carter, Jr., K. J. Nordquist, IEEE Electron Device Lett., vol. 15, no. 10, pp. 406–408 (1994).
42. K. Moore, DC, RF, and power performance of 6H-SiC JFETS, presented at WOCSEMMAD '95, New Orleans, February 1995; see also [17].
43. J. Nishizawa, T. Terasaki, and J. Shibata, IEEE Trans. Electron Devices, vol. ED-22, p. 185 (1975).
44. L. F. Eastman, R. Stall, D. Woodard, M. S. Shur, and K. Board, Ballistic electron motion in GaAs at room temperature, Electronics Letters, vol. 16, no. 13, p. 524, June (1980).
45. R. C. Clarke, R. R. Siergiej, A. K. Agarwal, C. D. Brandt, A. A. Burk, Jr., A. Morse, and P. A. Orphanos, in 15th Biennial IEEE/Cornell Conference on Advanced Concepts in High Speed Semiconductor Devices and Circuits Dig. Abstracts, August 7–9, 1995, p. 14.
46. Northrop-Grumman group report on the 1 kW multi transistor SIT module at the National Association of Broadcasters convention in Las Vegas, Nevada, April 1996.

47. R. R. Siergiej, S. Sriram, R. C. Clarke, A. K. Agarwal, C. D. Brandt, A. A. Burk, Jr., T. J. Smith, and P. A. Orphanos, High power 4H-SiC static induction transistors, in IEDM Tech. Dig., pp. 353–356 (1995).
48. G. L. Harris, K. Wongchotigul, K. Irvine, and M. G. Spencer, Proceedings of the Third International Conference on Solid State and Integrated Circuit Technology, China, October 18–24, 1992.
49. W. Xie, J. A. Cooper, Jr., and M. R. Melloch, Monolithic NMOS digital integrated circuits in 6H-SiC, IEEE Electron Device Letters, vol. 15, p. 455 (1994).
50. D. B. Slater et al., Trans. 3rd International High Temp. Elec. Conf., vol. 2, pp. XVI-27–XVI-32 (1996).
51. D. B. Slater, Jr., G. M. Johnson, L. A. Lipkin, A. V. Suvorov, and J. W. Palmour, Demonstration of 6H-SiC CMOS technology, 54th Device Research Conf. Dig., pp. 162–163, June (1996).
52. J. W. Palmour, R. Singh, L. A. Lipkin, and D. C. Waltz, Proceedings of 3d High Temperature Electr. Conf., Eds. D. B. King and F. V. Thorne, Sandia National Laboratories, Albuquerque, NM, p. XVI-9J (1996).
53. S. N. Vainshtein, V. A. Dmitriev, A. L. Syrkin, and V. E. Chelnokov, Sov. Phys. Lett., vol. no. 13, pp. 413–414 (1987).
54. V. A. Dmitriev, M. E. Levinshtein, S. N. Vainshtein, and V. E. Chelnokov, Electronics Letters, vol. 24, p. 1031 (1988).
55. J. W. Palmour, S. T. Alen, R. Singh, L. A. Lipkin, and C. H. Carter, in Int. Conf. SiC and Related Materials, 1995 Tech. Dig., Kyoto, Japan, Sep. 18–21, pp. 319–320 (1995).
56. N. Rammungul, T. P. Chow, M. Ghezzo, J. Kretchmer, and W. Hennessy, "A Fully Planarized, 6H-SiC UMOS Insulated-Gate Bipolar Transistor," presented at IEEE Device Research Conference, June 24–26, 1996, University of California, Santa Barbara.
57. J. A. Cooper, Jr., J. W. Palmour, C. T. Gardner, M. R. Melloch, and C. H. Carter, Jr., "Dynamic Charge Storage in 6H-Silicon Carbide: Prospects for High-Speed Nonvolatile RAM's," IEEE Device Research Conf., Boston, June 22–24, 1992.
58. J. A. Cooper, Jr., M. R. Melloch, W. Xie, J. W. Palmour, and C. H. Carter, Jr., "Progress and Prospects for Nonvolatile Memory Development in Silicon Carbide." Inst. Phys. Conf. Ser. no. 137, ch. 7, pp. 711–714 (1993).
59. W. Xie, J. A. Cooper, Jr., M. R. Melloch, J. W. Palmour, and C. H. Carter, Jr., A vertically integrated bipolar storage cell in 6H silicon carbide for nonvolatile memory applications, IEEE Electron Device Lett., 15, 212 (1994).
60. Y. Wang, J. A. Cooper, Jr., M. R. Melloch, S. T. Sheppard, J. W. Palmour, and L. A. Lipkin, Experimental characterization of electron-hole generation in silicon carbide, J. Elec. Materials, 25, 899 (1996).
61. P. Rabkin, R. Cottle, P. A. Blakey, and M. S. Shur, "2D Simulation of DC, AC, and Breakdown Characteristics of Bipolar and Unipolar Silicon Carbide Devices," in Proceedings of International Semiconductor Device Research Symposium, Charlottesville, VA, Dec. 1–3, pp. 569–572 (1993).
62. P. Rabkin, M. S. Shur, and P. Blakey, DC and AC Analysis of a Silicon Carbide HJFET, Simulation Standards, vol. 3, no. 5, pp. 2–3, October (1992).
63. G. Kelner, M. S. Shur, S. Binari, K. J. Sleger, and H. Kong, High transconductance 3C-SiC buried-gate JFET's, IEEE Trans. Electron Devices, vol. 36, no. 6, pp. 1045–1049 (1989).
64. D. Scharfetter and H. Gummel, Large-signal analysis of a silicon read diode oscillator, IEEE Trans. Electron Devices, vol. ED-16, pp. 64–67, January (1969).
65. R. F. Davis, G. Kelner, M. S. Shur, J. W. Palmour, and J. A. Edmond, Thin film deposition and microelectronic and optoelectronic device fabrication and characterization in micro-

crystalline alpha and beta silicon carbide, Proceedings of the IEEE, vol. 79, no. 5, pp. 677–701 (1991).
66. K. Lee, M. S. Shur, T. A. Fjeldly, and T. Ytterdal, "Semiconductor Device Modeling for VLSI," Prentice Hall, Englewood Cliffs, NJ (1993).
67. T. Fjeldly, T. Ytterdal, and M. S. Shur, "Introduction to Device and Circuit Modeling for VLSI," Wiley, 1998.
68. K. Park, C. Y. Lee, K. R. Lee, B. J. Moon, Y. Byun, and M. Shur, A unified charge control model for long channel n-MOSFETs, IEEE Trans. on Electron Devices, vol. 38, no. 2, pp. 399–406, Feb. (1991).
69. J. E. Meyer, MOS models and circuit simulation, RCA Review, vol. 32, pp. 42–63 (1971).
70. N. Arora, "MOSFET Models for VLSI Circuit Simulation: Theory and Practices," Springer-Verlag, Wien, New York (1993).
71. D. E. Ward, Charge based modeling of capacitance in MOS transistors, Ph.D. thesis, Stanford University (1981).
72. D. E. Ward and R. W. Dutton, IEEE J. Solid-State Circuits, SC-13, p. 703 (1978).
73. O. G. Johannessen, T. A. Fjeldly and T. Ytterdal, Unified capacitance modeling of MOSFETs, Physica Scripta, T54, pp. 128–130 (1994).
74. V. F. Tsvetkov, S. T. Allen, H. S. Kong, and C. H. Carter, in Int. Conf. SiC and Related Materials, 1995 Tech. Dig., Kyoto, Japan, Sep. 18–21, pp. 11–12 (1995).
75. R. J. Trew, J. Yan, and P. M. Mock, Proc. IEEE, vol. 79, no. 5, pp. 598–620 (1991).
76. G. L. Bibro, M. W. Shin, R. J. Trew, and R. C. Clarke, in Proc. 5th Int. Conf. on SiC and Related Compounds, Washington, DC, Nov. 1–3, 1993, Eds. M. G. Spencer, R. P. Devaty, J. A. Edmond, M. A. Khan, R. Kaplan, and M. Rahman, IOP Publishing, Bristol, UK, No. 137, pp. 699 (1994).
77. M. A. Khan, Q. Chen, M. S. Shur, B. T. Dermott, J. A. Higgins, J. Burm, W. Schaff, and L. F. Eastman, CW operation of short channel GaN/AlGaN doped channel heterostructure field effect transistors at 10 GHz and 15 GHz, IEEE Electron Device Lett., vol. 17, no. 12, pp. 584–585, Dec. (1996).
78. S. Kobayashi, T. Kimoto, and H. Matsunami, Jpn. J. Appl. Phys., vol. 35, Part 1, no. 6A, pp. 3331–333, June (1996).
79. D. A. Grant and J. Gowar, "Power MOSFETs," John Wiley and Sons, New York (1989).
80. J. A. Edmond, J. W. Palmour, and C. H. Carter, Jr., in Proc. Intern. Semicond. Device Research Symposium, Charlottesville, VA, December 1991.
81. Michael S. Shur and M. Asif Khan, "Wide Band Gap Semiconductors," Good Results and Great Expectations," in Proceedings of 22nd International Symposium on GaAs and Related Compounds, St. Petersburg, Russia, Sep. 22–28, 1996, Institute Conference Series, pp. 25–32, M. S. Shur and R. Suris, Eds. IOP Publishing, London (1997).
82. O. Kordina, J. P. Bergman, C. Hallin, and E. Janzen, Appl. Phys. Lett., vol. 69, no. 5, pp. 679–681 (1996).
83. G. L. Harris, Ed., SiC devices and ohmic contacts, in "Properties of Silicon Carbide," INSPEC, IEE, United Kingdom (1995).
84. M. E. Levinshtein, S. Rumyantsev, and M. S. Shur, Eds. "Handbook of Semiconductor Material Parameters," vol. 1, World Scientific (1996).
85. M. E. Levinshtein, S. Rumyantsev, and M. S. Shur, Eds., "Handbook of Semiconductor Material Parameters," vol. 2, World Scientific, 1998, to be published.
86. V. A. Dmitriev, K. G. Irvine, C. H. Carter, Jr., N. I. Kuznetsov, and E. V. Kalinina, Electric breakdown in GaN pn junctions, Appl. Phys. Lett., 68, pp. 229–231 (1996).
87. V. E. Chelnokov, A. L. Syrkin, and V. A. Dmitriev, "Overview of SiC Power Electronics," in Proceedings of The First ECSCRM, Greece, 1996.

88. B. J. Baliga, Trends in power semiconductor devices, IEEE Trans. Electron Devices, vol. 43, no. 10, pp. 1717–1731, Oct. (1996).
89. L. A. Lipkin and J. W. Palmour, J. Electronic Materials, vol. 25, p. 909 (1996).
90. J. W. Palmour, C. E. Weitzel, K. J. Nordquist, and C. H. Carter, Jr., SiC microwave power FETs, Inst. Phys. Pub., no. 137, pp. 495–498 (1994).
91. J. W. Palmour, Private communication (1997).
92. J. W. Palmour, J. A. Edmond, H. S. Kong, and C. H. Carter, Jr., Physica B, vol. 185, pp. 461–465 (1993).
93. K. Xie, J. H. Zhao, J. R. Flemish, K. Burke, W. R. Buchwald, G. Lorenzo, and H. Singh, IEEE Electron Device Lett. vol. 17, no. 3, pp. 142–144 (1996).

CHAPTER 5

SiC for Applications in High-Power Electronics

C. D. Brandt, R. C. Clarke, R. R. Siergiej, J. B. Casady, S. Sriram, A. K. Agarwal

NORTHROP GRUMMAN SCIENCE AND TECHNOLOGY CENTER
PITTSBURGH, PA

A. W. Morse

NORTHROP GRUMMAN ELECTRONIC SENSORS AND SYSTEMS DIVISION
BALTIMORE, MD

I.	INTRODUCTION	195
II.	BULK SiC GROWTH	198
III.	EPITAXIAL GROWTH OF SiC	202
IV.	ADVANTAGES OF SiC FOR HIGH-POWER RF SYSTEMS	204
V.	THE SiC MESFET: DESIGN CONSIDERATIONS	205
VI.	MESFET FABRICATION	207
VII.	6H-SiC MESFET RESULTS	209
VIII.	4H-SiC MESFET RESULTS	211
IX.	THE SiC SIT: DESIGN CONSIDERATIONS	212
X.	SIT FABRICATION	215
XI.	6H-SiC SIT RESULTS	217
XII.	4H-SiC SIT RESULTS	219
XIII.	450 W UHF SIT	221
XIV.	2.0 kW UHF MODULE	222
XV.	S-BAND SiC SITs	223
XVI.	S-BAND SIT DEVICE SCALE-UP	224
XVII.	SiC POWER SWITCHING DEVICES	226
XVIII.	CONCLUSIONS	231
	References	232

I. Introduction

Most traditional integrated circuit technologies using silicon devices are not able to operate at temperatures above 250°C, especially when high operating temperatures are combined with high-power, high-frequency, and high-radiation environments. Much attention has been given to silicon carbide (SiC), currently the most mature of the wide-bandgap ($2.0\,\text{eV} < E_g < 7.0\,\text{eV}$) semiconductors, as a material well-suited for high-temperature and efficient high-power operation. High-temperature circuit operation from 350 to 500°C is desired for use in aerospace applications (turbine engines and the more electric aircraft initiative), nuclear power instrumentation, satellites, space exploration, and geothermal wells [1–6].

As an example of one such application, the more electric aircraft concept offers substantial system-level benefits that are similar to the more electric tank, more electric ship, and more electric automobile concepts. By replacing bulky hydraulic-driven flight control actuators and engine-gearbox–driven fuel pumps with power electronics capable of high efficiency and high operating temperatures, tremendous weight savings can be induced. Currently, hydraulic-based systems also provide cooling to aircraft electronics, so the removal of the hydraulic systems will increase the operating temperature for on-board electronics. Thus, a distributed flight control system, which would allow the elimination of cooling systems, is envisioned using wide-bandgap devices [2]. Over 90% of closed-loop environmental control system cooling requirements on modern fighter aircraft are utilized for the cooling of electronics (including radar electronics). Reducing or eliminating this need for cooling would reduce aircraft weight, maintenance support, down time, and ground supplies, while increasing the aircraft's performance, flying range, efficiency, and reliability. Other advantages of distributed flight control include elimination of long and heavy wiring–shielding runs to actuators and sensors, reduction of the number of unreliable connector pins between control electronics and sensor–actuators, and a decreased chance of catastrophic failure, since the control electronics would be located in multiple positions as compared to one centralized location [2]. A pictorial summary of the more electric aircraft concepts is shown in Fig. 1.

In addition to these applications, SiC has potential for use in numerous other high power, high-frequency, and radiation-resistant applications [7–9]. SiC, aluminum nitride (AlN), gallium nitride (GaN), boron nitride (BN), diamond, and zinc selenide (ZnSe) are the primary wide-bandgap semiconductors now being developed for use in the aforementioned applications. Of these, SiC presently has several advantages, including commercial availability of substrates since ~1991 [e.g., 10, 11], known device processing techniques, and the ability to grow a thermal oxide for use as masks in processing, device passivation layers, and gate dielectrics (Tab. 1). In addition, SiC's high thermal conductivity (about 3.3 times that of Si at 300 K for 6H-SiC), high electric field breakdown strength (about 10 times that of Si for 6H-SiC), and wide bandgap (about 3 times that of Si for

5 SiC for Applications in High-Power Electronics

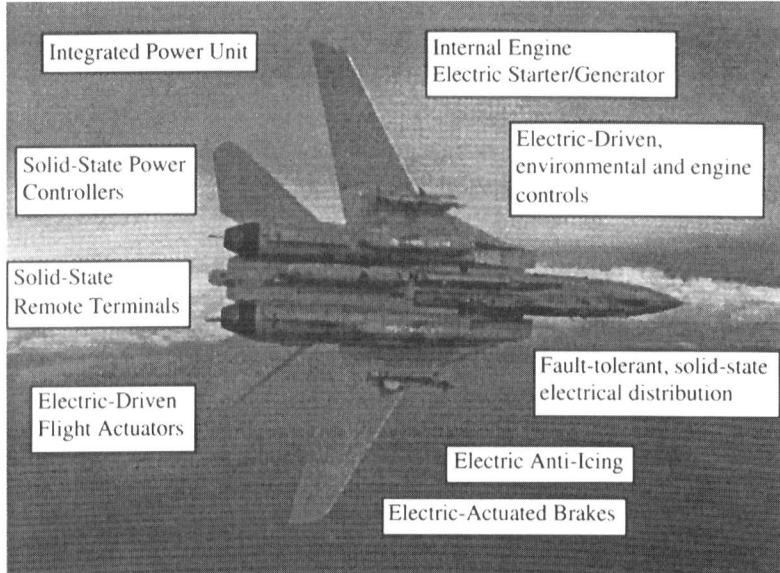

FIG. 1. Conceptual summary of the more electric aircraft concepts applied to a current-generation fighter aircraft.

TABLE I

Electronic and Physical Properties of Si, GaAs, GaN, and SiC

Material Property	Si	GaAs	GaN	6H-SiC	4H-SiC
Saturated electron velocity (10^7 cm/s)	1	2.0 peak 1.2 sat	2.5 peak 1.5 sat.	2	2
Breakdown field (MV/cm)	0.25	0.3	~3	~3	~3
Thermal conductivity (W/cm-K)	1.5	0.5	1.3 (on sapphire)	4.9	4.9
Bandgap (eV)	1.12	1.43	3.4	2.9	3.25
Electron mobility @ $N_d = 10^{16}$ cm^{-3} (cm^2/V·s)	1200	6500	900	400 ($\perp$ to c-axis) 60 ($\parallel$ to c-axis)	800 ($\perp$ to c-axis) 800 ($\parallel$ to c-axis)
Dielectric constant	11.8	12.8	9	9.7	9.7

4H-SiC and 6H-SiC) make it a material ideally suited for high-temperature, high-power, high-frequency, and high-radiation environments.

6H-SiC and 4H-SiC were the first polytypes of SiC to appear in bulk wafer form [10, 11], which has helped SiC to emerge as one of the relatively mature wide-bandgap semiconductor technologies. Lattice mismatches of only 1% for AlN [12] and 3% for GaN [13, 14] exist when these materials are grown on 6H-SiC substrates. Thus, SiC processing is often intimately linked with AlN and GaN electronic and optical device fabrication as well. For example, commercially available GaN blue (peak wavelength of 430 nm) LEDs manufactured on 6H-SiC substrates were released in 1995 [10]. The small lattice mismatches with AlN and GaN, as well as the abundance of polytypes in SiC, combine to make SiC a material with an immense potential for use in heterostructure electronic devices, which take advantage of the differing bandgaps, carrier mobilities and so on.

II. Bulk SiC Growth

Historically, bulk growth of SiC has been perhaps the most significant problem limiting the usefulness of SiC in electronic applications [15–31]. Single-crystal wafers of 6H-SiC have been available commercially only since 1991 (from Cree Research, Inc. [10]), and 4H-SiC wafers have been available only since 1994 (from Cree in 1994 and ATMI in 1995 [10, 11]). 3C-SiC wafers [32] and 15R-SiC [11] wafers have been produced in research environments, but not commercially. An excellent review of commercial SiC boule growth by seeded sublimation is given by Tsvetkov *et al.* [15]. Single-crystal boules grown from either a melt or solution would require excessive temperatures ($>3200°C$) and very high pressures ($>100,000$ atm), which precludes this growth method from being utilized. Thus, in the absence of other viable melt-growth techniques, physical vapor deposition via seeded sublimation is the most commonly used approach. The vapor phase of SiC (typically Si, Si_2C, SiC_2) is deposited upon an SiC seed crystal at high temperatures ($>2000°C$). Examples of a 4H-SiC boule grown via the sublimation process is shown in Fig. 2, and example wafers shown after slicing and polishing are shown in Fig. 3. Typical wafer diameters are 35 mm, although Cree Research is currently offering the sale of 2-in. (~ 50 mm) diameter wafers, beginning in 1998.

One major problem with the sublimation growth technique has been the formation of micropipe defects. Although defects such as micropipes are not found in Lely platelets of SiC, this type of growth results in irregular-shaped

FIG. 2. Polished 38-mm-diameter, ⟨0001⟩-oriented 4H-SiC boule grown via seeded sublimation technology shown prior to slicing. (Photo courtesy of Northrop Grumman.)

SiC substrates, which are unsuitable for commercial SiC device production [22, 27]. Micropipes are bulk defects (voids) that propagate the length of the boule from the seed crystal, and are also found to propagate through subsequent epitaxially grown SiC layers. Micropipes have hexagonal cross-sections with diameters from about 0.1 to 5 µm [15, 18]. Mechanisms causing the micropipes have not been clearly identified in the literature, but 13 possible thermodynamic, kinetic, and technological mechanisms have been identified [15]. A good discussion of various defects (hexagonal pits, micropipes, screw dislocations, hillocks, etc.) and possible causes is found in [28]. In physical vapor transport–grown 6H-SiC substrates, all micropipe defects were positioned along the lines of superscrew dislocations, with a Burgers vector of at least four times that of the c-axis lattice constant [18]. Micropipe defect density (MDD), found in densities of $1000/cm^2$ in the early 1990s, were reported as reduced to $3.5/cm^2$ at the research level on a 30-nm (1.18-in.) 4H-SiC wafer in 1995 [15], although typical commercial wafers currently possess MDDs ranging from 50 to $200/cm^2$. One fact often overlooked by device engineers is that the MDD is nonuniform and often locally clustered on the wafer. The nonuniform density is a real benefit to device engineers seeking large-area regions for power devices or complex circuits, since 1-cm^2 areas on wafers have often been found with zero micropipes [16]. Figure 4 illustrates an example of the nonuniform MDD found on a typical 4H-SiC substrate, similar to that reported in the

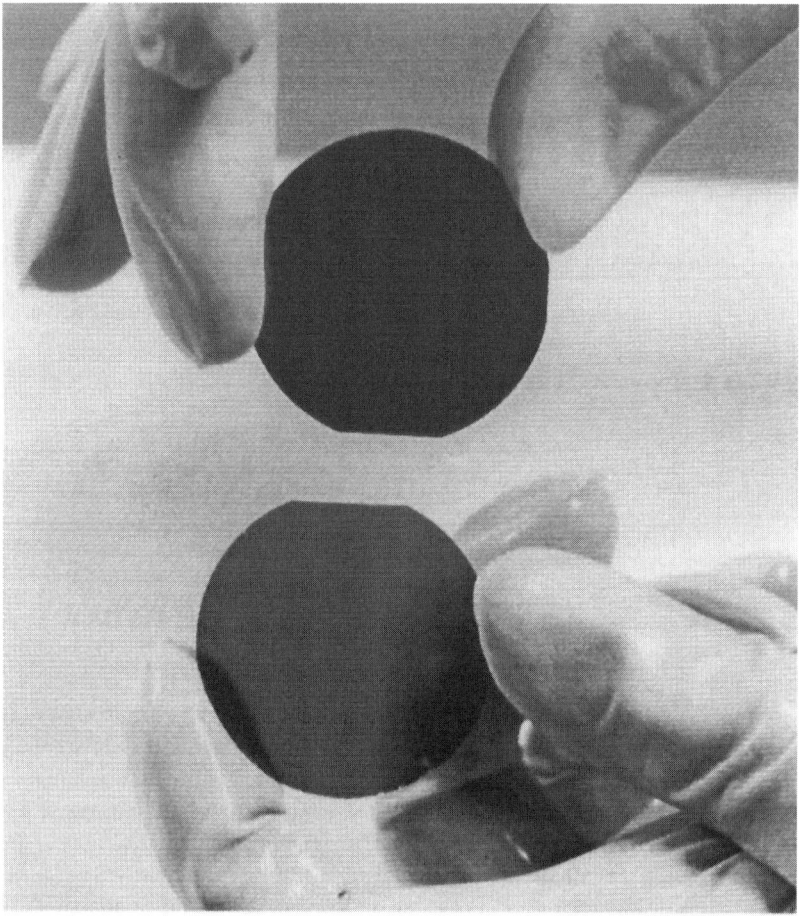

FIG. 3. Examples of sliced and polished SiC substrates. Note that SiC is optically transparent and that substrate color is dependent on doping type, concentration, and material polytype. (Photo courtesy of Northrop Grumman.)

literature [16]. While some areas are virtually defect-free, other areas have a large density of micropipe defects on the same wafer. Thus, quotes of average MDD across a wafer may not give a true representation of the substrate quality. Elimination of the micropipes found in bulk SiC is a critical issue for development of SiC power devices and larger area integrated circuits [17].

It should be noted that at least three different corporations in the United States (Northrop Grumman, Cree Research, and ATMI) are currently

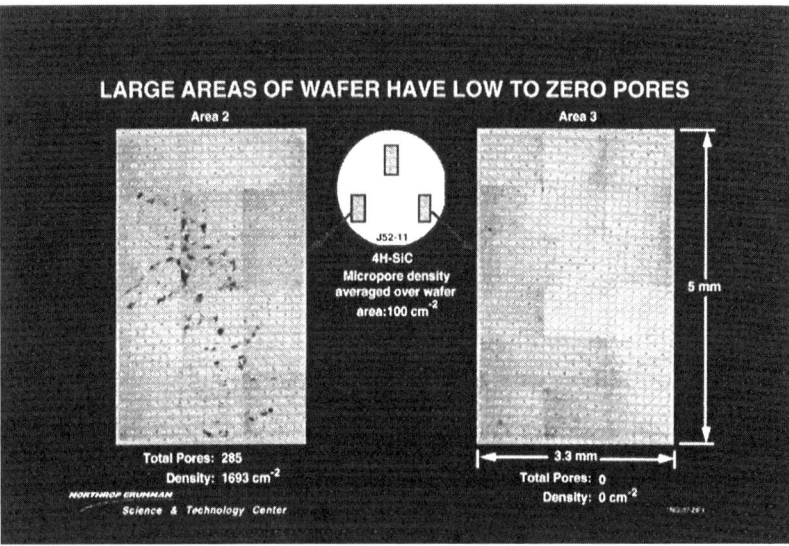

FIG. 4. Example of the nonuniform MDD on a typical 4H-SiC substrate. (Photo courtesy of Northrop Grumman.)

producing SiC wafers via seeded sublimation, with other companies in Russia, Japan, and Europe also producing wafers. A brief, noninclusive listing of other outstanding references to bulk growth of SiC are listed for the interested reader [15–32, 35]. Work by Hobgood *et al.* [19], Barrett *et al.* [24, 27], and Augustine *et al.* [16] provide superb discussions of SiC bulk growth using a sublimation-source physical vapor transport (PVT) system at Northrop Grumman, with results comparable to that of Cree's. Notable achievements include production of 6H-SiC boules up to 60 mm (2.36 in.) in diameter. It is estimated that 100-mm, high-quality wafers of reasonable cost will be required for high-power commercial SiC device production, while 50 to 75-mm wafers should suffice for low-power commercial products [15]. Growth was done at $\sim 2300°C$, while the oriented SiC seed crystal was held at a lower temperature ($\sim 2200°C$). The major crystalline defects reported in the 4H-SiC substrates grown by physical vapor transport were micropipes ($10\,\text{cm}^{-2}$ on best wafers) and dislocations ($10^4\,\text{cm}^{-2}$ range). Room-temperature electrical conductivity of the substrates could be varied from less than $1 \times 10^{-2}\,\Omega\text{-cm}$, n-type, to insulating ($> 10^{15}\,\Omega\text{-cm}$). Other research has focused on the use of tantalum-coated crucibles and containers for sublimation growth in place of the traditional graphite crucibles. For example, growth rates of 1.5 mm/hr have been achieved using Ta container material, with growth temperatures ranging from 1600 to 2100°C [29]. When comparing Ta and graphite crucibles for growth (using polycrystalline SiC source material and (0001) 6H-SiC Lely-grown platelets as seed), it was found that at low temperature gradients (<30 K/cm) the growth rate of material in Ta containers exceeded that of material grown in graphite containers. Of extreme importance for microwave applications is the development of semi-insulating substrates [30, 31]. Single-crystal 6H-SiC substrates grown by PVT have demonstrated high-resistivity (1–30 k$\Omega \cdot$cm), semi-insulating (>100 k$\Omega \cdot$cm), and even insulating (10^{11}–10^{12} k$\Omega \cdot$cm) type behavior. Achieving semi-insulating 6H-SiC has been accomplished by using undoped and vanadium-compensated substrates.

The resistivity of the undoped 6H-SiC semi-insulating substrates has a strong temperature dependence dominated by a single activation (ionization) energy of ~ 0.35 eV, identified as the B acceptor level [31]. The temperature dependence of the vanadium-compensated 6H-SiC material is also strong, but more complex, with at least three activation energies identified resulting from residual boron and the complex behavior of vanadium. Vanadium has been proposed as both a donor ($E_D = 1.35$ eV) and an acceptor ($E_A = 0.8$ eV). These temperature dependencies should obviously be taken into account when examining limits to high-temperature operation of high-frequency SiC devices fabricated on semi-insulating SiC substrates.

III. Epitaxial Growth of SiC

Doping in SiC for device fabrication is accomplished via epitaxially controlled doping and hot ion implantation. Temperatures required for diffusion are too high (>1800°C) for standard device processing because of the very high bond strength possessed by SiC. The two most common dopants used in SiC are nitrogen (n-type) and aluminum (p-type), as discussed previously, although boron is sometimes used for p-type implants. In the absence of diffusion, epitaxial and ion-implanted control of dopants is critical for the development of devices and integrated circuits. Numerous high-quality publications on epitaxial growth processes exist [e.g., 33, 34, 36, 37–52]. Silane and propane are typical source gases of Si and C, respectively. Typical growth rates for 6H-SiC homoepitaxy layers on Si-face n-type substrates are $\sim 3\,\mu$m/hr. As an example, a prototype horizontal-flow epitaxial growth reactor used at Northrop Grumman Science and Technology Center for homoepitaxial 4H-SiC and 6H-SiC growth is shown in Fig. 5. Increasing the growth rate while maintaining polytype control, good surface morphology, uniform thickness, and accurate control of dopant

FIG. 5. A horizontal vapor-phase epitaxy growth system used to grow SiC epitaxy at rates of $\sim 3\,\mu$m/hr at temperatures of 1500 to 1600°C. (Photo courtesy of Northrop Grumman.)

levels is critical in producing thick (50–100 μm) blocking layers necessary for high-voltage (>5 kV) power devices. Growth rates of up to 6 μm/hr have been demonstrated in vertical low-pressure chemical vapor disposition (CVD) systems manufactured by EMCORE, Inc [49]. In this system, high-speed rotation of the wafer carrier was employed to stabilize the gas flow in the reactor at a growth temperature of ~ 1500°C. The resulting epitaxial layers were characterized to have breakdown strengths of 2 MV/cm, electron mobility of 700 cm^2/V·s in lightly doped 4H-SiC epitaxy, and a background concentration of approximately 2×10^{14} cm^{-3} (boron). Doping levels were controlled over a three-order-of-magnitude range. Other work using hot-wall CVD reactors has produced impressive results, with 45-μm-thick epitaxial layers grown and used in 100-μm diameter 4.5-kV diodes [51].

For homoepitaxial growth of 6H-SiC, the substrates used are normally Si-face, 3.5-degree off-axis, while for homoepitaxial growth of 4H-SiC, Si-face 8-degree off-axis substrates are generally used. The reason for the difference in off-axis angle for the different polytypes is based on the greater step height in 4H-SiC. Using on-axis or nonoptimized off-axis wafers or nonoptimized growth conditions for growth of epitaxy results in the formation of triangular morphological defects caused by inclusions of 3C-SiC, which interrupt the step-flow mode of growth. These defects have been shown to cause high leakage currents in diodes. These and other surface morphology defects, such as Si-droplet formation, faceting, and hillocks, are reported in greater depth elsewhere [15, 50].

Nitrogen and aluminum doping have been investigated for many years; however, residual doping levels were too high for many devices. The discovery of the site-competition effect [40] for Si-face SiC epitaxy enabled

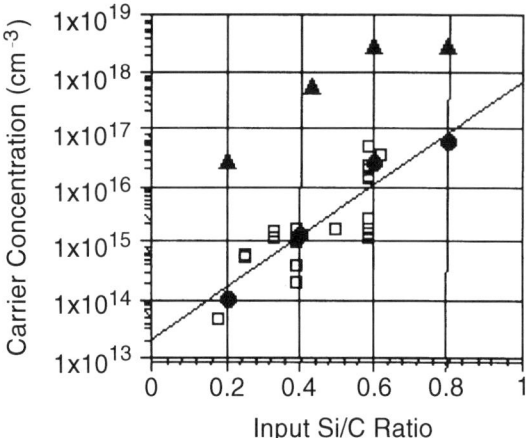

FIG. 6. Control of doping demonstrated by varying the Si/C gas ratio. (Taken from [75].)

reduction of residual doping levels to $10^{14}\,\text{cm}^{-3}$ and intentional incorporation over the entire range of possible concentrations. An example of controlling the nitrogen doping of SiC via gas Si/C gas ratio during growth is seen in Fig. 6 [52] for the same reactor pictured in Fig. 5. This discovery has opened the way for much of the device results shown in subsequent sections. The site-competition effect works by adjusting the Si/C source gas ratio in the growth reactor to control the amount of dopant incorporated into substitutional SiC crystal lattice sites [40]. The model is based upon N and C competition for C sites, with Al and Si competition for Si sites in the SiC lattice. This effect has also been observed for boron and phosphorous doping of SiC [38]. Growth on C-face substrates has behaved quite differently, and more work is needed to fully understand all growth mechanisms [38] and tie together behavior on both faces. Numerous industrial and university laboratories now produce homoepitaxial, device-quality growth of SiC.

IV. Advantages of SiC for High-Power RF Systems

SiC devices are emerging as very promising candidates for high-power microwave devices due the unique combination of high saturated electron velocity (2×10^7 cm/sec), high electric field breakdown strength (2.5×10^6 V/cm), and high thermal conductivity (4.9 W/cm-K). In addition, the wide bandgap of SiC allows high-power RF operation at high temperatures.

Significant progress has been achieved in recent years, and SiC metal semiconductor field-effect transistors (MESFETs) with cut-off frequencies of 42 GHz and power densities as high as 3.3 W/mm have been demonstrated. The power density is more than three times higher than that normally obtained with current GaAs technology. Similarly, static induction transistors (SIT) have also been demonstrated from VHF through S-bands. Power densities 2.5 times higher than silicon at UHF frequencies have been experimentally verified. In this section, we first review the advantages of SiC for high-power RF devices. We then describe the operating principles and the current state of the art for both SITs and MESFETs.

The advantages of SiC for high-power microwave devices arise due to its favorable electronic and thermal properties, as illustrated in Table I. The key parameters that are particularly important in high-power microwave devices are high breakdown strength, high saturated electron velocity, and high thermal conductivity. The high breakdown strength allows SiC devices to be operated at much higher voltages than are possible with Si or GaAs. At the same time, the high saturated velocity of SiC facilitates high device current density. The combination of high voltage and high current density results in very high power densities for SiC devices, an important consideration in high-power microwave devices, since the device size is limited to a fraction of the wavelength of operation. The high power density is also accessible in practical devices because of the thermal characteristics of the material. The high thermal conductivity of SiC provides superior conduction of the waste heat generated by the high power density, and its wide bandgap allows these devices to be operated at higher temperatures than are possible with Si or GaAs.

In addition to improving the maximum power output, the beneficial electronic properties of SiC also translate directly to important advantages in microwave circuit and system design. For example, the high-voltage operation of SiC raises the operating impedance closer to $50\,\Omega$ required in microwave circuits. The resulting simpler impedance transformation leads to wider bandwidth, to lower cost due to simpler circuits, and also to higher efficiency due to the lower losses in the matching circuits. The high-voltage, high-power operation of SiC devices also leads to considerable simplification in RF system design, resulting in lighter, lower cost, and more efficient systems.

V. The SiC MESFET: Design Considerations

The MESFET was first proposed by Mead in 1966 [53] and has been widely used in GaAs for both microwave and high-speed applications. The operation of SiC MESFETs is very similar to that of its GaAs counterparts.

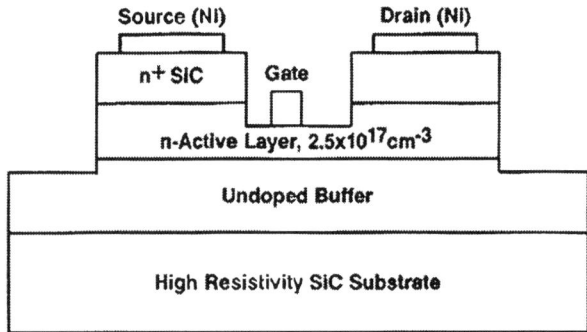

FIG. 7. The cross-section of an SiC MESFET.

Figure 7 shows the schematic cross-section of an SiC MESFET structure that consists of an n-type channel layer usually grown by epitaxy on top of a p-type or high-resistivity substrate. For microwave devices, the n-channel layer is typically 100 to 300 nm thick and is doped in the range of 10^{17} to $10^{18}/cm^3$. The n^+ layer on top of the channel layer is included to achieve low-resistance ohmic contacts, an important requirement in microwave devices. During operation, the current flowing between the source and drain ohmic contacts through the n-channel layer is controlled by the voltage on the Schottky gate.

The microwave performance of MESFETs is strongly influenced by the material properties, such as electron mobility, saturated electron velocity, the field at which velocity saturation is reached, the ionization efficiency of donors, and the electric field breakdown strength in SiC. Even though the electron mobility in SiC is much lower than in GaAs, good microwave performance can be obtained with submicron gate devices, which operate at high electric fields where saturation velocity is important. In these devices, the high saturated electron velocity in SiC (2×10^7 cm/sec), which is nearly twice that of GaAs, is an important advantage.

Drain-gate breakdown is another important factor that limits the maximum RF power output obtainable from a MESFET. The breakdown strength of SiC is nearly eight times higher than that of GaAs and enables breakdown voltages in excess of 100 V for SiC MESFETs, compared to about 20 V normally observed in GaAs-based devices. In addition to improving the power output, the high voltage capability allows SiC MESFETs to be operated at high impedance levels, leading to ease of impedance matching and improved RF performance of high-power microwave circuits.

Another parameter that impacts the microwave performance of SiC MESFETs is the relatively deep ionization energy of the commonly used nitrogen shallow donors in SiC [54]. This leads to incomplete ionization of the donors in the FET channel, resulting in increased parasitic resistances and lower drain currents for a given channel-doping level. For example, at a doping level of $5 \times 10^{17}/cm^3$ in 4H-SiC, the electron density can be calculated to be $2.8 \times 10^{17}/cm^3$. However, it may be pointed out here that such calculations are for thermodynamic equilibrium, whereas the substantial electric field present in the FET channel during its operation may lead to possible field-enhanced ionization of the donors. To the best of our knowledge, such effects are not well understood at the present time, and further work is neccessary to properly model the FET behavior.

Availability of high-resistivity or semi-insulating SiC substrates is another critical requirement to achieve good microwave performance. When MESFETs are fabricated on low-resistivity substrates, part of the RF power at the output is dissipated in the substrate, leading to lower gain and degraded RF performance.

VI. MESFET Fabrication

The fabrication process for SiC MESFETs is quite similar to that of its GaAs counterparts. Figure 8 shows a schematic of the SiC MESFET fabrication sequence. The first step in this process is mesa isolation, created by reactive ion etching (RIE) with Al as the mask. The channel region is next formed by etching the n^+ layer and part of the active layer using RIE. This is a critical step in the process, since small changes in the thickness of the channel under the gate will degrade the device performance due either to lowered device currents or to low breakdown voltages. After completion of the recess, ohmic contacts are formed using rapid thermal annealing (RTA) of nickel. Contact resistances of less than $3 \times 10^{-6}\,\Omega\text{-cm}^2$ are routinely obtained with this procedure. Submicron gates are defined next, usually by electron-beam lithography and lift-off procedures. A variety of metals, including Cr, Ni, Pt, Ti, and Au, can be used for the gates. In all cases, a thick top layer of gold should be added to reduce the gate resistance, an important consideration for improving gain in microwave devices. The remaining steps in the process are formation of thick metal for the source, drain, and pad regions of the device and interconnection of source fingers using plated gold air bridges similar to that used with GaAs MESFETs. Figure 9 shows a photograph of a processed 1.5-in. SiC wafer. A scanning electron microscopy (SEM)-image of an air-bridged 2-mm MESFET showing well-defined air bridges and gates can be seen in Fig. 10.

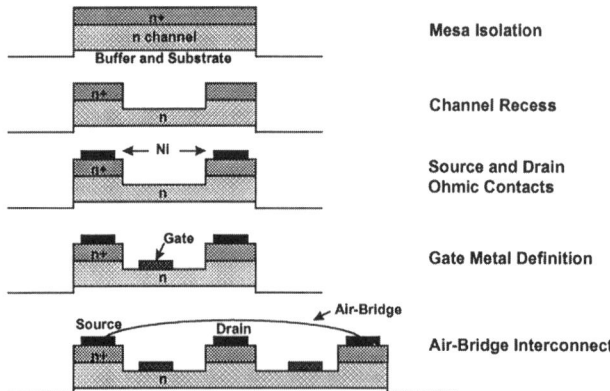

FIG. 8. The SiC MESFET fabrication sequence features mesa isolation and air-bridge interconnects.

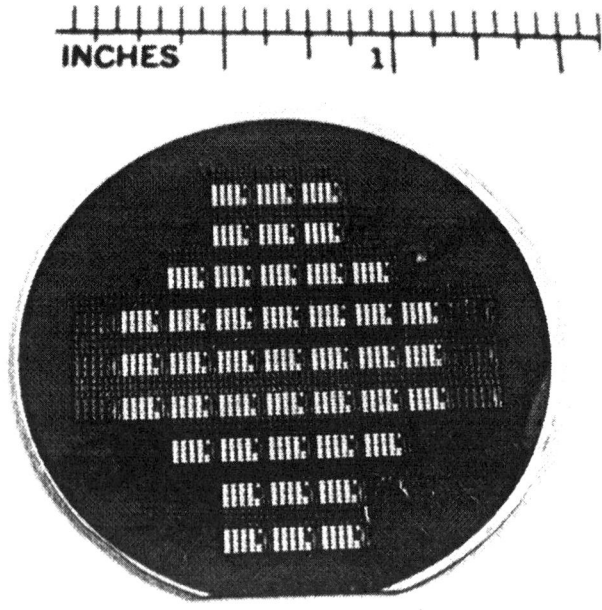

FIG. 9. A fully processed 1.5-in. diameter SiC wafer of air-bridged MESFETs.

FIG. 10. A fully air-bridged 6H-SiC power MESFET having gate lengths of 0.5 μm and a total gate periphery of approximately 2 mm.

VII. 6H-SiC MESFET Results

SiC MESFETs were first fabricated on 6H-substrates due to the wide availability of this material. The substrates were p-type, with resistivities usually less than 10 ohm-cm. The f_t and f_{max} for such devices with 0.7-μm gate lengths were 5.7 and 10.0 GHz, respectively [55]. Substantial improvement in RF performance was obtained by fabricating devices on high-resistivity substrates [56, 57]. The substrates used exhibited as-grown resistivities in the 1500 to 2000-Ω-cm range. The fabricated devices had a gate length of 0.5 μm and a gate width of 320 μm. The drain-gate breakdown for these devices exceeded 70 V, making them suitable for RF power operation. The f_t and f_{max} for these devices were 10 and 25 GHz, respectively (6H-MESFET curve in Fig. 11). This is the highest f_{max} reported to date for 6H-SiC MESFETs. These devices also showed small-signal RF gain of 8.5 dB at 10 GHz, making them suitable for X-band operation. It is noteworthy that these RF results were obtained at a high drain bias voltage of 40 V, which clearly demonstrates the simultaneously high voltage and frequency capabilities of SiC devices.

6H-SiC MESFETs fabricated on high-resistivity substrates have also shown excellent RF power performance [58, 59]. At 6 GHz, a power output of 3.5 W, with 45.5% power-added efficiency (PAE), was demonstrated for a 2-mm periphery 6H-SiC MESFET operated with a drain bias voltage of 40 V (Fig. 12). The corresponding power density is 1.75 W/mm and is more

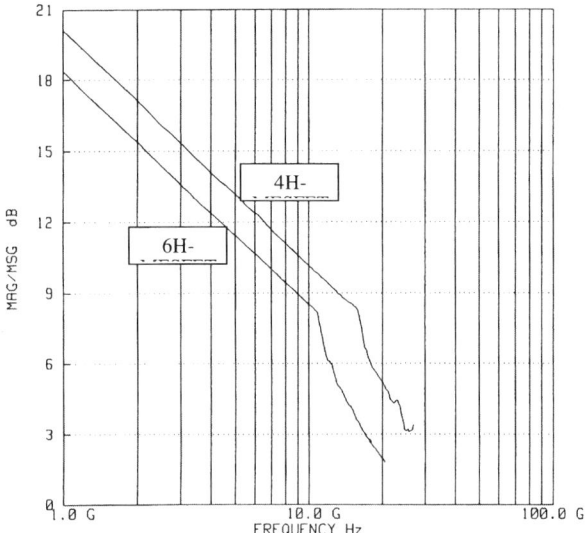

FIG. 11. The comparative advantage of a 4H-SiC MESFET over a 6H-SiC is seen in this small-signal result. The f_{max} values for the 4H and 6H MESFETs are 25 and 42 GHz, respectively.

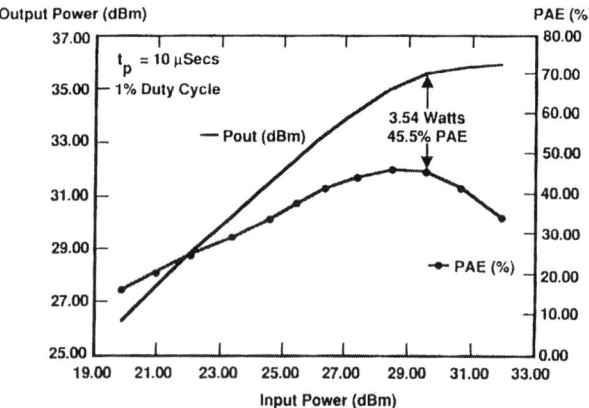

FIG. 12. The RF power and PAE for a 6H-SiC MESFET operating at 6 GHz ($W = 2$ mm, $L_g = 0.5\,\mu$m, $V_d = 40$ V).

than a factor of 3 higher than that normally obtained in GaAs. These results represent the highest power output and operating frequency reported for SiC MESFETs.

VIII. 4H-SiC MESFET Results

Significant improvements in both the power and frequency performance of SiC MESFETs can be obtained by using the 4H-polytype of SiC due to its higher electron mobility and lower donor ionization energy compared to 6H-SiC. With the development of both conducting and high-resistivity and/or semi-insulating 4H-SiC substrates, most of the microwave MESFET work is presently focused on the 4H material. The microwave performance data demonstrated to date are summarized next.

The first 4H-SiC MESFETs were fabricated on conducting substrates [60]. These devices with gate length of 0.7 μm and gate width of 332 μm showed maximum DC transconductance in the range of 38 to 42 mS/mm. The f_t and f_{max} values were 6.7 and 12.9 GHz, respectively, measured at a drain bias of 30 V. The high-frequency performance of these devices was limited due to the conducting substrates used. Substantial improvements in RF performance was obtained by using high-resistivity 4H-SiC substrates [61, 62]. The highest f_{max} reported to date using this material is 42 GHz for a device with gate dimensions of 0.5 × 200 μm [61]. These devices also showed a small-signal gain of 5.1 dB at 20 GHz (4H-MESFET curve in Fig. 11). It is noteworthy that these RF results were obtained at a high drain bias of 40 V. These 4H-SiC MESFETs also showed excellent DC characteristics, with maximum drain current of 500 mA/mm and gate breakdown voltage in excess of 100 V, clearly indicating the possibility of obtaining more than six times the power density achievable with GaAs MESFETs.

More recently, 0.4-μm MESFETs, also fabricated on semi-insulating 4H-SiC, demonstrated f_t and f_{max} values of 18 and 50 GHz, respectively [63]. These MESFETs had a gate width of 0.5 mm and a channel doping of $3.0 \times 10^{17}/cm^3$.

Excellent RF power performance has also been experimentally demonstrated for 4H-SiC MESFETs. Initially, a RF power density of 3.3 W/mm at 850 Mhz was reported [64]. This device had gate dimensions of 0.5 × 332 μm and delivered a maximum output power of 1.12 W under class A conditions, with a drain voltage of 50 V. In class B operation ($V_d = 40$ V, $I_d = 50\% I_{dss}$), the same device showed a PAE of 65.7%. The power output at this condition was 2.12 W/mm. At a higher frequency of 1.8 GHz, the same devices showed 2 W/mm, with 50% PAE in class B conditions, with

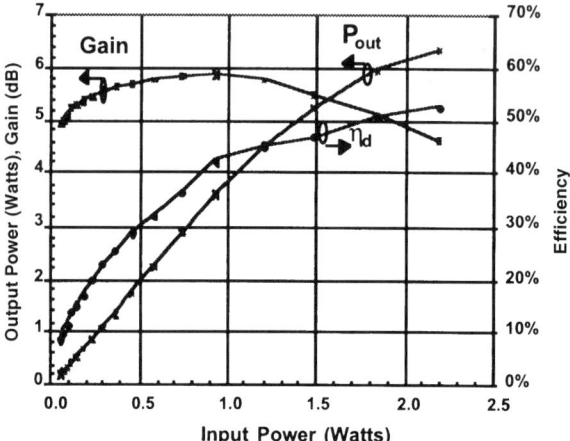

FIG. 13. Performance of SiC MESFET at 10 GHz.

drain biases at 40 V [65]. All of these power results were obtained for FETs fabricated on conducting substrates, which limited their operating frequency.

More recently [66], we have demonstrated high-power operation of 4H-SiC MESFETs at X-band. We have obtained power output of 6 W at 10 GHz using a 1.92-mm periphery MESFET with 0.5-μm gate length (Fig. 13). The associated gain, drain efficiency, and PAE were 5.1 dB, 50%, and 34.6%, respectively.

These results were obtained under class AB conditions with the gate biased at 10% I_{DSS}. A higher power output of 6.35 W (3.3 W/mm) was obtained at higher input drive level, with 4.6 db gain and 52% drain efficiency. These power results were obtained under pulsed conditions with 100-μs-wide pulses at 10% duty cycle and a drain bias voltage of 45 V. To our knowledge, the present results represent the highest power output obtained at X-band using wide-bandgap semiconductors. The power density at X-band is also the highest reported to date. Further improvements in performance can be expected by optimizing the device structure.

IX. The SiC SIT: Design Considerations

The SIT, invented by Watanabe and Nishizawa in 1950, consists of a multichannel, vertical structure [67]. It controls current flow by modulating the internal potential of a single channel, using a surrounding gate structure.

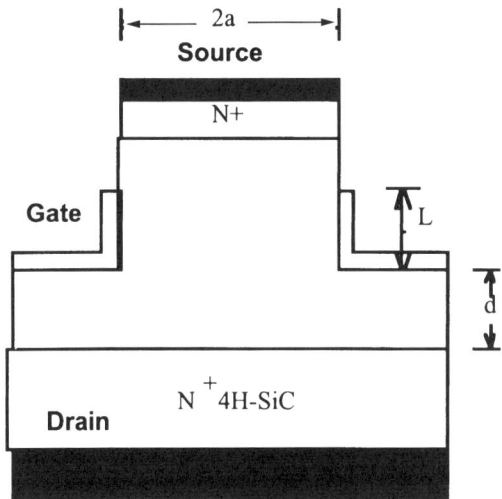

FIG. 14. The cross-section of a single finger element in an SiC SIT.

In 1975, experimental SITs were fabricated, and the source-drain current of this device was shown to follow a space-charge injection model [68]. More recently, very high power broad-band performance has been obtained from silicon SIT devices [69–78].

SITs are a class of transistors with a short-channel FET structure in which a current flowing vertically between source and drain is controlled by the height of an electrostatically induced potential energy barrier under the source. This electrostatic barrier develops at pinch-off when negatively charged opposing gate depletion layers converge to deplete the source drain channel of mobile charge carriers. Analogous to the vacuum triode, both the gate (grid) voltage and the drain (anode) voltage affect the source drain current. A cross-section of a single element of a SIT defining the gate length, L, and the channel length, d, is shown in Fig. 14.

Referring to the current–voltage characteristic of Fig. 15, the various regions of operation can be defined as (A) ohmic, (B) thermionic emission, (C) space-charge limited current (SCLC), and (D) velocity-saturated SCLC. Examination of the DC SIT characteristic of Fig. 15 under zero gate bias (point A) reveals conventional ohmic conduction of the doped channel dominated by the electron drift mobility, μ_n. This mode is described by

$$I_D = AqN_D\mu_n \frac{V_D}{L+d} \qquad (1)$$

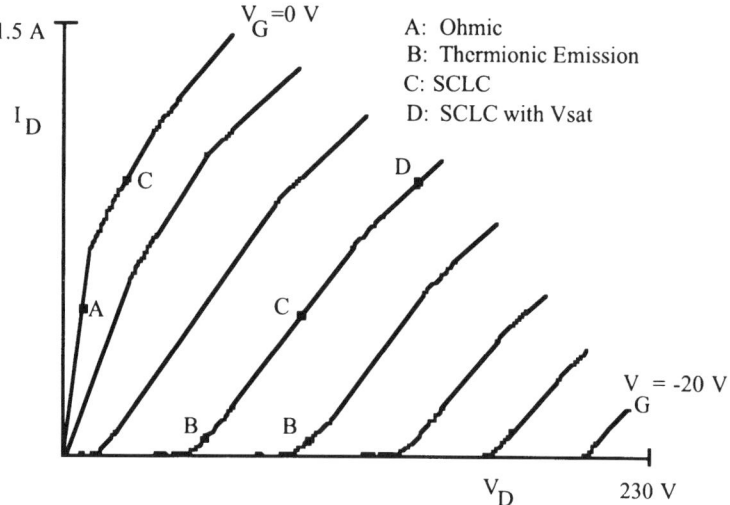

FIG. 15. Current versus voltage characteristic of a mixed-mode SIT.

where A is the cross-sectional area for current flow, q is the elemental charge, N_D is the doping, μ_n is the mobility, V_D is the applied drain bias, L is the gate height, and d is the length of the epilayer under the finger. With the application of a gate bias, the drain current moves toward pinch-off (point B) and the channel is depleted of carriers, as in a conventional MESFET. At this point, thermionic emission current flows, which is given by

$$I_D = AA^{**}T^2 \exp\left[-\frac{q}{kT}(\phi - \alpha V_D)\right] \quad (2)$$

where A^{**} is Richardson's constant, ϕ is the barrier height at zero drain bias, T is the temperature, and αV_D represents the barrier lowering due to the drain bias. When the device is pinched off, the application of further drain bias causes the induced barrier between the opposed gates to be lowered (point C), and SCLC begins to flow. SCLC, a signature of classic silicon-based SITs, is described by

$$I_D = \frac{9}{8}\mu_n\varepsilon_s \frac{V_D^2}{(L+W)^3} A \quad (3)$$

where ε_s is the permittivity of SiC, and W is the amount of depleted SiC below the finger. A further increase in drain bias raises the electron velocity in the drift region until velocity saturation occurs, and the drain current as a function of drain bias changes slope (point D). Assuming that the carriers move with saturated drift velocity throughout the channel, the current is given by

$$I_D = 2\varepsilon_s v_{\text{sat}} \frac{V_D}{(L+d)^2} A \qquad (4)$$

where v_{sat} is the saturated drift velocity for electrons.

In SIT devices, the gate-to-drain spacing, or drift region distance under the gate, comprises an undoped film several microns thick that can withstand the application of large gate-to-drain voltages and permit useful high blocking voltage in the transistors (200 V, for example). The maximum blocking voltage is a function of geometry and E_c, the electric field strength of the material. In classic SIT operation, the channel between the two gates is depleted of carriers when the gate is at zero bias, and application of drain bias is needed to turn the device on and produce source drain current. However, to maximize channel current and minimize the effect of knee voltage on device efficiency, mixed-mode SIT operation is usually employed. In mixed-mode operation, enough doping is added to the channel to provide ohmic currents at medium gate bias, while higher gate bias reverts to classic SIT operation, with its attendant large blocking voltage. Transconductance, transit time, unity gain frequency, and cut-off frequency are all strong functions of the electron mobility and saturated electron velocity.

This first-order discussion of SIT operation clearly identifies the material parameters most significant for high-frequency power performance: electron mobility, μ_n; saturated electron velocity, v_{sat}; dielectric constant, ε_r; and critical field strength, E_c. Relevant material parameters are compared in Table I for various semiconductor families. Based on first-order equations describing the SIT current–voltage relationship and the data of Table I, it may be deduced that SiC SITs, particularly the 4H-SiC polytype, will outperform SITs made from other materials with regard to power output, frequency response, and DC-to-RF conversion efficiency.

X. SIT Fabrication

SITs require two extremes of doping: a thick, approximately 5-μm-long, unintentionally doped drift layer of 1×10^{16} cm^{-3} carrier concentration and a very highly doped (1×10^{19}/cm^3), 0.2-μm-thick contact layer. The

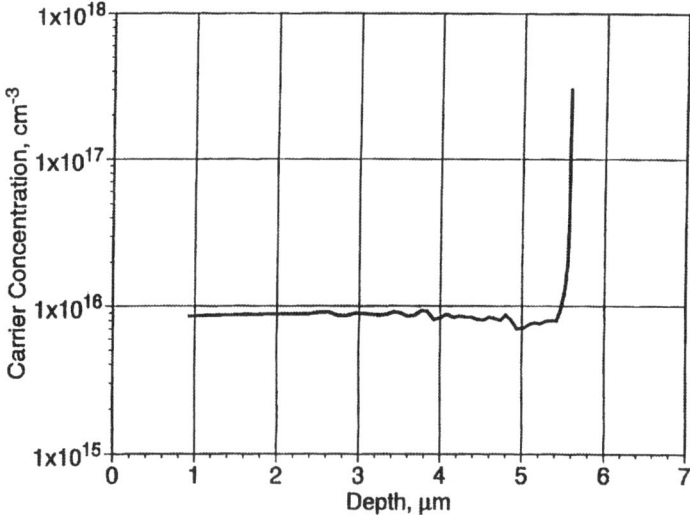

FIG. 16. A carrier concentration depth profile of a SIT drift region formed by vapor-phase epitaxy.

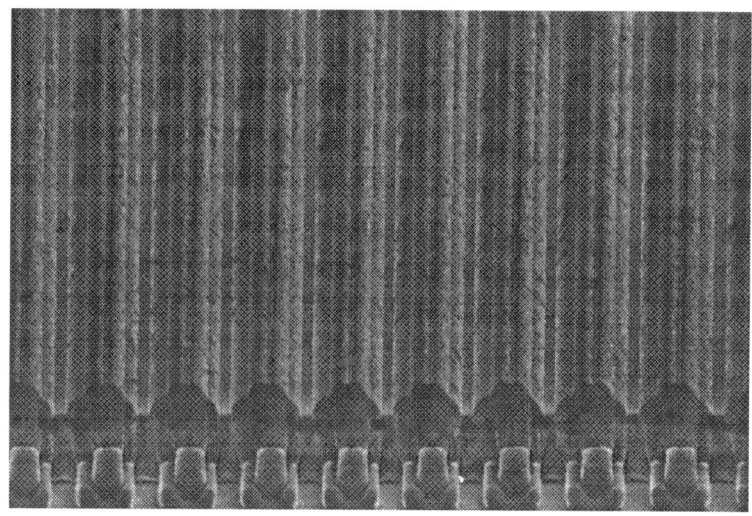

FIG. 17. Structure detail of an SiC SIT. Seen are the individual fingers and the top source electrode.

growth chemistry for SiC employs propane as a source of carbon and silane as a source of silicon, transported via hydrogen carrier gas to an inductively heated graphite susceptor (1450°C). SiC is grown homoepitaxially on c-axis-oriented n-type SiC substrates placed on the susceptor. A SIT carrier concentration versus depth profile is shown in Fig. 16. (The 0.2-μm-thick surface n$^+$ is added subsequently.)

Electron-beam lithography was used to define RIE masks and metal patterns to provide source ohmic contacts and Schottky metal gate electrodes. Dielectric passivation was applied, and thick metal was patterned to provide low-resistance source and gate interconnects to the surface of the device, with the drain electrode being the n$^+$ substrate. An SEM photograph of a completed SIT is shown in Fig. 17.

XI. 6H-SiC SIT Results

Like the MESFET, SiC SITs were first fabricated on 6H-SiC material owing to its wide availability. Low-wattage DC characteristics of a 3-cm source periphery 6H-SiC SIT are shown in Fig. 18. Measured data revealed a maximum channel current of 300 mA/cm, a transconductance of 30 mS/cm, and a voltage gain of 8. On-wafer small-signal measurements were conducted using cascade probes and a common drain configuration. A

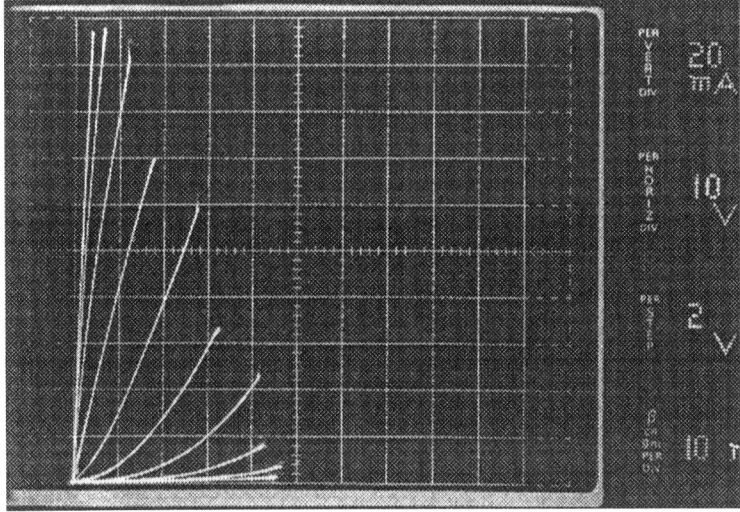

FIG. 18. DC characteristics of a 3-cm periphery SIT.

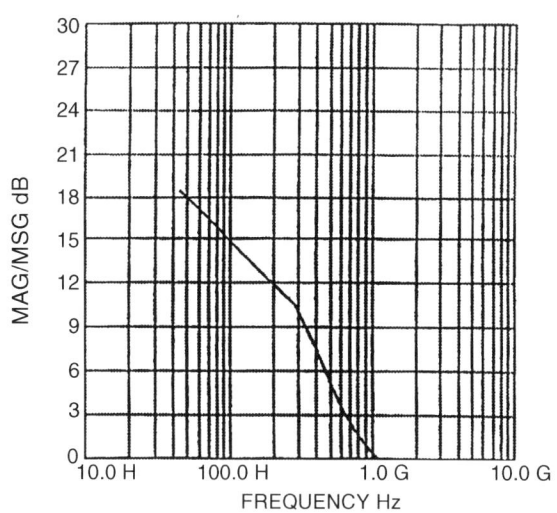

FIG. 19. Small-signal gain of a 1-cm periphery SIT.

small-signal gain of 12 dB at 200 MHz was measured, as shown in the gain frequency curve of Fig. 19. Monolithic SIT chips were screened for DC performance and wirebonded in parallel into a Kyocera package. Thirty-seven watts of pulsed output power was developed from an 11-cm periphery SIT at 175 MHz, with a PAE of 60%, as shown in Fig. 20. The associated gain was 10 dB.

Observed maximum currents in 6H-SiC SITs were nearly five times less than predicted by mathematical models. Reduced channel currents are a problem for power devices, because both power gain and PAE will be

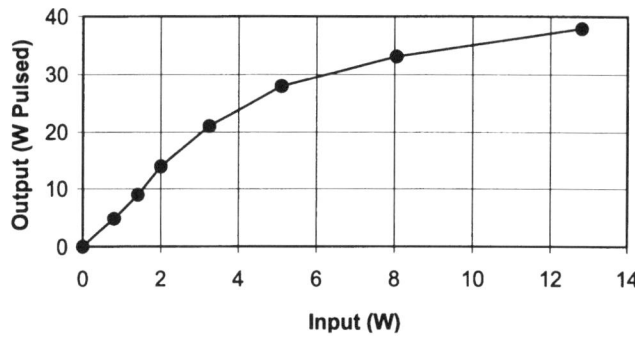

FIG. 20. Pulsed-power performance of a monolithic 11-cm periphery SIT measured at 175 MHz.

adversely impacted. The poor currents in 6H-SiC were found to be associated with anisotropic electron transport behavior. Currents traveling in a direction parallel to the c-axis (into the surface) of a 6H wafer were observed to be five times smaller than currents normal to the c-axis (along the surface) for the same applied voltage.

Experimental evaluation of current transport in SIT-like structures was conducted by measuring pulsed currents (to minimize heating effects). Extracted quantities for electron mobility and saturated drift velocity in 6H- and 4H-SiC parallel to the c-axis (in the SIT current direction) reveal a much larger electron mobility (6 times) in the 4H polytype than seen in 6H. The saturated electron drift velocity was determined to be $\sim 2.0 \times 10^7$ cm/s. Therefore, improved SIT characteristics are expected in the 4H-SiC polytype.

XII. 4H-SiC SIT Results

4H-SiC SITs showed a current density of 1 A/cm of source periphery, a voltage gain of 15, and a maximum transconductance of 75 mS/cm, as seen in Fig. 21. The device blocking voltage with a gate bias of -20 V was 200 V. Small-signal common drain measurements were performed on wafers using cascade probes. As can be seen in the computer-derived, common-source, gain-frequency curve of Fig. 22, a cut-off frequency, f_{max}, of 4 GHz was

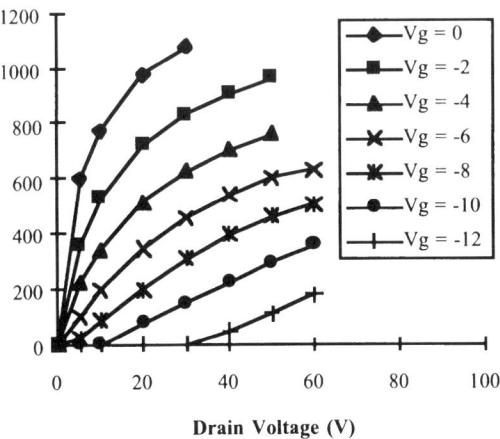

FIG. 21. Five-microsecond pulsed DC SiC SIT characteristic of a 1-cm source periphery device.

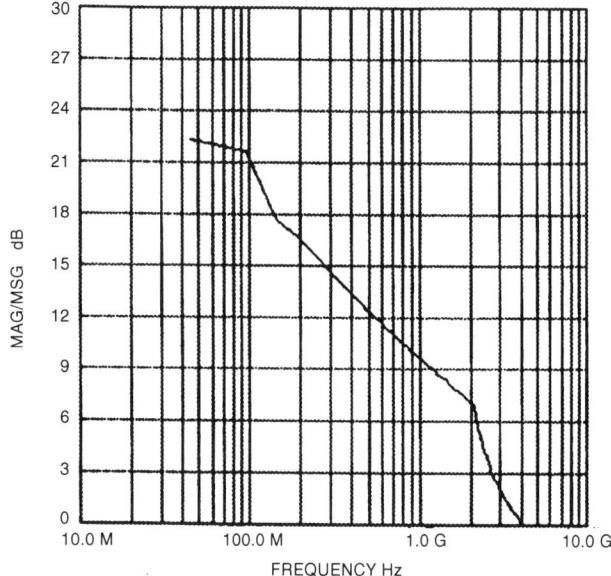

FIG. 22. Small-signal data of a 1.5-cm 4H-SiC SIT ($V_d = 20$ V, $V_g = -4$ V).

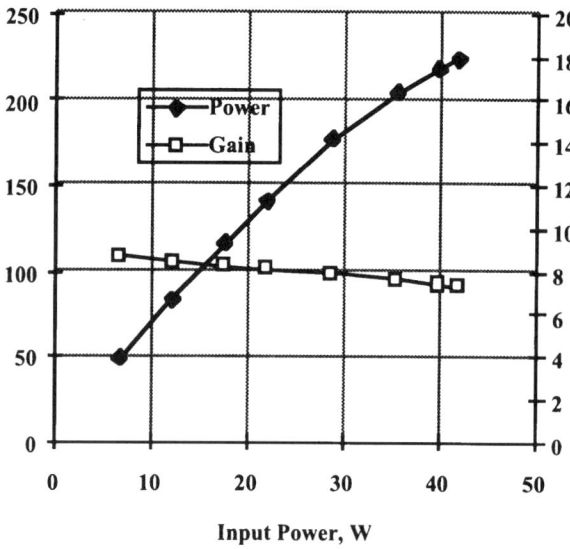

FIG. 23. Pulsed-power and gain curves for a 16.5-cm SIT measured at 600 MHz.

obtained. Automated wafer mapping was employed to identify candidate 1.5-cm SITs. Transistors of combined 16.5-cm periphery were cut from the wafer, soldered, and wirebonded into a Kyocera package. This package developed 225 W output power with an associated gain of 8.7 dB at 600 MHz (Fig. 23). The maximum PAE was 47%, and the power density was 13.5 W/cm. These results from 4H-SiC SIT devices are significantly better than those from silicon devices, in terms of power density, breakdown voltage, and frequency response.

XIII. 450 W UHF SIT

A total of 60 UHF SITs contained on three separate die were mounted into a standard 0.4-in. power transistor package [79]. Fondly referred to as "Godzilla," 23 of the 60 cells were found to pass a 250 V breakdown test and were subsequently wired in parallel for 34.5-cm total SIT periphery. The packaged device contained no internal matching components other than paralleled wirebonds, and is illustrated in Fig. 24.

The packaged impedance levels at 600 MHz were found to have magnitudes of approximately 2 ohms on the input and 4.1 ohms for optimum drain circuit load. These relatively high values of unmatched impedances reflect the high voltage swings of the cells, which are about 2.5 times higher than in similar structures available in silicon.

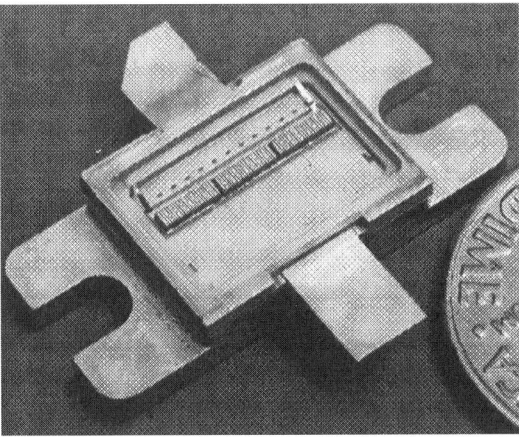

FIG. 24. A package containing 23 1-cm periphery cells of UHF SIT.

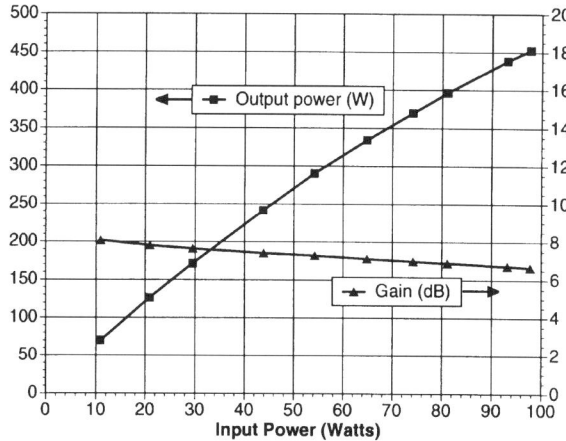

FIG. 25. 450-W UHF SIT RF power performance at 600 MHz.

The unmatched device was operated in a microstrip circuit at 90 V drain bias at 600 MHz, with the performance shown in Fig. 25. The power output density of the paralleled cells is 165 kW/cm^2. This value is substantially higher than the practical limit seen in silicon devices of 78 kW/cm^2 and is a direct result of the physical characteristics of SiC. This is the largest SiC power transistor yet reported.

XIV. 2.0 kW UHF Module

SiC power transistors were operated in a balanced configuration for improved intermodulation product performance and provided over twice the power available from conventional silicon devices. Remarkably, this 2 × power improvement was achieved with parts that were only one-third populated with transistor cells. The transistors were mounted in air-cooled heatsinks and assembled into a 2.0 kW module, seen in Fig. 26.

Due to the linear nature of the waveform and the high average power, the channel temperature within the SiC module was greater than 200°C. A demonstration UHF SiC power module was used at the National Association of Broadcasters Convention in 1996 for an over-the-air broadcast of high-definition television (HDTV). Figure 26 shows the module with a one-driving four-transistor architecture, which was used in the module to provide 200 W average (1 kW peak) output power out of the transmitter, with better than −30 dB third-order products. The linearity requirements

FIG. 26. SiC UHF TV module.

limited the module output power for the television application, but the module is actually capable of 400 W average (2.0 kW peak). These performance levels are achieved from four gemini push-pull output transistors utilizing only 30% of the transistor package capacity.

Under class C operation, the same 24-cell gemini transistor produces over 350 W peak output power at efficiencies greater than 50%, with greater than 10-dB gain. Since that time, the transistors have been operated under higher average power conditions at channel temperatures approaching 300°C, again with no resulting failures. Reliability testing of these parts continues.

XV. S-band SiC SITs

For the S-band SIT, a mixed-mode SIT design was used [80]. This construction maximizes channel current and transconductance under MESFET operation and voltage gain and blocking voltage under SIT conditions. The value of this optimization is maximum power with maximum gain and power density. In addition, the SIT finger spacing was reduced from the 4 μm of the UHF design to ~2.5 μm, to reduce capacitive parasitics.

In small-signal *on wafer* measurements, this device shows an excellent 9 dB of gain at 2.0 GHz, SITs were diced from the wafers, soldered into

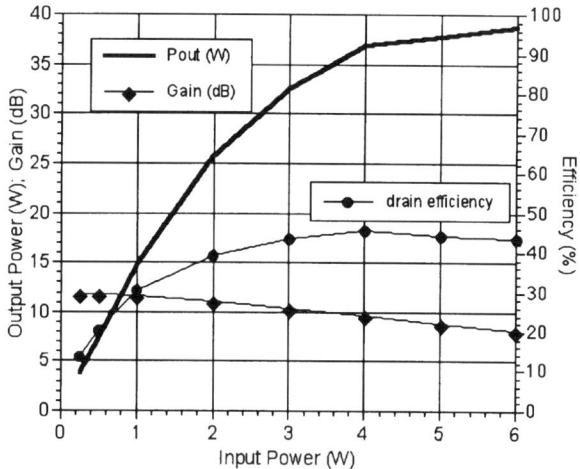

FIG. 27. Power, efficiency, and gain curves measured at 3.0 GHz for a 3-cm periphery common-gate SiC SIT.

commercial RF packages, and wirebonded for common gate biasing (which provides the highest frequency power gain). A 3-cm periphery SIT was operated at 90 V drain bias using 55-μs pulses for 1% duty cycle and delivered 38 W of output power at 3.0 GHz with 9.5 dB of associated gain (Fig. 27). The device operated with 43% PAE. The power density of this transistor is greater than 100 kW/cm^2, which is at least four times that obtainable from comparable S-band silicon power transistors.

XVI. S-band SIT Device Scale-up

SIT cells fabricated at the Northrop Grumman Science and Technology Center have previously shown proof of principle for high gain and power density at S-band. The S-band SIT process has been used at the Northrop Grumman Advanced Technology Laboratory pilot production SiC line.

Multiple 1-cm periphery cells were fabricated [66] on die measuring 800 × 3000 μm. Cell pitch was 250 μm, such that three die and 36 cells total could be packaged in a standard 0.4 × 0.5-in. Kyocera package pill.

An initial package was assembled with 19 cells wired in parallel. A single-step LC matching circuit was included on the input side, using bondwire inductance and a silicon metal-oxide semiconductor (MOS) film capacitor. The parallel summed output capacitance was measured at 13.3 pf

FIG. 28. Nineteen-cell, 200 W S-band SIT.

and was resonated in the package using a parallel inductance consisting of multiple bondwires and a MOS DC blocking capacitor. The input impedance level was measured as $3.0 + j6.55$ ohms at 2.9 GHz. The impedance of the optimum large signal load, referenced to the package interface, was measured at $6.6 + j0.3$ ohms at 2.9 GHz. Figure 28 shows the packaged device assembly.

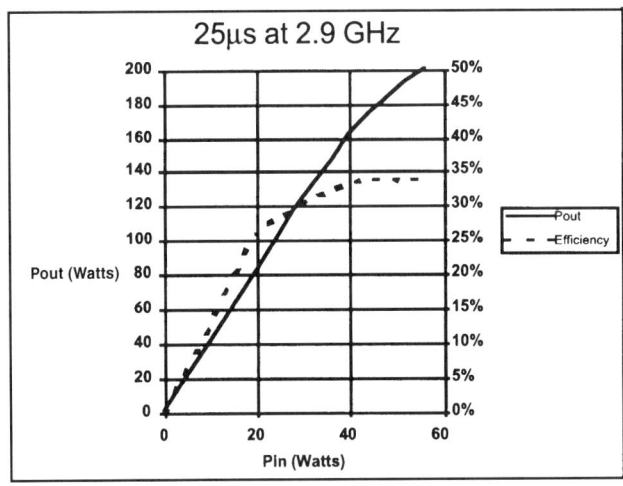

FIG. 29. Short-pulse operation of 19-cell SIT.

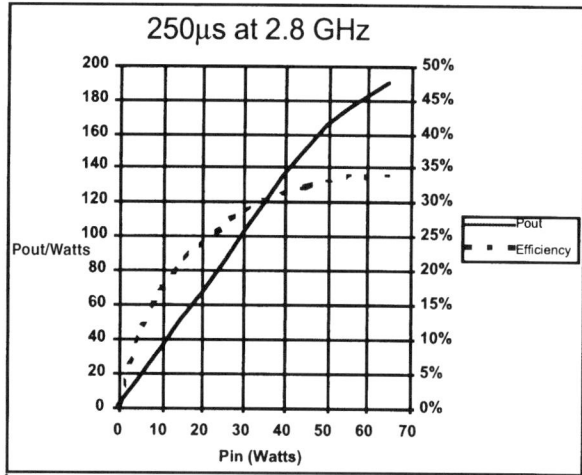

FIG. 30. Long-pulse operation of 19-cell SIT.

The transistor was operated in a class AB pulse mode with 90-V drain voltage and 500-mA quiescent current. Data were taken at both 25- and 250-μs pulse length, with performance summarized in Figs. 29 and 30. Very linear operation was obtained over a wide dynamic range, and a soft saturation characteristic was obtained, similar to that found in the UHF devices. Heating effects in the long-pulse case produced an 0.8-dB drop in gain as compared to the shorter pulse. This gain drop correlates with the calculated cell transient temperature and a gain sensitivity of 0.013 dB per degree centigrade. This is a similar temperature-versus gain sensitivity to that observed in Si and GaAs devices. No equivalent degradation was observed in the device drain efficiency.

XVII. SiC Power Switching Devices

Power MOSFETs with 260 V blocking capability and R_{on} of 18 m$\Omega \cdot$ cm^2 have been obtained [81] for vertical 4H-SiC UMOSFETs. Also, larger area devices, such as 175 V MOSFETs with a rated current of 2 A ($J = 200$ A/cm^2) at a forward drop of 2.65 V were also demonstrated. Operation of similar devices had previously been demonstrated to 300°C with short lifetimes because of gate oxide failure, but no high-temperature data were reported for these later developments [81]. Because of SiC's superior electric field strength, R_{on} has been affected by channel resistance, and not simply the resistance of the lightly doped drain-drift region, which is much thinner than that of comparable Si power MOSFETs. Improvement of channel mobility to lower R_{on} is critical. Poor channel mobility in SiC Power MOSFETs (e.g., $\sim 13\,\text{cm}^2/\text{V}\cdot\text{s}$ in 4H-SiC UMOSFET [81]) has thus far been generally attributed to poor interface quality between oxide and SiC in the channel region. Additionally, long-term stability of the insulator has been shown to be the limiting factor in SiC power UMOSFET structures, although use of a p^+ polysilicon gate reduces Fowler-Nordheim injection from the gate electrode, thereby increasing breakdown voltage. An excellent discussion of these issues is found elsewhere [82]. In an attempt to increase the channel mobility as well as blocking voltage, the first double implanted MOS (DIMOS or DMOS) FET fabricated in 6H-SiC was reported by a group at Purdue University [83]. Blocking voltage operation up to 760 V was demonstrated with 10-μm blocking layers doped at n-type at 1.7×10^{16} and $6.5 \times 10^{15}\,\text{cm}^{-3}$. A peak inversion-layer channel mobility of $26\,\text{cm}^2/\text{V}\cdot\text{s}$ was obtained, showing some improvement over the UMOSFETs.

Later 4H-SiC UMOSFETs were fabricated by Northrop Grumman, with a 12-μm-thick drain-drift layer ($N_D \sim 2 \times 10^{15}\,\text{cm}^{-3}$) and a 90-nm gate oxide. The gate oxide consisted of a thin layer of thermally grown SiO_2 followed by a thicker layer of deposited and densified SiO_2 to ensure a more uniform gate oxide across the trench [84]. Nickel was used for the drain-source metals, with transmission line method measurements yielding a specific contact resistance of 5 to $7 \times 10^{-6}\,\Omega\cdot\text{cm}^2$. Channel length was $4\,\mu$m. Blocking voltages of 1.0 to 1.2 kV were measured at room temperature by probing bare die under Fluorinert. In Fig. 31, the blocking voltage for a typical device is shown. Under a moderate gate bias of 32 V (3.55 MV/cm) at room temperature, the effective channel mobility (μ_{eff}) was only 1.5 $\text{cm}^2/\text{V}\cdot\text{s}$, while increasing the temperature to 100°C increased μ_{eff} to $7\,\text{cm}^2/\text{V}\cdot\text{s}$ at a gate bias of only 26 V (2.9 MV/cm). Specific on-resistance was $74\,\text{m}\Omega\cdot\text{cm}^2$ at 100°C. Increasing channel mobility with temperature may be explained by deep-level interface traps located at the SiC-oxide interface, which are thermally ionized at higher temperatures, resulting in increased conduction for a given bias. This explanation would account for the typical observed increased in I_{DS} with temperature, which opposes the expected $T^{-3/2}$ temperature dependence of μ_{eff} resulting from increased acoustic phonon scattering. 4H-SiC DMOS structures were also reported on in the same time frame [84] using a lightly doped 10-μm-thick drain-drift layer and a 50-nm-thick oxide-nitride gate dielectric. Multiple ion implants (high-dose nitrogen and aluminum implants for source-body contact regions, and high-energy aluminum or boron implants for formation of the p-wells) were used to fabricate a planar power MOSFET in SiC. Initial

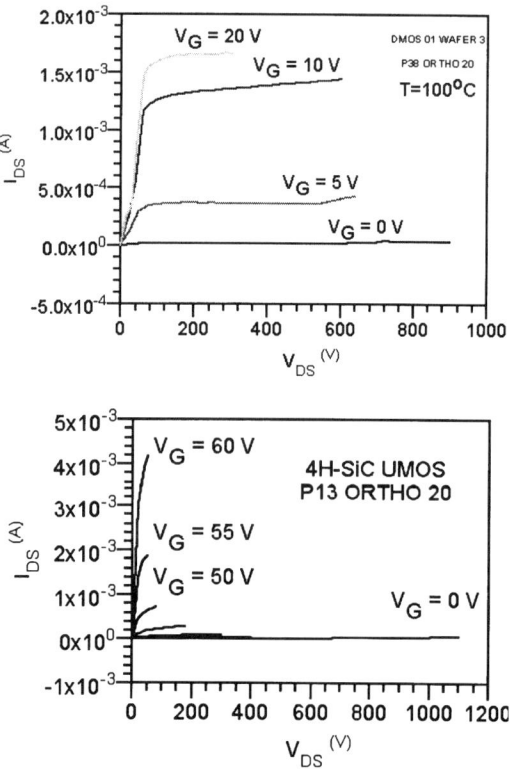

FIG. 31. Output characteristics for a 4H-SiC DMOS exhibiting 900-V blocking voltage at 100°C (above) and similar characteristics for an 1100 V 4H-SiC UMOSFET at room temperature (below). (Taken from [244].)

results have resulted in 900 V devices being demonstrated (see Fig. 31), although the μ_{eff} still shows the *same* temperature effects of interface traps as were found for the UMOSFET case.

Other high-temperature, high-power devices, such as thyristors, have been demonstrated to 500°C [81]. At room temperature, this same device had a blocking voltage of 600 V with a 1.8 A forward current at a voltage drop of 3.7 V. A 10 A, 200 V thyristor with a forward voltage drop of 3.6 V, a current density of 620 A/cm^2, and an R_{on} of 1.2 mΩ·cm^2 was also demonstrated [81]. Other thyristors with 100 V blocking voltage, pulse-switched current density of 5200 A/cm^2 (1.8 A) have also been fabricated [85]. The turn-on rise time was 43 ns, and a fall time of less than 100 ns was measured.

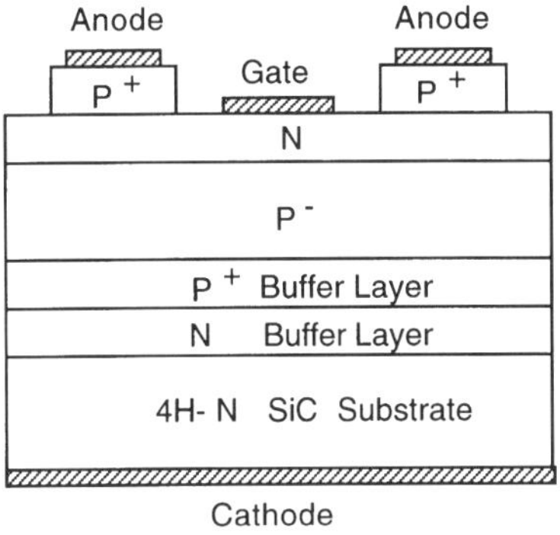

5 SiC for Applications in High-Power Electronics

FIG. 32. Cross-section of a 4H-SiC asymmetrical GTO. (Reproduced from [4], with permission.)

Silicon carbide (4H-SiC), asymmetrical gate turn-off thyristors (GTOs) have also been fabricated and tested with respect to forward voltage drop (V_F), forward blocking voltage, and turn-off characteristics [86]. The GTO structure (Fig. 32) has advantages over the conventional thyristor in that it can be turned off with a gate pulse, allowing it to be used in DC power switching since it does not require the negative AC cycle for turn-off. The asymmetrical GTO structure also has the advantage of easier turn-off in comparison to symmetrical devices. Additionally, the asymmetrical GTO structure is less susceptible to open-base transistor breakdown, which generally determines the breakdown voltage in conventional thyristors. Devices were tested from room temperature to 350°C in the DC mode. Forward blocking voltages ranged from 600 to 800 V at room temperature for the devices tested. V_F of a typical device at 350°C was 4.8 V at a current density of 500 A/cm^2. Turn-off time was less than 1 μs. Although no beveling or advanced edge-termination techniques were used, the blocking voltage represented approximately 50% of the theoretical value when tested in an air ambient. Also, four GTO cells were connected in parallel to demonstrate 600 V, 1.4 A performance. The p-type buffer layer shown in Fig. 32 was epitaxially deposited to form a punch-through structure and also allow easier turn-off of the device by reducing the injection efficiency of the bottom npn bipolar transistor. The blocking voltage is supported across a lightly doped ($\sim 5 \times 10^{14}$ cm^{-3}) 6.0-μm-thick p-type base layer. Finally, n-type gate and p-type anode epitaxial layers were grown. Interdigitated anode and gate fingers were of equal lengths and widths for these nonoptimized, first-generation devices. Typical finger lengths and widths were 250 and

(a)

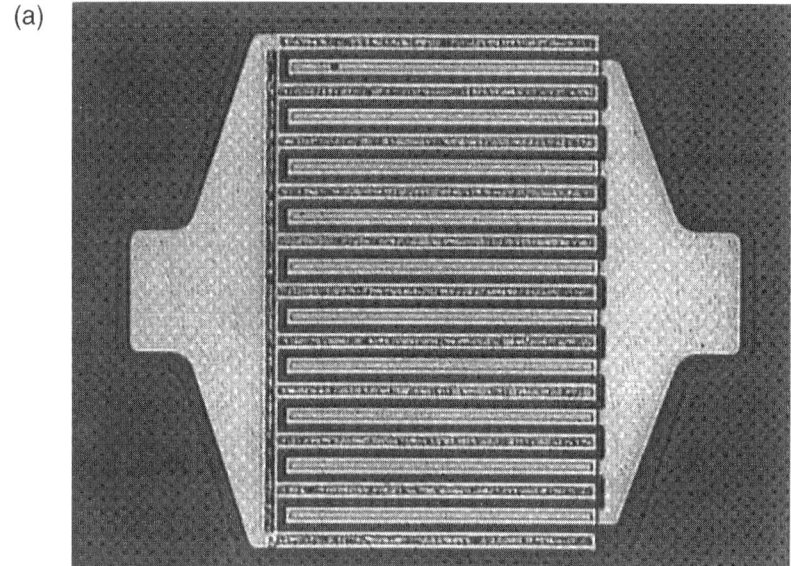

(b)

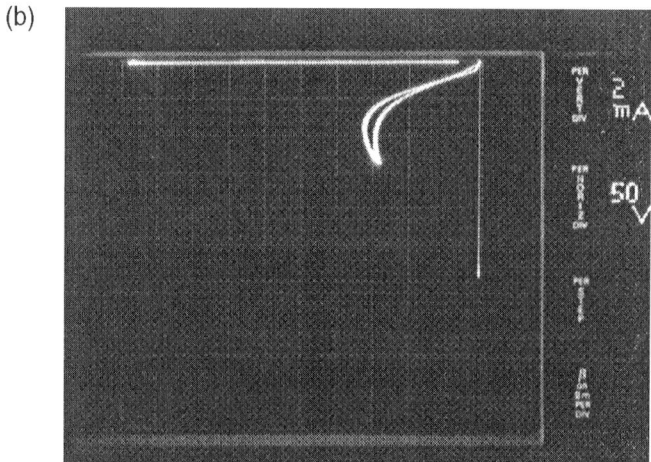

FIG. 33 (a) Top view of a 4H-SiC GTO with interdigitated gate and anode fingers. The cathode (backside) is not shown. (Photo courtesy of Northrop Grumman.) (b) Typical I-V characteristics of same device showing off-state blocking, turn-on, and forward conducting states. (Photo courtesy of Northrop Grumman.)

15 μm, respectively. The top view of one of these devices is shown in Fig. 33, and typical I–V characteristics obtained using a curve tracer in the pnp mode are shown in Fig. 33 as well.

This blocking voltage corresponded to a maximum electric field (E_m) of approximately 1.2 MV/cm across the blocking layer for a nominal 700 V blocking device when assuming negligible voltage drop across the anode, substrate, contacts, and buffer layers. This value of E_m is about 50% of the theoretical critical electric field possible in 4H-SiC. Emissions of light at the periphery of the device structure in breakdown were observed under high-power (50×) magnification in the dark, consistent with edge breakdown. Further improvements in edge termination should improve the breakdown characteristics of the devices. For comparison purposes, it is notable that for punch-through structures in Si, such as a p-i-n diode, a 45-μm-thick blocking layer of equal doping would be necessary to achieve the same 600 V blocking as with the 6-μm SiC blocking layer used here. The dramatic reduction in the required blocking-layer thickness for this device structure utilizing SiC's high electric breakdown strength should allow for much faster power switching than conventional silicon devices.

XVIII. Conclusions

This chapter has reviewed the current state of the art in high-frequency, high-power SiC MESFET and SIT technology. It has been described that the advantageous properties of SiC, namely high electric field breakdown strength (2.5×10^6 V/cm), high saturated electron velocity (2×10^7 cm/sec), and high thermal conductivity (4.9 W/cm-K) synergistically combine to produce high-frequency, high-power density devices that far exceed the performance of their GaAs and Si counterparts.

The highest reported f_{max} for a 0.5-μm MESFET made in semi-insulating 4H-SiC is 42 GHz. Power densities as high as 3.3 W/mm at 850 MHz have been demonstrated. This is more than three times the power density obtained with current GaAs technology. The largest SiC power transistor reported is a 450 W SIT operating at 600 MHz. UHF transistors have shown about 2.5 times the power density compared to silicon SITs. Other UHF SITs have been combined in a 2.5 kW module useful for HDTV transmission. S-band SITs have been reported, with a 3-cm package developing 38 W of output power and a power density four times that of S-band Si-based transistors.

Most researchers believe better performing microwave power devices made from SiC are in the offing. Such improvements will lead not only to higher power, higher gain devices performing at frequencies described in this chapter, but also to devices operating through and beyond 10 GHz. These improvements will be a direct result of continuing advances in substrate and epitaxial purity in addition to increased understanding of device physics and fabrication technology.

ACKNOWLEDGMENTS

The authors would like to acknowledge all of the people responsible for the success of the SiC devices reported in this article. SiC wafers were supplied by G. Augustine, V. Balakrishna, and G. T. Dunne. L. B. Rowland and A. A. Burk, Jr., provided the epitaxial films. R. J. Bojko provided e-beam lithography engineering support. Microwave packaging was performed by J. Ostop, and power testing was performed by T. J. Smith and P. M. Esker. The technical staff of the electronic and system support department at the Northrop Grumman Science and Technology Center (Pittsburgh, PA) and the staff of the Advanced Technology Lab (Baltimore, MD) are acknowledged for fabrication of the SiC devices.

REFERENCES

1. D. E. Cusack, W. M. Glasheen, and H. R. Steglich, "Flame Indication Sensors Operate at Temperatures Above 500°C," Trans. of 2nd International High Temperature Electronics Conference (HiTEC), Charlotte, NC, Session III, pp. 17–22, June 5–10, 1994.
2. K. C. Reinhardt and M. A. Marciniak, "Wide Bandgap Power Electronics for the More Electric Aircraft," Trans. of 3rd International High Temperature Electronics Conference (HiTEC), Session I, pp. 9–14, June 9–15, 1996.
3. P. L. Dreike, D. M. Fleetwood, D. B. King, D. C. Sprauer, and T. E. Zipperian, "An Overview of High Temperature Electronic Device Technologies and Potential Applications," *IEEE Trans. on Comp., Hybrids, Man. Tech. A*, **17**, 4, pp. 594–609, Dec. 1994.
4. A. K. Agarwal, G. Augustine, V. Balakrishna, C. D. Brandt, A. A. Burke, L. S. Chen, R. C. Clarke, P. M. Esker, H. M. Hobgood, R. H. Hopkins, A. W. Morse, L. B. Rowland, S. Seshadri, R. R. Siergiej, T. J. Smith, Jr., and S. Sriram, "SiC Electronics," *Tech. Dig. Int'l. Elect. Dev. Meeting*, pp. 9.1.1–9.1.6, Dec. 1996.
5. M. Tajima, "Research Activity on High Temperature Electronics and Its Future Application in Space Exploration in Japan," Trans. of 2nd International High Temperature Electronics Conference (HiTEC), Charlotte, NC, Session I, pp. 29–34, June 5–10, 1994.
6. D. M. Fleetwood, "High Temperature Silicon-on-Insulator Electronics for Space Nuclear Power Systems: Requirements and Feasibility," *IEEE Trans. Nucl. Sci.*, **NS-35**, pp. 1099–1112, 1988.
7. J. W. Palmour and C. H. Carter, Jr., "4H-Silicon Carbide Devices for High Power and High Frequency," *Proceedings of 1993 International Semiconductor Device Research Symposium*, Omni Charlottesville Hotel, pp. 695–696, December 1–3, 1993.

8. R. J. Trew, J. B. Yan, and P. M. Mock, "The Potential of Diamond and SiC Electronic Devices for Microwave and Millimeter-Wave Power Applications," *Proceedings of the IEEE*, **79**, pp. 598–620, May 1991.
9. J. M. McGarrity, F. B. McLean, W. M. DeLancey, J. W. Palmour, C. H. Carter, Jr., and J. A. Edmond, "Silicon Carbide JFET Radiation Response," *IEEE Trans. on Nuclear Science*, **39**, No. 6, pp. 1974–1981, 1992.
10. Cree Research, Inc., 2810 Meridian Parkway, Durham, NC 27713.
11. Advanced Technology Materials, Inc., 7 Commerce Drive, Danbury, CT 06810-4169.
12. T. George, W. T. Pike, M. A. Khan, J. N. Kuznia, and P. Chang-Chien, "A Microstructural Comparison of the Initial Growth of AlN and GaN Layers on Basal Plane Sapphire and SiC Substrates by Low Pressure Metalorganic Chemical Vapor Deposition," *J. of Electronic Materials*, **24**, 4, pp. 241–247, April 1995.
13. W. M. Yim, E. J. Stofko, P. J. Zanzucchi, J. I. Pankove, E. Ettenberg, and S. L. Gilbert, "Epitaxially grown AlN and its optical bandgap," *J. Appl. Phys.*, **44**, 1, pp. 292–296, Jan. 1, 1973.
14. J. Edmond, H. Kong, V. Dmitriev, G. Bulman, and C. Carter, Jr., "Blue/UV Emitters from SiC and Its Alloys," *Inst. Phys. Conf. Ser.* 137, pp. 515–518m IOP Publishing, Ltd., Bristol, 1994.
15. V. F. Tsvetkov, S. T. Allen, H. S. Kong, and C. H. Carter, Jr., "Recent Progress in SiC Crystal Growth," *Inst. Phys. Conf. Ser.* 142, pp. 17–22, IOP Publishing, Ltd., Bristol, 1996.
16. G. Augustine, H. McD. Hobgood, V. Balakrishna, G. Dunne, and R. H. Hopkins, "Physical Vapor Transport Growth and Properties of SiC Monocrystals of 4H Polytype," *Phys. Stat. Solidi*, submitted Sept. 1997.
17. P. G. Neudeck and J. A. Powell, "Performance Limiting Micropipe Defects in Silicon Carbide Wafers," *IEEE Elect. Dev. Lett.* **15**, pp. 63–65, Feb. 1994.
18. J. Giocondi, G. S. Rohrer, M. Skowronski, V. Balakrishna, G. Augustine, H. M. Hobgood, and R. H. Hopkins, "The Relationship Between Micropipes and Screw Dislocations in PVT Grown 6H-SiC," MRS Symposium Proc., **423**, pp. 539–544, 1996.
19. H. M. Hobgood, D. L. Barrett, J. P. McHugh, R. C. Clarke, S. Sriram, A. A. Burk, J. Greggi, C. D. Brandt, R. H. Hopkins, and W. J. Choyke, "Large Diameter 6H-SiC for Microwave Device Applications," *J. Crystal Growth*, **137**, pp. 181–186, 1994.
20. A. Udding, H. Mitsushashi, and T. Uemoto, "Investigation of Deep Levels and Residual Impurities in Sublimation-Grown SiC substrates," *Jpn. J. Appl. Phys.* **33**, p. L908–L911, 1994.
21. Yu. M. Tairov and V. F. Tsvetkov, "Investigation of Growth Processes of Ingots of Silicon Carbide Single Crystals," *J. Cryst. Growth,* **43**, pp. 209–212, 1978.
22. Yu. M. Tairov and V. F. Tsvetkov, "General Principles of Growing Large-Size Single Crystals of Various Silicon Carbide Polytypes," *J. Cryst. Growth,* **52**, pp. 146–150, 1981.
23. G. Ziegler, P. Lanig, D. Theis, and C. Weyrich, "Single Crystal Growth of SiC Substrate Material for Blue Light Emitting Diodes," *IEEE Trans. Elect. Dev.* **ED-30**, pp. 277–281, 1983.
24. D. L. Barrett, R. G. Seidensticker, W. Gaida, R. H. Hopkins, and W. J. Choyke, "SiC Boule Growth by Sublimation Vapor Transport," *J. Cryst. Growth* **109**, pp. 17–23, 1991.
25. R. F. Davis, C. H. Carter, Jr., and C. E. Hunter, U.S. Patent No. 4,866,0005, Sept. 12, 1989.
26. T. Nakata, K. Koga, Y. Matsushita, Y. Ueda, and T. Niina, *Proc. of 2nd Int'l. Conf. on Amorphous and Crystalline Silicon Carbide II & Related Mat.,* Santa Clara, CA, Dec. 15–16, 1988, *Springer Proc. Phys.* **43**, pp. 26–34, 1989.
27. D. L. Barrett, J. P. McHugh, H. M. Hobgood, R. H. Hopkins, P. G. McMullin, R. C. Clarke, and W. J. Choyke, "Growth of Large SiC Single Crystals," *J. Cryst. Growth* **128**, pp. 358–362, 1993.

28. F. Takahashi, M. Kanaya, and Y. Fujiwara, "Sublimation Growth of SiC Single Crystalline Ingots on Faces Perpendicular to the (0001) Basal Plane," *J. Cryst. Growth* **135**, pp. 61–70, 1994.
29. D. Hofmann, S. Y. Karpov, Y. N. Makarov, E. N. Mokhov, M. G. Ramm, M. S. Ramm, A. D. Roenkov, and Y. A. Vodakov, "Use of Ta Container Material for Quality Improvement of SiC Crystals Grown by the Sublimation Technique," *Inst. Phys. Conf. Ser.* 142, pp. 29–32, IOP Publishing, Ltd., Bristol, 1996.
30. H. McD, Hobgood, R. C. Glass, G. Augustine, R. H. Hopkins, J. Jenny, M. Skowronski, W. C. Mitchel, and M. Roth, "Semi-insulating 6H-SiC Grown by Physical Vapor Treatment," *Appl. Phys. Lett.* **66**, 11, pp. 1364–1366, Mar. 13, 1995.
31. G. Gradinaru, T. S. Sudarshan, S. A. Gradinaru, W. Mitchell, and H. M. Hobgood, "Electrical Properties of High Resistivity 6H-SiC Under High Temperature/High Field Stress," *Appl. Phys. Lett.* **70**, 6, pp. 735–737, Feb. 10, 1997.
32. V. Shields, K. Fekade, and M. G. Spencer, "A Process for the Growth of Beta-SiC Substrates," *Inst. Phys. Conf. Ser.* 137, pp. 21–24, IOP Publishing, Ltd., Bristol, 1994.
33. I. Golecki, F. Reidinger, and J. Marti, "Single Crystalline Epitaxial Cubic SiC Thin Films Grown on [100] Si at 750°C by Chemical Vapor Deposition," presented at the Workshop on High Temperature Power Electronics for Vehicles, Fort Monmouth, NJ, April 26–27, 1995.
34. J. Anthony Powell and D. J. Larkin, "Controlled Thin-Film Growth of Silicon Carbide Polytypes," NASA Tech. Briefs, pp. 58–59, May, 1995.
35. M. A. Tischler, N. Hamaguchi, S. Choi, A. Powell, and P. Dobrilla, "Progress in Bulk and Thin Film Growth of Silicon Carbide," presented at the Workshop on High Temperature Power Electronics for Vehicles, Fort Monmouth, NJ, April 26–27, 1995.
36. D. J. Larkin, P. G. Neudeck, J. A. Powell, and L. G. Matus, "Site-Competition Epitaxy for Controlled Doping of CVD Silicon Carbide," *Inst. Phys. Conf. Ser.* 137, pp. 51–54, IOP Publishing, Ltd., Bristol 1994.
37. Y. C. Wang, R. F. Davis, and J. A. Edmond, "In Situ Incorporation of Al and N and p-n Junction Diode Fabrication in Alpha (6H)-SiC Thin Films," *Journal of Electronic Materials* **20**, pp. 289–294, April 1991.
38. D. J. Larkin, "Site-Competition Epitaxy for n-Type and p-Type Dopant Control in CVD SiC Epilayers," *Inst. Phys. Conf. Ser.* 142, pp. 23–28, IOP Publishing, Ltd., Bristol, 1996.
39. D. J. Larkin, S. G. Sridhara, and R. P. Devaty, "Hydrogen Incorporation in Boron-Doped 6H-SiC CVD Epilayers Produced Using Site-Competition Epitaxy," *J. of Electronic Materials*, 24, pp. 289–294, April 1995.
40. D. J. Larkin, P. G. Neudeck, J. A. Powell, and L. G. Matus, "Site-Competition Epitaxy for Superior Silicon Carbide Devices," *Appl. Phys. Lett.* **65**, p. 1659, 1994.
41. A. A. Burk, Jr., L. B. Rowland, A. K. Agarwal, S. Sriram, R. C. Glass, and C. D. Brandt, "Vapor Phase Homoepitaxial Growth of 6H and 4H Silicon Carbide," *Inst. Phys. Conf. Ser.* 142, pp. 201–204, IOP Publishing, Ltd., Bristol, 1996.
42. T. Kimoto and H. Matsunami, "Two-Dimensional Nucleation and Step Dynamics in Crystal Growth," *Inst. Phys. Conf. Ser.* 137, pp. 55–58, IOP Publishing, Ltd., Bristol, 1994.
43. H. J. Kim and R. F. Davis, "Theoretical and Empirical Studies of Impurity Incorporation into β-SiC Thin Films During Epitaxial Growth," *J. Electrochem. Soc.* **133**, pp. 2350–2357, Nov. 1986.
44. M. Ikeda, H. Matsunami, and T. Tanaka, "Site Effect on the Impurity Levels in 4H, 6H, and 15R SiC," *Phys. Rev. B* **22**, 6, pp. 2842–2845, Sept. 15, 1980.
45. O. Kordina, A. Henry, J. P. Bergman, N. T. Son, W. M. Chen, C. Hallin, and E. Janzen, "High Quality 4H-SiC Epitaxial Layers Grown by Chemical Vapor Deposition," *Appl. Phys. Lett.* **66**, 11, pp. 1373–1375, Mar. 13, 1995.

46. A. Hoh, H. Akita, T. Kimoto, and H. Matsunami, "High Quality 4H-SiC Homoepitaxial Layers Grown by Step-Controlled Epitaxy," *Appl. Phys. Lett.* **65**, 11, pp. 1400-1402, Sept. 12, 1994.
47. H. S. Kong, J. T. Glass, and R. F. Davis, "Chemical Vapor Deposition and Characterization of 6H-SiC Thin Films on Off-Axis 6H-SiC Substrates," *J. Appl. Physics* **64**, 5, pp. 2672–2679, Sept. 1, 1988.
48. R. F. Davis, "Deposition and Characterization of Diamond, Silicon Carbide, and Gallium Nitride Thin Films," *J. Cryst. Growth* **137**, pp. 161–169, 1994.
49. R. Rupp, P. Lanig, J. Vilkl, and D. Stephani, "Silicon Carbide CVD Approaches Industrial Needs," *Mat. Res. Soc. Symp. Proc.* **423**, pp. 253–263, 1996.
50. A. A. Burk, Jr., L. B. Rowland, G. Augustine, H. M. Hobgood, and R. H. Hopkins, "The Impact of Pregrowth Conditions and Substrate Polytype on SiC Epitaxial Layer Morphology," *Mat. Res. Soc. Symp. Proc.* **423**, pp. 275–280, 1996.
51. E. Janzen and O. Kordina, "Recent Progress in Epitaxial Growth of SiC for Power Device Applications," *Inst. Phys. Conf. Ser.* 142, pp. 653–658, IOP Publishing, Ltd., Bristol, 1996.
52. L. B. Rowland, A. A. Burk, Jr., R. C. Clarke, R. R. Siergiej, S. Sriram, G. Augustine, H. M. Hobgood, and M. C. Driver, "Vapor Phase 6H and 4H SiC Epitaxy for High-Speed Devices," *Proc. IEEE/Cornell Conf. Advanced Concepts in High Speed Semiconductor Devices and Circuits,* Aug. 7–9, 1995.
53. C. A. Mead, *Proc. IEEE* **54**, p. 307, 1966.
54. G. Pensl and W. J. Choyke, *Physica B* **185**, p. 264, 1993.
55. J. W. Palmour, C. E. Weitzel, K. J. Nordquist, and C. H. Carter, Jr., *Inst. Phys. Conf. Ser.* 137, p. 495, 19xx.
56. H. M. Hobgood, D. L. Barrett, J. P. McHugh, R. C. Clarke, S. Sriram, A. A. Burk, Jr., J. Greggi, C. D. Brandt, R. H. Hopkins, and W. J. Choyke, *J. Cryst. Growth,* **137**, p. 181, 1994.
57. S. Sriram, R. C. Clarke, A. A. Burk, Jr., H. M. Hobgood, P. G. McMullin, P. A. Orphanos, R. R. Siergiej, T. J. Smith, C. D. Brandt, M. C. Driver, and R. H. Hopkins, *IEEE Electron Dev. Lett.* **15**, p. 458, 1994.
58. H. M. Hobgood, R. C. Glass, G. Augustine, R. H. Hopkins, J. Jenny, M. Skowronski, W. C. Mitchel, and M. Roth, *Appl. Phys. Lett.* **66**, p. 1364, 1995.
59. S. Sriram, R. R. Barron, A. W. Morse, T. J. Smith, G. Augustine, A. A. Burk, Jr., R. C. Clarke, R. C. Glass, H. M. Hobgood, P. A. Orphanos, R. R. Siergiej, C. D. Brandt, M. C. Driver, and R. H. Hopkins, *53rd Dev. Res. Conf. Digest,* Charlottesville, VA, p.104, 1995.
60. C. E. Weitzel, J. W. Palmour, C. H. Carter, Jr., and K. J. Nordquist, *IEEE Electron Dev. Lett.* **15**, p. 406, 1994.
61. S. Sriram, G. Augustine, A. A. Burk, Jr., R. C. Glass, H. M. Hobgood, P. A. Orphanos, L. B. Rowland, T. J. Smith, C. D. Brandt, M. C. Driver, and R. H. Hopkins, *IEEE Electron Dev. Lett.* **17**, p. 369, 1996.
62. S. T. Allen, J. W. Palmour, V. F. Tsvetkov, S. J. Macko, and C. H. Carter, Jr., *53rd Dev. Res. Conf. Digest,* Charlottesville, VA, p. 102, 1995.
63. S. T. Allen, private communication.
64. K. E. Moore, C. E. Weitzel, K. J. Nordquist, L. L. Pond, J. W. Palmour, S. A. Allen, and C. H. Carter, Jr., *IEEE Electron Dev. Lett.* **18**, 1997.
65. S. T. Allen, J. W. Palmour, C. H. Carter, Jr., C. E. Weitzel, K. E. Moore, K. J. Nordquist, and L. L. Pond, *IEEE MTT-S Dig.*, p. 681, 1996.
66. A. W. Morse et al., *IEEE MTT-S Digest*, 1997 (in publication).
67. Y. Watanabe and J. Nishizawa, Japanese patent 205 068, published No. 28-6077, application date, Dec. 1950.

68. J. Nishizawa, T. Terasaki, and J. Shibata, *IEEE Trans. on Elect. Dev.* **22**, p. 185, 1975.
69. B. J. Baliga, *IEEE Trans. on Elect. Dev.* **29**, p. 1560, 1982.
70. A. I. Cogan, R. J. Regan, I. Bencuya, S. J. Butler, and F. Rock, *IEDM Dig.*, p. 221, 1983.
71. R. J. Regan, A. I. Cogan, S. J. Butler, I. Bencuya, and P. Haugsjaa, paper presented at 14th European Microwave Conf., Aug. 1984.
72. M. G. Kane and R. Frey, Microwave Systems News **14**, 1984.
73. I. Bencuya, A. I. Cogan, S. J. Butler, and R. J. Regan, *IEEE Trans. on Elect. Dev.* **32**, pp. 1321, 1985.
74. R. J. Regan, I. Bencuya, S. J. Butler, S. Stites, and W. Harrison, Microwaves & RF **24**, April 1985.
75. R. J. Regan and S. J. Butler, Proc. RF Tech. Exp., 1985.
76. S. J. Butler and R. J. Regan, Proc. RF Tech. Exp., 1986.
77. R. J. Regan, Proc. Southwest Semi. Elect. Exp., Oct. 1986.
78. R. J. Regan, S. J. Butler, E. Bulat, A. Varallo, M. Abdollahian, and F. Rock, paper presented at the RF Exposition East, Nov. 1986.
79. A. W. Morse, P. Esker, R. C. Clarke, C. D. Brandt, R. R. Siergiej, and A. K. Agarwal, MTT-S Dig., 1996.
80. R. C. Clarke, A. K. Agarwal, R. R. Siergiej, C. D. Brandt, and A. W. Morse, *54th Dev. Res. Conf.*, Santa Barbara CA, p. 62, 1996.
81. J. W. Palmour, S. T. Allen, R. Singh, L. A. Lipkin, and D. G. Waltz, "4H-Silicon Carbide Power Switching Devices," *Inst. Phys. Conf. Ser.* 142, pp. 813–816, IOP Publishing, Ltd, Bristol, 1996.
82. A. K. Agarwal, R. R. Siergiej, M. H. White, P. G. McMullin, A. A. Burk, L. B. Rowland, C. D. Brandt, and R. H. Hopkins, "Critical Materials, Device Design, Performance and Reliability Issues in 4H-SiC Power UMOSFET Structures," *Mat. Res. Soc. Symp. Proc.* **423**, pp. 87–92, 1996.
83. J. N. Shenoy, J. A. Cooper, Jr., and M. R. Melloch, "High-Voltage Double-Implanted Power MOSFET's in 6H-SiC," *IEEE Electron Dev. Lett.* **18**, 3, pp. 93–95, Mar. 1997.
84. J. B. Casady, A. K. Agarwal, L. B. Rowland, W. F. Valek, and C. D. Brandt, "900 V DMOS and 1100 V UMOS 4H-SiC Power FETs," *IEEE 55th Annual Device Research Conference*, June 23–25, 1997.
85. K. Xie, J. H. Zhao, J. R. Flemish, T. Burke, W. R. Buchwald, G. Lorenzo, and H. Singh, "A High-Current and High-Temperature 6H-SiC Thyristor," *IEEE Electron Dev. Lett.* **17**, 3, pp. 142–144, Mar. 1996.
86. A. K. Agarwal, J. B. Casady, L. B. Rowland, S. Seshadri, W. F. Valek, and C. D. Brandt, "700 Volt Asymmetrical 4H-SiC Gate Turn-Off (GTO) Thyristors," *IEEE Electron Device Letters*, submitted Feb. 11, 1997.

CHAPTER 6

SiC Microwave Devices

R. J. Trew

U.S. DEPARTMENT OF DEFENSE
WASHINGTON, DC

I.	INTRODUCTION	237
II.	BACKGROUND	239
III.	SEMICONDUCTOR MATERIAL AND CONTACT PROPERTIES	241
IV.	SEMICONDUCTOR DEVICE MODELS	247
V.	TEMPERATURE EFFECTS	250
VI.	RF ACTIVE DEVICES	252
	1. *MESFETs*	253
	2. *SiC Static Induction Transistors*	264
	3. *Bipolar Transistors*	267
	4. *IMPATT Diodes*	272
VII.	SUMMARY	279
	References	280

I. Introduction

The utilization wide-bandgap semiconductors such as SiC for fabrication of high-frequency and microwave electronic devices has been of interest to device physicists and researchers for many years. Based on the electronic and thermal properties of the semiconductor material, it has been predicted that devices fabricated from selected SiC crystal polytypes will have performance superior to that of present-day devices [1], and this has motivated significant development effort. The 6H-, 4H-, and 3C-SiC polytypes, in particular, have been used to fabricate a variety of devices with good dc and RF performance. There is interest in the use of SiC devices for applications such as microwave power amplifiers that can be used in phased-array radars, base station transmitters for mobile communications, high-efficiency and broadband radar transmitters, and other applications.

Although the interest in SiC-based device development has intensified in recent years, serious work with SiC dates at least to the early 1960s. As the material parameters became known, it became apparent that these materials possess a combination of parameters that, in many respects, make them ideal for various applications, including high-frequency power devices for use at elevated temperature and in harsh environmental conditions. Desirable properties of SiC include wide bandgap ($Eg \sim 3\,\text{eV}$), high electron saturation velocity ($v_s \sim 2 \times 10^7\,\text{cm/sec}$), high breakdown field ($Ec > 10^6\,\text{V/cm}$), and high thermal conductivity ($\kappa \sim 4°\text{C/W}$). These properties offer the potential to fabricate a new generation of electronic devices with improved dc and RF performance.

The early attempts to utilize SiC material were hindered by technological problems related to crystal growth, purity, and doping, and to problems associated with the development of suitable ohmic and rectifying contacts. Although rapid progress has been made in recent years on solutions to these problems, growth and device fabrication technology is still primitive compared to that for Si and GaAs devices, and the necessary SiC-based technology has not yet developed to the point where high-performance devices can be easily fabricated. Nonetheless, solutions to the early problems are being determined and reports of prototype devices with significantly improved dc and RF performance are increasing. RF and microwave devices that can be fabricated from SiC include metal semiconductor field-effect transistors (MESFETs), static induction transistors (SITs), bipolar junction transistors (BJTs) and heterojunction bipolar transistors (HBTs), and impact avalanche transit-time (IMPATT) diodes. Blue light–emitting diodes (LEDs) have also been fabricated from SiC and are commercially available, and various types of high-temperature SiC diodes and transistors are suitable for use at dc and low frequency for sensing and power-conditioning applications.

In this chapter, the operation and performance of various RF and microwave electronic devices fabricated from SiC are investigated. The investigation makes use of circuit-oriented RF performance simulators that include physically based models of the devices. The simulators permit the operation of the devices to be examined and optimum device structures to be determined. This, in turn, permits performance capability and limitations to be investigated. The results of the simulations are compared to experimental measurements where possible, and excellent agreement between the simulated and measured data is obtained.

II. Background

SiC is the only compound in the SiC system that exists in the solid state, but it can occur in many polytype structures [2, 3]. More than 170 polytypes have been identified. The lone cubic polytype crystallizes in the zinc-blende structure and is denoted 3C-SiC (sometimes called β-SiC). The C indicates the cubic structure and the 3 indicates the atomic repeat sequence. In 3C-SiC, the atomic layers in the crystal structure repeat every third layer. The cubic (C) polytype along with the hexagonal (H) and rhombohedral (R) polytypes are often collectively referred to as α-SiC. The most common polytypes are 3C, 6H, and 4H; however 15R and 2H have also been identified in crystalline form, but are rare. Most of the polytypes are extremely stable, except that 2H is unstable and can transform to other polytypes at temperatures as low as 400°C.

SiC does not exist in significant quantities in nature. The first reported synthesis was accidental: the result of attempts by Berzelius to make diamond. The development of the Acheson process [4] in 1891 brought SiC production to commercial scale in the abrasives industry. Initial SiC research was generally conducted using crystals that were occasional by-products of this process. In 1955, Lely [5] developed a laboratory version of the industrial sublimation process and was able to produce purified 6H-SiC single crystals. Due to the growth temperature of about 2500°C, only 6H-SiC polytypes were produced in the Lely process. Growth of 3C-SiC was sometimes observed during the cool-down phase. The doping level of the SiC crystals depends strongly on the impurity content of the starting material, the quality of the argon atmosphere, as well as temperature and duration of degassing cycles used in this process. The success of the Lely process led to significant research effort during the 1960s directed toward development of SiC.

The electron mobility of 3C-SiC over the temperature range of 27 to 730°C was predicted from theoretical calculations to be significantly greater that can be obtained from 6H-SiC, due to reduced phonon scattering in the cubic material. For this reason, there was early interest in the growth of thin films of 3C-SiC for device applications. However, most recent work has concentrated on the use of 4H-SiC, which has an electron mobility essentially the same as 3C-SiC and twice that of 6H-SiC.

SiC can be doped n- or p-type by diffusion [6], epitaxial growth [7, 8], and ion implantation [9–11]. The diffusion of dopants into SiC requires temperatures around 1900°C, and special precautions are necessary to prevent sublimation of the bulk crystal at this temperature. Moreover, this temperature is considerably above the melting point of SiO_2, and no commercially attractive alternative diffusion masking material has yet been found. Therefore, doping during epitaxial growth or ion implantation with boron (B) or aluminum (Al) for p-type and phosphorus (P) or nitrogen (N) for n-type is preferred. The most commonly used acceptor and donor atoms are Al and N. Using epitaxial growth, the dopant gases of N_2 (or NH_3), PH_3, B_2H_6, and $AlCl_3$ (or $Al(CH_3)_3$ carried in H_2) are incorporated directly into the primary gas stream during chemical vapor deposition (CVD). For ion-implanted crystals, damaged or amorphous regions are annealed at a temperature between 1400 and 1800°C [11]. Unintentionally doped 6H-SiC epilayers are usually n-type [12] with electron concentrations and mobilities of mid 10^{16} cm^{-3} and 500 to 700 cm^2/V-sec, respectively.

Historically, SiC can be considered one of the first known semiconductors, and electroluminescence was reported by Round [13] in 1907. Blue LEDs have been fabricated and are now commercially available. Bipolar transistors fabricated from material grown by a similar technique have also been reported [8]. MESFETs fabricated from 6H-SiC were first reported by Muench et al. [8, 14] in 1977. In this early work, a thin layer of n-type SiC doped with nitrogen to a concentration of about 10^{16} cm^{-3} was deposited by liquid-phase epitaxy (LPE) on a p-type SiC crystal. The ohmic contacts were formed from an Al-Si alloy, and the gate was formed by depositing thin layers of titanium and gold. Current saturation was observed, but the device gain was very low and limited by resistive losses. The maximum transconductance was reported to be only 1.75 mS/mm for a device with a gate length of 10 μm. The first 3C-SiC MESFET was fabricated by Yoshida et al. [15]. An Al-doped p-type 3C-SiC layer was epitaxially grown on a p-type Si substrate, followed by growth of an unintentionally doped n-type layer. The device had high channel resistance, and a maximum transconductance of only 90 μS/mm was obtained. Improved devices with transconductances of 1.7 mS/mm and 0.15 mS/mm at room temperature and 400°C were later reported by the same researchers [16]. Junction-gate field-effect transistors [17–19] and depletion-mode metal-oxide semiconductor field-effect transistor (MOSFETs) have also been fabricated from 3C-SiC grown by CVD on a 3C-SiC substrate [10, 11]. A 3C-SiC MESFET with a 1-μm gate length demonstrated a maximum room-temperature transconductance of 25 mS/mm and a current gain cut-off frequency (f_T) of about 3 GHz [20]. Although this device has a significantly improved transconductance relative to previously reported devices, its performance is still limited by high parasitic resistances. Improved device fabrication techniques yield lower parasitics and devices with impressive dc and RF performance have been reported. For example, 6H-SiC MESFETs that can operate at frequencies up to X-band (10 GHz) have been produced with RF output power for a class A amplifier in the range of 2.5 W/mm and power-added efficiency (PAE) on the order of 45% at 6 GHz [21]. MESFETs with excellent RF performance have also been fabricated from 4H-SiC with RF output power on the order of 2.8 W/mm at 1.8 GHz [22], and 2.27 W/mm with 65.7% for a class B amplifier reported at 850 MHz [23, 24]. A 4H-SiC MESFET with an f_{max} of 42 GHz has been reported [25]. This device produced 5.1-dB gain at 20 GHz. SIT's also look very promising, and a 4H-SiC SIT with 38-W RF output power, 9.5 dB of gain, and 45% drain efficiency at 3 GHz has been reported [26, 27]. The device was operated under pulse bias and is useful for radar applications [28]. A two-stage amplifier with 1-kW RF output power using these devices was reported and is a commercial product for the high-definition television market. HBTs fabricated from SiC may also be possible, although the low mobility of p-type material required for the base region may limit the frequency performance of the device [1, 29]. The most commonly employed acceptor is Al, which has an activation energy of Ea $\sim$ 220 meV in SiC. This results in low activation at normal operating temperature, and very high impurity density must be used to produce low-resistivity material. Typically, the acceptor density is about two orders of magnitude greater than the free-hole density at room temperature. Due to the high acceptor density, impurity scattering dominates and the hole mobility is very low, generally in the range of 20 to 50 cm^2/V-sec, and it is very difficult to produce low-resistance base regions. This may limit the operation of bipolar transistors to S-band (4 GHz) [1] or less, unless techniques can be found to produce significantly reduced base resistances. It has been determined that contact resistance to the p-type base must be reduced to about Rc $\sim 10^{-6}$ Ω-cm^2 in order to obtain X-band (i.e., 8–12 GHz) performance [29]. The p-type contact resistance must be reduced one to two orders of magnitude lower than currently obtained. HBTs using GaN as the emitter and SiC for the base and collector have been proposed and initial results reported [30]. DC current gain as high as 100,000 was reported. Since SiC has very high breakdown fields, however, devices with excellent high-power performance are possible at lower frequencies.

III. Semiconductor Material and Contact Properties

The RF performance of electronic devices is determined by both the structural design of the particular device and the electronic, breakdown, and thermal characteristics of the material from which the device is fabricated. In addition, the manufacture of devices requires that low-resistance ohmic

TABLE I

MATERIAL PROPERTIES FOR SEVERAL SEMICONDUCTORS

Material	E_g (eV)	ε_r	κ ($W/°$K-cm)	E_c (V/cm)
Si	1.12	11.9	1.5	3×10^5
GaAs	1.43	12.5	0.54	4×10^5
InP	1.34	12.4	0.67	4.5×10^5
4H-SiC	3.2	10.0	4.0	3.5×10^6
6H-SiC	2.86	10.0	4.0	3.8×10^6
GaN	3.4	9.5	1.3	2×10^6
Diamond	5.6	5.5	20–30	5×10^6

contacts be fabricated between the semiconductor and external metal conductors. Rectifying contacts are also required for many devices in order to establish potential barriers for the control of currents within device structures. In general, contact technology is difficult on wide-bandgap semiconductors, especially for ohmic contacts.

A summary of the semiconductor material properties for several semiconductors most important to electronic device performance is provided in Table I. A large energy gap is desirable because it results in low intrinsic concentration and resistance to ionization from radiation. A wide energy bandgap also permits operation at elevated temperature, and SiC can be operated at temperatures in excess of 500°C without excessive thermal intrinsic charge generation. The bandgap of SiC is almost three times that of Si and over twice that of GaAs. The bandgap is also instrumental in determining the resistance of the device to radiation. Minority carrier lifetimes are important when designing bipolar devices such as transistors and diodes, where switching speed from a conductive to a nonconductive state is of importance.

It is desirable that the dielectric constant for the semiconductor be low. The dielectric constant is a measure of the capacitive loading of the device, and a low dielectric constant produces reduced device impedances. This means that for the same device impedance, a larger device area can be used, and this, in turn, will permit higher RF power levels to be developed. The dielectric constant for SiC is about 20% less than for Si or GaAs.

High thermal conductivity is extremely important since many of the electronic devices of interest are intended for high-power applications. The dc and RF performance of these devices depends on the ability to extract the heat due to dissipated power. Device operational efficiency is directly dependent on operating temperature, and efficient operation requires that dissipated energy have an effective outlet from the device. The unconverted dc power produces a temperature rise in the device, which, in turn, degrades the electronic charge carrier transport characteristics. Since the charge transport characteristics determine the current density that can flow in the device, an increase in operating temperature produces reductions in device current. This, in turn, reduces the RF power that can be generated by the device. The thermal conductivity of SiC is excellent and a factor of three higher than for Si and a factor of eight higher than for GaAs.

At typical $N_d \sim 10^{17}$ cm^{-3} doping levels, the electron mobility in 6H-SiC and 4H-SiC is on the order of $\mu_n \sim 250$ cm^2/V-sec and 500 cm^2/V-sec, respectively. The hole mobility in both materials is very low and on the order of $\mu_p \sim 20$ to 50 cm^2/V-sec. The higher electron mobility of 4H-SiC is one of the reasons this material is preferred over 6H-SiC for device fabrication. Low mobility results in increased parasitic resistance, increased losses, and reduced gain. The problem is, of course, worse as operating frequency is increased. The low value for hole mobility severely limits the use of p-type SiC in RF transistors that are to operate above about 2 to 3 GHz. Microwave transistors with very good dc and RF performance can be fabricated with the conducting fabricated from majority carrier n-type semiconductor, without the need to use lossy p-type material in the primary current path. The two most attractive transistor structures of this type for fabrication in SiC are the MESFET and the SIT. Both of these transistors can be fabricated with n-type conducting channels with the need to use no or minimal p-type regions.

The dc and RF currents that flow through a device are directly dependent on the charge carrier velocity versus electric field transport characteristics of the semiconductor material. Generally, high charge carrier mobility and high saturation velocity are desirable. A comparison of the electron velocity–electric field (v-E) characteristics for several semiconductors is shown in Fig. 1. The v-E characteristic is described in terms of charge carrier mobility defined from the slope of the v-E characteristic at low electric field and the saturated velocity defined when the carrier velocity obtains a constant, field-indendent magnitude. A primary disadvantage of fabricating transistors from SiC is the relatively low values for the charge carrier mobilities. In general, the wide-bandgap semiconductors such as SiC have relatively low mobility but very high saturation velocity. For typical device doping density ($N_d \sim 2 \times 10^{17}$ cm^{-3}), the electron mobility for 6H- and 4H-SiC is about 250 cm^2/V-sec and 500 cm^2/V-sec, respectively. The factor of 2 increase in mobility for 4H-SiC compared to 6H-SiC is one of the major reasons that the 4H polytype is preferred for device applications, and most current device development effort is directed toward use of the 4H polytype. Also, the mobility of 6H-SiC is anisotropic and has different magnitude

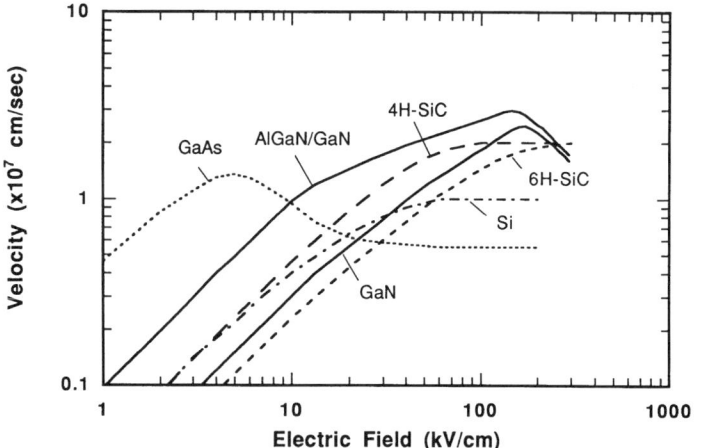

FIG. 1. Electron velocity versus electric field for several semiconductors at $10^{17}\,\text{cm}^{-3}$.

depending on crystal orientation. The 4H polytype does not demonstrate this phenomenon to a major extent. The electron saturation velocity in both 6H- and 4H-SiC is $v_s \sim 2 \times 10^7\,\text{cm/sec}$, which is a factor of 2 higher than for Si ($v_s \sim 1 \times 10^7\,\text{cm/sec}$) and a factor of 4 higher than for GaAs ($v_s \sim (0.5$–$0.6) \times 10^7\,\text{cm/sec}$). The magnitude of electric field that produces saturated charge velocity is also important because the device must be able to develop the saturation field to obtain maximum RF performance and high-frequency operation. The saturation fields for 4H- and 6H-SiC are about $E_s \sim 60\,\text{kV/cm}$ and $E_s \sim 200\,\text{kV/cm}$, respectively, which are high compared to the comparable values of $E_s \sim 3\,\text{kV/cm}$ and $E_s \sim 35\,\text{kV/cm}$ for GaAs and Si. Operation under velocity saturation conditions permits high dc and RF currents to be developed and should permit efficient RF operation through the microwave region and into the low millimeter-wave region. The hole mobility in SiC is very low (on the order of 20–50 cm^2/V-sec), and it is very difficult to observe saturation effects. The extremely low mobility requires very high saturation fields, which approach the critical field for avalanche breakdown. The low hole mobility presents serious problems for use of p-type material in devices.

The critical electric field for breakdown of the semiconductor is fundamental in determining the maximum power handling capability of the device. The breakdown field places a limit on the total voltage that can be applied to a device and is therefore instrumental in determining the RF power that can be converted from the dc bias. The wide bandgap of SiC results in high critical electric field for breakdown, which is on the order of 3 to 4 MV/cm and almost an order of magnitude greater than for Si and GaAs. High breakdown voltage permits high drain bias voltages to be applied, which is necessary to obtain high RF output power. The high breakdown field also permits the electric field internal to the device to achieve the value necessary to produce charge carrier velocity saturation. This latter factor is fundamental to successful development of SiC electronic devices because the charge carrier mobility is low and high velocity saturation fields result. High critical field for breakdown permits the saturation fields to be achieved, and this permits device operation in the velocity saturation region that is necessary for high performance operation. Avalanche breakdown can be described in terms of the ionization integral, written as

$$\int_0^W \beta_p \exp\left[-\int_0^x (\beta_p - \alpha_n)dx'\right]dx = 1 \tag{1}$$

where β_p and α_n are the hole and electron ionization coefficients and W is the avalanche region. The ionization rates for 6H-SiC have been found to be strongly anisotropic with respect to avalanche breakdown [31]. Although breakdown in SiC is not well understood, it has been suggested [31] that the process of impact ionization is influenced by superstructure splitting in the conduction band and that holes dominate the carrier generation. The avalanche generation by electrons is considered insignificant. Other reports [32] also find the avalanche breakdown to be anisotropic and dominated by hole generation but suggests that the avalanche mechanism involves deep-level states corresponding to residual impurities and not the conduction-band superstructure. For modeling purposes, the ionization rates can be expressed as

$$\alpha, \beta = A_{n,p} \exp\left(-\frac{B_{n,p}}{|E|}\right) \tag{2}$$

where the constants are determined by comparison to experimental data. Epitaxial 6H-SiC is typically grown in the c-crystal-axis direction, and when the electric field is parallel to the c-axis, the ionization rates listed in Table II result. These values were determined by comparison to breakdown data for 6H-SiC p-n junction diodes and yield results that agree with experimental data to within a few percent. The a-axis breakdown voltage is about one-third of that for c-axis orientation for equivalent doping levels. These values are suitable for use in device design models.

TABLE II

INDIRECTLY DETERMINED IONIZATION COEFFICIENTS FOR 6H-SiC

Parameter	c-Axis	a-Axis
A_n(cm^{-1})	1.0×10^6	1.6×10^6
A_p(cm^{-1})	5.1×10^7	5.8×10^7
B_n(V/cm)	2.4×10^7	1.7×10^7
B_p(V/cm)	1.8×10^7	9.6×10^6

Both ohmic and rectifying contacts are required in device fabrication. Contact technology, however, is difficult on wide-bandgap semiconductors. When metals are placed on these materials, rectifying behavior is generally obtained. Good rectifying contacts to SiC have been formed by deposition of Ti/Au and thermal evaporation of Au. Gold, which is the most successful Schottky material for SiC, has several disadvantages, including poor adherence to SiC and reaction with SiC above 400°C. There is interest in refractory metals and refractory metal silicides (e.g., PtSi$_x$) for use as Schottky barriers on SiC. Rectifying p-n junctions in SiC are formed by *in situ* doping during growth or by ion implantation. Typical p-type and n-type dopants are B or Al and P or B, respectively. Diodes typically exhibit high reverse voltages with low leakage current, high current-carrying capability, and fast switching speed. High-temperature operation is possible, and no significant degradation in junction characteristics are observed, at least to 350°C. The built-in potential obtained from gold on SiC is in the range of 1.4 to 1.7 eV, compared to 0.7 to 1.0 eV obtained with GaAs and Si. Ideality factors for SiC junctions are generally about 1.4 to 2.0, indicating a significant generation–recombination current mechanism. The absence of a pure diffusion current (which would be indicated by an ideality factor approaching unity) is consistent with the low intrinsic density of $n_i \sim 10^{-5}$ cm^{-3} characteristic of SiC. Due to the high critical field, SiC diodes have high breakdown voltage (e.g., $V_B > 400$ V are easily achieved). Reverse leakage currents can be very low due to the wide bandgap, and reverse-saturation current densities in the range $J_s \sim 10^{-14}$ A/cm^2 have been obtained.

Ohmic contacts to SiC can be formed by deposition of metals such as Ni, Ag, Ta, W, Mo, and Ti. Sputtered TaSi$_2$ has provided a good ohmic contact to n-type SiC; however, e-beam evaporation of elemental Ta has produced ohmic contacts on this material, with contact resistivities of Rc $\sim 10^{-4}$ Ω-cm^2. Annealed Ni followed by Au plating has produced ohmic contacts to heavily doped SiC with contact resistivities approaching the range of Rc $\sim 10^{-6}$ Ω-cm^2, which is sufficient for fabrication of low-resistance contacts for use in high-performance microwave devices.

IV. Semiconductor Device Models

Semiconductor device models can be formulated by solution of the basic semiconductor device equations applied to a particular device structure. The equations are solved to determine expressions for the current flowing through the device subject to the applied dc and RF voltages. The fundamental set of semiconductor device equations provides the basis for development of a device model, which can be either small-signal or large-signal. The equations consist of the current-density equations, the continuity equations, Poisson's equation, and Faraday's law. These equations must be solved simultaneously with the appropriate boundary conditions in order to obtain an accurate representation of the device operation. For devices such as bipolar transistors, where both electrons and holes are important to the operation of the device, two sets of equations are required, one for each type of charge carrier. These equations form the drift–diffusion approximation that is applicable to conditions where the mobile charge carriers can be considered to be in thermal equilibrium with the crystal lattice. These conditions are generally valid when the device dimensions are large compared to the wavelength of the operation frequency, or when the RF period is long compared to the charge carrier relaxation time. When these conditions are not applicable, non-equilibrium (i.e., hot-electron) effects can be significant, and the drift–diffusion approximation is not valid. Under these conditions, alternate modeling approaches must be employed. For the majority of devices that operate under large-signal conditions, the drift–diffusion approximation is valid and the current density equation can be separated into terms that describe current conduction by drift and diffusion mechanisms. For electrons, the current-density equation can be written as

$$\vec{J}_n = q\mu_n n \vec{E} + qD_n \nabla n \tag{3}$$

where the first term indicates that electrons drift in response to an applied electric field, with current density proportional to the electric field according to the electron mobility μ_n. The second term indicates that electron conduction also occurs by a diffusion mechanism, where current density flows in response to a gradient in the electron density. For the latter case, the proportionality constant is the electron diffusion coefficient D_n (units of cm^2/sec). For nondegenerate semiconductors, the diffusion coefficient and the mobility are related by the Einstein relation

$$D_n = \mu_n \frac{kT}{q} \tag{4}$$

The current density equation, as expressed in Eq. (3) applies to relatively low electric field operation where electron flow is linear. In this region, the electron velocity is directly related to the magnitude of the electric field, according to the relation

$$v = \mu_n E \qquad (5)$$

For large electric fields, as are encountered under large-signal operating conditions, the electron velocity saturates and becomes a nonlinear function of electric field.

The continuity equation for electron is

$$\frac{\partial n}{\partial t} = G_n + \frac{1}{q}\nabla \cdot \vec{J}_n \qquad (6)$$

This equation states that the time rate of change of charge within a volume is equal to the rate of flow of charge out of the volume. Charge within the volume may be generated by some mechanism (G_n), such as optical excitation or avalanche ionization. Poisson's equation and Faraday's law complete the basic set of equations required for device analysis. Poisson's equation is

$$\nabla \cdot \vec{E} = \frac{q}{\varepsilon}(N_d - n) \qquad (7)$$

where N_d represents the density of ionized donor impurities and n is the free electron density (both have units of cm^{-3}). Faraday's law is

$$\nabla \times \vec{E} = -\frac{\partial \vec{B}}{\partial t} \qquad (8)$$

Magnetic effects (B) can usually be ignored so that the right-hand side of Eq. (8) is zero. This, in turn, allows the electric field to be determined from the spatial gradient of the electric potential, according to the expression

$$\vec{E} = -\nabla V \qquad (9)$$

Simultaneous solution of these equations permits a physically based model for any device to be developed. The model will be a time-dependent current–voltage description of device operation. The model can be solved directly in the time domain for dc or pulse operation. For small- and large-signal RF operation, the model is generally embedded into a circuit simulator that operates in the frequency domain. Harmonic balance techniques can be applied to interface the time-domain device model with the frequency-domain RF circuit simulator. An integrated simulator, such as is shown in Fig. 2 results. This approach has proved to provide accurate simulation of practical circuits, and the technique is widely used in practice.

Since device nonlinearities are inherent in the basic semiconductor equations, it is not necessary to make *a priori* assumptions as to the form or identity of model nonlinearities. These physics-based models can be applied,

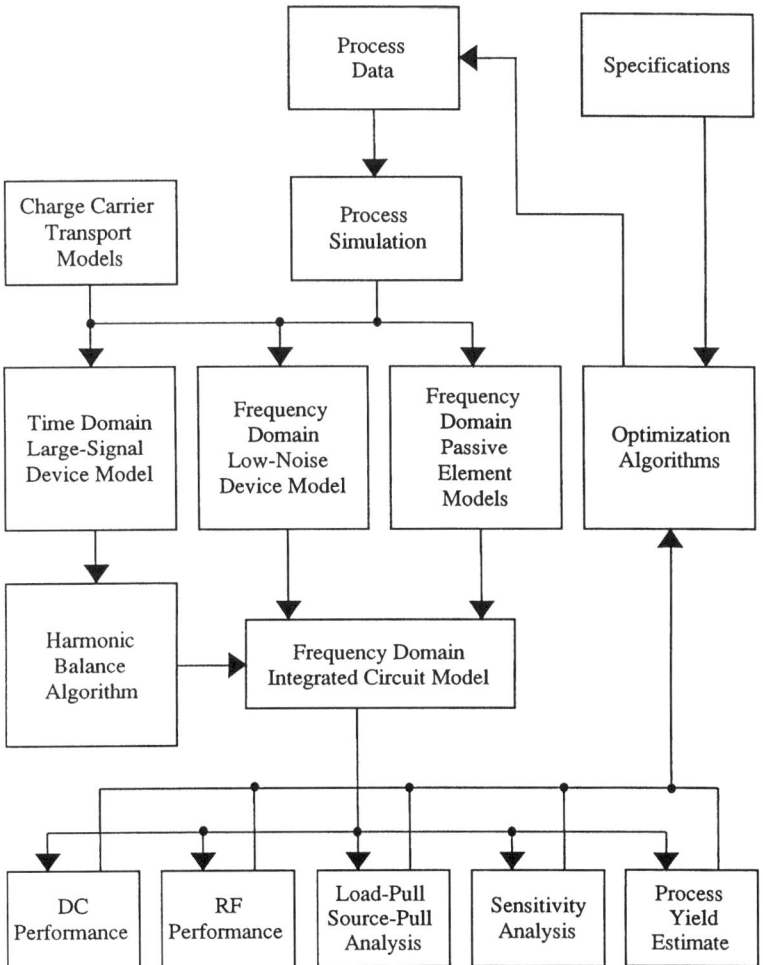

FIG. 2. Block diagram for an integrated microwave circuit simulator.

therefore, in oscillator and mixer applications where multifrequency performance is to be investigated, as well as in amplifier applications. This type of simulator is well suited for both device and RF circuit optimization studies. Application of the simulators to RF amplifier performance optimization is possible, and large numbers of simulations can be efficiently performed.

V. Temperature Effects

Temperature has a significant effect on the operation of SiC electronic devices and influences both the density of free charge and the charge transport characteristics. The free-electron density in the active region of the device depends on the dopant density and also on the activation. In materials such as n-type GaAs, the donor activation energy is so small (e.g., Si is a shallow donor in GaAs with an activation energy of 5.8 meV) that essentially all donors are activated at room temperature. Therefore, in the commonly used semiconductors, such as GaAs, Si, InP, and so on, activation is not an issue. However, in the various polytypes of SiC, the donors are at significantly deeper energy levels in the bandgap, and full activation will require elevated temperature. Activation energy is a function of crystal perfection, and poor crystal quality generally produces increased activation energy. The early work with SiC often demonstrated relatively deep donor levels. However, as crystal quality improved, reduced activation energies were obtained. Nitrogen is the most commonly used donor impurity in SiC, and, currently, n-type 6H-SiC and 4H-SiC demonstrate activation energies of about 85 and 70 meV, respectively. Acceptors in SiC are significantly deeper in the bandgap, and the most commonly used acceptor, Al, has an activation energy of Ea ~ 220 meV. The high activation energy requires high-temperature operation to activate charge carriers.

For device model calculations, the free-electron density is essentially equal to the density of activated donors and can be calculated from the expression

$$N \cong N_{d_{\text{act}}} = \frac{N_d}{1 + g_c \exp\left[\dfrac{E_a - E_f}{kT}\right]} \tag{10}$$

where g_c is the degeneracy factor and has a value of 2, E_a is the activation energy, and the other terms have their usual meaning. The electron transport characteristics degrade with temperature, and the mobility demonstrates a significant temperature effect and scales inversely according to T^{-s},

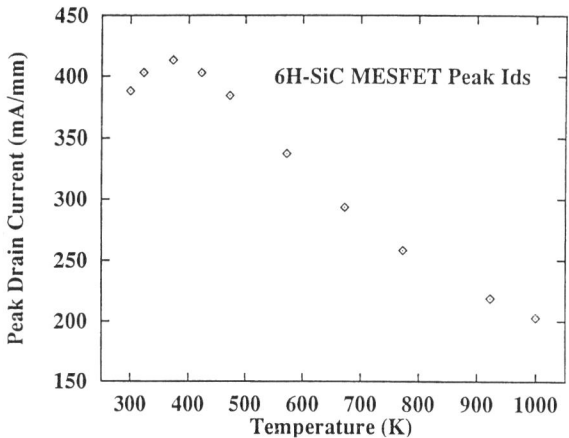

FIG. 3. Open-channel current as a function of temperature for a 6H-SiC MESFET.

where $s \sim 2.1$ for SiC. The hole mobility degrades slightly faster, with temperature with $s \sim 2.5$. The saturated charge carrier velocity does not degrade as rapidly, typically demonstrating $s \sim 1.1$ to 1.6.

The current density that can flow through a semiconductor device is proportional to the product of the density of free charges and the charge carrier transport characteristic. These two factors vary with temperature in an opposite manner, and this results in a maximum device current at some temperature, dependent on the activation energy of the impurity that is used in the particular semiconductor. For example, the open-channel current density in a 6H-SiC MESFET is shown in Fig. 3. The donor impurity is nitrogen with an assumed activation energy of 100 meV. The current density is a maximum at slightly greater than 350 K. Channel current decreases below this temperature due to a reduction in the number of activated carriers and above it due to charge transport degradation due to increased lattice scattering. This type of behavior is often observed in experimental SiC transistors [33].

The temperature at which maximum channel current occurs is directly related to the activation energy. For activation energies below about 80 to 90 meV, essentially all impurities are activated at room temperature. Significant reduction in the density of free carriers at room temperature occurs as activation energy exceeds about 100 to 150 meV. For activation energy of about 220 meV, typical of p-type SiC, only about 1 to 5% of the impurities are activated at room temperature, and operating temperature must reach greater than about 300°C before the majority of the acceptors in bulk material are activated. This means that for room temperature operation and to achieve low resistivity p-type material, an impurity density about 20 to 100 times greater than the desired free-hole density must be used. Excessive impurity scattering occurs, and it is for this reason that hole mobilities are generally very low, with typical values of about $\mu_p \sim 50\,\text{cm}^2/\text{V-sec}$. For this reason, it is desirable to minimize the use of p-type regions in devices fabricated from these materials.

VI. RF Active Devices

Virtually all RF systems require active circuit elements for use as oscillators, amplifiers, and so forth. These elements permit conversion of energy from dc bias sources to RF bands, where the energy can be used to provide useful gain at specified frequencies. Although active elements can be constructed from both diodes and transistors, transistors are preferred since they are three-terminal devices. This permits isolation between input and output ports and significantly simplifies circuit design. A variety of bipolar and field-effect transistor types are possible. The choice of a specific device type depends on factors such as frequency of operation, desired operating power level, and so on. In this section, various SiC transistors and the IMPATT diode are described, and performance principles and limitations are presented.

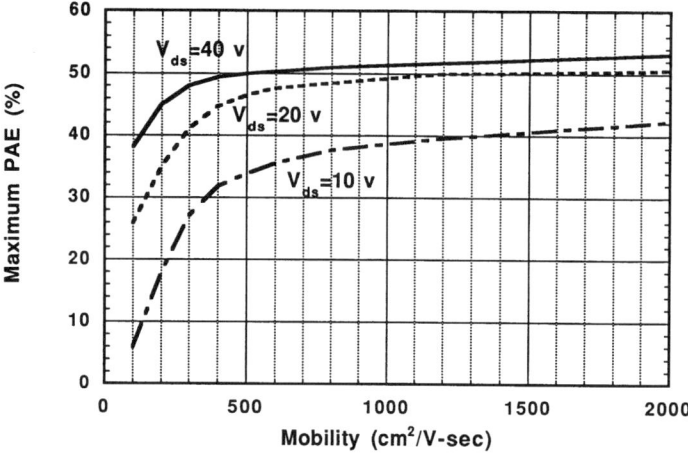

FIG. 4. Maximum PAE for a class A amplifier as a function of semiconductor mobility and drain bias.

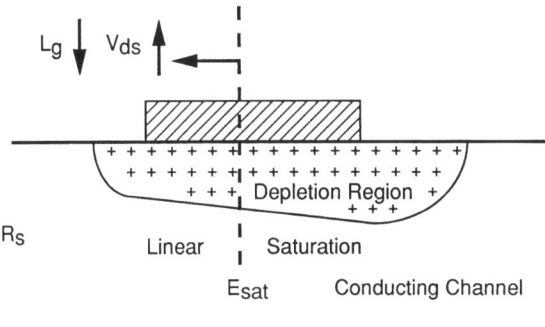

FIG. 5. Conducting channel under the gate in a MESFET.

There has been concern that the low mobility of the wide-bandgap semiconductors would severely limit transistor performance. This is investigated in Fig. 4, which shows the maximum PAE for a MESFET class A amplifier as a function of electron mobility for three values of drain-source voltage. A critical value for mobility exists at which optimum performance is obtained. For mobility below the critical value, device performance is severely degraded, but for mobility above the critical value, essentially no improvement in performance occurs. The reason for this effect is shown in Fig. 5, which shows the conducting channel under the gate. The critical mobility value is achieved when the conducting channel under the gate region operates in velocity saturation. The critical value is a function of gate length and drain voltage because these parameters affect the electric field magnitude under the gate. Reductions in gate length or increases in drain bias will force the electric field toward the saturation value. Once the field is sufficiently high to saturate the electron velocity, the mobility only affects source resistance, and this effect is minimized for power devices by wide gate width. In general, for power MESFETs, the electron mobility needs to be about 500 cm^2/V-sec to obtain near-ideal RF performance. Since 6H-SiC has mobility lower than this value, degraded performance is expected. However, both 4H- and 3C-SiC have mobility adequate for fabrication of high-performance transistors.

1. MESFETs

One of the most promising devices for fabrication in SiC is the MESFET [34]. A cross-section of a typical device is shown in Fig. 6. The MESFET is a planar device fabricated by growth of a thin and doped epitaxial layer

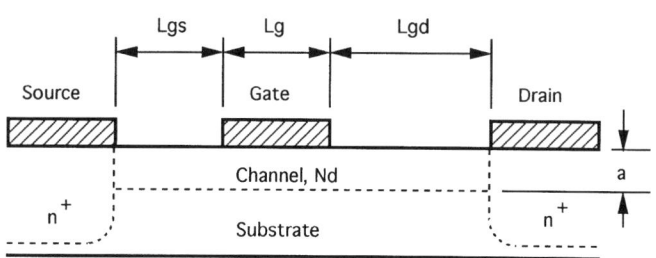

Fig. 6. Cross-section for a MESFET.

located on a semi-insulating substrate. Generally, high-resistivity substrates are desirable and result in improved dc and RF performance for the transistor by confining electrons to the doped epitaxial layer, which is generally thin and on the order of 0.1 to 0.3 μm in thickness. This layer forms a conducting channel. Current is passed through the conducting channel by means of two ohmic contacts (the source and drain), which are separated by some distance, usually in the range of 2 to 10 μm. The actual dimensions are dependent on the operating frequency, and smaller dimensions are used as operating frequency is increased. A rectifying Schottky contact (the gate) is located between the two ohmic contacts. Typically, the gate length is on the order of 0.1 to 2 μm for modern microwave devices. Shorter gate lengths results in improved transconductance (i.e., gain) but lower power handling ability due to reduced breakdown. Typically, the gate-drain breakdown voltage is inversely proportional to gate length. The width of the device is scaled with frequency and typically ranges from about 50 μm for millimeter-wave devices to 1 to 10 mm for power microwave devices. Devices with larger gate periphery are possible at lower frequencies. RF output power, of course, scales with device area. In operation, the drain contact is biased at a specified potential (positive drain potential for an n-channel device), and the source is grounded. The flow of current through the conducting channel is controlled by negative dc and superimposed RF potentials applied to the gate. The RF signal modulates the channel current, thereby providing RF gain. The operation of the transistor is determined by the ability of the gate signal to effectively modulate and control the current in the conducting channel. For this reason, electronic charge that leaks into the substrate or through the gate electrode will produce performance degradation. The availability of high-resistivity, low-leakage substrates and low-leakage Schottky gates in SiC makes high-performance devices possible.

The dc and RF performances of both experimental 6H-SiC and 4H-SiC MESFETs were simulated and investigated. The parameters for the 6H-SiC MESFET [21] are listed in Table III. The simulated and measured dc I–V

TABLE III
EXPERIMENTAL 6H-SiC MESFET PARAMETERS

Parameter	Value
Gate length, L_g	0.5 μm
Gate-source spacing, L_{gs}	0.5 μm
Gate-drain spacing, L_{gd}	1.0 μm
Channel thickness, a	0.21 μm
Channel doping, N_d	2.5×10^{17} cm^{-3}
Unit finger length	160 μm
No. gate fingers	6
Activation energy, E_a	0.10 eV
Electron mobility, μ_n	190 cm^2/V-sec
Saturated velocity, v_s	2×10^7 cm/sec
Energy gap, E_g	2.86 eV
Relative permittivity, ε_r	9.66
Contact resistance, R_c	8×10^{-6} Ω-cm^2
Gate metal work function, W_f	5.7 eV
Gate barrier height, ϕ_b	1.35 eV

characteristics are shown in Fig. 7. As indicated, good agreement between the simulated and measured characteristics is obtained. The device was operated in a class B amplifier circuit at 6 GHz, and the simulated and measured RF performances are shown in Fig. 8. At a drain bias of 45 V, the device produces about 3.5-W RF power, 45% PAE, and about 6-dB linear

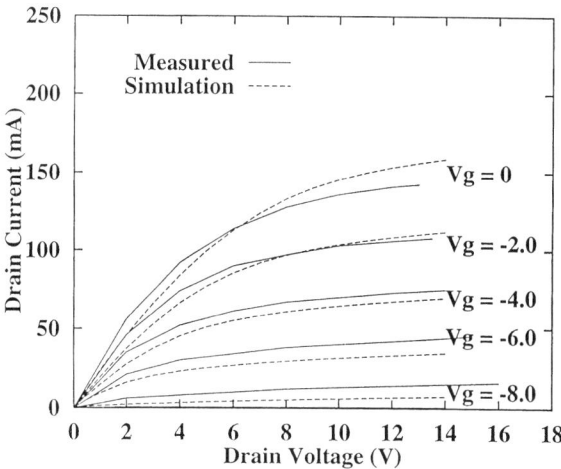

FIG. 7. Measured and simulated dc I–V characteristics for a 6H-SiC MESFET.

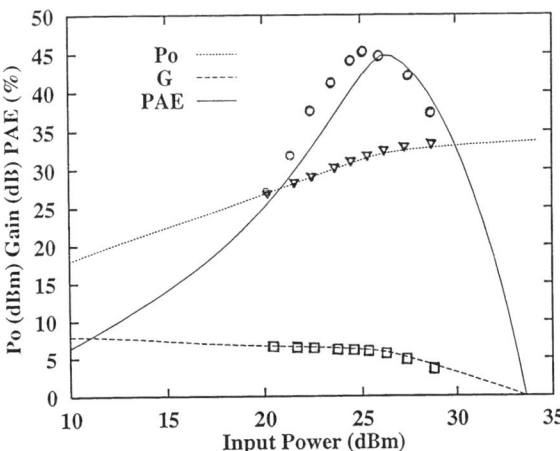

FIG. 8. Measured and simulated RF performance versus RF input power for a 6H-SiC MESFET class B amplifier operating at 6 GHz.

gain. The dc drain curent as a function of RF input power is shown in Fig. 9a and demonstrates typical class B characteristics. That is, as the device is driven into saturation by increasing RF input power, the dc drain current increases due to rectification effects and RF voltage and current waveform clipping. The terminal RF current waveform is clipped on the negative portion of the RF cycle, which results in an increase in the dc component of the current. The dc gate current as a function of RF input power is shown in Fig. 9b. The increase in gate current with RF input power indicates that the device is being driven into saturation by waveform clipping due to forward conduction of the gate electrode. In general, saturation can occur due to forward or reverse conduction of the gate electrode. Reverse conduction occurs when the reverse-biased Schottky gate breaks down and negative dc current in the gate circuit results. The lack of negative gate current for this device indicates that breakdown is not occurring. An experimental 4H-SiC MESFET [22-24] was also simulated, and the measured and simulated dc I–V characteristics are shown in Fig. 10. Again, good agreement between the measured and simulated data was obtained. The device parameter values are listed in Table IV. The MESFET produces a maximum channel current of about 300 mA/mm of gate periphery. The ability of the device to modulate current, which is the measure of the gain of the device, is given by the device transconductance, which for this device is about $g_m = 42$ mS/mm. The ability of the gate bias to turn the device off and on is good, as indicated by the channel current with zero and high

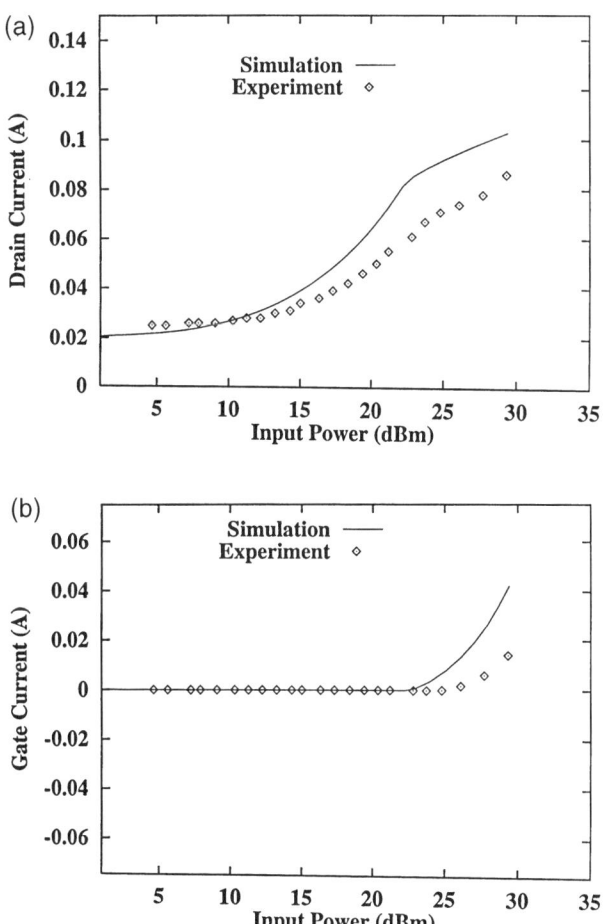

FIG. 9. (a) Measured and simulated dc drain current versus RF input power for a 6H-SiC MESFET class B amplifier operating at 6 GHz. (b) Measured and simulated dc gate current versus RF input power for a 6H-SiC MESFET class B amplifier operating at 6 GHz.

reverse gate bias applied. Very good turn-off characteristics are observed for reverse bias of slightly greater than -8 V.

The measured and simulated large-signal RF performance of the 4H-SiC MESFET operated in a class A amplifier and tuned for maximum RF output power is shown in Fig. 11. The device was biased with a drain voltage of Vds = 50 V, a drain current of Ids = 40 mA, and was operated at F = 850 MHz. As shown, excellent agreement between the simulated and

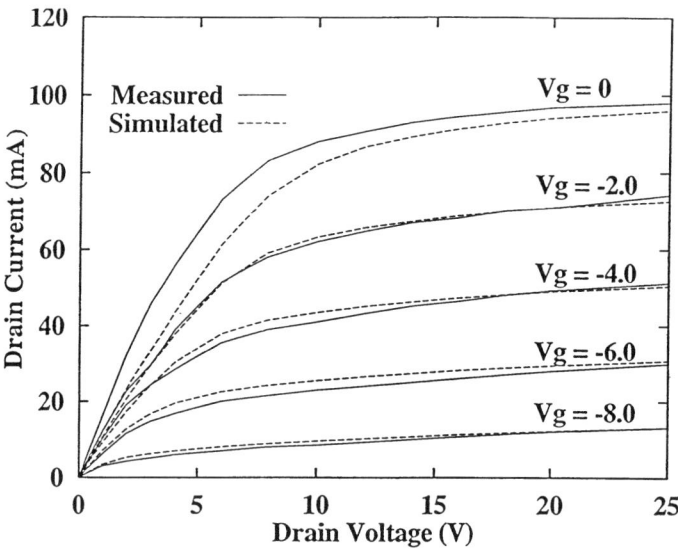

FIG. 10. Measured and simulated dc I–V characteristics for a 4H-SiC MESFET.

measured data was obtained. The amplifier produced a maximum RF output power of 1.12 W, which yields a normalized RF output power density of 3.37 W/mm. The amplifier produced a peak PAE of 38.9% and a linear gain of about 17.5 db. The RF output power obtained is greater than a factor of 3 higher than that obtained from comparable GaAs MESFETs.

To investigate the RF performance capability of SiC MESFETs in microwave power amplifiers, the experimental device designs were modified until optimized performance was obtained. Design optimization investigations were performed for both 6H-SiC and 4H-SiC. The resulting parameter values are listed in Table V.

TABLE IV
EXPERIMENTAL 4H-SiC MESFET PARAMETERS

Parameter	Value
Gate length, L_g	0.7 μm
Gate width, W	332 μm
Source-drain spacing, L_{sd}	1.8 μm
Channel doping, N_d	1.7×10^{17} cm^{-3}
Channel thickness, a	0.26 μm

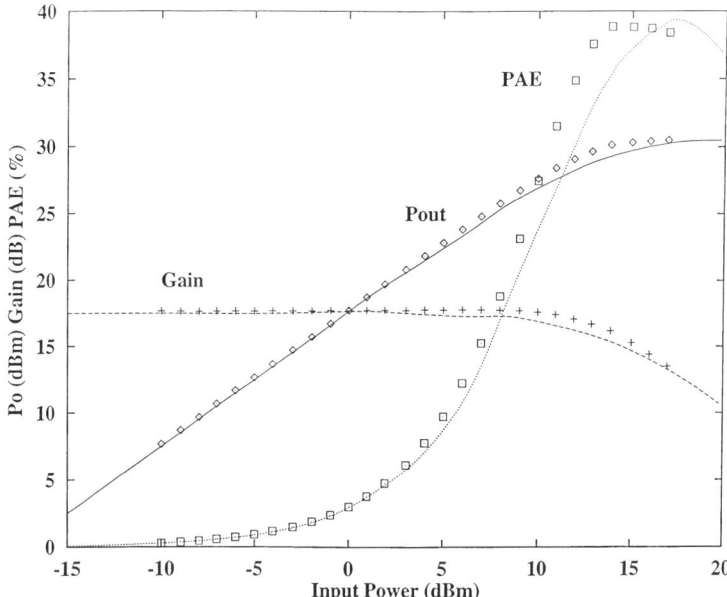

FIG. 11. Measured and simplified RF performance versus RF input power for a 4H-SiC MESFET class A amplifier operating at 850 MHz.

The dc I–V characteristics for the 6H-SiC MESFET at room temperature indicate a maximum channel current of $I_{dss} = 300\,\text{mA/mm}$ and a maximum device transconductance of $g_m = 30\,\text{mS/mm}$. When the device is heated to 500°C, the maximum channel current and device transconductance are reduced to $I_{dss} = 75\,\text{mA/mm}$ and $g_m = 10\,\text{mS/mm}$, respectively. The RF output power, PAE, and gain for the 6H-SiC device operated in a class B

TABLE V
OPTIMIZED SiC MESFET DESIGN PARAMETERS

Parameter	6H-SiC	4H-SiC
Gate length, L_g	0.5 μm	0.5 μm
Gate width, W	1 mm	1 mm
Channel doping, N_d	$5 \times 10^{17}\,\text{cm}^{-3}$	$5 \times 10^{17}\,\text{cm}^{-3}$
Channel thickness, a	0.15 μm	0.15 μm
Electron mobility, μ_n	175 cm^2/V-sec	350 cm^2/V-sec
Saturated velocity, v_s	2×10^7 cm/sec	2×10^7 cm/sec

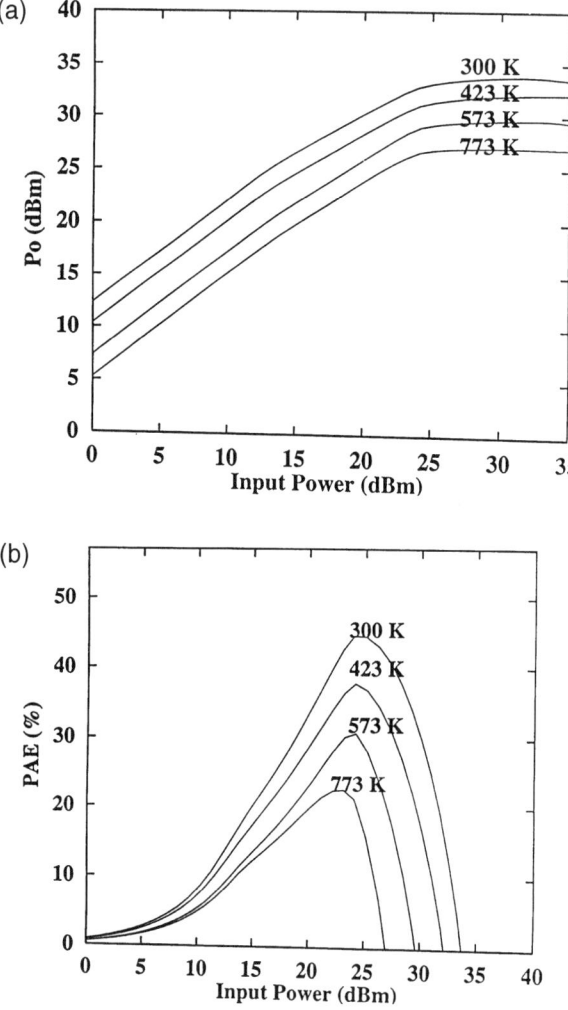

Fig. 12. (a) Simulated RF output power versus RF input power as a function of temperature for an optimized 6H-SiC MESFET class B 8-GHz amplifier. (b) Simulated power-added efficiency versus RF input power as a function of temperature for an optimized 6H-SiC MESFET class B 8-GHz amplifier. (c) Simulated gain versus input power as a function of temperature for an optimized 6H-SiC MESFET class B 8-GHz amplifier.

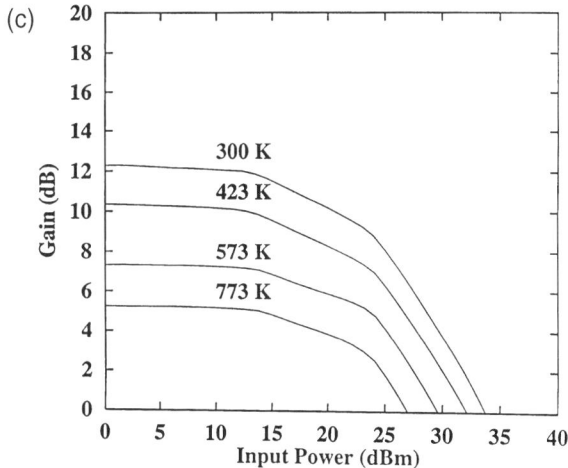

FIG. 12. (Continued).

amplifier are shown in Fig. 12. The calculations are shown for operating temperature, extending from room temperature to 500°C. The amplifier was biased with a drain voltage of $V_{ds} = 40\,\text{V}$ and operated at $F = 8\,\text{GHz}$. At room temperature, the amplifier produces 34 dBm (2.5 W), 45% PAE, and about 12-dB linear gain. At 500°C, the performance degrades to 26.5 dBm (0.45 W), 22% PAE, and slightly over 5-dB linear gain. The dc I–V characteristics for the 4H-SiC MESFET indicated higher dc current, with a maximum channel current of $I_{dss} = 490\,\text{mA}$ and a maximum device transconductance of $g_m = 50\,\text{mS/mm}$. Heating the device to 500°C resulted in a reduction in the maximum channel current to $I_{dss} = 135\,\text{mA/mm}$ and a reduction in the maximum transconductance to $g_m = 18\,\text{mS/mm}$. When operated in a class B amplifier at 8 GHz, the results shown in Fig. 13 were obtained. The RF output power, PAE, and gain are $P_0 \sim 36\,\text{dBm}$ (4 W), 60% PAE, and about 15-dB linear gain, respectively. At 500°C, the amplifier produces about $P_0 = 30\,\text{dBm}$ (1 W), PAE = 32%, and a linear gain of 8.5 dB.

To investigate the RF performance capability of SiC MESFETs in microwave power amplifiers at higher frequencies, the 4H-SiC MESFET amplifier was operated in a class A amplifier, and the performance was simulated as a function of frequency. The RF performance shown in Fig. 14 results. The amplifier performance is good through X-band, with gain above 10 dB and RF output power about 4 to 5 W/mm. Amplifier performance degrades at frequencies above X-band but is still good as high as 30 GHz,

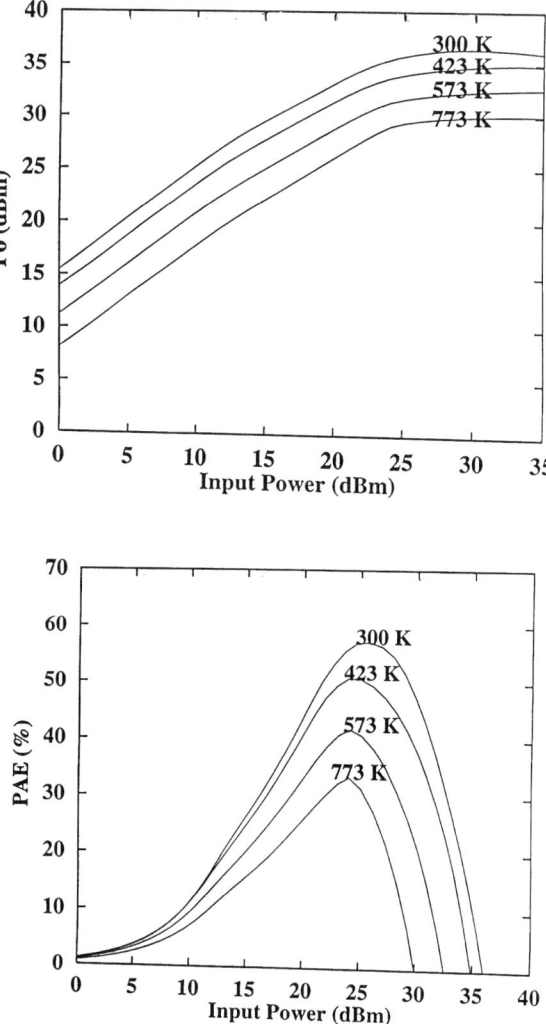

FIG. 13. (a) Simulated RF output power versus RF input power as a function of temperature for an optimized 4H-SiC MESFET class B 8-GHz amplifier. (b) Simulated PAE versus RF input power as a function of temperature for an optimized 4H-SiC MESFET class B 8-GHz Amplifier. (c) Simulated gain versus RF input power as a function of temperature for an optimized 4H-SiC MESFET class B 8-GHz amplifier.

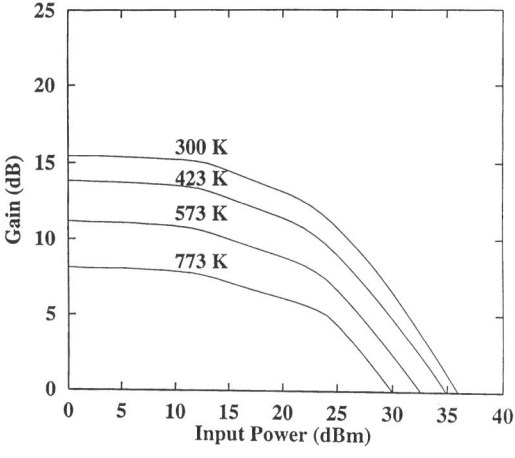

FIG. 13. (*Continued*).

where the amplifier produces 4-W/mm RF output power, 26% PAE, and 4-dB gain. Performance degrades with increasing frequency due to resistive losses that result from the relatively low magnitude for the electron mobility. A 6H-SiC MESFET amplifier produces limited RF performance compared to the 4H-SiC MESFET amplifier due to the reduced electron mobility of

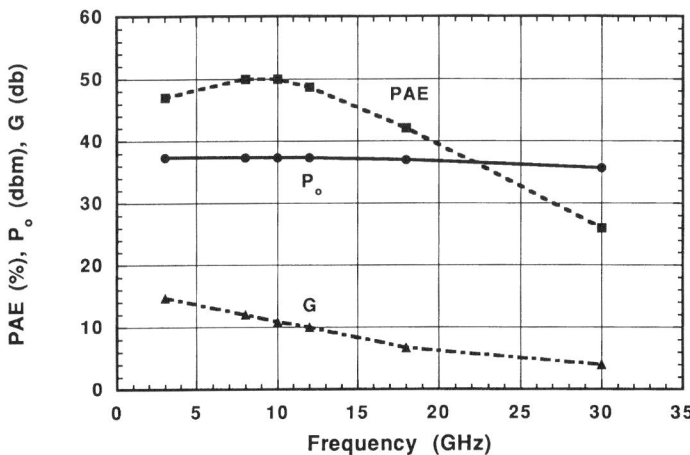

FIG. 14. Simulated RF performance as a function of frequency for a 4H-SiC MESFET class A amplifier.

the 6H material. These calculations indicate that 4H-SiC MESFETs have excellent microwave RF performance potential up to X-band and are capable of producing good RF performance as high as 30 GHz.

2. SiC Static Induction Transistors

SITs are useful as high-power RF sources and amplifiers at UHF and microwave frequencies. The device was originally proposed by Nishizawa *et al.* [35] as a power amplifier for applications such as audio amplifiers. A cross-sectional diagram for the transistor is shown in Fig. 15. The device has a structure very similar to a vacuum triode, with source and drain contacts separated by a certain distance. A potential diagram for the conducting region between the source and drain is shown in Fig. 16a, and the potential as a function of distance is shown in Fig. 16b. Electrons are emitted from the source, which is generally at ground potential, and are accelerated to the drain, which is biased at a positive potential, where they are collected. A grid structure is located in the space between the source and drain electrodes so that the charge carriers can be externally modulated. The grid and drain bias voltages establish a saddle point in the potential in the conducting channel between the grids over which electrons must be injected in order to have current flow between source and drain. The RF gain of the device is determined by the efficiency at which the modulation is affected. The grid structure can be fabricated using p-n junctions or metal Schottky barriers. When the latter structure is used, the device is called the permeable base transistor [36], and these transistors have been fabricated in Si and GaAs

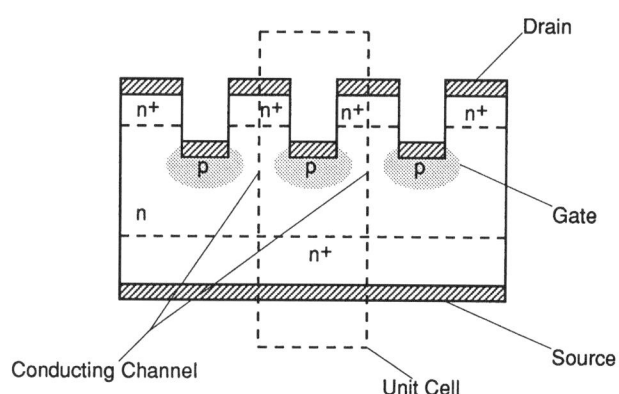

Fig. 15. Cross-section for a SIT.

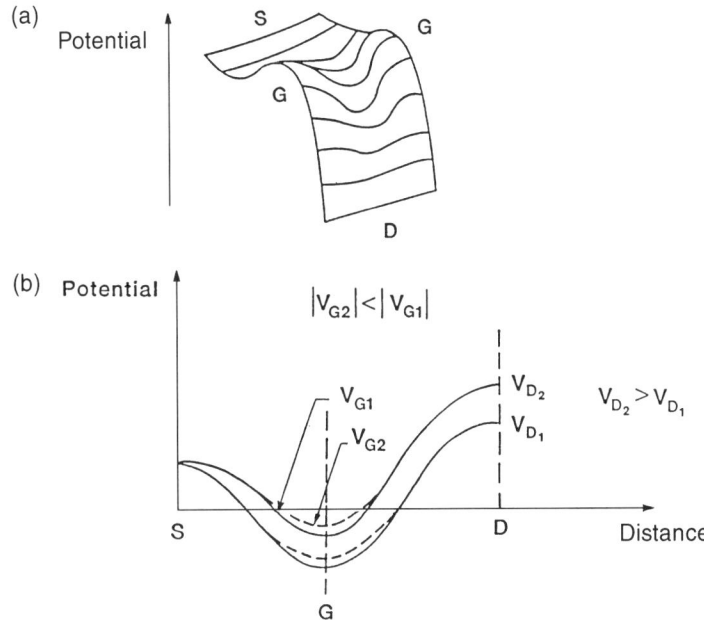

FIG. 16. (a) Potential diagram for a SIT. (b) Potential versus distance for a SIT.

with excellent RF power gain well into the microwave region [37]. Generally, SIT devices are capable of high voltage gain but limited current gain. However, since the device has good impedance characteristics, high power gain is available and the device can be used effectively at frequencies significantly in excess of the f_T for the device.

The device can be operated in various modes, depending on the device dimensions and bias conditions. For example, if the grid junctions are closely spaced and are forward biased, minority charge will be injected into the device from the grid bars. Under these conditions, the minority charge injected from each bar can overlap creating a "virtual base." The device will operate as a bipolar transistor. If the grid bars are reverse biased, depletion regions surrounding the grid bars will be created. If the bars are widely spaced, the depletion regions will work together to control current flow and the device will operate as a vertical field-effect transistor. The saturated dc I–V characteristics of a MESFET, as shown in Fig. 17a, will be obtained. However, if the grid bars are closely spaced, the depletion regions will overlap and a space charge region will be created in the space between the grid bars. The nonsaturated dc I–V characteristics, shown in Fig. 17b and

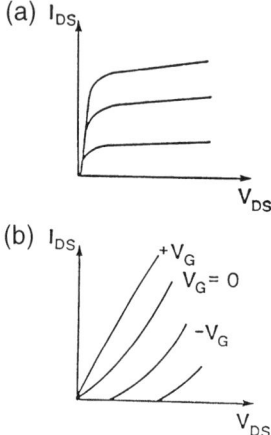

FIG. 17. (a) Saturating dc I–V characteristics for a SIT. (b) Non-saturating dc I–V characteristics for a SIT.

typical of a space-charge limited current device, are obtained. Since the potential in the conducting regions between the grid bars is less than that applied to the bars, a saddle point will be created. Modulation of the saddle point by signals applied to the grids allows grid control. Also, since the saddle potential is affected by the potential applied to the drain, a feedback effect is created. This effect is indicated in Fig. 16b and is the source of the name *static induction transistor*, which indicates that there is a drain dependence to the current flow. This is observed in the I–V characteristics, which have a positive slope, as shown in Fig. 17b.

A limitation to the gain of SIT devices is due to a relatively limited current density that can flow under space-charge-limited operation. Space-charge-limited conduction conditions dominate when the electric field within the conducting channel is suppressed by the space-charge of the injected carriers. A negative feedback is established where further increases in charge density result in increased suppression of the electric field, and a net decrease in current density occurs. This mechanism creates a fundamental limit to the current density that can flow through the device and also a limit to the frequency response of the device. The use of SiC enhances the performance of a SIT because the saturated velocity of electrons in SiC is very high (i.e., $v_s = 2 \times 10^7$ cm/sec). Although the high saturation velocity makes a high channel current possible, the device is still dominated by space-charge-limited operation. The current density in an SiC SIT will be limited by the density of carriers that can be injected.

The relatively low mobility of SiC creates another limitation to the performance of SiC SITs because the electrons must, in general, be accelerated from ground potential at the source to the saturation field. The low electron mobility creates a significant series resistance at the source that degrades the current density that can flow through the device. A possible solution to this problem is to use hot-injection contacts at the source. This may permit the injection of hot electrons directly into the control region of the device where they would travel at saturated velocity. The hot-contact structure is attractive because the hot-injection contacts are located away from the grid region and do not interfere with the modulation and control function. The result would be increased current-carrying capacity and improved current gain. The hot-injection contacts can be fabricated using a heterojunction that has a band discontinuity in the conduction band so that electron energy is higher in the source region. As bias is applied, the electrons would be emitted from the contact material into the SiC and could be injected with sufficient energy to achieve velocity saturation conditions. High RF output power and gain with good PAE should result.

The use of wide-bandgap semiconductors, such as SiC, offers improved RF performance of SIT. In fact, SiC SIT devices with record RF output power in S-band, on the order of 38 W per cell with 45% drain efficiency and 9.5 dB of gain, have been reported [26, 27]. These devices have been used to fabricate a two-stage amplifier with RF output power of 1 kW at 3 GHz [28]. High-performance SiC SITs will, most likely, be limited in frequency response to less than about 10 GHz. These devices, however, are attractive for applications such as RF power amplifiers for cellular system base station transmitters and radar transmitters.

3. BIPOLAR TRANSISTORS

A bipolar transistor is fabricated from a three-layer device structure, as shown in Fig. 18. The device performance is dependent on minority carrier diffusion through the base region, and since electrons diffuse faster than holes, the n-p-n structure is preferred for high-frequency transistors. The necessity of using a p-type base region presents problems for SiC bipolar transistors because the resistivity of p-type SiC is high due to the low activation of Al acceptors, as previously discussed. Also, the contact resistivity to p-type SiC is relatively high and on the order of $Rc \sim 10^{-4}\,\Omega\text{-cm}^2$. This makes it difficult to easily fabricate low-resistance base contacts. Base resistance directly affects device performance, and the base resistance must be reduced as much as possible. Early development work with SiC bipolar transistors [8] produced a transistor with current gains in the range of $\beta \sim 4$

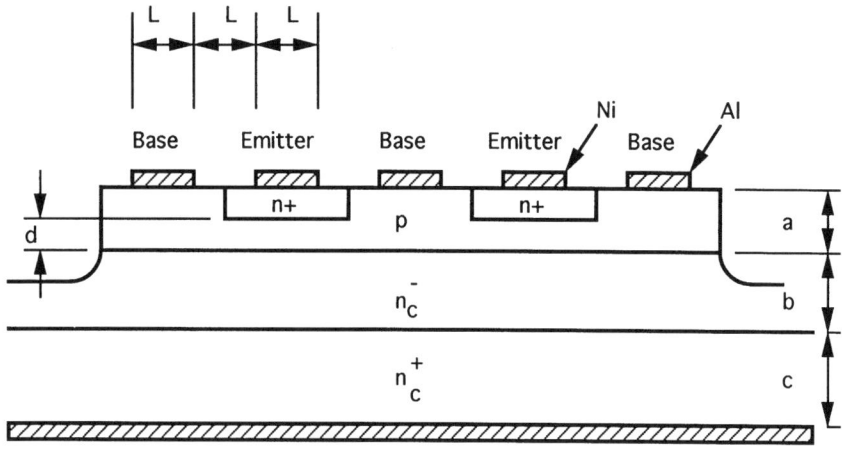

FIG. 18. Cross-section for a bipolar transistor.

to 8. More recent work [30] has produced an HBT using GaN as the emitter and SiC for the rest of the device. Early work with this structure has produced dc common-source current gain of $\beta \sim 100{,}000$. This structure is promising for high-performance and high-power transistors.

The performance of SiC bipolar transistors was simulated in order to determine the capability of these devices [1]. The design of the bipolar transistor was accomplished using an iterative procedure beginning with assumed geometry and doping concentrations. For this investigation, a transistor geometry and doping levels were selected based on physical operation principles. An initial estimate for the device geometry can be determined from impedance matching considerations, subject to distributed effects limitations. A multiple-emitter finger structure was selected. The cross-section of the final device design is shown in Fig. 18 and the design parameters are listed in Table VI.

The device was designed for 10-GHz operation and has a total of eight emitter fingers. The number of emitter fingers was selected based on impedance matching and emitter current density considerations. As the base-emitter junction area is increased, the input impedance decreases. BJTs fabricated from SiC, however, will have a relatively large base resistance, and this factor will ultimately dominate the input impedance. Under these conditions, the device area will be limited by emitter current density and output port impedance-matching considerations. It is desirable to keep the

TABLE VI

SiC BIPOLAR TRANSISTOR DIMENSIONS

Parameter	Value
L	2 μm
a	0.5 μm
b	5 μm
c	50 μm
d	0.2 μm
n^+	2×10^{19} cm^{-3}
p	3×10^{18} cm^{-3}
n^-	8×10^{15} cm^{-3}
n_c^+	4×10^{17} cm^{-3}
$R_{c(E)}$	5×10^{-5} Ω-cm^2
$R_{c(B)}$	7×10^{-4} Ω-cm^2
$R_{c(C)}$	10^{-3} Ω-cm^2

device output impedance in the range of 25 to 50 Ω and the emitter current density in the range of 20 to 30 kA/cm^2.

The most critical design considerations were directed toward the base and collecter region. Base region design issues include the conflicting effects of base region resistance and base region transit time. Collector region design issues include base–collector region capacitance (charge storage) and base–collector depletion region transit time. The base region design involves a calculation of the current gain, base resistance, and base region transit time. The current gain is calculated from consideration of minority carrier transport across the base region. In the common-base configuration, the dc current gain is defined as α_0 and is given by the expression

$$\alpha_0 = \frac{1}{\cosh\left(\frac{W_B}{L_B}\right) + \frac{D_{pB} L_B N_B}{D_{nB} L_E N_E} \sinh\left(\frac{W_B}{L_B}\right)} \tag{11}$$

where the various terms are calculated from the design dimensions listed in Table VI. Base region transit time τ_B is also an important factor, and this parameter is generally defined in terms of the alpha cut-off frequency for the device, defined as

$$f_\alpha = \frac{1}{2\pi\tau_B} = \frac{D_{nB}}{\pi W_B^2} \tag{12}$$

The dc current gain will degrade with frequency according to the expression

$$\alpha = \frac{\alpha_0}{1 + j\dfrac{f}{f_\alpha}} \exp(-j\omega\tau_c) \qquad (13)$$

where the various terms are indicated in Table VII. These expressions indicate the trade-offs between base region and base–collector region transit times in determining the current gain for the device.

The base thickness is $W_B = 0.2\,\mu$m. For this base thickness, a value of $\alpha_0 = 0.894$ is obtained, which results in a common emitter current gain of $\beta_0 = 8.4$. This is low according to Si BJT standards, where α_0 and β_0 are typically greater than 0.95 and 20.0, respectively. The parameters listed in Table VI yield a base resistance value of $29.7\,\Omega$ for the eight-emitter finger device. Selection of the base region thickness and doping concentration are critical design factors for the device. As base region thickness is reduced, α_0 increases and base region transit time decreases, enhancing performance, but base resistance increases, thereby degrading performance. An optimum thickness for the base region exists. An increase in base region doping reduces base resistance but decreases α_0 and base region transit time.

TABLE VII

SiC Bipolar Transistor Element Values

Parameter	Value
No. emitter fingers	8
V_{ce}	150 v
I_{ce}	1 A
BV_{ce}	346 v
R_B	29.7 Ω
R_{BE}	0.05 Ω
C_{BE}	15.92 pF
$C_{BE(0)}$	11.93 pF
R_E	2.5 Ω
C_{BC}	0.184 pF
$C_{BC(0)}$	1.5 pF
R_{BC}	250 Ω
R_C	6.4 Ω
α_0	0.894
β_0	8.4
f_α	23.7 GHz
τ_c	23 pS
$\tau_{\text{(minority lifetimes)}}$	10 nS

Trade-offs involved in collector region design are directed toward base–collector region capacitance, C_{BC}, and base–collector depletion region transit time, τ_C. An increase in collector doping decreases the base–collector depletion region and corresponding transit time but increases collector capacitance. An increased collector capacitance lowers output impedance, thereby limiting device area. The base–collector depletion region transit time introduces an inductive delay that degrades RF performance.

The large-signal equivalent circuit model for the BJT used in this work is shown in the common emitter configuration in Fig. 19. This is a standard model for the bipolar transistor and contains elements of most significance to the RF operation of the device. The calculated parameter values are listed in Table VII. Package and lead parasitic elements were added to make the simulations more physical. For this work, the common emitter configuration and class A operating conditions were chosen. Power devices can be operated either in common emitter or common base configurations. A common base configuration is generally used when the device is limited in RF performance by breakdown voltage considerations. Since SiC has a large critical field for breakdown, collector breakdown voltage limitations are not

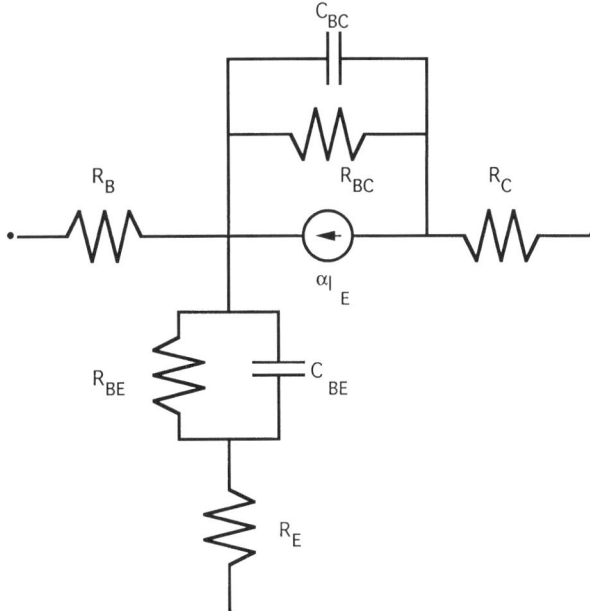

FIG. 19. Large-signal bipolar transistor equivalent circuit.

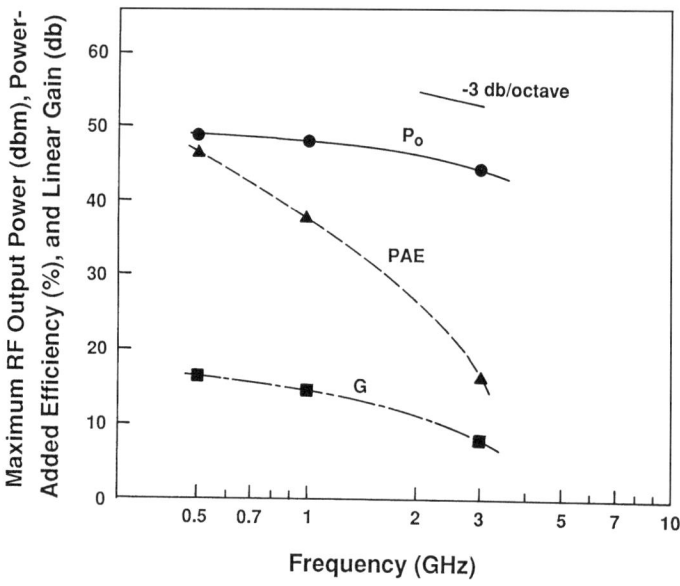

Fig. 20. RF performance versus frequency for a 6H-SiC bipolar transistor.

expected to be a factor, and for this reason, the more desirable common emitter configuration is selected.

The RF performance as a function of frequency for the SiC BJT is shown in Fig. 20. At frequencies above about 1.5 GHz, the RF output power and gain degrade at a -3-dB/octave rate. The PAE degrades rapidly with frequency, and the device will not produce useful power above approximately 4 GHz. Below 1.5 GHz, the RF output power of the device is essentially constant, indicating that the device design is probably not optimum for low-frequency operation. At these frequencies, the device area could possibly be increased to increase output power. Attempts to design such a device, however, were not successful due to impedance-matching problems introduced by increased collector capacitance and conductance with large-area devices.

4. IMPATT DIODES

The IMPATT diode has proven to be a useful device for the generation and amplification of RF energy from the microwave to the high millimeter-wave spectrum. Although in recent years the GaAs MESFET and the

HEMT and HBT fabricated from AlGaAs/GaAs and GaInAs/InP have taken over many of the systems applications in the microwave spectrum, IMPATTs are still used at millimeter-wave frequencies. In general, IMPATT diodes produce more RF power per unit area than do transistors and they are much easier to fabricate for millimeter-wave applications. They are easily integrated into resonant cavities to produce high-performance oscillators and amplifiers, although the amplifiers require use of a non reciprocal device such as a circulator to separate the input and output signals. In operation, IMPATT diodes are thermally limited, rather than electronically limited, as are MESFETTs and HEMTs, their main competitors for high-frequency applications. The thermal limitation permits much higher RF output power to be developed under pulse bias operation, where thermal effects can be minimized. An IMPATT diode will typically produce an order of magnitude greater RF output power under pulse bias than under CW operation, whereas MESFETs and HEMTs, since they are electronically limited by breakdown of the gate electrode, do not produce significantly greater RF output power when operated under pulse bias conditions. This makes IMPATT diodes attractive for use in transmitters in pulsed radar systems.

Although IMPATT diodes can be fabricated in single-drift or double-drift structures, the double-drift structure is generally preferred for millimeter-wave applications because more RF output power results. The basic structure for a double-drift IMPATT diode is shown in Fig. 21. The device consists of a p-n junction sandwiched between two low-doped "drift" regions. In operation, the device is biased into avalanche breakdown of the p-n junction. The electron and hole densities are driven by the electric field and travel in opposite directions through the corresponding drift regions to the device contacts. The avalanche process produces approximately a 90° phase shift in the RF current relative to the RF voltage. The delay through the drift regions causes an additional inductive-phase delay, which, when added to that due to the avalanche process, results in a total delay exceeding

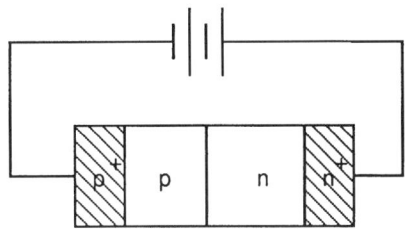

Fig. 21. IMPATT diode.

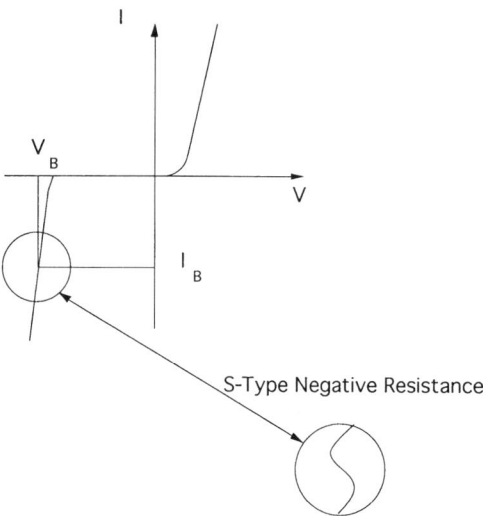

FIG. 22. DC I–V characteristics for an IMPATT diode showing s-type negative resistance.

90°, thereby generating an S-type negative resistance, as shown in Fig. 22. Typical RF voltage and current waveforms for a 6H-SiC IMPATT diode operating at 60 GHz are shown in Fig. 23. The waveforms labeled J_{av}, J_{RF} and V_{RF} are the avalanche region current density, terminal RF current density, and RF voltage, respectively. The S-type negative resistance only exists above a certain frequency, called the avalanche frequency, which is determined by the details of the avalanche region of the device under operation conditions. The avalanche frequency is given by the expression

$$f_{av} = \frac{1}{2\pi}\sqrt{\frac{2(\partial\alpha/\partial E)v_s J_0}{\varepsilon_r}} \tag{14}$$

where f_{av} is the avalanche frequency, α is the ionization coefficient, v_s is the saturated velocity, and J_0 is the current density at which the diode is operated. The avalanche frequency increases as the square root of the current density, and increasing the bias current through an IMPATT oscillator will increase its frequency of oscillation and provide a means for electronic tuning. There is a lower bound to the avalanche frequency, and the IMPATT diode produces no negative resistance or active characteristics at dc. An equivalent circuit for the device is shown in Fig. 24, where the S-type negative resistance is modeled as a negative resistance in series with

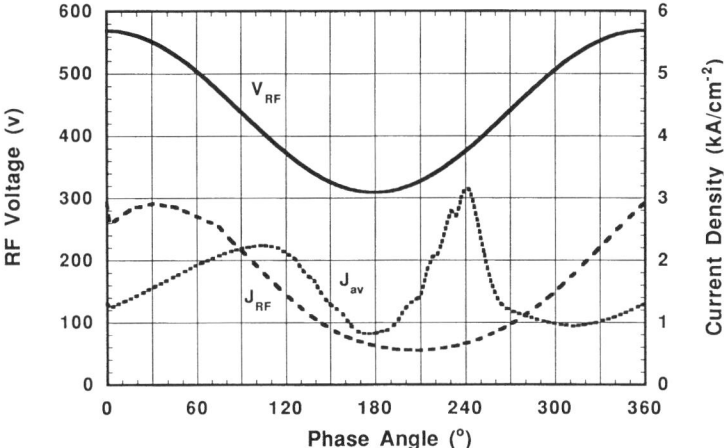

FIG. 23. RF terminal current and voltage waveforms for a 6H-SiC IMPATT diode oscillator.

an inductor. The active region of the diode is shunted by the parallel plate capacitance of the reverse-based junction, indicated as C_d in Fig. 24. When placed in a resonant circuit, the device will oscillate when the bias current is raised above the critical magnitude that results in the diode becoming active in a frequency region near the resonant frequency of the cavity. When this occurs, the oscillation frequency will be primarily determined by the resonant cavity characteristics.

The double-drift structure operates basically as two back-to-back diodes. Device impedance levels are increased, thereby permitting larger area devices and higher output power to be obtained. A disadvantage of the double-drift structure is that the most significant dc power dissipation

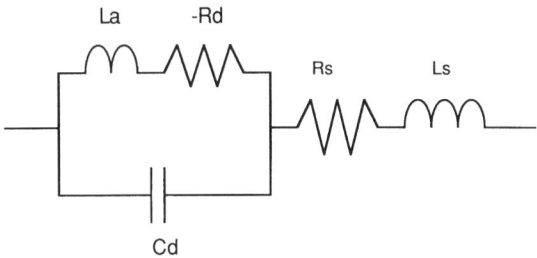

FIG. 24. IMPATT diode equivalent circuit.

occurs in the avalanche region, which is located inside the device. The problem is alleviated for millimeter-wave devices due to thin drift regions, which scale inversely with frequency. When properly designed, the device will operate with the electric field within the device above that required to achieve charge carrier velocity saturation. The low field mobility for materials such as SiC is very low and can affect device performance, because the magnitude of the charge carrier mobility determines if velocity saturation conditions can be achieved. If charge carrier mobility is too slow, the semiconductor may reach dielectric breakdown conditions before the saturation field can be achieved. If velocity saturation conditions are not maintained throughout the RF cycle, performance degradation occurs. The parasitic resistances due to the necessary bulk semiconductor and contact regions are also directly proportional to the charge carrier mobility. These regions produce a series resistance, shown as R_s in Fig. 24, that will affect RF performance. Since the negative resistance generated by an IMPATT diode is only about $R_d \sim -1\,\Omega$ to $R_d \sim -10\,\Omega$, it is necessary to reduce parasitic resistances as much as possible, and less than the magnitude of the negative resistance.

Optimum IMPATT operation is obtained when the avalanche region is restricted to a small portion of the total diode length. When this becomes difficult, more complex doping profiles, such as high-low or low-high-low doping profiles, can be used. These structures make use of alternating regions of high and low impurity doping to tailor and shape the electric field within the diode so that high breakdown fields exist in the avalanche region, which is designed to be small, while maintaining sufficiently high electric field in the remainder of the diode to saturate the charge carrier velocity. In this manner, the avalanche region can be confined to a minimum portion of the diode. These complex diode structures generally produce greater dc-to-RF conversion efficiency compared to simple, uniformly doped structures, since the total diode voltage can be reduced, thereby decreasing the dc power into the diode without any loss of RF output power.

To investigate the potential of SiC IMPATT diodes for millimeter-wave oscillator and amplifier applications, the dc and RF operation of a series of diode structures was simulated using a physical model simulator. The IMPATT diode model solves the semiconductor device equations under large-signal RF drive conditions and calculates the dc and RF performance of the device as a function of device design, bias, and RF operating conditions. The various diodes were optimized by adjusting dopant levels and layer thicknesses until a peak dc-to-RF conversion efficiency for each diode at each frequency of interest was obtained. Operation at 35, 44, 60, and 94 GHz was considered. The material parameters used in the simulations are listed in Table VIII and the diode dimensions are listed in Table IX.

TABLE VIII
MATERIAL PARAMETERS FOR 6H-SiC IMPATT SIMULATIONS

Parameter	Value
Hole mobility (μ_p)	50 cm^2/V-s
Hole saturation velocity (v_{psat})	5.4 × 10^6 cm/s
Hole diff. coef. (D_p)	0.55 cm^2/s
Electron mobility (μ_n)	250 cm^2/V-s
Electron saturation velocity (v_{nsat})	2.0 × 10^7 cm/s
Electron diff. coef. (D_n)	9.88 cm^2/s
Ionization constants	
(α_p)	4.65 × 10^6 cm^{-1}
(α_n)	4.65 × 10^4 cm^{-1}
($\beta_{n,p}$)	1.2 × 10^7 V/cm
Dielectric constant (ε_r)	9.7
Thermal conductivity (κ_T)	5 W/cm-°C

The CW RF output power results predicted by the simulations for the 6H-SiC IMPATT oscillators are shown in Fig. 25, and these results are compared to comparable GaAs and Si IMPATT oscillators. The RF output power and dc-to-RF conversion efficiency as a function of frequency and operating conditions are listed in Table X. As shown in Fig. 25, the SiC IMPATT oscillators produce RF output power slightly higher than the GaAs or Si IMPATT oscillators, but the conversion efficiency is lower.

TABLE IX
OPTIMIZED PROFILES FOR SiC DOUBLE-DRIFT IMPATT DIODES

Freq. (GHz)	p-Doping (10^{17} cm^{-3})	p-Width (μm)	n-Doping (10^{17} cm^{-3})	n-Width (μm)
35	1.50	1.00	0.63	2.30
44	1.80	0.90	1.00	1.70
60	2.20	0.75	1.15	1.40
94	2.50	0.60	1.50	1.10

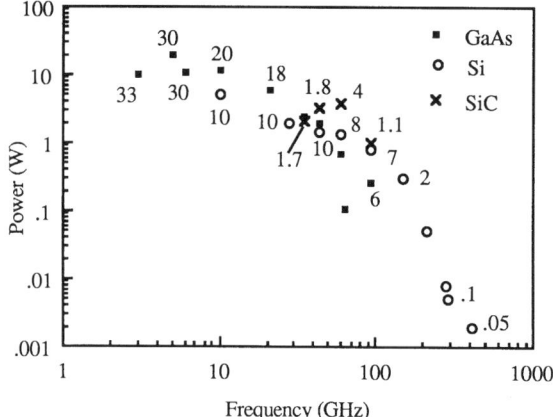

FIG. 25. Simulated SiC IMPATT diode oscillator RF output power compared to measured GaAs and Si IMPATT diode performance as a function of frequency. (Numbers indicate conversion efficiency.)

Conversion efficiency is generally in the range of 1 to 5%, and RF output power up to about 3 to 4 W can be obtained. These calculations were performed with full consideration of thermal effects, which limited the current that could be used.

The RF performance of IMPATT diodes fabricated from 4H-SiC would yield higher conversion efficiency because the electron mobility is twice that of the 6H-SiC. The higher mobility would produce smaller parasitic series

TABLE X
SiC Diode CW RF Performance Characteristics

Freq. (GHz)	Area (cm^{-2})	J (kA/cm^2)	I (mA)	Vdc (V)	PRF (W)	η (%)
35	7.0×10^{-5}	4.0	280	512	2.10	1.7
35	2.3×10^{-5}	6.0	138	517	0.92	1.6
44	1.1×10^{-4}	4.0	440	436	3.26	1.8
44	2.0×10^{-5}	8.0	160	425	0.25	2.1
44	2.0×10^{-6}	15.0	30	426	0.36	3.2
60	3.3×10^{-5}	3.0	100	356	0.21	0.6
60	5.6×10^{-5}	5.0	280	352	0.84	1.88
60	2.9×10^{-5}	8.5	247	373	3.92	4.28
94	3.8×10^{-5}	10.5	400	243	1.03	1.1
94	2.1×10^{-6}	28.6	60	236	0.42	3.0

resistances. Also, the use of more advanced doping profile structures, such as the high-low or low-high-low designs, would produce improved RF performance. Improved structure design should yield RF output power on the order of watts, with dc-to-RF conversion efficiency in the range of 10 to 15% at millimeter-wave frequencies. One final improvement would be use of one-sided devices, rather than the double-drift devices described here. Minimization or elimination of the lossy p-type material would be expected to improve the RF performance, even though the advantages of the double-drift design would be lost. These results indicate that SiC IMPATT diodes with good-to-excellent RF power performance should be possible. These devices may prove useful for applications such as high-power pulse radar transmitters.

VII. Summary

Rapid progress in development of SiC bulk and epitaxial growth technology and the commercial availability of high-quality epitaxial layers are producing significant interest in exploitation of this material for a variety of device applications. In particular, microwave power amplifiers and oscillators fabricated from SiC devices offer the potential for significantly improved dc and RF performance compared to devices fabricated from traditional semiconductors, such as Si and GaAs. The combination of wide energy bandgap, high carrier velocity, high thermal conductivity, and high breakdown field indicates that devices fabricated from the various polytypes of SiC are capable of high-power generation with high PAE in harsh environments and at high temperatures. Theoretical predictions indicate that devices fabricated from SiC have microwave power capability at room temperature about a factor of 4 greater than comparable devices fabricated from GaAs or Si. These theoretical predictions are being verified in experimental prototype devices. In addition, the wide-bandgap devices are capable of operation at temperatures at least to 500°C, where comparable devices fabricated from GaAs or Si cannot be operated. The most attractive polytypes of SiC are the 3C-, 4H-, and 6H-SiC, with 4H-SiC being the single most attractive material for electronic device applications.

A variety of devices, including MESFETs, BJTs and HBTs, SITs, and IMPATT diodes all have significant potential for improved RF performance when fabricated from 4H-SiC. The MESFET, in particular, is very attractive for use in high-power, high-frequency amplifiers because the MESFET is structurally simple and one of the easiest devices to fabricate. These devices are rapidly approaching the commercialization stage. The SIT devices also look very promising, particularly for lower frequency operation. The inherent channel current limitation of these devices will likely limit their use to C-band, or lower operation. The SIT, however, is capable of high RF output power and good efficiency performance. These devices are also rapidly approaching the comercialization stage, and commercial amplifiers have already been announced. The HBT and IMPATT have not received as much attention, although they too offer improved performance, particularly for high RF output power. The HBT, in particular, is likely to find use.

The greatest limitation to development of SiC electronic devices is the high resistivity of p-type material, which is due to the low activation of the acceptors (generally Al) in SiC and the need to use extremely high doping density to produce low-resistivity material. The acceptor doping is generally 1 to 2 orders of magnitude greater than the free-charge density, and extremely low mobility due to high impurity scattering results. If techniques to improve activation cannot be found, the use of p-type material will be limited. This will hinder the development of devices that require p-type regions, such as p-n junctions, bipolar transistors, IMPATT diodes, and so forth. Majority carrier devices, such as MESFETs and SITs, that can be fabricated completely from n-type material are very promising and will find many applications.

REFERENCES

1. R. J. Trew, J. B. Yan, and P. M. Mock, "The Potential of Diamond and SiC Electronic Devices for Microwave and Millimeter-Wave Power Applications," *Proc. IEEE*, **79**, pp. 598–620, May 1991.
2. H. Jagodzinski and H. Arnold, "The Crystal Structure of Silicon Carbide," *Silicon Carbide, A High-Temperature Semiconductor*, edited by J. R. O'Connor and J. Smiltens, New York: Pergamon, 1960, pp. 136–145.
3. N. W. Jepps and T. F. Page, "Crystal Growth and Characterization of Polytype Structures," *Progress in Crystal Growth and Characterization*, **7**, p. 259, 1983.
4. A. G. Acheson, British Patent, No. 17911, 1892.
5. A. Lely, "Darstellung von Einkristallen von Silizium Carbid und Bekerrschung von Art and Meng der Eingebauten Verunreinigunger," *Ber. Deut. Keram. Ges.*, **32**, pp. 229–231, 1955.
6. L. J. Kroko and A. G. Milnes, "Diffusion of Nitrogen into Silicon Carbide Single Crystals Doped with Aluminum," *Solid-State Electronics*, **9**, pp. 1125–1134, 1966.
7. G. Kelner, S. Binari, K. Sleger, and H. Kong, "β-SiC MESFETs and Buried-Gate JFETs," *IEEE Electron Dev. Lett.* **EDL-8**, pp. 428–430, Sep. 1987.
8. W. V. Muench, P. Hoeck, and E. Pettenpaul, "Silicon Carbide Field-Effect and Bipolar Transistors," *1977 IEEE International Electron Device Meeting Digest*, pp. 337–339.
9. S. Shibahara, T. Saito, S. Nishino, and H. Matsunami, "Fabrication of Inversion-Type n-Channel MOSFETs Using Cubic-SiC on Si(100)," *IEEE Electron Dev. Lett.* **EDL-7**, pp. 692–693, Dec. 1986.
10. J. W. Bumgarner, H. S. Kong, H. J. Kim, J. W. Palmour, J. A. Edmond, J. T. Glass, and R. F. Davis, "Monocrystalline β-SiC Semiconductor Thin Films: Epitaxial Growth,

Doping and FET Device Development," *1988 Proc. of the 38th Electronics Components Conf.*, pp. 342–349.
11. R. F. Davis, "Epitaxial Growth and Doping of and Device Development in Monocrystalline β-SiC Semiconductor Thin Films," *Thin Solid Films*, **181**, pp. 1–15, Dec. 1989.
12. M. Yamanaka, H. Daimon, E. Sakuma, S. Misawa, and S. Yoshida, "Temperature Dependence of Electrical Properties of n- and p-type 3C-SiC," *J. Appl. Phys.* **61**, pp. 599–603, Jan. 1987.
13. H. J. Round, "A Note on Carborundum," *Electrical World*, **49**, p. 309, Feb. 1907.
14. W. V. Muench and E. Pettenpaul, "Saturated Electron Drift Velocity in 6H Silicon Carbide," *J. Appl. Phys.* **48**, pp. 4823–4825, Nov. 1977.
15. S. Yoshida, H. Daimon, M. Yamanaka, E. Sakuma, and K. Endo, "Schottky-Barrier Field-Effect Transistors of 3C- SiC," *J. Appl. Phys.* **60**, pp. 2989–2991, Oct. 1986.
16. H. Daimon, M. Yamanaka, M. Shinohara, E. Sakuma, S. Misawa, K. Endo, and S. Yoshido, "Operation of Schottky-Barrier Field-Effect Transistors of 3C-SiC up to 400°C," *Appl. Phys. Lett.* **51**, pp. 2106–2108, Dec. 1987.
17. K. Furukawa, A. Hatano, A. Uemoto, Y. Fujii, K. Nakanishi, M. Shigeta, A. Suzuki, and S. Nakajima, "Insulated-Gate and Junction-Gate FETs of CVD-Grown β-SiC," *IEEE Electron Dev. Lett.* **EDL-8**, pp. 48–49, Feb. 1987.
18. K. Furukawa, A. Hatano, A. Uemoto, Y. Fujii, K. Nakanishi, M. Shigeta, A. Suzuki, and S. Nakajima, "Insulated-Gate and Junction-Gate FETs of CVD-Grown β-SiC," *IEEE Electron Dev. Lett.* **EDL-8**, pp. 48–49, Feb. 1987.
19. G. Kelner, M. S. Shur, S. Binari, K. J. Sleger, and H. S. Kong, "High-Transconductance β-SiC Buried-Gate JFETs," *IEEE Trans. Electron Dev.* **36**, pp. 1045–1049, June 1989.
20. J. W. Palmour, "α-SiC MESFETs," *1990 WOCSEMMAD Conference*, San Francisco, CA.
21. S. Sriram, R. R. Barron, A. W. Morse, T. J. Smith, G. Augustine, A. A. Burk, R. C. Clarke, R. C. Glass, H. M. Hobgood, P. A. Orphanos, R. R. Siergiej, C. D. Brandt, M. C. Driver, and R. H. Hopkins, "High Efficiency Operation of 6H-SiC MESFETs at 6 GHz," *53rd Device Research Conference*, Charlottesville, VA, 1995.
22. C. Weitzel, J. Palmour, C. Carter, and K. Nordquist, "4H-SiC MESFET with 2.8 W/mm Power Density at 1.8 GHz," *IEEE Electron Dev. Lett.* **15**, 10, Oct. 1994.
23. K. E. Moore, C. E. Weitzel, K. J. Nordquist, L. L. Pond, J. W. Palmour, S. Allen, and C. H. Carter, "4H-SiC MESFET with 65.7% Power-Added Efficiency at 850 MHz," *IEEE Electron Dev. Lett.* **18**, pp. 69–70, Feb. 1997.
24. K. Moore, C. Weitzel, K. Nordquist, L. Pond, J. Palmour, S. Allen, and C. Carter, "Bias Dependence of RF Power Characteristics of 4H-SiC MESFET's," *Proc. IEEE/Cornell Conference on Advanced Concepts in High Speed Semiconductor Devices and Circuits*, Ithaca, NY, Aug. 1995, pp. 40–46.
25. S. Sriram, G. Augustine, A. A. Burk, R. C. Glass, H. M. Hobgood, P. A. Orphanos, L. B. Rowland, T. J. Smith, C. D. Brandt, M. C. Driver, and R. H. Hopkins, "4H-SiC MESFET's with 42 GHz fmax," *IEEE Electron Dev. Lett.* **17**, pp. 369–371, July 1996.
26. R. R. Siergiej, R. C. Clarke, A. K. Agarwal, C. D. Brandt, A. A. Burke, A. Morse, and P. A. Orphanos, "High Power 4H-SiC Static Induction Transistors," *IEDM Digest*, pp. 353–356, Washington, DC, Dec. 1995.
27. R. C. Clarke, A. K. Agarwal, R. R. Siergiej, C. D. Brandt, and A. W. Morse, "The Mixed Mode 4H-SiC SIT as an S-Band Microwave Power Transistor," *Device Research Conference Digest*, pp. 62–63, Santa Barbara, CA, June 1996.
28. A. W. Morse, P. M. Esker, R. C. Clarke, C. D. Brandt, R. R. Siergiej, and A. K. Agarwal, "Application of High Power Silicon Carbide Transistors at Radar Frequencies," *1996 IEEE MTT-S Digest*, pp. 677–680, San Francisco, CA.
29. G.-B. Gao, J. Sterner, and H. Morkoc, "High Performance of SiC Heterojunction Bipolar Transistors," *IEEE Trans. Electron Dev.* **41**, pp. 1092–1097, July 1994.

30. J. Pankove, S. S. Chang, H. C. Lee, R. J. Molnar, T. D. Moustakas, and V. Van Zeghbroeck, "High-Temperature GaN/SiC Heterojunction Bipolar Transistor with High Gain," *Technical Digest of the 1994 International Electron Device Meeting*, pp. 389–392.
31. A. P. Dmitriev, A. O. Konstantinov, D. P. Litvin, and V. I. Sankin, "Impact Ionization and Superlattice in 6H-SiC," *Sov. Phys. Semicond.* **17**, pp. 686–689, June 1983.
32. M. M. Anikin, M. E. Levinshtein, I. V. Popov, V. P. Rastegaev, A. M. Strel'chuk, and A. L. Skyrkin, "Temperature Dependence of the Avalanche Breakdown Voltage of Silicon Carbide p-n Junctions," *Sov. Phys. Semicond.* **22**, pp. 995–998, Sep. 1988.
33. H. S. Kong, J. W. Palmour, J. T. Glass, and R. F. Davis, "Temperature Dependence of the Current-Voltage Characteristics of Metal-Semiconductor Field-Effect Transistors in n-type β-SiC Grown via Chemical Vapor Deposition," *Appl. Phys. Lett.* **51**, pp. 442–444, Aug. 1987.
34. R. J. Trew and M. W. Shin, "High Frequency, High Temperature Field-Effect Transistors Fabricated from Wide Bandgap Semiconductors," *International Journal of High Speed Electronics and Systems*, **6**, 1, pp. 211–236, Mar. 1995.
35. J. I. Nishizawa, T. Terasaki, and J. Shibata, "Field-Effect Transistor Versus Analog Transistor (Static Induction Transistor)," *IEEE Trans. Electron Dev.* **ED-22**, pp. 185–197, April 1975.
36. C. O. Bozler and G. D. Alley, "Fabrication and Numerical Simulation of the Permeable Base Transistor," *IEEE Trans. Electron Dev.* **ED-27**, pp. 1128–1141, June 1980.
37. W. R. Frensley, "An Analytic Model for the Barrier-Limited Mode of Operation of the Permeable Base Transistor," *IEEE Trans. Electron Dev.* **ED-30**, pp. 1624–1628, Dec. 1983.

CHAPTER 7

SiC-Based UV Photodiodes and Light-Emitting Diodes

*J. Edmond, H. Kong, G. Negley, M. Leonard,
K. Doverspike, W. Weeks, A. Suvorov,
D. Waltz, and C. Carter, Jr.*

CREE RESEARCH, INC.
DURHAM, NC

I.	INTRODUCTION	283
II.	SiC BLUE LEDs	286
	1. *Epitaxy and Device Fabrication*	286
	2. *Device Performance*	287
III.	SiC GREEN LEDs	288
	1. *Epitaxy and Ion Implantation*	288
	2. *Device Performance*	289
IV.	UV PHOTODIODES	290
	1. *Device Fabrication*	290
	2. *Electrical Characteristics*	292
	3. *Optical Responsivity*	294
	4. *Applications*	296
V.	GROUP III-NITRIDES ON 6H-SiC	297
	1. *Epitaxial Growth, Characterization, and Device Fabrication*	297
	2. *Hall Effect*	298
	3. *Photoluminescence*	300
	4. *GaN: SiC Blue LEDs*	302
	5. *Electrical Static Discharge Survivability*	305
VI.	SUMMARY	305
	References	306

I. Introduction

SiC has been investigated as a short-wavelength optoelectronic material since the early years of semiconductor development. As a result of the 3.0-eV bandgap of 6H-SiC, any color light-emitting diode (LED) in the visible spectrum can be achieved in this material. Despite its indirect bandgap, which makes light emission relatively inefficient, many scientifically interesting and some commercially successful emitters have been developed. Dmitriev et al. [1] demonstrated a red-green-blue LED display made from a 1 × 1-cm single-crystal wafer of 6H-SiC. Ion implanatation and epitaxial techniques were employed to achieve the three LED colors. No information on the efficiency of the devices was given. Vodakov et al. [2] demonstrated a green LED in 6H-SiC by implanting Al into an n-type epilayer 10 to 30 μm thick and grown by sublimation epitaxy. The Al implant produced the p-type region of the p–n junction and simultaneously generated luminescence centers. The resulting LEDs emitted green with a peak wavelength of 530 nm. The output power was in the range of 5 to 15 μW.

Development of 6H-SiC blue LEDs was for many years a major research effort at Siemens Research Laboratory, various labs in the former Soviet Union, and a number of Japanese labs, in particular Sanyo Electric Co., Ltd. Several of these groups released blue LEDs as products. The highest external quantum efficiency (η_{ext}) reported by these groups was achieved by Hoffmann et al. [3] (Siemens) using a high-resistivity transparent p-type substrate with contacts made to via-etched topside p^+ and light-emitting compensated n-type epilayers. The epitaxial layers were deposited via liquid-phase epitaxy (LPE). This nontraditional two-topside contact device design resulted in a maximum η_{ext} of $\sim 2 \times 10^{-4}$. However, typical commercial devices with a vertical device configuration exhibited much lower values: $\eta_{ext} \sim 0.3 - 0.4 \times 10^{-4}$ with excessive forward voltage (4–8 V) as a result of the low optical transparency and high electrical resistivity of the p-type substrates employed. Koga et al. [4] of Sanyo introduced a device also grown by LPE utilizing an n-type substrate. A light-emitting compensated n-layer was grown, followed by a high-resistivity p-layer with poor current and therefore light-spreading characteristics. To compensate for this problem, the chip was inverted from a traditional junction up-configuration with the p-contact face down and a polished n-substrate face up. This provided for uniform light emission in the shape of the p-contact imaged through the chip. The lower resistivity and higher optical transparency of the n-substrate yielded an improvement over the Siemens device with regard to lower forward voltage, 4 V, and higher η_{ext} of $\sim 1 \times 10^{-4}$ for commercial devices. This nontraditional chip configuration, however, made it difficult to package with standard silver epoxy mounting without shorting the junction.

The wide bandgap of SiC also make it an excellent ultraviolet (UV) detector. The earliest published research on SiC UV photodiodes utilized diffusion of Al into n-type 6H substrates [5]. This process resulted in high-leakage devices with corresponding low quantum efficiency. Glasow et al. [6] improved on this initial design by utilizing N ion implantation to form a very shallow (0.05 μm) n^+–p junction in a 5-μm p-type epitaxial layer grown on a p-type substrate. This research resulted in devices with a maximum quantum efficiency of 75% at a peak wavelength (λ_p) of 280 nm at room temperature. However, the reverse-bias dark current density was excessive, on the order of 10^{-5} A/cm^2 at -10 V at room temperature.

A related and very rapidly growing application for SiC in optoelectronics is its use as a substrate for group III-nitride LED and laser structures. Higher efficiency blue and green LED emission has been achieved with the direct bandgap alloy system AlN-InN-GaN, which has a structural and chemical compatibility with SiC substrates. To date, most researchers have employed sapphire as a substrate, which also has a structural and chemical compatibility with GaN, but has two major shortcomings: (1) sapphire is electrically insulating and (2) it has a 16% lattice mismatch with GaN. This requires two topside contacts to pass curent through the p–n junction and results in a high dislocation density in the device structure, on the order of 10^9 cm^{-2}. SiC has a lattice mismatch of 3.5% with GaN and is conductive, which allows for the fabrication of a vertical device structure for nitride-based LEDs. Cree Research uses 6H-SiC substrates for commercial production of nitride-based LEDs, while the other two commercial sources of super-bright blue LEDs employ sapphire.

Nakamura [7] of Nichia Chemical has demonstrated InGaN single quantum well (SQW) LEDs grown on sapphire emitting from violet to orange. The external quantum efficiency of these devices are the highest recorded for the InGaN system. As the indium fraction of the InGaN SQW increases, the efficiency of the LED decreases from 10% at 400 nm ($\sim 14\%$ In) to 1.2% at 600 nm ($\sim 77\%$ In). Based on this research, Nichia Chemical released blue and green SQW LEDs into production in 1996. The blue LED with an active layer composition of $In_{0.36}Ga_{0.64}N$ has a typical output of 2.5 mW at a peak wavelength of 470 nm and a bandwidth of 30 nm. The green version, with an active-layer composition of $In_{0.53}Ga_{0.47}N$, has a typical output of 1.5 mW at a peak wavelength of 525 nm and a bandwidth of 40 nm.

Koike et al. [8] of Toyoda Gosei Co. demonstrated asymmetric double heterostructures (ADHs) of AlGaN/InGaN/GaN blue and InGaN/GaN MQW UV LEDs. The active layer of the blue LED is $In_{0.08}Ga_{0.92}N$ co-doped with Zn and Si, which results in an emission peak of 450 nm and a bandwidth of 70 nm. In production, this device has a typical output of 2 mW at 20 mA. The multiple quantum well emitter employed five wells of $In_{0.08}Ga_{0.92}N$/GaN bounded by AlGaN.

This chapter describes the development, fabrication, characterization, and some applications of 6H-SiC homojunction blue and green LEDs and UV photodiodes at Cree Research, Inc. In addition, electrical and optical properties of nitride layers grown on SiC substrates will be presented along with the current status of blue LEDs fabricated from this material system.

II. SiC Blue LEDs

1. EPITAXY AND DEVICE FABRICATION

Cree Research produces the only commercially available SiC blue LED. This device utilizes an n-type 6H substrate with a resistivity of 0.02 to 0.04 Ω-cm. Prior to epitaxial growth, the wafer is cut and polished to a thickness of $\sim 300\,\mu$m, with an orientation of $\sim 3.5°$ tilted off-axis along the $(11\bar{2}0)$ direction. The diameter of commercial wafers is presently 30 mm. Epitaxial growth is achieved via chemical vapor deposition (CVD) as opposed to LPE. This provides for greater control of dopant profiles and layer thicknesses. An n-type layer is first grown, followed by a compensated

FIG. 1. SEM micrograph of a typical Cree 6H-SiC blue LED.

p-type layer doped with N and Al. The blue light is generated via radiative recombination between N donors and Al acceptors in this layer. It was determined by the present authors that p-compensated material is more efficient for light emission than n-compensated, as employed by previous researchers. This may be a result of the higher mobility and thus recombination lifetime of electrons in SiC. A p^+-layer with a resistivity of 1 to 2 Ω-cm is deposited on the p-compensated layer, which serves both as a contact and as a current-spreading layer. The p–n junction is delineated via mesa etching with NF_3 in a reactive ion etch mode. The ohmic contact to this top layer is Ti/Al, followed by Ti/Au as a bond pad. The contact to the n-substrate is Ni. Both contacts are sintered to the SiC at 850 to 900°C.

Figure 1 shows a scanning electron microscopy (SEM) micrograph of Cree's typical SiC blue LED chip. The chip size is 225 × 225 μm, with a thickness of 250 μm. The top p^+-layer is 20 to 30 μm thick, which increases the escape probability and current spreading in the device. To further increase the light output, the surfaces of the chip are also roughened, as shown in the micrograph.

2. DEVICE PERFORMANCE

As mentioned previously, the 6H polytype of SiC has an indirect bandgap of ~ 3.0 eV. The proposed light generation mechanism for Cree's LEDs is phonon-assisted donor–acceptor (D-A) pair recombination between nitrogen donors and aluminum acceptors. This doping scheme results in a device that emits light with a peak wavelength of ~ 470 nm, with a spectral halfwidth ($\Delta \lambda_p$) of ~ 70 nm. Figure 2a shows the relative spectral emission versus wavelength for a Cree blue LED operating at a forward current of 20 mA measured in an integrating sphere. As shown, the peak emission is ~ 470 nm with $\Delta \lambda_p \sim 70$ nm and a dominant wavelength of ~ 481 nm. The resultant spectral purity is $\sim 82\%$. The radiant flux or optical power output of this device is ~ 34 μW. The typical output power is between 25 and 35 μW for a forward current of 20 mA at ~ 3.2 V. This represents an external quantum efficiency of 0.05 to 0.07%. With respect to photometric units, the die luminous intensity is ~ 0.5 millicandela (mcd), for a radiant flux output of 34 μW. The corresponding luminous intensity of a T $1\frac{3}{4}$-packaged lamp with a viewing angle of 16° is ~ 35 mcd.

The effect of increasing forward current to 50 mA on the radiant flux and corresponding value of the external quantum efficiency is shown in Fig. 2b. As shown, the radiant flux increases in a sublinear fashion with increasing current. This translates into a decreasing external efficiency, also shown in this illustration. This sublinear behavior is most probably due to the onset

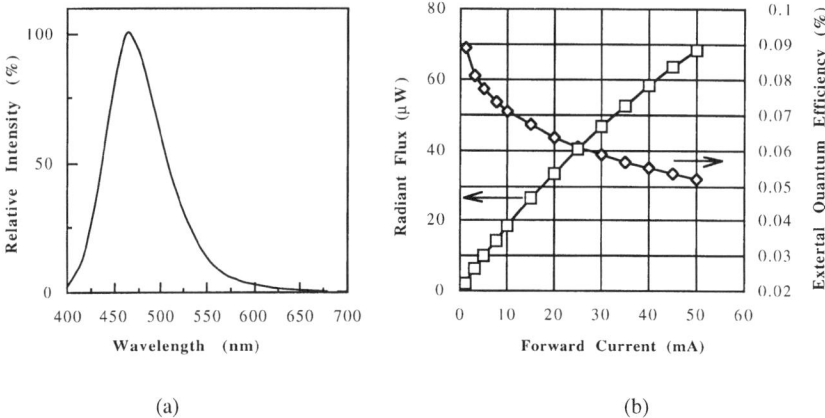

FIG. 2. Typical (a) emission spectra and (b) current-dependent output of 6H-SiC blue LEDs.

of decreasing D-A pair recombination or pair saturation and an increase in lower efficiency recombination events at higher energy. By increasing the D-A pair density, it is believed that a more linear relationship between light output and drive current can be achieved, as well as an overall increase in the efficiency of the device.

Another important aspect of SiC blue LEDs is their stability in light output over time. Using a stress current of 50 mA at room temperature, the typical 24-hr degradation is $\sim 1\%$. After 10,000 hr, the degradation is typically ~ 10 to 15%—better than most commercial green and red LEDs produced today.

III. SiC Green LEDs

1. Epitaxy and Ion Implantation

N-type 6H substrates with a resistivity of 0.02 to 0.04 Ω-cm were employed as the base substrate for epitaxial growth and ion implantation. An n-type epitaxial layer was grown via CVD to a thickness of 3 to 5 μm and a carrier concentration of 10^{15} to 10^{16} cm^{-3}. To form the p-type layer, ion implantation of either Al or Ga was then performed. The typical implant conditions were a dose of 2 to 5×10^{16} cm^{-2} and energy of 40 to 90 keV. During implantation, the wafer was maintained at a temperature of 1700 to 1800°C. This prevented the SiC from becoming amorphous from the high-dose implant. After implantation, the surface had a thin C-rich layer, which was easily removed via ashing with an oxygen plasma. A p^+-layer with a resistivity of 1 to 2 Ω-cm and a thickness of 20 μm was epitaxially grown on the p-type Al-implanted layer as a contact and a current-spreading layer. The chip fabrication process was the same as described for the blue LED.

2. DEVICE PERFORMANCE

As a result of the high-temperature implant, Al diffuses into the SiC and thus the p–n junction is located ~1 μm below the implanted surface. This is nearly an order of magnitude deeper than the Al implant range at 90 keV. This enhanced diffusion was investigated by Suvorov et al. [9] for Al implanted at 40 keV and a dose of 2×10^{16} cm^{-2} into SiC at 1800°C. Their work concluded that the radiation-enhanced diffusion coefficient was 2 orders of magnitude higher than the thermally activated diffusion coefficient as a result of an Si vacancy-rich near-surface layer formed by this implant condition.

Defects generated within a few diffusion lengths of the p–n junction electroluminescence (EL) in the green region of the visible spectrum. Figure 3a shows the relative spectral emission versus wavelength for this device operating at a forward current of 20 mA, measured in an integrating sphere. As shown, the peak emission is ~530 nm, with a bandwidth of ~85 nm. As

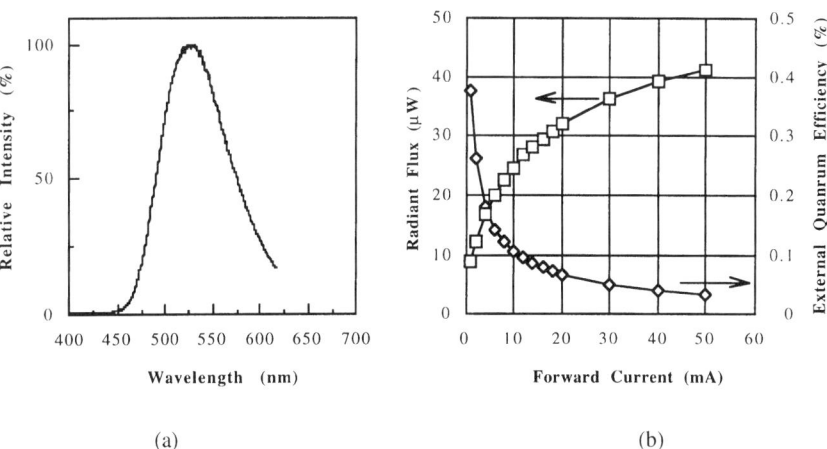

FIG. 3. Typical (a) emission spectra and (b) current-dependent output of 6H-SiC green LEDs.

a result of this broad emission peak, the spectral purity is very low ($\sim 45\%$). The typical output power is about the same as for Cree's 6H-SiC blue LED—between 25 and 35 μW for a forward current of 20 mA at ~ 3.2 V. With respect to photometric units, however, the die luminous intensity is a factor of 2 higher than the blue LED (1.2 mcd), for a radiant flux output of 33 μW. The corresponding luminous intensity of a T $1\frac{3}{4}$-packaged lamp with a viewing angle of 16° is ~ 85 mcd.

The effect of increasing forward current to 50 mA on the radiant flux and corresponding value of the external quantum efficiency is shown in Fig. 3b. As shown, the radiant flux at 1-mA forward current is nearly 10 μW, a factor of 5 higher than for the SiC blue LED at this current. The value of η_{ext} is 0.4% at this operating current, which is very high for an indirect gap material. If this efficiency were maintained to 20 mA, the output would be nearly 200 μW. However, as shown, the efficiency drops drastically with increasing current. This behavior is most probably due to a small recombination cross-section produced by the ion-implantation process. If the volume of centers could be increased, the saturation behavior would not be so severe. This could be done in two ways. The easiest method would be to increase the chip size and thus reduce the current denisty. If the chip had a junction area 6 times the area employed in this research, the brightness would indeed be nearly 200 μW at 20 mA. However, a 550 × 550-μm LED chip is neither practical nor economical. The other method to increase the recombination volume would be to increase the thickness of the active region of the device. A method to achieve this is not obvious because the exact mechanism for the EL in this device is not well understood.

IV. UV Photodiodes

1. Device Fabrication

The most common SiC-based UV photodiode today utilizes a p-type 6H substrate with a resistivity of 1 to 10 Ω-cm, which is prepared for epitaxy as previously described for Cree's blue LED. The epitaxial junction is produced by first growing a predominantly Al-doped p-type layer followed by a predominantly nitrogen-doped n-type layer. The doping and thickness of these layers can be varied to alter the responsivity of the devices. The doping in the background p-layer is typically between 1 and 5×10^{16} cm^{-3}, with a thickness of 0.4 to 5.0 μm. The n-layer is heavily doped to $\sim 10^{19}$ cm^{-3}, with a thickness of ~ 0.1 to 0.35 μm. The devices are fabricated using a mesa geometry to delineate the p–n junction. Typical sizes vary from 0.3×0.3 mm^2 to 3×3 mm^2, depending on the signal requirements of the application. The junction is passivated with dry thermally grown SiO$_2$.

7 SiC-Based UV Photodiodes and Light-Emitting Diodes   291

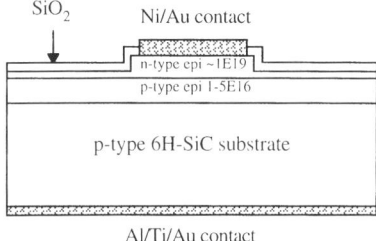

Fig. 4. Schematic showing the structure of a typical 6H-SiC UV detector.

Contact materials for the *p*-side and *n*-side are Al and Ni, respectively. The metals are sintered to the SiC at 850 to 900°C to obtain ohmic contacts.

Figure 4 shows a schematic of a commercial SiC UV detector produced by Cree Research. As shown in this illustration, the top *n*-type layer is thinned to ~ 0.1 μm from the 0.35-μm as-grown layer. This is important for

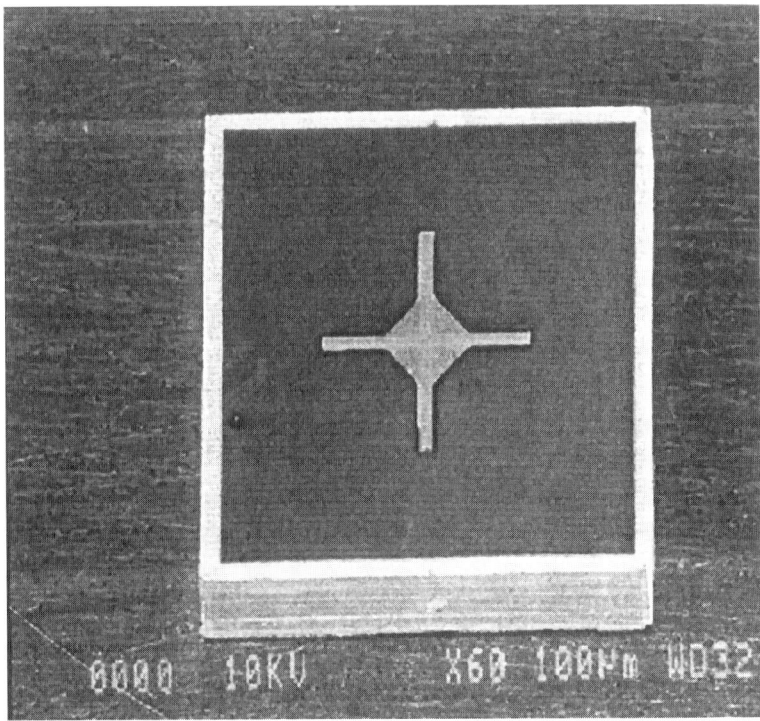

Fig. 5. SEM micrograph of a typical 6H-SiC UV detector.

two reasons. First, Ni when sintered as described, reacts to a depth of ~0.25 μm, shorting the top *n*-type to the underlying *p*-type region. Therefore, the top *n*-layer must exceed 0.25 μm. Second as a result of the high optical absorbance and surface recombination velocity of the n^+-epilayer, the responsivity of the device is enhanced by reducing its thickness from 0.35 to 0.1 μm [10]. To accomplish both conditions, the layer is selectively thinned beyond the region of the Ni contact centered on the top layer via reactive ion etching, using NF_3 as shown in Fig. 4.

Figure 5 is an SEM image of the aforementioned chip, showing the structure described previously. The cross-shaped center Ni contact reduces the active region of the device by ~2%. The center square metallization is the wire bond pad, which consists of a metal stack of Ti/Pt/Au with a thickness of 0.1, 0.1, and 2 μm, respectively. This metal scheme provides for a reliable high-temperature contact with excellent adhesion and bondability. The backside metallization employed is also Ti/Pt/Au, which can be easily brazed to an Au-plated header for hermetic packaging.

2. Electrical Characteristics

a. Dark Current

As mentioned, an advantage of SiC for UV detection is its extremely low reverse dark current, even at elevated temperatures. Figure 6a shows the reverse-bias dark-current density versus voltage as a function of temperature for a typical 6H-SiC UV photodiode. The active junction is square with an area of 0.04 cm^2. For many applications, a photodiode is operated at a reverse bias of −1.0 V, taking advantage of the fact that reverse current increases almost linearly with the increase in incident radiation. Since the sensitivity of the device depends on the level of background or dark current, it should be kept to a minimum for a high signal-to-noise level. As shown in Fig. 6a, the dark current is extremely low at −1.0 V, being $\sim 10^{-11}$ A/cm^2, even at a temperature of 473 K. At the highest voltage measured (−10 V), dark current density increased from $\sim 10^{-9}$ to $\sim 3 \times 10^{-8}$ A/cm^2 when increasing the temperature from 473 to 623 K. This relationship is plotted as a function of temperature in Fig. 6b. Here, the dark-current density at −10 V as a function of temperature is a straight-line function, with an activation energy of 0.60 eV. Brown *et al.* [10] showed similar values of reverse dark current for devices in their research, using epitaxially grown junctions. Data for the ion-implanted detectors [6] is included in Fig. 6b for comparison purposes. As shown, the devices reported herein exhibit orders of magnitude lower leakage than implanted UV detectors in 6H-SiC.

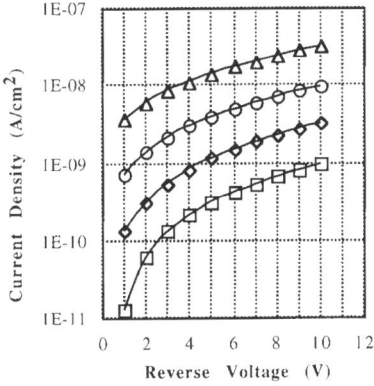 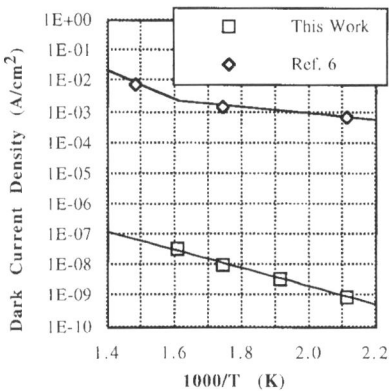

FIG. 6. Logarithmic dark-current denisty versus (a) reverse voltage at ☐ 473 K, ◇ 523 K, ○ 573 K, △ 623 K, and (b) inverse temperature for 6H-SiC UV detectors.

b. Power Generation Characteristics

Another important electrical characteristic to consider for a photodiode is the ability to produce power during illumination. This mode of operation is typically utilized in solar cell operation and is referred to as the photovoltaic mode. At zero bias, the current generated is termed the *short-circuit current* (I_{sc}) because it represents the photocurrent generated in the absence of an external load. As a load is applied to the device, a voltage and current will be output until the junction potential is reached. At this point, the reverse current decreases until no current flows through the device. The voltage at this point is termed the *open-circuit voltage* (V_{oc}). Within the I–V region, there is a maximum region of power output referred to as the maximum power rectangle. The parameters I_m and V_m correspond to the current and voltage, respectively, for the maximum power output P_m $(= I_m V_m)$.

Figure 7 shows the I–V characteristics of a 3×3-mm^2 SiC UV power photodiode illuminated with radiation in the 200 to 350-nm range and a power of 7×10^{-3} W/cm^2. At 0 V, I_{sc} was approximately 0.47 mA. This curve was generated by measuring the voltage across a variable load resistor and varying the resistance while keeping the illumination constant. The output current remained relatively level, to about 0.5 V, where it then started to decrease. This phenomenon is typically the result of a high internal device resistance. Since the series resistance of the device is on the order of 10 to 20 Ω, an additional internal resistance dominates the device in this region. It may be a result of deep trap levels in the bandgap activating at a particular voltage. At $I_1 = 0$, V_{oc} is approximately 2.3 V. The value of P_m

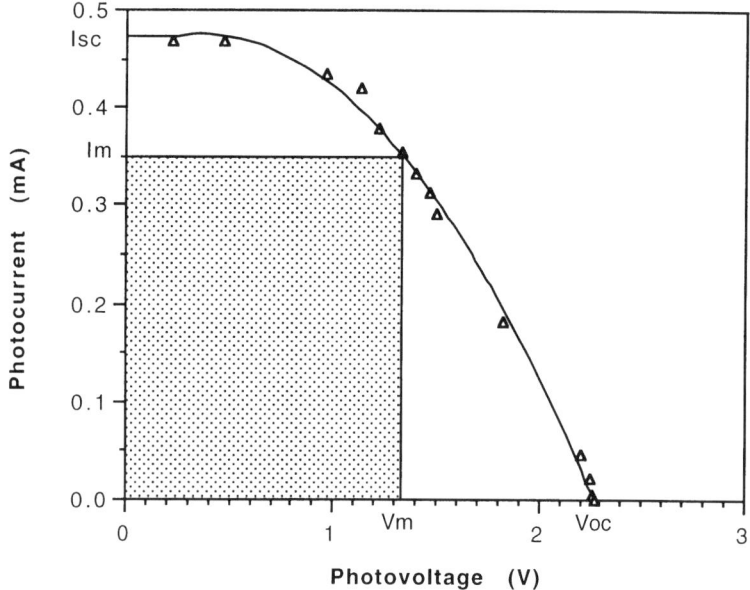

FIG. 7. Relationship between the photocurrent generated versus photovoltage output for a $3 \times 3 \, \text{mm}^2$ 6H-SiC photodiode. The incident light energy was $7 \times 10^{-3} \, \text{W/cm}^2$, with a wavelength in the range of 200 to 350 nm.

occurs at $I_m = 0.35$ mA and $V_m = 1.35$ V, or 0.47 mW. The maximum power rectangle is indicated by the shaded area in Fig. 7.

3. OPTICAL RESPONSIVITY

a. Effect of Epilayer Thickness

In addition to the thickness of the n^+ cap layer discussed earlier, two factors that can contribute to the responsivity of photodiodes are the thickness of the underlying p-layer of an n^+–p device and the temperature of operation. The depletion region of the detector is predominantly in the lower doped p-layer. When UV light is absorbed, electron–hole pairs are generated within the depletion region itself and in the p-layer bulk. Those carriers, in this case electrons, within a diffusion length, L_n, of the depletion region are swept by the large electric field of the p–n junction. These carriers are collected in a given circuit and generate current or voltage.

Brown et al. [10] employed modeling techniques to determine L_n, fitting experimental results of various SiC detectors to theory. From their results, the diffusion length of electrons in p-type SiC is $1.8 \pm 0.4\,\mu m$. This is smaller than the estimate made by Glasow et al. [6], whereby a value for L_n of $3\,\mu m$ was calculated from a lifetime of 20 ns for electrons and holes and an electron mobility of $200\,cm^2/V \cdot s$. Brown's value does not depend on any assumptions regarding carrier lifetime or mobility.

The effect of p-layer thickness on the responsivity and quantum efficiency of SiC UV detectors is shown in Fig. 8. The p-layer thicknesses employed were 0.4, 1.2, and $3.0\,\mu m$, with a junction depth of $\sim 0.1\,\mu m$. For all of the devices, the peak quantum efficiency was $\sim 80\%$. The effect of p-layer thickness is to shift the peak response and overall long-wavelength response. This is important if the objective is to produce nearly solar-blind UV detectors. For the 3-μm-thick p-layer, the peak response occurs at $\sim 280\,nm$. This response curve is very similar to that observed by Brown et al. [10], for a p-layer thickness of $5\,\mu m$, which exhibited a peak response at 275 nm. Decreasing the layer to 1.2 and $0.4\,\mu m$ shifts the peak response to $\sim 260\,nm$ and $\sim 250\,nm$, respectively. In doing so, the long-wavelength response is greatly reduced.

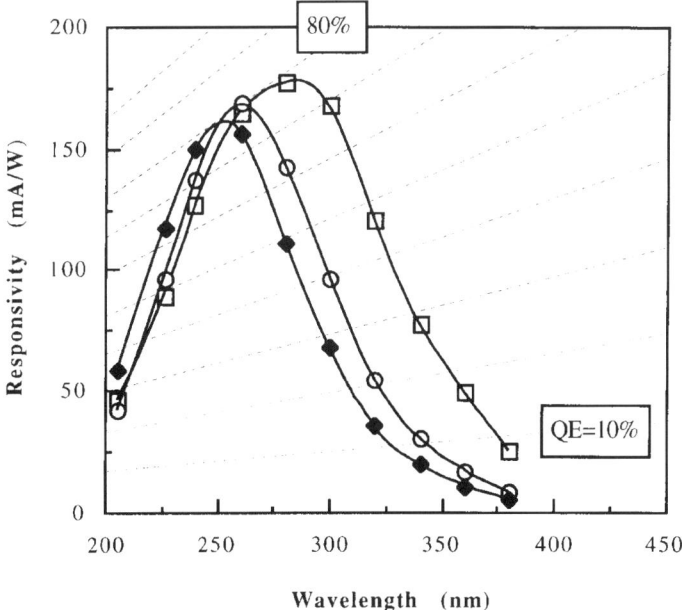

FIG. 8. The effect of p-layer thickness on the responsivity and quantum efficiency of SiC UV detectors. The p-layer thicknesses are ◆ $0.4\,\mu m$, ○ $1.2\,\mu m$, and ☐ $3.0\,\mu m$.

The reduction in longer wavelength response with decreasing p-layer thickness can be understood when considering diffusion length and the dependence of the absorption coefficient on wavelength in SiC. As determined by Choyke and Patrick [11], the absorption coefficient of SiC increases sharply with increasing energy, or decreasing wavelength, of radiation above the optical bandgap energy of ~ 3 eV. For p-layers thicker than L_n, carriers generated as far away from the depletion layer edge as L_n contribute to the photoresponse of the device. Since the absorption coefficient is smaller at longer wavelengths just below the absorption edge, carriers generated by these wavelengths have a chance to contribute. When the thickness is decreased to less than L_n, in this case 1.2 μm, fewer carriers are generated at the longer wavelengths (>250 nm) and thus a lower signal results. Below 250 nm, the absorption coefficient appears to dominate, which results in nearly identical photoresponse curves for all thicknesses of p-layers.

b. *Effect of Operating Temperature*

The responsivity and quantum efficiency of UV detectors as a function of operating temperature are shown in Fig. 9. The temperatures tested were 223, 300, 498, and 623 K. As shown, there is a shift in the peak and long-wavelength response to higher responsivity values at longer wavelengths. The peak response increases from ~ 268 nm at 223 K to ~ 299 nm at 623 K. This corresponds to a energy decrease from 4.63 to 4.15 eV over this temperature range, which corresponds to 1.2×10^{-3} eV/K. This is a result of the shift in bandgap with temperature. The indirect bandgap decreases from ~ 3.03 eV at 223 K to ~ 2.88 eV at 623 K, corresponding to a rate of 3.8×10^{-4} eV/K [10]. This is a factor of ~ 3 lower than experimentally observed from the shift in the peak response. Therefore, the peak response does not change at the same rate as the indirect bandgap. The indirect bandgap could be measured via the long-wavelength cut-off. Unfortunately the necessary narrow bandpass filters to measure this were not available at the time of this experiment. For all of the measurement temperatures, the peak quantum efficiency was between 82 and 96%. These results show the effectiveness of 6H-SiC as a UV photodiode, even to 623 K.

4. APPLICATIONS

High-temperature SiC UV photodiodes may have a significant impact in many application areas. Improvement in combustion control is anticipated with the ability to sense flames in aircraft engines, building boiler systems,

7 SiC-Based UV Photodiodes and Light-Emitting Diodes

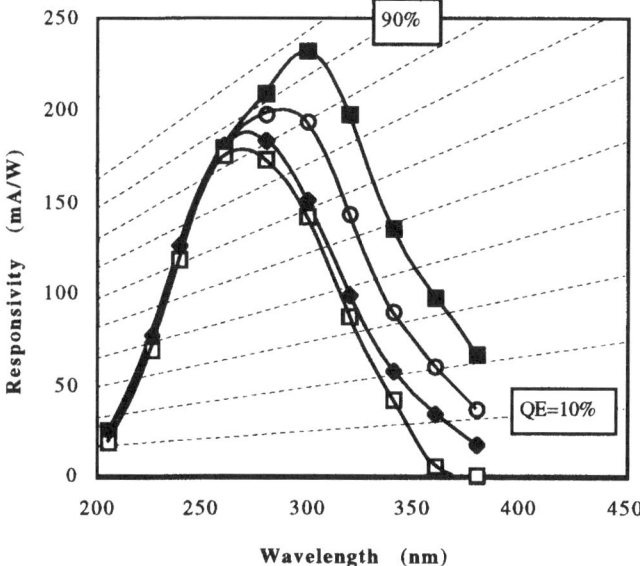

FIG. 9. The effect of measurement temperature on the responsivity and quantum efficiency of SiC UV detectors. The temperatures tested were ❑ 223 K, ◆ 300 K, ○ 498 K, and ■ 623 K.

and industrial processes. Air quality monitoring and UV dosimetry for industrial processes are other important application areas. A small, low-power, temperature-insensitive UV detector would enable the construction of rugged, self-contained, UV detector systems for use in remote-sensing applications. The availability of an array detector, sensitive to the near and middle UV, could be important to imaging applications, including atmospheric UV remote sensing, combustion control, and, potentially, missile plume detection and tracking.

V. Group III–Nitrides on 6H-SiC

1. Epitaxial Growth, Characterization, and Device Fabrication

High-quality GaN, AlGaN, and InGaN alloy thin films were epitaxially grown on the Si-face of 6H-SiC substrates with a diameter of 41 mm via metal-organic chemical vapor deposition (MOCVD). The substrates were n-type with a typical resistivity of 0.02 to 0.04 Ω-cm. The non-indium-containing films were grown in the temperature range of 1000 to 1050°C and decreased to 700 to 900°C for the InGaN layers. N-type and p-type doping was achieved with the introduction of Si and Mg, respectively.

Room-temperature photoluminescence (PL) was employed to optically characterize undoped and doped GaN, AlGaN, and InGaN layers. A 10-mW, 325-nm He-Cd laser with a spot size of $\sim 50\,\mu$m in diameter was used as the excitation source. These layers were combined in device structures to form LEDs.

Following growth of either an insulating or conducting buffer layer, Si-doped GaN was grown as a base layer for the double heterojunction (DH) structure. The DH light-emitting region was formed using a GaN active layer bordered by p- and n-AlGaN with an Al composition of 8 to 12%. A p-type GaN contact layer was deposited as the capping layer. The LEDs were fabricated with a p-type Au contact centered on the chip topside for wire bonding. The backside ohmic contact to the SiC was Ni sintered at 900 to 1000°C in an inert atmosphere. All devices were fabricated using a mesa geometry and passivated with SiO_2.

Room-temperature EL measurements were performed on blue LEDs using a Photoresearch Model PR703A Spectrascan Spectroradiometer System from 390 to 730 nm. The total optical output power of the devices was determined with the use of an integrating sphere attachment. The external quantum efficiency as a function of drive current was also determined. The current–voltage characteristics were measured with a Tektronix 370 Curve Tracer. A room-temperature Hall effect using Al and Au ohmic contacts for the respective n- and p-type layers was also performed. The dislocation density in the films ($\sim 10^8\,cm^{-2}$) was determined by transmission electron microscopy measurements.

2. Hall Effect

GaN epilayers grown on 6H-SiC substrates were controllably doped in the carrier concentration ranges of 10^{15} to $10^{19}\,cm^{-3}$ for n-type, and $\sim 10^{15}$ to $6 \times 10^{17}\,cm^{-3}$ for p-type, respectively, as determined by Hall and current-voltage measurements. Figures 10a and 10b show temperature-dependent Hall effect data for a 2-μm-thick n-type GaN layer doped with Si grown on 6H-SiC. Figure 10a shows the variation in electron concentration and also resistivity as a function of reciprocal temperature. At 300 K, the carrier concentration and resistivity values were $1.35 \times 10^{18}\,cm^{-3}$ and 0.016 Ω-cm, respectively. At lower temperatures, the carrier concentration saturated due to impurity band conduction. Figure 10b shows the Hall mobility as a function of temperature. The 300-K mobility of 293 cm^2/V-sec is comparable to the highest value reported at this carrier concentration and indicates the high quality of the epitaxial layer.

7 SiC-Based UV Photodiodes and Light-Emitting Diodes

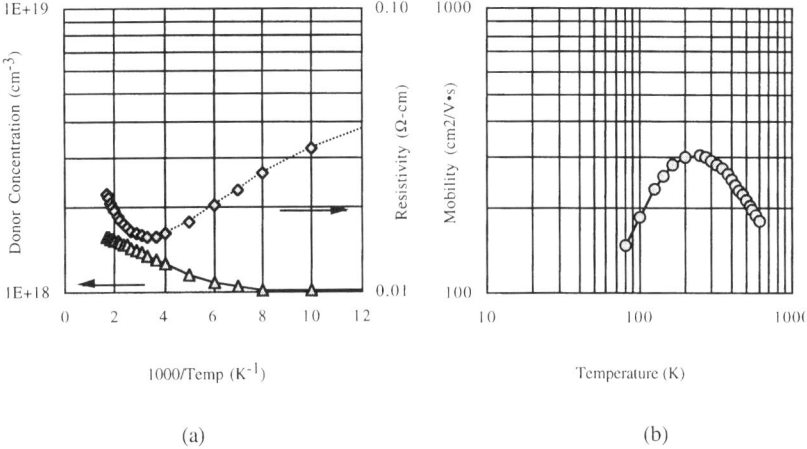

Fig. 10. Temperature-dependent Hall effect data for an Si-doped GaN layer grown on SiC. The (a) donor concentration and resistivity at 300 K were 1.35×10^{18} cm^{-3} and 0.016 Ω-cm, respectivley, with (b) a mobility of 293 cm^2/V·s.

Figures 11a and 11b show Hall data for a typical Mg-doped p-GaN top-contact layer. At 300 K, the carrier concentration and resistivity values for p-type GaN were 6×10^{17} cm^{-3} and 1.3 Ω-cm, respectively. The mobility of this layer was 8 cm^2/V-sec, which is a typical value reported by other researchers for p-type GaN.

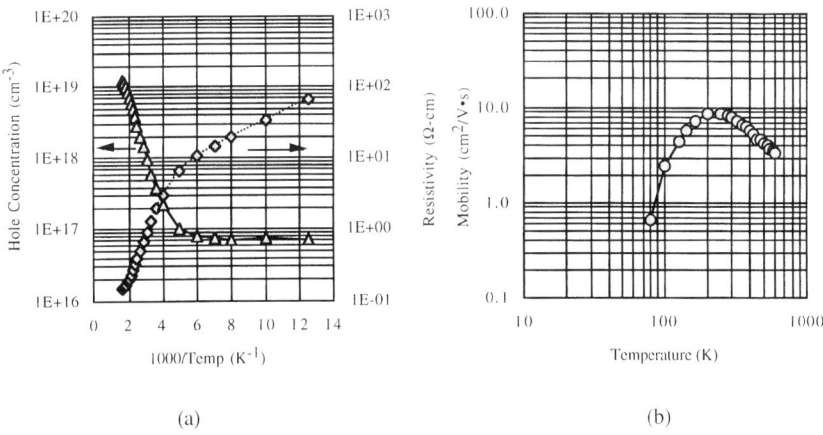

Fig. 11. Temperature-dependent Hall effect data for a Mg-doped GaN layer grown on SiC. The (a) hole concentration and resistivity at 300 K were 6.1×10^{17} cm^{-3} and 1.3 Ω-cm, respectively, with (b) a mobility of 8 cm^2/V·s.

3. Photoluminescence

Room-temperature PL spectra of undoped and Mg-doped GaN films were measured. PL spectra of undoped GaN epilayers exhibit an emission dominated by the band-edge exciton of 3.41 eV (Fig. 12a). The defect luminescence at 2.2 eV was not observed. PL spectra of a Mg-doped GaN epilayer are shown in Fig. 12b. The peak emission is at 435 nm. The peak emission wavelength changes with Mg doping concentration, as observed on Mg-doped GaN films grown on sapphire substrates by other researchers [12].

Figure 13 shows room-temperature PL spectra for undoped $Al_xGa_{1-x}N$ layers with an Al composition in the range of 0 to 10%. As shown, room-temperature PL of pure GaN on SiC has a characteristic peak at 366.4 nm and an FWHM of ~ 30 to 40 meV. As the percentage of Al was increased to 5, 7, and 10% in the epilayer, the peak emission shifted to 351.5, 346.2, and 337.5 nm, respectively. The emission halfwidth also increased slightly with increasing Al composition, which is typically an indication of degradation in material quality. Growth conditions were optimized to reduce the 2.2-eV luminescence and was thus not observed in any of the layers. In addition to these, AlGaN layers with Al composition up to 22% have been achieved.

Figure 14 shows room-temperature PL spectra for undoped $In_xGa_{1-x}N$ layers, with an In composition in the range of 15 to 48%. As shown, the

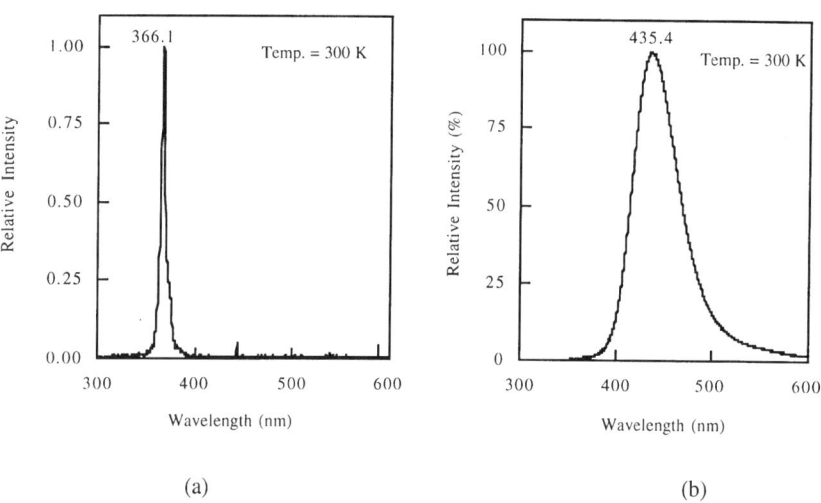

FIG. 12. Room-temperature PL spectra of (a) undoped GaN and (b) Mg-doped GaN.

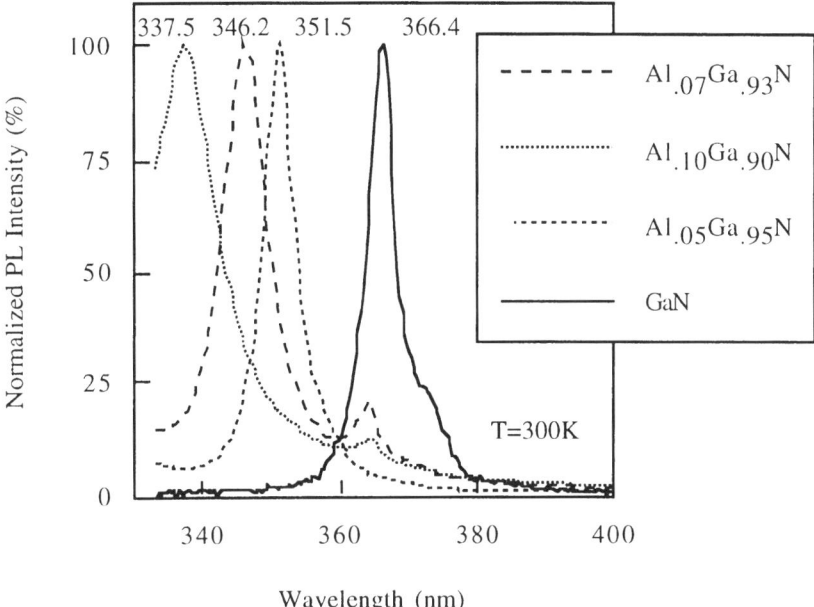

FIG. 13. Room-temperature PL spectra for undoped $Al_xGa_{1-x}N$ layers with Al compositions in the range of 0 to 10%.

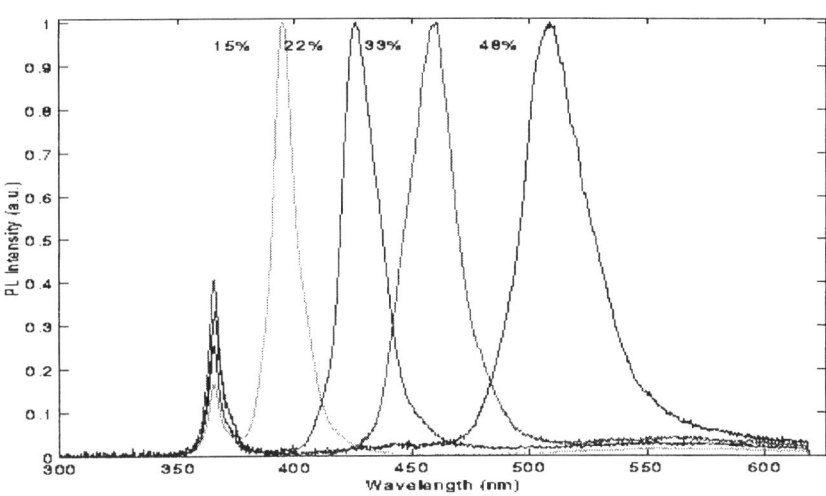

FIG. 14. Room-temperature PL spectra for InGaN layers with an In composition in the range of 15 to 48%.

underlying GaN on SiC has a characteristic peak at 366.4 nm. At a composition of $In_{0.15}Ga_{0.85}N$, a peak in the UV at 395 nm was observed, with an FWHM of 11 nm. As the indium concentration was increased to 22, 33, and 48%, the emission peak red shifted from deep blue to green. The corresponding peak wavelengths were 426, 460, and 509 nm, while the bandwidth of the emission was 18, 24, and 32 nm, respectively. This increase in bandwidth is typically an indication of degradation in material quality. As can be seen in the PL spectra, the defect luminescence near 550 nm was weak compared to the luminescence from the InGaN layers.

4. GaN: SiC Blue LEDs

Two types of LED structures were grown and compared for light emission and forward operating voltage. Cree's initial production of blue LEDs fabricated on SiC substrates was released in mid-1995 and employed a highly resistive buffer layer consisting of a high concentration of Al. To bypass this resistive layer between n-SiC and n-GaN, a shorting ring was applied that allowed for vertical current flow from the top p-type GaN layer to the chip backside. The most recent epitaxy structure and chip design takes full advantage of the high conductivity of SiC by employing a conducting AlGaN buffer layer with a low Al percentage.

Figure 15 shows a schematic diagram of the two types of blue LEDs discussed: (a) insulating buffer and (b) conductive buffer. As shown in both structures, the AlGaN/GaN/AlGaN DH is grown on n-GaN separated from the n-SiC substrates by a strain-relieving buffer layer. In Fig. 15a, the buffer layer is substantially AlN and electrically isolates the GaN from the SiC. This requires the use of a shorting ring, which spans the n-GaN to an Ni

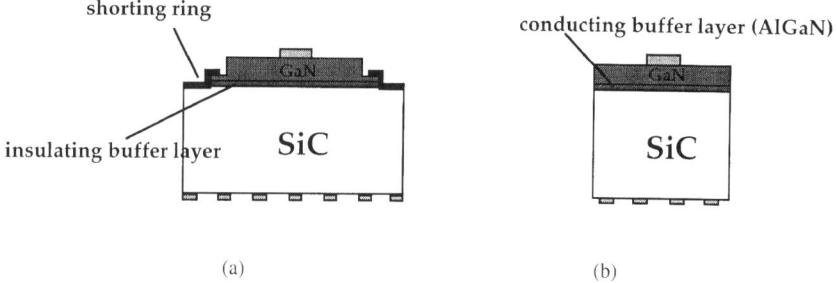

(a) (b)

Fig. 15. Schematic showing the difference between the (a) insulating and (b) conducting buffer layer GaN: SiC LEDs.

ohmic contact on the perimeter of the topside of the SiC chip. This ring not only complicates the chip fabrication process by adding four photoresist steps, but also absorbs light being emitted by the chip itself. As shown in Fig. 15b, this shorting ring structure is eliminated when an AlGaN conductive buffer is employed. This results in a vertical chip that is smaller in size, more efficient, and easier to fabricate.

Figure 16 is an SEM image of the two chip types. The insulating buffer chip shown in Fig. 16a is $300 \times 300\,\mu$m with a square bond pad in the center. The shorting ring can be seen on the chip periphery extending in

FIG. 16. SEM images showing the (a) insulating and (b) conducting buffer layer GaN:SiC blue LEDs.

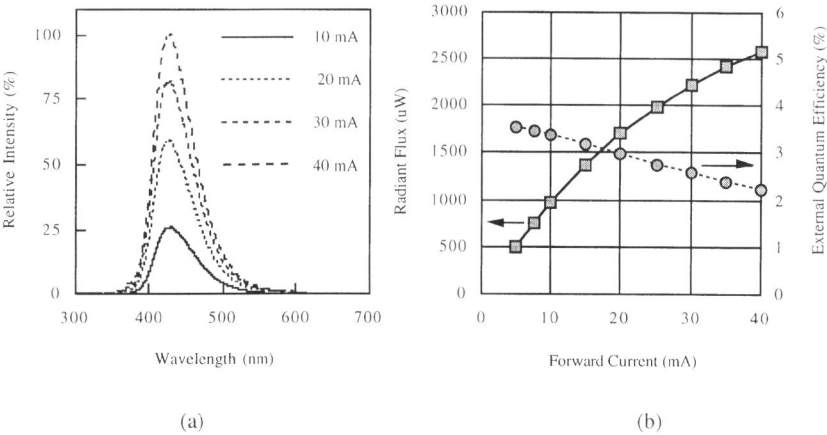

(a)

(b)

FIG. 17. Optical characterization of GaN:SiC blue LEDs showing (a) a peak emission wavelength of 430 nm from 10 to 40 mA drive current and (b) an output of 1.7 mW at 20 mA, corresponding to an external quantum efficiency of 3%.

~20 μm. By eliminating this ring, as shown in Fig. 16b, the chip size is reduced to 260 × 260 μm. In this design, a round bond pad was employed, which is more of an industry standard.

The output characteristics of a blue LED utilizing a high-quality AlGaN/GaN/AlGaN double heterojunction on SiC are shown in Figs. 17a and 17b. The peak emission from this device is 430 nm, with a bandwidth of 65 nm (Fig. 17a), characteristic of a deep blue. The maximum output power observed for the insulating buffer LED is 1.7 mW at 20 mA, which corresponds to an external quantum efficiency of 3% (Fig. 17b). Presently, the output power is typically 800 μW at 20 mA for production devices. For the same epilayer emission efficiency, the output of the conductive buffer chip increases 25 to 50% as a result of eliminating the shorting ring contact.

The I–V characteristics of both chip types are shown in Fig. 18. As shown for the insulating buffer LED, the forward operating voltage (V_f) using the shorting ring geometry is as low as 3.45 V at 20 mA. The dominant resistance is from the p-type contact. From the forward-bias I–V characteristics, the total resistance of the chip is 8 to 10 Ω. Attributing all the resistance to the specific contact resistance (ρ_c), the value of ρ_c for the p-type contact was <3 to 4×10^{-3} Ω-cm² for this device. The operating voltage of

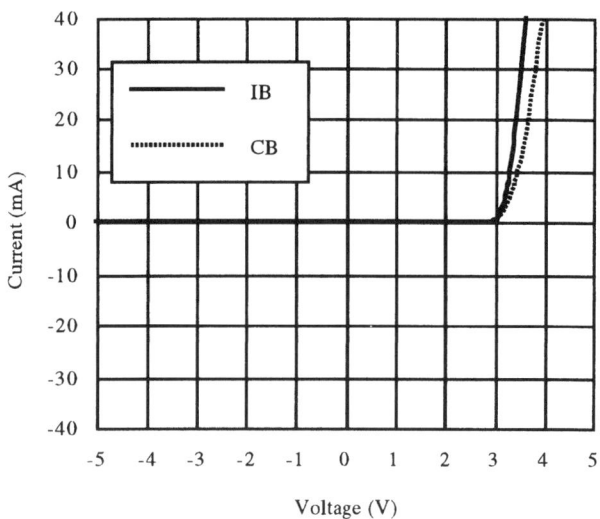

FIG. 18. The current–voltage characteristics of an insulating (IB) and conducting (CB) buffer layer GaN:SiC blue LED.

the conductive buffer chip was slightly higher than the insulating buffer device: 3.66 V at 20 mA. This may be a result of the heterobarrier between the n-SiC substrate and n-AlGaN buffer layer. Work is being performed to reduce this heterobarrier, and thus overall V_f, of the device.

5. ELECTRICAL STATIC DISCHARGE SURVIVABILITY

Another important parameter that has been a reliability issue in nitride LEDs is electrostatic discharge (ESD) survivability. Both of the commercial DH LED products on sapphire include a warning of the sensitivity of these devices. Other types of LED materials, including GaP, AlGaAs, AlInGaP, and SiC, show no sensitivity to ESD. In this study, GaN:SiC LEDs were tested for ESD sensitivity by applying a reverse-bias pulse up to 2000 V and measuring any change in the reverse-bias leakage current at -10 V. This type of test simulates the discharge from a human body and is commonly employed for ESD testing.

DH LEDs on sapphire typically failed the ESD test at <500 V. The GaN:SiC LEDs were tested to 6000 V, at which point the plastic lamp encapsulant broke down. Earlier GaN:SiC LEDs showed sensitivity similar to those grown on sapphire. By optimizing the MOCVD growth parameters, nearly 100% of the GaN:SiC chips will now withstand >2000 V, which exceeds the human body model. The improved ruggedness of GaN on SiC is likely due to the lower defect density of this structure as compared to GaN on sapphire.

VI. Summary

While both blue and green 6H-SiC LEDs have been developed to the point of commercial viability, with the blue being a relatively successful product, the brightness and efficiency of these devices is far below that of the group III-nitride-based blue and green LEDs developed by Nichia Chemical, Cree, and others in the past few years. As a result, the future of emitters that use SiC as the active structure appears to be quite limited. However, 6H-SiC is an excellent substrate for heteroepitaxial growth of III-nitrides, and Cree's line of G·SiC LEDs is based on nitride structures grown on 6H-SiC. As such, SiC will continue to play a major role in future super-bright visible emitters and UV detectors.

Acknowledgments

This research was performed at Cree Research, Inc., and partially funded by the Defense Advanced Research Projects Agency under contract numbers N00014-95-C-0038, MDA972-95-C-0016, and F19628-96-C-0066. The authors would like to thank Mark Edmond for aiding in producing the SEM images of the LED chips.

References

1. Dmitriev, V., Morozenko, Y., Popov, I., Suvorov, A., Syrkin, A., and Chelnokov, V. (1986). "Three-Color Blue-Green-Red Display Made from One Single Crystal." *Sov. Tech. Phys. Lett.* **12**, 221.
2. Vodakov, Y., Vol'fson, A., Zaritskil, G., Mokhov, E., Ostroumov, A., Roenkov, A., Semenov, V., Sokolov, V., Syralev, V., and Udal'tsov, V. (1992). "Efficient Green-Emitting SiC Diodes." *Sov. Phys. Semicond.* **26**, 59.
3. Hoffman, L., Ziegler, G., Theis, D., and Weyrich, C. (1982). "SiC Blue LEDs with Improved External Quantum Efficiency." *J. Appl. Phys.* **53**, 6962.
4. Koga, K., Nikata, T., and Niina, T. (1985). "Single Crystal Growth of 6H-SiC by a Vacuum Sublimation Method and Blue LEDs." *Extended Abstr. 17th Conf. Solid State Devices and Materials* (Tokyo) p. 249.
5. Campbell, R. and Chang, H. (1967). "Detection of UV Radiation Using SiC $p-n$ Junctions." *Sol. St. Electron.* **63**, 949.
6. Glasow, P., Ziegler, G., Suttrop, W., Pensl, G., and Helbig, R. (1987). "SiC UV Detectors." *SPIE* **868**, 40.
7. Nakamura, S. (1996). InGaN Light-emitting diodes with quantum-well structures. In "Gallium Nitride and Related Materials" (F. Ponce, R. Dupuis, S. Nakamura, and J. Edmond, eds.) *Mat. Res. Soc. Symp. Proc.* Vol. 395, pp. 879–887. Materials Research Society, Pittsburgh.
8. Koike, M., Shibata, N., Yamasaki, S., Nagai, S., Asami, S., Kato, H., Koide, N., Amano, H., and Akasaki, I. (1996). Light emitting devices based on GaN and related compound semiconductors. In "Gallium Nitride and Related Materials" (F. Ponce, R. Dupuis, S. Nakamura, and J. Edmond, eds.) *Mat. Res. Soc. Symp. Proc.* Vol. 395, pp. 889–895. Materials Research Society, Pittsburgh.
9. Suvorov, A., Usov, I., Sokolov, V., and Suvorova, A. (1996). Enhanced diffusion of high-temperature implanted Al in SiC. In "Ion-Solid Interactions for Materials Modification and Processing" (D.B. Poker, D. Ila, Y-S. Cheng, L. Harriot, and T. Sigmon, eds.) *Mat. Res. Soc. Symp. Proc.* Vol. 396, pp. 239–242. Materials Research Society, Pittsburgh.
10. Brown, D., Downey, E., Gezzo, M., Kretchner, J., Saia, R., Liu, Y., Edmond, J., Gati, G., Pimbley, J., and Schnieder, W. (1992). "SiC UV Photodiodes." *IEEE Trans. Electron Devices* **ED-40**, 325.
11. Choyke, W. and Patrick, L. (1968). "High Absorption Edges of 6H-SiC." *Phys. Rev.* **172**, 769.
12. Amano, H., Kitoh, M., Hiramatsu, K., and Akasaki, I. (1990). "Growth and Luminescence Properties of Mg-Doped GaN Prepared by MOVPE." *J. Electrochem. Soc.* **137**, 1639.

CHAPTER 8

Beyond Silicon Carbide! III–V Nitride-Based Heterostructures and Devices

Hadis Morkoç

DEPARTMENT OF ELECTRICAL ENGINEERING AND PHYSICS
VIRGINIA COMMONWEALTH UNIVERSITY
RICHMOND, VA

I. INTRODUCTION 307
II. STRAIN AND STRUCTURAL DEFECTS 309
   1. *Effect of Strain and Lattice Mismatch on Crystal Structure* 309
   2. *Dislocations* 310
   3. *Stacking Fault Defects* 311
   4. *Point Defects* 313
   5. *P-Type Doping by Mg vis-à-vis Defects* 320
   6. *Defect Analysis by Deep-Level Transient Spectroscopy* 330
   7. *Defect-Aided Current in Unintentionally Doped GaN* 333
III. OPTICAL MANIFESTATION OF DEFECTS 335
   1. *Yellow Band* 335
   2. *Defects Caused by P-Type Doping in GaN* 339
IV. APPLICATIONS 342
   1. *Field-Effect Transistors* 342
   2. *Blue, Green, and Yellow LEDs* 373
   3. *LEDs by MBE* 378
   4. *Lasers in Semiconductor Nitrides* 380
   5. *UV Detectors* 385
V. CONCLUSIONS 389
   References 390

I. Introduction

GaN-based materials are promising candidates for short-wavelength emitters and detectors and for high-power–high-temperature electronic devices. Consequently, a great deal of effort is being expended on synthesis and characterization of these materials and on processes necessary to develop devices based on them, as well as exploiting applicable devices [1–3]. It appears imminent that digital storage technology will especially benefit from short-wavelength coherent sources (lasers) based on III–V nitrides, because the minimum pit size in disks that can be sensed is diffraction-limited, and optical storage density can be increased quadratically as the read laser wavelength is reduced. Light-emitting diodes (LEDs) are essential for full-color displays and for new applications such as in signal and illumination and for sign applications [3]. Although blue-green lasers based on ZnSe heterostructures have been demonstrated, their short lifetimes and the fragile nature of the material are of considerable concern despite the continuous improvements. GaN and/or its alloys are potentially free of these shortcomings and are promising candidates to fill the green, blue, and ultraviolet (UV) region of the spectrum [4].

Both GaN lasers and LEDs have expanded remarkably in terms of brightness and range of wavelength. These LEDs are very reliable and have a wide range of applications [1–3]. Commercialization of bright blue and green LEDs by Nichia Chemical [4] paved the way for full-color displays. In other potential applications, if these blue, green, and already available red LEDs are used in place of incandescent light bulbs, they would not only provide compactness and longer lifetime (tens of thousands of hours for LEDs as opposed to ~ 2000 hr for incandescent bulbs), but also consume about 10 to 20% of the power for the same luminous flux output. Another attractive approach for producing white light utilizes a blue LED pumping a medium containing green and red phosphors. This has also been done by Nichia Chemical. As for colored LEDs, field tests are already underway in many municipalities around the world to install LED-based traffic lights in place of the filtered incandescent bulbs. The energy savings in about 3 to 4 years alone would offset the cost of installation, particularly in locations where the energy cost is very high.

Though in an embryonic state, electronic devices such as heterojunction field-effect transistor (FETs) built in a GaN environment have shown remarkable performance. Extrinsic transconductances of about 220 mS/mm have been obtained in devices with 1.5-μm gate lengths, rivaling even their august GaAs counterparts, with the added advantage that much larger breakdown fields and a large thermal conductivity pave the way for high-power applications. Output microwave power levels of about 1.5 W/mm in inverted modulation doped FETs with 2-μm gate lengths have been obtained at 4 GHz, which is remarkable.

Due to a lack of lattice-matched substrates, early stages of development, the wurtzitic nature of the III-nitride semiconductor materials, and possibly some yet to be undetermined reasons, GaN and related materials are unfortunately rich in structural and electrical defects. Extremely small native substrates prepared in a high-pressure and high-temperature environment have been reported [5], but their use is limited to a few experiments, and thus any impact they may have will probably not be felt for quite some time. However, ZnO, which is isomorphic with GaN, with a lattice mismatch of under 2%, looks as though it may provide the much-needed reduction in defects until such time as native substrates are available and/or other yet-unproved techniques are developed. Techniques commonly applied to the characterization of defects are increasingly being applied to this material system in order to gain an understanding about the nature of these defects as well as the pathways for their formation. In this chapter, structural and electrical defects and their signature, electrical and optical properties of GaN and GaN-based heterosturcutres, and electronic and optoelectronic devices based on the same will be discussed.

II. Strain and Structural Defects

1. Effect of Strain and Lattice Mismatch on Crystal Structure

Lack of native substrates and associated lattice mismatch causes the epitaxial layer grown to have structural defects. In addition, if the stacking order of the epitaxial layer differs from that of the substrate, steps on the substrate surface would cause rotation and inversion domain boundaries, meaning the epitaxial layers on adjacent terraces would be rotated with respect to another, referred to as stacking mismatch boundaries. There is also the thermal mismatch in heteroepitaxy, which manifests itself as either additional structural defects created during cool down from the growth temperature and/or strain if any defect formed during cool down is not sufficient to cause complete relaxation. Strain causes modification to band structure and, if controlled, can be used to advantage. GaN is a hard material, and a sufficiently thick film can actually crack the substrate. That this is true at least for a number of substrates is evident from an observation of Chu [6] on SiC and of Grimmiss and Monemar [7] on sapphire. Transmission electron microscopy (TEM) has been used to directly observe structural defects in GaN films on GaAs substrates caused by thermal strain from a comparatively thin GaN film.

Amano et al. [8] studied the effect of the thermal mismatch between GaN and sapphire substrates. The GaN was strained under biaxial compression by an amount that depended on the substrate orientation. From the shift in the lattice constant and the photoluminescence (PL) peak energy, the deformation potential for GaN was found to be 12 eV. Naniwae et al. [9]

studied the strain as a function of epitaxial layer thickness for GaN grown on sapphire. The perpendicular lattice constant was observed to be $c = 5.187$ Å up to GaN thicknesses of 50 mm. Thicker films showed a gradual relaxation to $c = 5.185$ Å as the thickness increased to 300 mm. The PL peak energy was observed to have a similar downward trend in energy as a function of thickness. The authors demonstrated reduced residual strain in a GaN film that was grown on a thick GaN layer previously prepared on a sapphire substrate. Because of the poor thermal match between the nitrides and their substrate materials, and the rather high growth temperature used in chemical vapor deposition (CVD) growth, and to a lesser extent in molecular beam epitaxy (MBE) growth processes, nearly all of the nitrides that have been reported were probably under varying amounts of strain. This fact is beginning to get deserved attention [10].

Attention has also been paid to the optimization of the initial nitride overgrowth in order to minimize the defect density resulting from the substrate lattice mismatch. Yoshida et al. [11, 12] were the first to observe an improvement in GaN grown on sapphire when an AlN buffer layer was used. Amano et al. [13] and Akasaki et al. [14] extensively studied the effect of the AlN buffer layer, whose incorporation reduced the background electron concentration by 2 orders of magnitude, while increasing the electron mobility by a factor of 10. The near-bandgap PL was 2 orders of magnitude more intense, and the x-ray diffraction peak width was 4 times smaller in layers having AlN buffer layers. GaN grown directly on sapphire nucleated in tiny microcrystallites, which led to hexagonal islands on the surface of the GaN films. When a buffer layer was used, the AlN was more highly oriented, and the growth became two-dimensional more quickly, allowing improved GaN morphology to be obtained. An optimum AlN thickness of 500 Å was determined with the GaN tending to become polycrystalline when grown on thicker AlN layers. Kistenmacher and Bryden [15] have noted a similar improvement in the morphology and electrical characteristics of InN grown on sapphire when an AlN buffer layer is used. However, Nakamura [16] has been able to grow GaN of extremely high quality directly on sapphire substrates without any AlN buffer layer. Instead, the growth was initiated with several hundred angstroms of GaN deposited at a low temperature.

2. DISLOCATIONS

The use of non-ideal substrate for the growth of thin-film nitrides accompanies a large number of dislocations, which are formed in the epitaxial layer to alleviate the lattice mismatch and the strain of postgrowth cooling. These dislocations result also from thermal expansion mismatches propagated through all layers. Ponce *et al.* [17] have studied the crystallinity of AlGaN films in the neighborhood of the substrate. The film was grown on (0001) sapphire substrate by metal-organic chemical vapor deposition (MOCVD) using an AlN buffer layer. Transmission electron lattice images indicate that the sapphire/AlN interface exhibits a uniform array of misfit dislocations. The misfit dislocations lie along (110) AlN directions, one of which coincides with the direction of lattice projection in the image. They are separated from each other by about 2 nm, which is close to the 2.03 nm expected for a 12.46% lattice mismatch between AlN and Al_2O_3. When AlN is grown on Al_2O_3, eight AlN planes match nine planes at the interface, and the termination of the extra {110} Al_2O_3 plane results in misfit dislocation. The critical thickness for generation of this misfit dislocation is about one monolayer of the film. The exact nature of chemical bonding responsible for this dislocation is yet to be known. One possibility is that the Al atoms at the interface bond with three oxygen atoms in the sapphire direction and to two nitrogen atoms in the AlN direction. Dislocation at the $AlN/Al_{0.5}Ga_{0.5}N$ interface occurs along $\langle 1\bar{1}0 \rangle$ directions. The measured distance of separation between these misfit dislocations is about 22 nm, as compared to 22.6 nm expected for bulk values. The experimental position of these dislocations is also not very distinct. Unlike the dislocations at the AlN/Al_2O_3 interface, the dislocation plane for these dislocations can vary by six or more monolayers (± 1.5 nm).

3. STACKING FAULT DEFECTS

Stacking faults are a common form of strain relief in face-centered cubic crystal structures because their formation energy is fairly low. The schematic diagram of a typical stacking fault is shown in Fig. 1. From this figure it may be seen that, when crystal growth is not in the stacking direction, the formation of a stacking fault requires the termination of an atomic plane at a dislocation. In lattice-mismatched heteroepitaxy of group III nitrides, stacking faults serve as a form of strain relief. If the substrate orientation is chosen in such a way that the growth direction is parallel to the stacking direction, the polytype boundary can be coherent (see, e.g., for (111) 3C and (0001) 2H GaN). Davis *et al.* [18] grew zinc-blende GaN on 3C-SiC and through TEM analysis observed many defects, mainly microtwins and stacking faults propagating into the epitaxial layer. Likewise, films grown on other substrates are not free of these defects. Large quantities of stacking faults have been observed [19] in GaN films grown on sapphire, GaAs, 6H-SiC, and ZnO, although the overall density of these stacking faults vary

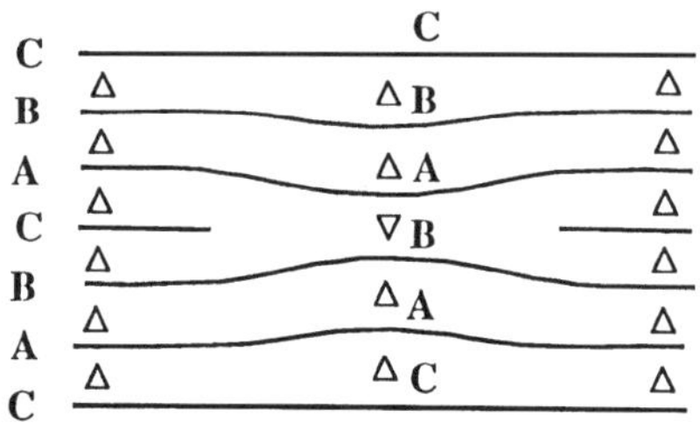

```
C ─────────────  C  ─────────────
B  △          △ B            △
A  △          △ A            △
   △             △           △
C  ─────────  ▽ B  ──────────
B  △          △              △
A  △          △ A            △
C  △          △ C            △
   ─────────     ──────────
```

FIG. 1. Schematic diagram of a typical stacking fault defect. When growth is not along the stacking direction, the formation of a stacking fault requires the termination of an atomic plane at a dislocation.

from substrate to substrate. For example, probably due to the better lattice match, the densities of stacking faults were significantly lower in layers grown on 6H-SiC and ZnO. Lei et al. [20] grew GaN films onto (001) Si by ECR-assisted MBE. GaN buffer layer was first grown at low temperatures, which was followed by the growth of epitaxial GaN layer at high temperature. The buffer layer appeared to be highly faulted, with some very weak texturing visible, which is more apparent in the selected area diffraction pattern. Single-crystal reflections expected from the [001] zone axis show that the buffer layer consists of highly defective single-crystal and, to a lesser extent, highly textured and possibly twinned polycrystalline material. Into the thicker region of the specimen, the single-crystal nature of the specimen was more evident. This resulted from numerous zero-layer reflections due to faulting on the (110) planes. A high-resolution TEM study by Strite et al. [21] of as-grown n-GaN on (001) GaAs substrate showed a large degree of roughness at the GaN/GaAs interface. Twins and stacking faults propagating along the {111} planes appeared to be the major form of strain relief, because a few threading dislocations were seen.

Several investigations have attempted to elucidate the role of stacking faults in group III nitrides. Based on observation of a small zinc-blende component in a bulk wurtzitic film, Seifert and Tempel [22] suggested the existence of these faults in GaN. Powell [23] has since reported the existence of small regions of wurtzitic GaN in bulk zinc-blende GaN crystals. Lei et al. [24] used x-ray diffraction to suggest that wurtzitic and zinc-blende polymorphs co-exist widely in GaN thin films. It was concluded that all of the films except those on Si (001) have domain sizes along the surface. These domains had their (111) axis parallel to the (0001) wurtzitic axis, and the zinc-blende domains were nucleated at a stacking fault, which allowed the wurtzitic stacking to shift to the zinc-blende stacking. This view was verified by TEM images showing the nucleation of a wurtzitic InN domain from a stacking fault in zinc-blende InN [21]. The impact of stacking faults on carrier mobility is not precisely known. Away from the fault edges, their primary effect would most probably be to cause a local variation in the energy bandgap.

Almost all epitaxial GaN-based structures have so far been grown on substrates that do not have matching stacking order. As a result, inversion and rotation domain boundaries, which will be generically referred to as stacking mismatch defects (SMDs), form at each step unless each terrace has the same bonding configuration. In short, unless the steps in 6H-SiC are six double steps high, SMDs would be formed. Consequently, it is highly desirable to deposit GaN on substrates with a common stacking order with nitrides. ZnO is one such substrate, with the added advantage that the lattice mismatch between GaN and ZnO is $\varepsilon = 0.017$, which leads to a range of critical thickness between 80 and 120 Å, which implies that coherently strained layers of GaN could be grown with thicknesses up to about 100 Å. Some compositions of the InGaAlN lattice match ZnO. Although, opinion is divided, GaN grown on the O face of (0001) ZnO exhibited emission characteristics similar to the best ones on (0001) sapphire. More work is underway to harness the possibilities offered, including easy cleavage planes that align with GaN.

4. Point Defects

Point defects, also known as native defects or intrinsic defects, are the common defects occurring in semiconductors. As in all semiconductors, these defects play an important role in the electrical and optical properties of nitride semiconductors. For example, the carrier lifetime is strongly dependent on traps and plays a pivotal role in radiative the quantum efficiency and longevity of GaN-based lasers and LEDs. Experimental evidence and calculations support the premise that N vacancies form a shallow donor level in InN and GaN, getting deeper as the Al mole fraction is increased in AlGaN or the pressure on GaN is increased, as shown schematically in Fig. 2. Such donor and trap levels determine the carrier concentration and whether controlled doping in GaN-based device structures can be obtained.

There are three basic types of point defects: vacancies, self-interstitials, and antisites. Vacancies are the atoms missing from lattice sites. Self-interstitials are the additional atoms in between the lattice sites. Antisites are the cations sitting on anion sites in compound semiconductors. As the formation of native defects results when the bonds in a semiconductor are either broken or distorted, these defects give rise to deep levels within the

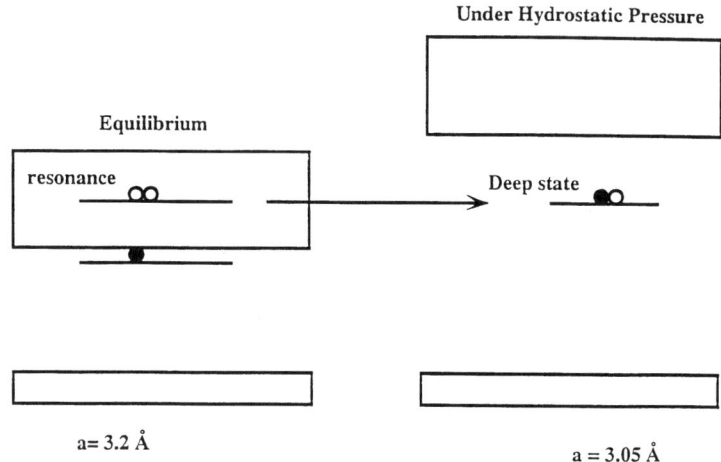

FIG. 2. N-vacancy level in GaN and AlGaN or GaN under sufficient hydrostatic pressure, showing the deepening of the N-vacancy level as the lattice constant is reduced. After Jenkins et al. [29].

energy bandgaps. The Fermi level, and thus occupation of the level, determines the charge state of the defect. Depending on the location of the levels in the energy bandgap and the charge state of the defects, they can be either donor-type, acceptor-type, or amphoteric.

Defects in general and point defects in particular have been with GaN research since the first reports of epitaxial growth. Characteristically, the early films suffered from large n-type background. Maruska and Tietjen [25] argued that autodoping is due to native defects, probably a nitrogen vacancy, because impurity concentration was at least 2 orders of magnitude lower than electron concentration in their samples. Consequently, "unintentional n-type doping in GaN is probably due to N-vacancies" became the crutch statement in nitride-related papers, because there had not been an unequivocal experiment to prove or disprove it. Neugebauer and Van de Walle [26] suggested that the formation of N vacancy in n-type material is highly improbable, based on their first-principles calculations. Instead, they pointed out that some contaminants, such as silicon or oxygen, may be responsible for the large electron concentration. On the other hand, the first-principle calculations of Perlin et al. [27] argue in favor of the N vacancy.

In one study, Wetzel et al. [28] attempted to determine the position of the localized states in GaN, assumed to be nitrogen vacancy related, with respect to the band edge employing infrared reflection and Raman spectroscopy analyses under large hydrostatic pressure. An independent observation of carrier localization by different experimental techniques was also reported. From the observed strong reduction of the free carrier concentration to 3% at 27 GPa with respect to the concentration at ambient pressure, they concluded that the defect concentration was as high as $10^{19}\,\text{cm}^{-3}$. This defect is strongly localized and has a bandgap state of $126 \pm 20\,\text{meV}$ below the conduction band at 27 GPa. This is responsible both for the high electron concentration at ambient pressure and for the capture and localization at 27 GPa. A resonant level of the neutral localized level was predicted to be $0.40 \pm 0.10\,\text{eV}$ above the conduction band edge at ambient pressure.

Jenkins et al. [29] calculated the energies of the N and Ga vacancies and antisite defects in GaN by using a tight binding approach. For the calculations, the wave functions of strongly localized defects, like nitrogen vacancies, were built up from contributions of the whole Brillouin zone. The calculations showed small variations with respect to alloy composition. They indicated that nitrogen vacancies lie roughly 40 meV below the conduction band and that Ga vacancies form shallow acceptors. Both types of antisite defects lie deep within the energy bandgap. In the case of GaN, the doubly occupied A_1 state was approximately 0.11 eV below the conduction band, and a T_2 level was about 0.61 eV above the conduction band edge. In the case of AlN, the A_1 and T_2 levels were predicted to lie 1.60 eV and 0.68 eV, respectively, below the conduction band edge.

Using a state-of-the-art supercell approach with 32 atoms per cell, a plane-wave basis set with an energy cut-off of 60 Ry, and soft Troullier-Martins pseudopotentials, Neugebauer and Van de Walle [30] investigated defect formation energies and electronic structures for native defects in wurtzitic and cubic GaN. It was predicted that 3d electrons are important for both the formation energy and the atomic relaxation. The breaking of the additional bonds between the Ga 3d orbitals and the N orbitals to create a vacancy costs more energy. This leads to a significant increase in the formation energy. Taking the Ga 3d electrons as core electrons results in a large relaxation. The introduction of 3d electrons causes the system to be stiffer. As a result, the outward displacement of the surrounding nitrogen atoms is reduced to 0.1 Å and the energy gain to 0.26 eV. The 3d electrons thus prevent the GaN bond length from becoming too short.

Another yet extensive theoretical study of native defects in hexagonal GaN carried out by Boguslawski et al. [31] contradicts the conclusions of Neugebauer and Van de Walle. For this study, the effects of cation and anion vacancies, antisites, and interstitials were considered. The computations were carried out using an ab initio molecular dynamics approach. Supercells contained 72 atoms. The results suggested that the residual

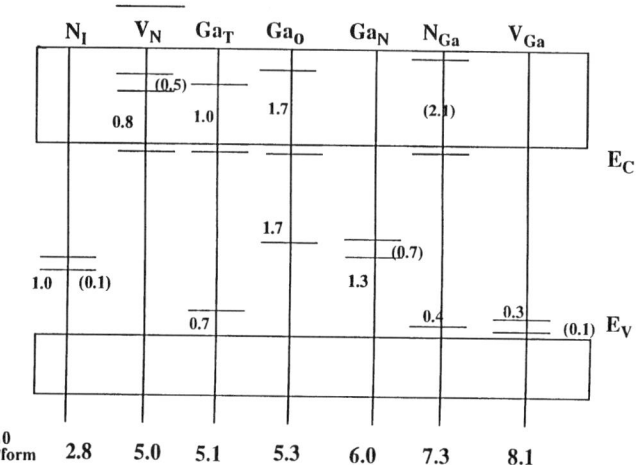

FIG. 3. Energy levels of neutral native point defects in GaN. After Boguslawski et al. [31] with permission.

donors responsible for the *n*-type character of as-grown GaN are N vacancies. However, the concentration of Ga interstitials under equilibrium conditions in the usual Ga-rich material can become comparable to that of the vacancy. Both *n*- and *p*-type doping efficiencies are substantially reduced by the formation of gallium vacancies, N vacancies, and gallium interstitials. In the zinc-blende structure, a substitutional defect (e.g., a vacancy) has four equivalent nearest neighbors. In the wurtzitic structure, the atom along the *c*-axis relative to the defect becomes inequivalent to the three remaining neighbors. This leads to a lowering of point symmetry, causing the defect states, which are threefold degenerate in the zinc-blende structure, to split into singlet and doublet pairs in the wurtzitic structure. The resulting energy levels are shown in Fig. 3.

As indicated previously, there have been various arguments about the origin of *n*-type conductivity of the *as-grown* undoped bulk and heteroepitaxial GaN. One argument is that the N vacancy [25, 27, 28] is responsible for autodoping (referred hereafter as the N-vacancy argument). The impurities [26], such as Si and O, responsible for autodoping represent the other argument (referred to hereafter as the impurity argument). Undoped GaN films with different ammonia flow rates were grown, keeping other experimental conditions the same in an attempt to unveil the origin of autodoping [32]. Hall effect measurements and secondary ion mass spectroscopy (SIMS) measurements were carried out to obtain background

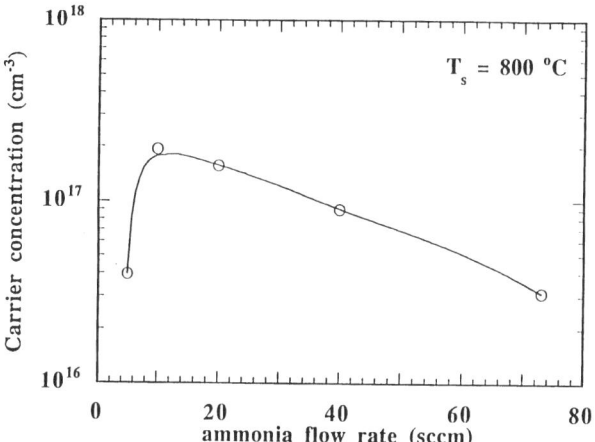

FIG. 4. The variation of the background carrier concentration for a series of undoped GaN films grown with different ammonia flow rates. The growth rate and film thickness were kept at 1.2 μm/hr and 2.4 μm, respectively. After Kim et al. [32].

doping and impurity levels, respectively. Figure 4 shows the change in background doping level for the samples grown with different ammonia flows rates (V/III ratio). In general, the background electron concentration decreases as ammonia flow increases, except for the point with the lowest ammonia flow rate. It is expected that the concentration of N vacancy in the film should decrease as the V/III ratio increases. Thus, the trend in Fig. 4 can be deduced such that the background doping level and N-vacancy level follow one another, supporting the *N-vacancy argument*. To verify the relationship between the autodoping level and the impurity level, SIMS measurements were carried out for two samples grown with 10 (#5465) and 73 (#5455) sccm of ammonia flow rates, as shown in Fig. 4. The background doping levels for the first two samples were about 2×10^{17} and 3×10^{16} cm^{-3}, respectively. Figure 5 shows the depth profile of O, H, and Si impurities for those two samples measured by SIMS. Shown in Fig. 5 at the same time is the depth profile of such impurities for the sample (#5169) grown with a higher ammonia injector temperature (600°C) and single ammonia purifier (ammonia flow rate = 4 sccm), which shows an unintentional electron level of 2×10^{18} cm^{-3}. The growth condition for sample 5169 is even closer to the N-deficient condition than that for sample 5465. As one can see from Fig. 5a, the Si impurity levels are lower than 10^{17} (mostly in the 10^{16} atoms/cc range, which is close to the lower detection limit of SIMS for Si) for all three samples, regardless of different background

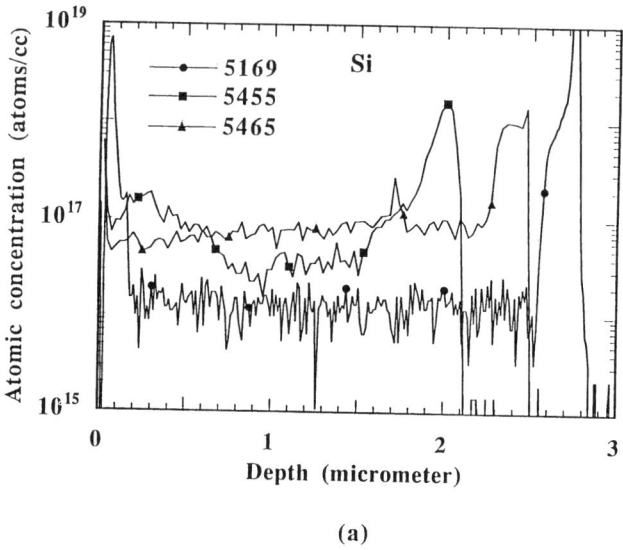

(a)

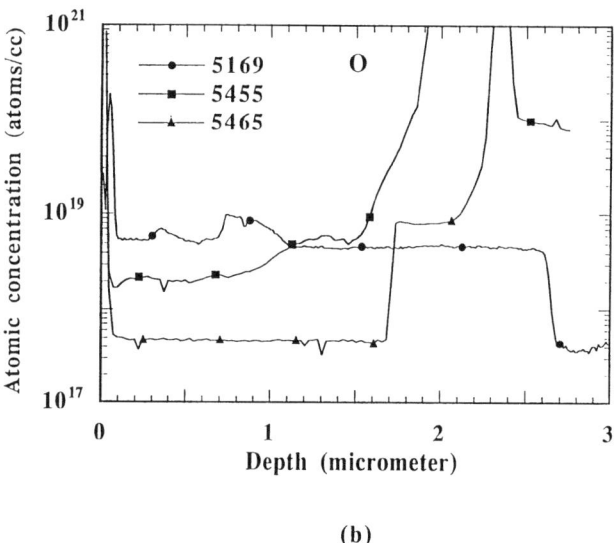

(b)

FIG. 5. The depth profile of (a) Si, (b) O, and (c) H measured by SIMS for various ammonia fluxes. Samples 5169, 5455, and 5465 were grown with 4, 73, and 10 sccm ammonia flow rates, respectively. A single ammonia purifier was used for sample 5169, and double purifiers, tandem, were used for the other two samples. Moreover, the injector temperature was 600°C for sample 5169 and 300°C for the others. After Kim et al. [32].

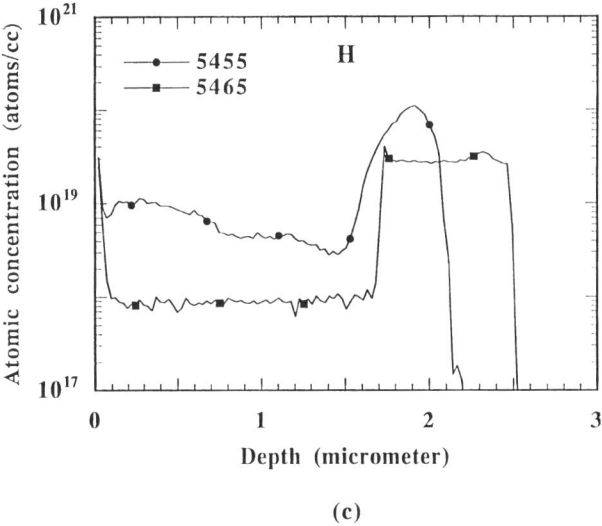

FIG. 5. (*Continued*).

doping levels. It is thus reasonable to assert that Si impurity is not mainly responsible for the autodoping of these undoped GaN samples.

Considering the lower ammonia flow rate for sample 5169, the 1 order of magnitude difference in O concentration can be attributed to the difference in the ammonia purifying state; that is, ammonia was purified in a single stage for sample 5169, while a two-stage purification was employed for sample 5465. The background O level from the system itself was checked by carrying out SIMS measurements for an undoped GaAs layer grown in the same system. The O concentration in this sample was about 6×10^{16} atoms/cc, assumed to be the instrumental detection limit in the GaAs matrix. Thus, it is concluded that the O contamination in the GaN samples is mainly due to ammonia.

It has been suggested that oxygen forms a reasonably "shallow" donor in GaN [33]. The oxygen level in samples 5169 and 5465 are comparable in terms of electrical activity, as shown in Fig. 5b. Thus, the autodoping can be attributed to O. At the same time, however, both samples were grown under the environment closer to N-deficient conditions than sample 5455, and thus these particular set of samples are not sufficient to rule out the O contamination. To delineate the role of O, we prepared a GaN sample with high ammonia flow rate, thus high O background. Figure 5b also shows the O concentration for sample 5455, which was grown with 73 sccm of ammonia flow rate. The O concentration for this sample is higher than that of sample 5465, as expected, due to the higher ammonia flow rate. The background doping level for this sample was about 3×10^{16} cm^{-3}, which is about 1 order of magnitude lower than that of sample 5465. Consequently, the N-vacancy argument is a more plausible source of autodoping in GaN, barring any active compensation processes being in effect during N-rich growth conditions.

Figure 5c shows the depth profile of H atoms for samples 5455 and 5465. The overall H content in sample 5455 is higher than that in sample 5465, which is consistent with the ammonia flow rate used for these two films, the former having been grown with a larger ammonia flow rate. It is predicted from the first-principles calculation that H incorporation in n-type GaN is much lower than that in p-type films [34], due to the higher formation energy, while there are two contradictory experimental reports about deuterium incorporation after hydrogenation for both n- and p-GaN [35, 36]. The H concentrations in our n- and p-type GaN films are almost the same. Although a detailed study about the incorporation behavior of H into n- and p-type GaN films is necessary, we suggest that the H concentration in our GaN films is determined by the amount of ammonia during the growth, regardless of carrier type, unless there are additional impurity incorporation factors, such as high density of defects involved.

5. P-Type Doping by Mg vis-à-vis Defects

Though a good understanding is lacking, sucessful p-type doping has been achieved with Mg by using both MOCVD and MBE techniques. As-grown GaN:Mg samples by the MOCVD technique are highly resistive but are made p-type after a thermal annealing in nitrogen or electron beam irradiation [37, 38]. However, thermal annealing in an NH$_3$ atmosphere reverts the samples to their high-resistivity state (10^6 Ω-cm) [39]. Van Vechten et al. [40] suggested that hydrogen passivates Mg, which is also supported by the calculations of Neugebauer and Van de Walle [34]. Moreover, Neugebauer and Van de Walle demonstrated that hydrogen is beneficial for p-type doping by Mg when compared to the hydrogen-free case, since the Mg concentration increases and N-vacancies concentration decreases in the presence of hydrogen. Although it has been mentioned frequently that Mg and H form a complex in Mg-doped GaN films, the mechanism of formation and release of H upon postgrowth treatment has not been definitively clarified. Also, the predictions about the position of the H atom by first-principles calculation are not consistent. For example, in

Neugebauer and Van de Walle's calculation [41], H is stable at the N antibonding (AB_N) site, while it is stable at bond center (BC) site according to Okamoto et al. [42]. Unlike the MOCVD-grown layers, Mg doped GaN layers grown by reactive molecular beam epitaxy (RMBE) exhibit p-type conductivity without any postgrowth treatment [43].

It is believed that H electrically passivates Mg in Mg-doped GaN film, as is the case for other III–V semiconductors [44]. This has been supported by the improvement in conductivity of MOCVD-grown Mg-doped GaN films when they undergo a low-energy electron beam irradiation (LEEBI) or thermal anneal in an N environment. Also, the first-principles calculations show that the same amount of Mg and H incorporate into GaN films when they are grown under an H-containing growth condition, such as MOCVD. Furthermore, the calculations predict that more Mg can be incorporated into GaN film when there are more H atoms.

As-grown p-GaN films grown by RMBE have shown p-type conductivity, even though these films were grown using H-containing ammonia [43]. Note that both the RMBE and MOCVD growth environments contain copious amounts of H and that MOCVD-grown Mg-doped GaN films [45] show p-type conductivity only after the thermal annealing. Actually, the amount of H in a p-GaN film in the study by Kim et al. [43] as well as in our samples, shown in Fig. 6, is much lower than the amount of Mg. This indicates that there are a good deal of Mg atoms that do not form complexes with H in the film if one-to-one Mg-H complex formation occurs. Figure 6 shows the depth profile of H for the two Mg-doped samples discussed in Fig. 4 of our previous report [43]. The Mg concentration for these two samples were about 5×10^{18} and 3×10^{19} atoms/cc, respectively. As one can see, the H level in RMBE-grown p-GaN is not negligible. It is noticeable that the H concentration in these two as-grown films is almost the same, although there is a distinct difference in Mg concentration. It was previously reported [43] that the thermal annealing, or rapid thermal annealing (RTA) process, did not change the electrical characteristics of these films. But, the amount of H decreased by about three times after RTA treatment for 2 min at 900°C, as shown in curve 2 of Fig. 6. Thus, the release of H atoms from as-grown GaN:Mg did not improve the conductivity of the already conducting RMBE-grown films. Although a further study is essential to delineate the nature of as-grown p-type conductivity in RMBE-grown GaN:Mg films, the following two explanations may be justified. The first has to do with the *in situ* annealing effect during the cooling procedure after the film growth. When the film growth is completed, the ammonia flux is normally shut off right away. Consequently, the film is vacuum-annealed, albeit for a short time, during the cooling process, during which H may be

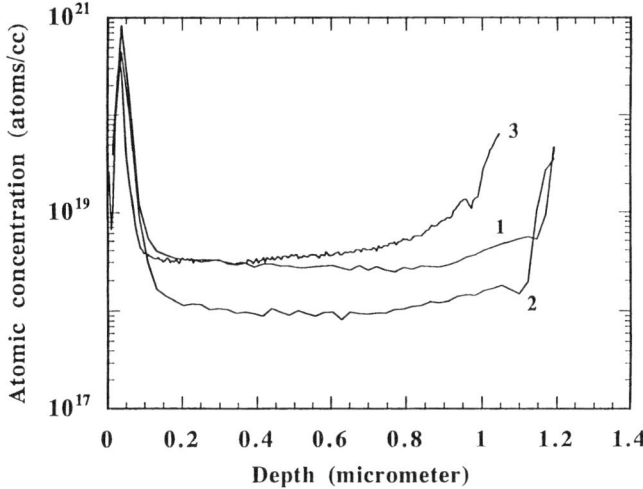

FIG. 6. The H profile measured by SIMS for two p-GaN films grown with different Mg fluxes. The Mg profiles for these two samples are shown in Fig. 4 of Kim et al. [43]. An ammonia flow rate of 37 sccm was used for both of the samples. 1, the H profile for the as-grown sample with low Mg flux; 2, the H profile for the same sample shown in curve 1, but after RTA treatment at 900°C for 2 min; 3, the H profile for the as-grown sample with high Mg flux. After Kim et al. [32].

released, which might not be the case for MOCVD-grown films. It has been reported that annealing in an ammonia atmosphere may passivate the conductive (after postgrowth treatment) p-GaN film, and the conductivity can be restored by annealing the same sample in an N atmosphere [39]. It is not clear, however, whether the activation and deactivation processes for already conductive (after postgrowth treatment) p-GaN film are the same as those for highly resistive as-grown GaN:Mg film.

The second and more plausible explanation is that the release of H might not be the direct cause of the recovery of p-type conductivity. Rather, the Mg atoms in GaN film can be at some metastable state due to the unique growth reaction or somewhat ionic nature of GaN:Mg. Now they can be activated by energy from either electron irradiation or thermal annealing, and the release of H can be just a side effect of this process. This idea can be supported by the fact that the electrical property is not affected by the RTA treatment, while the H concentration decreased about three times, as shown in Fig. 6. Even though there is some evidence supporting these two ideas, further research on the behavior of Mg and H atoms in atomic scale and their effect on the electrical properties of the Mg-doped GaN film is necessary.

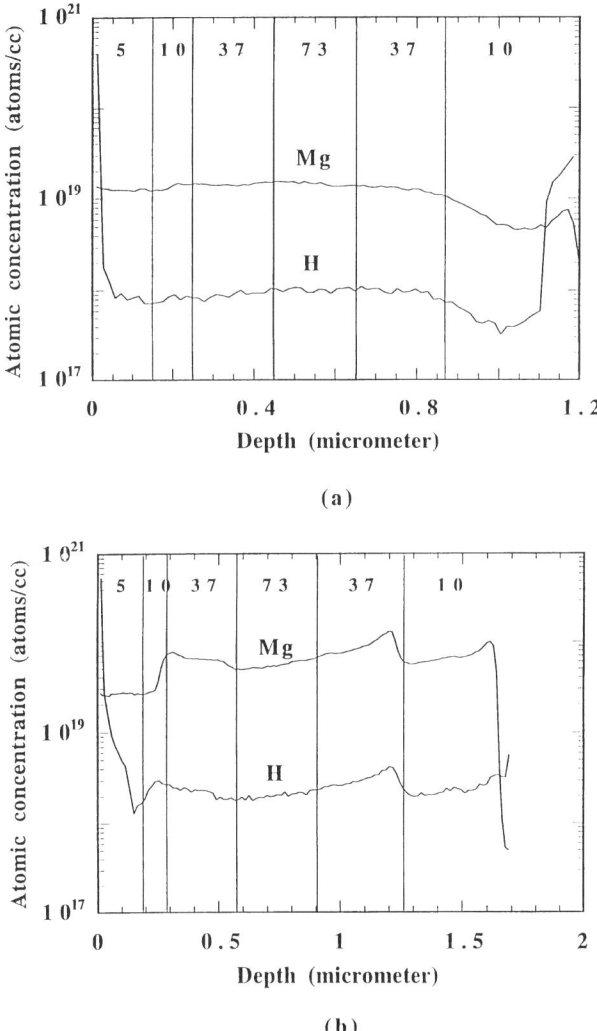

FIG. 7. The Mg and H profiles measured by SIMS for two Mg-doped samples grown with different Mg fluxes. The ammonia flow rate was changed in several steps throughout each growth. The numbers in each segment represent the ammonia flow rate. (a) Low Mg flux; (b) high Mg flux. After Kim et al. [32].

Two Mg-doped GaN films were grown on c-plane sapphire substrates with different Mg flux. The flow rate of ammonia was changed in several steps throughout each growth to observe the effect of ammonia flow rate on the incorporation of Mg into the film. The depth profiles of H and Mg atoms for these two samples were analyzed by SIMS. Figures 7a and 7b show the depth profiles of H and Mg for these two samples. The numbers appearing on the top portion of each growth segment indicate the ammonia flow rate for each period. No clear steps are seen in Fig. 7a, while two clear steps are seen in Fig. 7b when the ammonia flow rate was changed between 10 and 37 sccm for both H and Mg. It is noticeable that the Mg and H concentrations change at the same position. Also, the Mg concentration is more than 1 order of magnitude higher than the H concentration, as mentioned earlier. It is tempting to conclude that a portion of Mg atoms incorporated into the film is passivated by H, while the majority of the Mg atoms incorporate without such passivation by the accompanying H. And the ratio of passivated Mg may or may not be constant, which means that the H profile may or may not follow that of Mg. In fact, the H profile did not match the Mg profile when the same experiment was carried out on 6H-SiC substrate, as shown in Fig. 8. At this point, it is not clear whether there is a difference in growth characteristics in terms of H and Mg incorporation when SiC substrates are used. In fact, the H concentration, in general, was about 1 order of magnitude higher than the previous two samples, which might be due to the growth characteristics related to SiC substrates—for instance, diffusion of impurities from the SiC surface to the growing film. The H and O concentrations are about 1 order of magnitude higher in the part of the film near the film–substrate interface, as compared

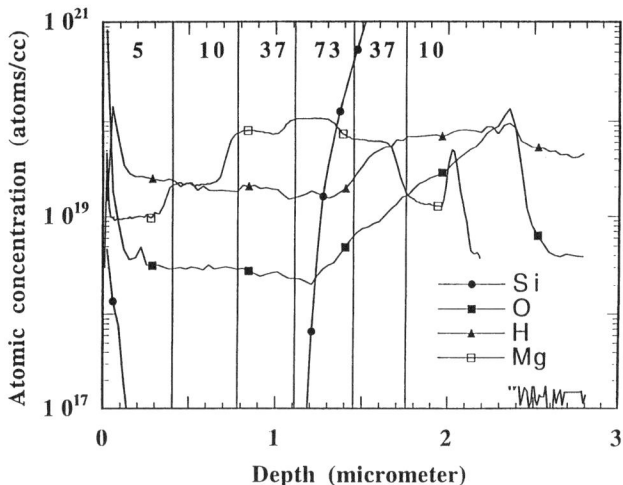

FIG. 8. The SIMS profile for Si, O, H, and Mg for the sample grown on 6H-SiC with the same scheme used in Fig. 5. The steps in Mg concentration are clearly seen for each segment. After Kim et al. [32].

to the rest of the film. This might be related to the out-diffusion of impurities from SiC substrate, as alluded to earlier, because the location where the concentration starts to increase approximately matches the point where the Si signal (of SiC) starts to increase drastically. In any case, Fig. 8 supports the latter idea, because there is no such trend in the H profile, as seen in Fig. 7b, while steps in Mg concentration are still seen. It appears that "more H during the growth" enhances the incorporation of Mg, as was expected by Neugebauer and Van de Walle [34], when we consider only the sample shown in Fig. 7b. However, the following argument having to do with site selection can explain the overall trend of Mg incorporation with changing ammonia flow rate: if films are grown with more active N species, the concentration of Ga vacancy in the film will be increased, and, consequently, it will be more likely for Mg atoms to incorporate into those sites, provided that there is sufficient Mg. In fact, "more H" also means "more active N species" in our experimental environment, which is consistent with the expected behavior of Mg.

It is suggested that the steps not seen in Fig. 7a are due to the Mg flux not being sufficient to saturate the available Ga-vacancy sites provided by the growth condition with a 5-sccm ammonia flow rate. The same argument can be applied for the 73-sccm flow rate portion in Fig. 7b. However, this does not exclude the presence of Mg atoms at sites other than the Ga sites and that we cannot keep a reasonable crystalline quality with very high Mg concentration. It has not been determined whether Mg atoms still locate substitutionally at the Ga sites when the concentration is relatively high, while a first-principles calculation predicts that the Ga substitutional site is energetically most stable [41] without considering the effect of concentration.

Based on the foregoing observations, we can assume that better p-GaN can be grown within a certain limit of Mg concentration for a given amount of active N species (in our case, ammonia flow rate). It still has to be verified if the upper limit, within which better quality p-GaN can be grown, is proportional to the amount of active N species. Actually, p-GaN films, with much-enhanced quality over the previous ones, were grown with 73-sccm ammonia flow rate using RMBE (about 2 times higher than the flow rate used for the sample in the study by Kim *et al.* [43]). Moreover, the injector temperature for the samples of Kim *et al.* was 600°C, possibly leading to less active N species on the growing surface when compared to the lower injector temperature case for a given ammonia flow rate. The best sample has a doping level, mobility, and resistivity of about $8 \times 10^{17} \, \text{cm}^{-3}$, $26 \, \text{cm}^2/\text{Vs}$, and $0.3 \, \Omega \, \text{cm}$, respectively. Figures 9a and 9b show the room-temperature PL spectra and Arrhenius plot for the hole concentration measured by the van der Pauw method. The thermal activation energy obtained from the

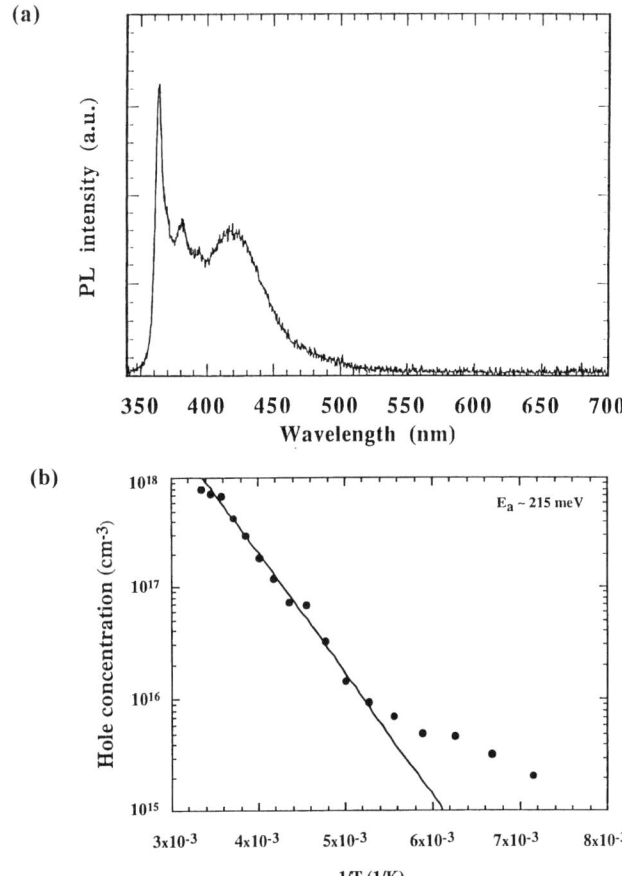

FIG. 9. (a) Room-temperature PL spectra for the p-GaN sample, which shows a doping level of 8×10^{17} cm^{-3} with 26 cm^2/Vs Hall mobility. (b) Arrhenius plot for the hole concentration measured by the van der Pauw method, giving about 215-meV thermal activation energy. After Kim et al. [32].

slope of the Arrhenius plot is about 215 meV, which is different from that previously reported [43]. These two different activation energies might reflect the multiple acceptor levels found in the deep-level transient spectroscopy (DLTS) study by Götz et al. [46]. Astonishingly, the Mg concentration in this film is about 8×10^{18} cm^{-3}, which means about 10% of activation ratio, which remains unexplained at this point. The enhanced quality is not surprising when this sample is compared to the two samples shown in Fig. 3 of the article by Kim et al. [43]. The Mg concentration in the latter sample (was grown with a 37-sccm ammonia flow rate and high injector temperature, which gave more than 2 times less active ammonia on the growing surface) was about 5×10^{18} cm^{-3} and the doping level was 3×10^{17} cm^{-3} with mobility of 3.7 cm^2/Vs. Even after assuming the same Mg concentration, it is not difficult to infer that higher ammonia flow rates (or active nitrogen species) can produce better p-GaN with other conditions being the same. The same experiment was repeated for Si-doped and undoped GaN films to see the effect of ammonia flow rate on the incorporation of Si and O. No particular steps in the Si and O profile were seen. The Si concentration was only about 2×10^{17} cm^{-3} for the Si-doped sample, and this level of Si concentration might be low enough to produce the same effect seen in Fig. 7a. An exhaustive study on the Si-doped samples, especially for the higher doping level, is underway.

As is the case with ZnSe [47], p-type GaN exhibits persistent photoconductivity (PPC), as detailed by Li et al. [48], who investigated the nature of Mg impurities in GaN in Mg-doped p-GaN grown by RMBE. They catalogued the PPC buildup and decay transients and the dependence of the

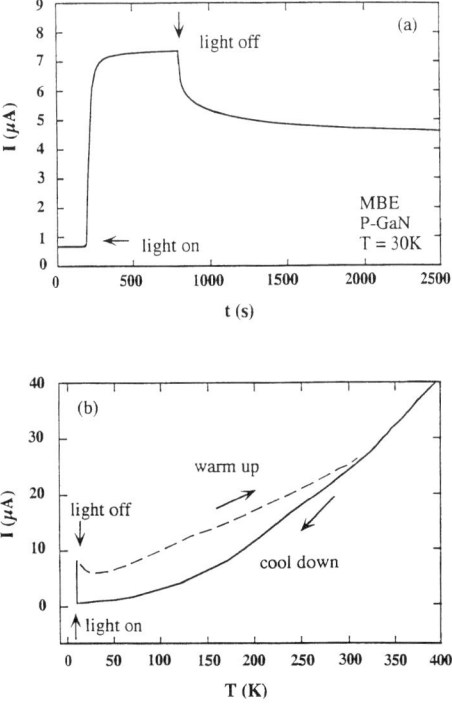

FIG. 10. (a) A typical behavior of PPC in Mg-doped p-GaN grown by RMBE. (b) The dark conductivity as a function of temperature. The bottom curve (solid dots) represents data taken with the sample cooling down in the dark, while the top curve (open triangles) represents data taken with the sample illuminated at 10 K for about 10 min and then warming up in the dark. After Okamoto et al. [42].

PPC decay time constant on the PPC buildup time and formulated their premise in the context of lattice-relaxed Mg impurities (or AX centers). The results demonstrated conclusively that there is a thermal energy barrier of about 130 meV that prevents free hole capture by ionized Mg impurities and that there is a lattice relaxation associated with Mg impurities in GaN. A typical low-temperature ($T = 30$ K) PPC behavior of a p-GaN sample grown by MBE is shown in Fig. 10a, which shows that the conductivity increases by several orders of magnitude after exposure to light and that the light-enhanced conductivity persists for a very long period. The PPC effect is also seen in MOCVD GaN to an extent that is about 5% less than that in MBE films. Since we are in the early stages of development and understanding, this assertion may have to be re-evaluated. Figure 10b shows the dark conductivity as a function of temperature. The bottom curve (solid line) represents data taken with the sample cooling down in the dark, while

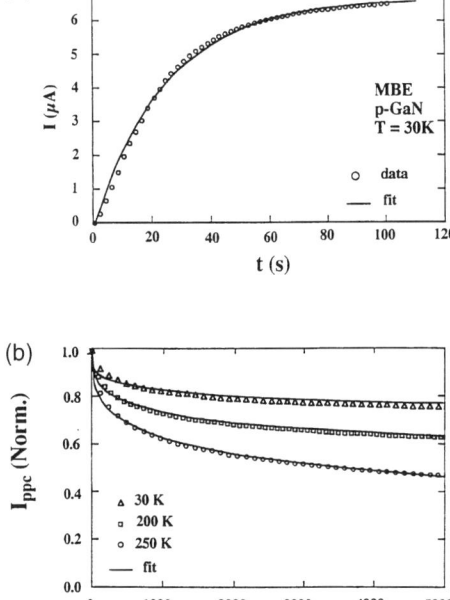

FIG. 11. (a) PPC buildup transient recorded for the first 100 S at 30 K. (b) PPC decay obtained at three representative temperatures for relatively low buildup levels. Each decay curve is normalized to a unity at $t = 0$, the moment at which the light excitation is terminated and the dark level has been subtracted. The solid curves are the least squares fit of data by stretched exponential functions of Eq. (4) in Okamoto et al. [42]. After Okamoto et al. [42].

the top curve (dashed curve) represents data taken with the sample illuminated at 10 K for about 10 min and then warmed up in the dark. Figure 10b clearly illustrates that the MBE sample exhibits PPC effect below 310 K. The observed hysteresis of the dark conductivity is the hallmark of DX centers (or AX centers) in semiconductors. A crucial feature that further reveals the AX nature of Mg impurities in GaN is the kinetics of PPC buildup and decay, shown in Fig. 11. As seen conclusively in Fig. 12, there is a lattice relaxation associated with Mg impurities as well as an energy barrier for photoexcited hole capture in Mg-doped GaN, possibly due to a broken bond along the c-axis. This is illustrated schematically in the inset of Fig. 12a. Such a feature has been theoretically predicted for dopants in several materials. As mentioned earlier, AX centers have also been experimentally observed in II–VI wide-bandgap semiconductors. The observation of AX centers in GaN seems to suggest that lattice relaxation

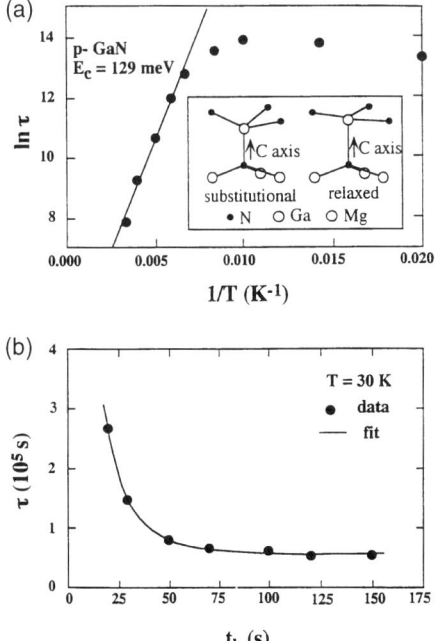

FIG. 12. (a) The Arrhenius plot of the PPC decay time constant τ (ln τ versus $1/T$). The capture barrier (E_c) obtained from data taken above 150 K is about 129 meV. The inset shows schematic views of (i) normal substitutonal sites and (ii) the possible broken-bond configuration giving rise to AX centers in Mg-doped GaN. (b) PPC decay time constant g as a function of PPC buildup time τ_b. The solid curve is the least squares fit of data with Eqs. (9) and (10) in Okamoto et al. [42]. After Okamoto et al. [42].

is a rather common phenomenon associated with dopants in wide-bandgap semiconductors. It is clear that Mg impurities are responsible for the PPC effect observed in p-GaN grown by both methods and that impurities (or H-Mg complexes) are responsible for the non-equilibrium process of the dark conductivity in MOCVD-grown p-GaN. It must be noted, however, that recent p-type samples grown by MBE do not show any PPC effect at all, indicating that properly grown samples may be void of lattice-relaxation stretched bonds, which were present in earlier samples.

6. Defect Analysis by Deep-Level Transient Spectroscopy

To gain some insight into the electrical character of point defects, the DLTS method is traditionally used. In this method a *pn* junction or a Schottky barrier structure is biased into depletion, which causes the defects above the Fermi level (for *n*-type samples) to release, to a first extent, their electrons. A charging pulse, which reduces the reverse bias across the junction, is applied to fill some of the previously emptied traps, increasing the junction capacitance. On the removal of the charging pulse, the electrons trapped at the deep level(s) would be freed at a rate that depends on the energy level of the trap. This is also related to the rate at which the capacitance would decrease toward the equilibrium value. In DLTS, the change in capacitance in an appropriate time window is plotted as a function of sample temperature, from which the activation energy of the trap can be determined. Alternatively, the time rate of change in the capacitance as the traps empty can be plotted versus time, from which pertinent information about the nature of the centers can be deduced. This technique is called isothermal capacitance transient spectroscopy (ICTS), which has also been employed for analyzing nitrides.

In very wide gap semiconductors, the really deep states may not effciently release their electrons, particularly at low temperatures. To circumvent this problem, optical excitation in conjunction with measurement of the capacitance transients can be employed. Similar to thermal excitation of electrons, when the electrons are optically excited into the conduction band in a reverse-biased junction, the charge in the depletion region increases following the resonant optical excitation, causing the capacitance to increase [49]. Each time the photon energy resonates with the deep level, there is a marked increase in the capacitance. However, the defect-state energies measured by this technique (photoemission spectroscopy) can be different from those discerned by DLTS, as the thermal processes are not involved in the former. Nevertheless, very useful information can be gleaned because DLTS cannot reach very deep states.

GaN samples prepared by MOCVD on sapphire substrates have been investigated by DLTS by several groups [50–52]. Defects states termed E_1, E_2, and E_3, have been observed by various investigators, with E_1 and E_2 appearing in ICTS [50]. The energies of the aforementioned defects have been determined to be in the range of, in order, 0.264 to 0.269 eV, 0.563 to 0.59 eV, and 0.662 to 0.686 eV. The concentrations of the same defects, again in order listed, are in the range of 2.6×10^{15} to 6.6×10^{16}, 3.0×10^{13} to 1.3×10^{14}, and 1.6×10^{14} to 3.8×10^{14} cm^{-3}. When the GaN layers were grown on sputtered or thick ZnO, concentration of the E_1 and E_3 defects are reduced to 1.4×10^{15} and 5.1×10^{13} cm^{-3}, respectively. The observed trap levels by DLTS and photoemission transient capacitance methods, with the spread in their measured energies, are shown in Fig. 13. Deep levels observed in GaN by DLTS and by various groups are shown in Fig. 14, as compiled by Hacke *et al.* [50]. The shallow level to the right on the E_1 level appears in films grown by a TMGa source. A notorious problem with wide-bandgap nitrides is that attempts to accomplish *p*-type doping often results in the creation of structural and electrical defects. Minimization of these defects is imperative and determines whether *p*-type doping can be successfully attained. Shown in Fig. 15 are the DLTS spectra obtained in films with and without Mg doping. As can be seen, the E_2 line increases remarkably with addition of Mg. Similar effects are seen for the other defects, albeit not to the same extent. The photoemission transient

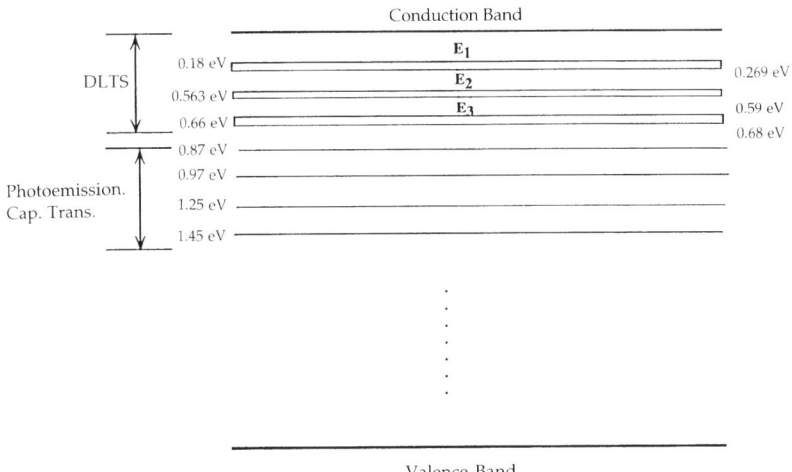

FIG. 13. Various deep levels within the gap energy of GaN as observed by DLTS and photoemission transient spectroscopy.

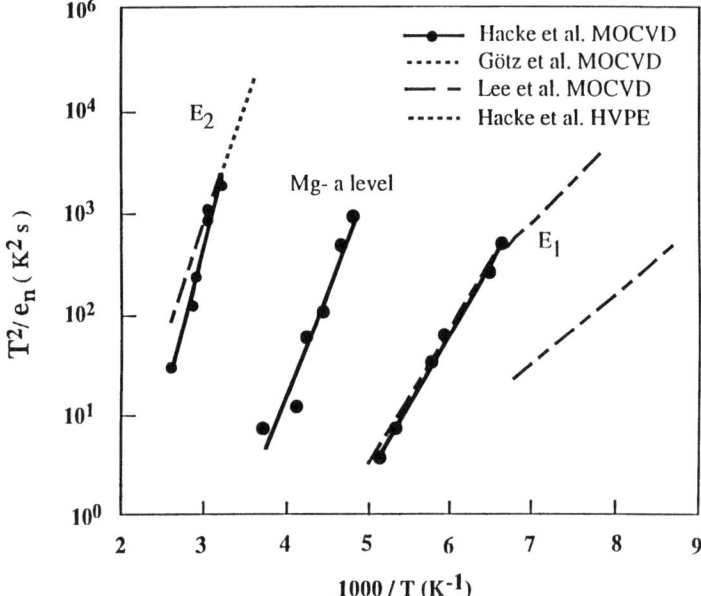

FIG. 14. Deep levels observed in GaN by DLTS. After Hacke *et al.* [50].

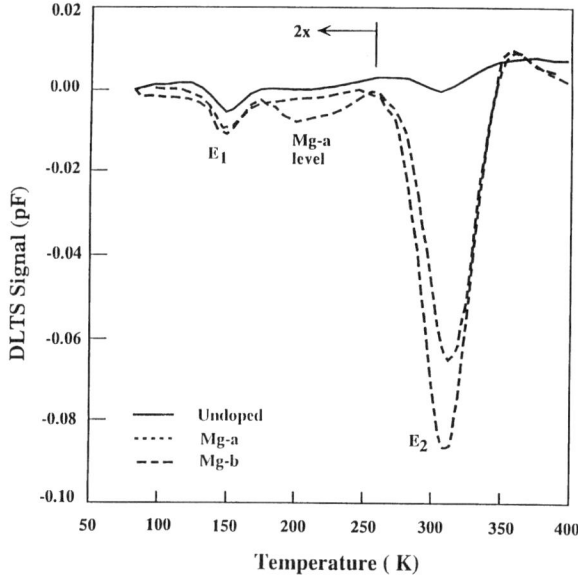

FIG. 15. DLTS spectra of undoped and Mg-doped GaN layers. After Hacke *et al.* [50].

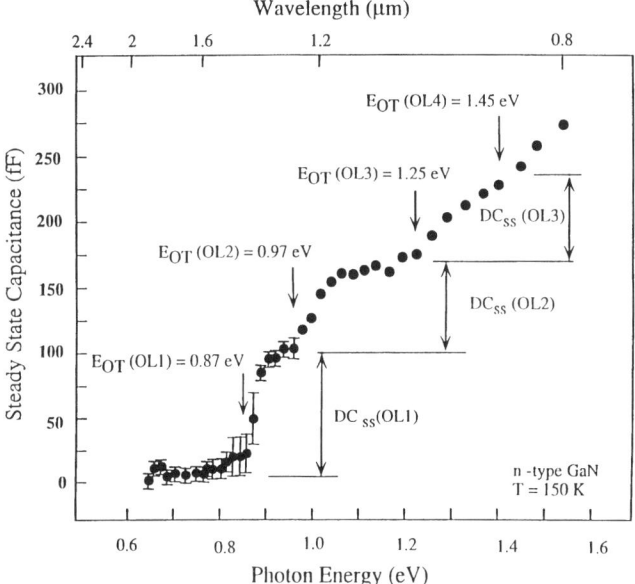

FIG. 16. Capacitance versus optical excitation energy in GaN Schottky diodes. After Götz et al. [49], with permission.

capacitance technique resulted in the observation of much deeper defects, as shown in Fig. 16. Caution must be exercised, as optical and thermal activation energies of defect vary substantially.

7. Defect-Aided Current in Unintentionally Doped GaN

Despite great strides in recent years, little is known about the compensation and conduction mechanisms in GaN. For example, most of the undoped material, both bulk and epitaxial—particularly early varieties—have strong n-type conductivity, with electron concentrations ranging from 10^{17} to 10^{20} cm^{-3} [53, 54]. As mentioned earlier, the dominant donor producing these large concentrations is thought to be the N vacancy V_N [29, 31], although oxygen contamination may also be important [34, 55, 56]. Besides n-type and p-type layers, which have been produced and used for LEDs and lasers, it is desirable to have semi-insulating (SI) material, especially for electronic applications. There are very few reports [53, 54, 57] and no detailed analyses of SI GaN in the literature. In this work, we show that increased N flux in MBE growth changes the electrical properties from highly conductive to SI.

The conduction mechanism in GaN has been investigated by Look et al. [58] and is very dependent on the layers themselves. For example, the layer grown with about 1.5×10^{15} cm^{-2} s^{-1} N flux (with about 1-μm/hr growth rate) shows a room-temperature electron concentration of 1.1×10^{18} cm^3 and is quite representative of typical GaN samples discussed in the literature. The data indicate that above 140 K, the dominant electrical transport mechanism results from conduction by electrons thermally excited from shallow donors into the conduction band. Below 140 K, the dominant mechanism is due to electrons "frozen out" in a band formed by these same shallow donors. This shallow donor band is probably formed from the hydrogenic-type wave functions of electrons loosely bound to N vacancies.

As for resistivity, the trend among all the samples studied is that as the nitrogen flux is increased, the measured Hall mobility decreases, eventually becoming immeasurable. For example, a sample grown with large N flux exhibits a much higher resistivity than those grown with lower N flux, and no measurable Hall effect ($\mu < 0.5$ cm^2/Vs), even at 400 K. For conduction-band transport, such a small μ would require $N_I > 5 \times 10^{20}$ cm^{-3}, and $n \ll N_D$, N_A (to minimize free-carrier screening), which is unlikely. Instead, phonon-assisted hopping among localized defect centers at first might appear to be a more plausible mechanism. Such hopping will, indeed, produce a very small or vanishing Hall coefficient, in agreement with observations. Basically, hopping involves two limiting factors [59]: (1) the energy $\varepsilon_3 \sim (\varepsilon_3 N_D^{1/3})$ necessary to hop from an occupied defect to one that is unoccupied and (2) wave-function overlap. At high enough temperatures, the energy will not be a limiting factor, and nearest-neighbor hopping will dominate. At low enough temperatures, hopping between more distant neighbors, which are closer together in energy, will begin to dominate. However, the defect concentrations required are inconsistent with the PL data, which yield extremely clean spectra with only the free excitons with the excited states of A and B exciton observable. We therefore conclude that the variable range v_r, hopping model, which requires $N_D \approx 1 \times 10^{17}$ cm^{-3}, is reasonable and consistent with the good PL results. Thus, variable-range hopping seems to be the best model to explain the low or vanishing mobilities in high-resistivity samples 5175 and 5069. It is not clear, however, why nearest-neighbor hopping does not begin to dominate at the higher temperatures, because sufficient thermal energy should be available for large-energy hops.

In short, we can conclude that conduction in samples grown with low nitrogen flux is dominated by conduction-band and impurity-band conduction at high (>140 K) and low ($T < 140$) temperatures, respectively. On the other end of the spectrum, the samples grown with excess nitrogen flux exhibit conduction consistent with variable-range hopping. If one assumes that the N vacancies are responsible for the donor-like states, then it is plausible that increased N flux reduces the N vacancies, thus donor-like states, giving rise to increased resistivities.

III. Optical Manifestation of Defects

Defects manifest as extrinsic transitions in optical emission and absorption, if the contrast is large enough for the latter. It is therefore warranted to succinctly discuss the optical signature of defects, as it provides important clues to the processes involved that are essential precursors to their minimization if not total elimination.

1. Yellow Band

Many wide-bandgap semiconductors (GaN is no exception) suffer from emission near the midgap (yellow-green in the case of GaN). The situation is exacerbated by the enhanced response of the human eye to this region of the spectrum as compared to blue. In LEDs, this yellow emission causes a serious deviation from the saturated blue and inhibits the achievement of all colors. Consequently, a great deal of attention has been paid to the causes and nature of this emission. There are two arguments for the origin of this transition. One suggests transitions from shallow donors to deep acceptor, while the other attributes this transition from deep donor states to shallow acceptor states (Fig. 17). Exhaustive studies conducted at the Naval Research Laboratory [60] culminated in the conclusion that the yellow emission is due to transitions from deep donor-like states to shallow acceptors. In contrast, the pressure dependence of this yellow transition is such that it seems to follow the conduction band, accounting for the fact that the transition should have the character of about half the bandgap of GaN, thus invoking the notion that it must be from a shallow donor-like state to a deep acceptor-like state [61]. Reynolds *et al.* [62] reported on the similarities between the green-band luminescence band in ZnO and the yellow luminescence band in GaN and drew the conclusion that the genesis of these bands is of the same type. Experiments by Hofmann *et al.* [63] suggest that the yellow band in GaN results from the recombination between a shallow door and a deep level. They propose that the deep level

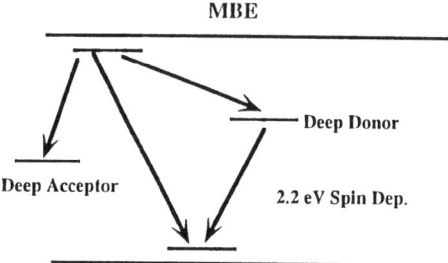

FIG. 17. Schematic representation of the origin of the yellow-band emission in GaN. In one model, the said transition is attributed to that between a deep donor and a shallow acceptor. In the other model, the same is attributed to that between a shallow donor and a deep acceptor.

may be a double donor, although an acceptor cannot be ruled out. In the case of ZnO, a typical green band obtained from a platelet-type crystal (at 2 K) is shown in Fig. 18 as a solid curve. A modulated structure is observed on the high-energy side of the band. The modulated structure can be explained by the model shown in Fig. 19.

The PL emission results from the recombination between the shallow donor level and the deep level. Hot electrons in the conduction band are pumped up by the He-Cd excitation source. Peaks in the PL emission band occur whenever the energy of the PL peak coincides with the sum of the energies of the donor level plus an integral multiple of a principal optical phonon energy. At adjacent energy values, an equilibrium number of electrons will arrive at the donor level and thus take part in the recombination with the deep level. The deep level will also have accompanying excited states due to interaction with local vibrational modes as well as lattice modes. It would be expected that the dominant transition would occur between the shallow donor and the ground state of the deep level. Transitions will also occur between the shallow donor and the excited states of the deep level, with reduced oscillator strengths. This model agrees with the model proposed by Hofmann *et al.* [63] for GaN and has the added advantage that it can explain the width of the emission band. The width of the yellow band in GaN and the green band in ZnO is extremely broad and would not be explained by the width of the impurity levels. The phonon involved in the green band in ZnO is the longitudinal optical phonon, which has an energy of 0.072 eV, corresponding to the energy separation of the PL peaks on the high side of the green band. It is noted in Fig. 20 that the modulated energy structure does not occur on the low-energy side of the green band. This would be expected because the phonons that are involved in cascading hot electrons from the conduction band to the donor level are

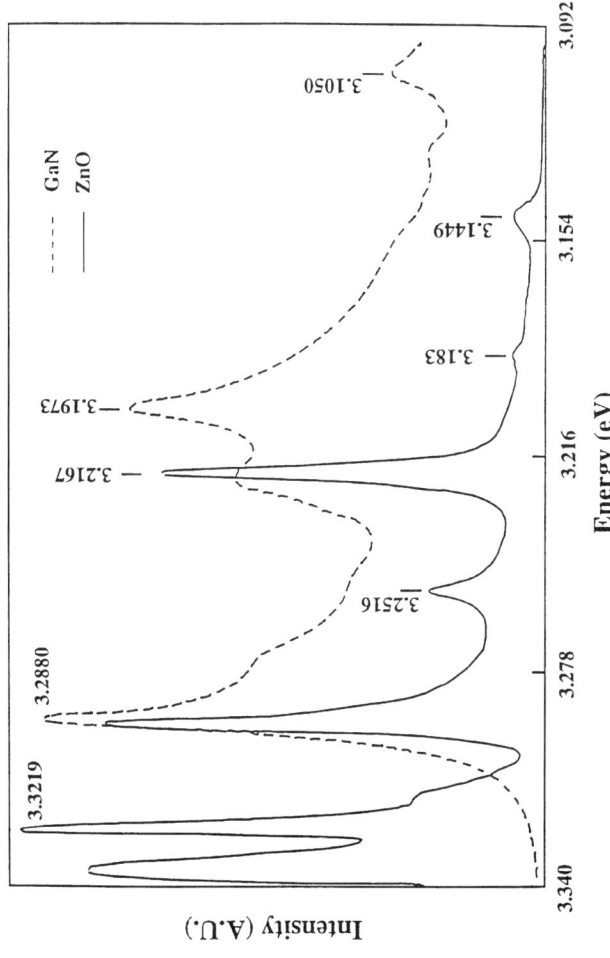

FIG. 18. Donor–acceptor pair transitions in GaN (dashed curve) and in ZnO (solid curve). The zero phonon transitions along with their phonon replicas are identified in the text. After Reynolds et al. [62].

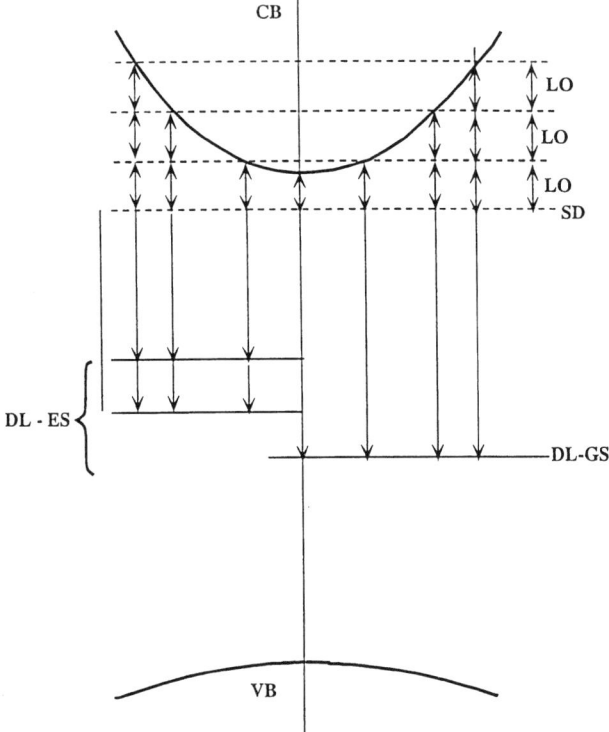

Fig. 19. Model to explain the green-band emission in ZnO. The band model in K-space shows the conduction band, the shallow donor level (SD), the deep-level ground state (DL-GS), and the deep-level excited states (DL-ES). The longitudinal optical phonons are designated LO. The energies of these states are all related to the valence band. After Reynolds et al. [62].

not involved with the low-energy emission. This emission is accounted for by recombination of donor electrons with excited states of the deep level. The deep level may be a complex center whose excited states consist of both local vibrational modes and lattice modes. These excited states are so distributed that they do not produce a resolvable modulated structure on the low-energy side of the green band.

If one now assumes that there is a relationship between the green band in ZnO and the yellow band in GaN, then a similar model would apply to the latter. This would support the model of Hofmann et al. [63] and would also explain the very large width of the yellow band. The modulated structure observed for the green band in ZnO could not be accounted for by the alternative model, proposed by Kennedy et al. and Glaser et al. [60],

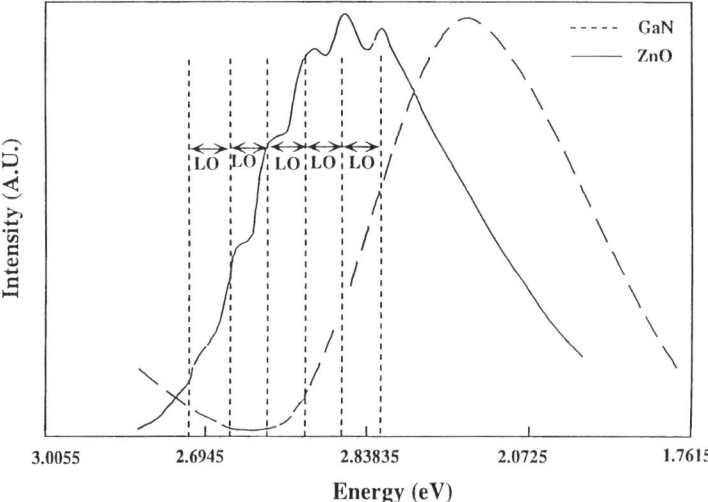

FIG. 20. The yellow emission band (dashed curve) observed in GaN and the green emission band (solid curve) observed in ZnO. The structured peaks occurring on the high-energy side of the green band are separated in energy by that of the longitudinal optical phonon in ZnO. After Reynolds et al. [62].

to explain the yellow band in GaN. In the Glaser model, the electrons transfer from a shallow donor to the deep state in a nonradiative process, followed by radiative decay from the deep state to a shallow acceptor.

2. DEFECTS CAUSED BY P-TYPE DOPING IN GaN

As in any wide-bandgap semiconductor, p-type doping in GaN and related materials is rather complex. In GaN, for example, while the effective mass–like acceptor is about 200 meV from the valence band, Mg-doped GaN exhibits emission at centers that are about 0.5 eV above the valence band when the Mg concentration exceeds a certain level. In optical spectra, two broad-emission bands of about 290 (dominant for $T < 150$ K) and 550 meV (dominant for $T > 150$ K) below the bandgap appear. A typical continuous-wave (CW) PL spectrum of p-type GaN layers at 10 K is dominated by a band at about 3.21 eV, which nearly disappears for $T > 150$ K. As the temperature is increased above 150 K, a weak emission band at ~2.95 eV appears. Moreover, the peak position of the lower energy emission red shifts considerably as the Mg doping level is increased. At room temperature, the peak position of this lower energy emission band can be varied from 430 to about 700 nm.

To explore the physical origin of the observed emission lines, their dynamic behavior has been studied [64]. At low temperatures, PL decay is nonexponential but can be approximated by two-exponential decay having fast and slow components. The typical lifetime of the fast component, which contributes 90% of the PL signal, is about 0.6 ns, and the lifetime of the slow component is about 5.0 ns. In the temperature region $T < 150$ K, where the 3.21-eV emission band dominates, the recombination lifetime decreases progressively from 0.6 to 0.3 ns as temperature increases from 10 to 140 K. This behavior can be accounted for by an increased nonradiative recombination rate at higher temperatures, caused by the nonradiative carrier transfer to the lower energy recombination channels. This is consistent with the observation of the thermal quenching of the 3.21-eV emission line and the subsequent increase in the emission intensity of the lower energy band at 2.95 eV with temperature.

In the higher temperature region ($T > 150$ K), where the lower energy emission band (~ 2.95 eV at $T < 150$ K) dominates, the fast decay component contributes nearly 95% of the PL signal, and consequently the decay kinetics of PL are nearly single exponential. The temperature dependence of the recombination lifetime of the lower energy emission band indicates an increase with temperature reaching 0.3 ns at room temperature. This is due to the carrier transfer from the 3.21 eV recombination channel, as discussed.

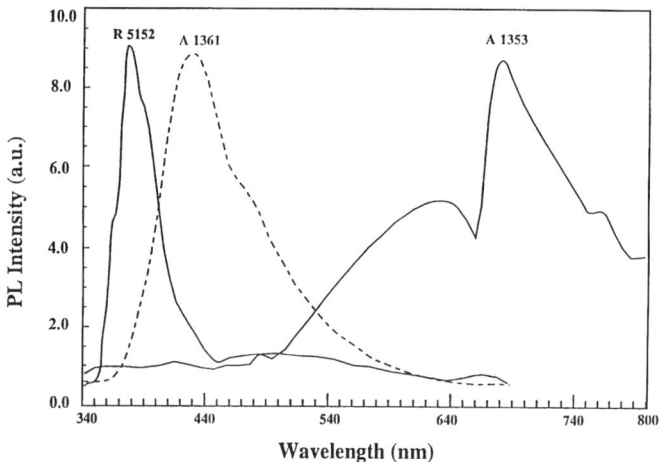

Fig. 21. PL spectra of Mg or Be doped GaN. With increasing concentration, the spectrum red shifts.

The observed subnanosecond PL recombination lifetimes suggest that the band-edge emissions in Mg-doped p-GaN result predominantly from the conduction band-to-impurity recombination, involving substitutional shallow Mg acceptors at low temperatures ($T < 150$ K) and Mg-related deep-level centers, or deep-level centers generated most likely during growth by the presence of large concentrations of Mg on the surface of the film, at high temperatures ($T > 150$ K). In such a context, the quenching of the 3.21-eV emission line is due to either thermal ionization of shallow neutral Mg acceptors or hole transfer from the shallow to the deep impurities as temperature increases.

As shown in Fig. 21, the transitions associated with the deep levels can extend into the band several electron volts above the valence band. It should also be pointed out that while defect generation is aided by the presence of Mg on the growth surface, there is no evidence that any Mg complex is formed in the process. Until such time as a correlation is made, we propose that Mg induces native defects when introduced in high concentrations and or under conditions not optimum for substitutional incorporation on the Ga sites. To depict this phenomenological picture, we present Fig. 22, where the shallow band is associated with the substitutional Mg; a band can be formed when the Mg concentration is high; and the deeper band, which is associated with the defects, is caused by the presence of large concentrations of Mg. Similar phenomena are observed with Be doping, which suggests that the processes involved are not Mg-specific, but rather p-type dopant-specific [65].

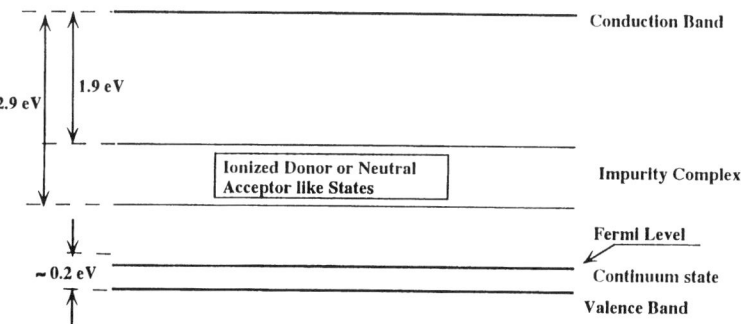

FIG. 22. Schematic representation of the acceptor level and/or band and deep levels caused by Mg impurities. Reprinted with permission from *Journal of Electronic Materials*, Vol. 21, pg. 437 (1992), a publication of the Minerals, Metals and Materials Society, Warrendale, Pennsylvania 15086.

Improved *p*-type Mg-doped samples prepared by reactive MBE in the author's laboratory do not show any PPC effect. This is true even for samples in which the optical signature indicates defect states in addition to the expected acceptor level. It can be concluded that the lattice distortion, to the extent required for the PPC effect to appear, must not be taking place in the more recent samples. The samples in question have been prepared taking into account the notion that Mg on the Ga site is a stable site.

IV. Applications

As alluded to earlier, the impetus behind the interest in GaN and its alloys stems from their applictions to green, blue, and UV portions of the spectrum as emitters and detectors and as high-temperature and/or high-power electronic devices. Following is a succinct treatment of the subject matter, limited to devices deemed to hold this potential.

1. FIELD-EFFECT TRANSISTORS

GaN-based FETs are projected to be highly useful for amplification and switching in high-power and/or high-temperature environments. This optimism is justifiably fueled by the calculated large electron velocity and the robustness of the material. Other pertinent parameters include, but are not limited to, large thermal conductivity of GaN, type I heterojunctions, large-band discontinuities with resultant large interface carrier concentrations, and large breakdown voltage. Even though the research activity is still in its embryonic state, a few reports have been put forth that describe fabrication and electrical characterization of some GaN FETs. The results are extraordinarily encouraging and the FET research is on a rapid rise. Among the applications are solid-state replacement parts for ground stations in telecommunication and compact radar. As in the case of the conventional III–V area, the modulation-doped FETs (MODFETs) based on this new heterostructure system are expected to exhibit the best performance. Consequently, only the MODFETs will be covered here.

a. Theoretical Method

Electronic properties of modulation-doped structures based on the III-nitride semiconductor system have been theoretically treated by Stengel *et al.* [66]. The structure considered for this particular study was a wurtzitic $Al_xGa_{1-x}N$/GaN-layered normal MODFET structure. For source and drain contacts, a scheme by which the metal contact is deposited on $Al_xGa_{1-x}N$ was considered, with the well-justified assumption that contact metal penetrates down to the GaN layer that hosts the two-dimensional electron gas (2DEG). Because of conduction-band discontinuity, the electrons diffusing from the larger bandgap AlGaN into the smaller bandgap GaN form a triangular quantum well, the $Al_xGa_{1-x}N$/GaN interface. In this well, electrons accumulate, forming a 2DEG. In the next section, the calculation of the band-bending 2DEG and its variation with gate bias will be described. Also to be discussed are transconductance and current–voltage characteristics of MODFET devices.

b. *The Two-Dimensional Electron Gas*

To obtain the 2DEG concentration at the $Al_xGa_{1-x}N$/GaN interface, Poisson's equation [Eq. (1)] and Schrodinger's equation [Eq. (2)] are solved self-consistently. In reduced units (energy units being Rydberg and the length unit being Bohr radius), these equations are given by

$$\frac{d^2 E_c}{dz^2} = -8\pi n(z) \tag{1}$$

$$-\frac{\hbar^2}{2m^*}\frac{d^2 \Psi_i}{dz^2} + E_c(z)\Psi_i = E_i \Psi_i \tag{2}$$

where z is the coordinate in a direction perpendicular to the channel, $E_c(z)$ is the conduction-band position, Ψ_i is the wave function of electrons, m^* is the effective electron mass in GaN, $n(z)$ is the electron concentration at the point z, and $\hbar = h/2\pi$ with h being Planck's constant. When unintentionally doped, GaN shows a donor-implanted behavior. However, these donors do not affect the 2DEG concentration, because most of them are not ionized at the interface, owing to the Fermi level lying above the conduction band. If, for the sake of convenience, the origin of the energy is chosen at the edge of the conduction-band level of GaN at the $Al_xGa_{1-x}N$/GaN interface, and if the unintentional donor concentration in the bulk of GaN is accounted for, then the first boundary condition may be written as

$$E_c(z \to +\infty) = E_F + \frac{1}{2} E_g \tag{3a}$$

or

$$E_c(z \to +\infty) = E_F + k_B T \ln\left(\frac{N_c}{N_d}\right) \tag{3b}$$

where N_c is the effective electron density of states, N_d is the donor impurity density in the bulk GaN, E_F is the Fermi level, E_g is the energy bandgap (all for GaN), k_B is the Boltzmann constant, and T is the absolute temperature. Note that, in practical situations, this boundary condition merely affects the 2DEG concentration. The two other boundary conditions are

$$\left(\frac{dE_c}{dz}\right)_{z=0} = 8\pi n_{2D} \tag{4a}$$

$$\Psi_i(z \to \pm \infty) = 0 \tag{4b}$$

where n_{2D} is the density of 2DEG. The value of the electron concentration at point z, $n(z)$, is given by

$$n(z) = \sum_{i=1}^{\infty} N_i |\Psi_i(z)|^2 \tag{5}$$

where

$$N_i = \frac{k_B T}{2\pi} \ln\left[1 + \exp\left(\frac{E_F - E_i}{k_B T}\right)\right] \tag{6}$$

Equations (1) and (2) are solved using trial wave functions of the form

$$\phi_n(z) = (z + z_0)e^{-b_a z} \quad \text{for} \quad z \geq 0 \tag{7a}$$
$$\phi_n(z) = z_0 e^{(1/z_0 - b_a)z} \quad \text{for} \quad z < 0 \tag{7b}$$

The wave function is obtained by Eq. (8), with C_n^i as the variational parameter.

$$\Psi_i = \sum_{n=1}^{\infty} C_n^i \phi_n \tag{8}$$

Although, theoretically, the summation series of Eqs. (5) and (8) can extend up to ∞, a value of $n = 10$ is sufficient for self-consistency. Correlation and exchange terms are also considered in this self-consistent calculation [67]. This calculation, for a given 2DEG concentration, yields the conduction band, the Fermi level (which is also the quasi-Fermi level of the electrons), the energy levels of the electrons in the triangular well, and the electron concentration as a function of z. Another important parameter extracted from these calculations is $\Delta E_{Fi} = E_F - E_c$ at the $Al_xGa_{1-x}N/GaN$ interface.

c. *The $Al_xGa_{1-x}N$ Layer*

Poisson's equation and the current continuity equation for the $Al_xGa_{1-x}N$ region may be given by

$$\frac{d^2 E_c}{dz^2} = 8\pi [N_D^+(z) - n(z)] \qquad (9)$$

$$\frac{dE_F}{dz} = \frac{J}{e\mu n(z)} \qquad (10)$$

where J is the constant current density, μ is the constant mobility of electrons, $N_D^+(z)$ is the ionized donor concentration, and $n(z)$ is the electron concentration, all in $Al_xGa_{1-x}N$. For the solution of Poisson's Eq. (9) and the current continuity Eq. (10), which involve the quasi-Fermi level of the electrons [68], the origin of energy is taken at the Fermi level at the $Al_xGa_{1-x}N/GaN$ interface.

The boundary conditions necessary for the solutions of Eqs. (9) and (10) are

$$E_c(0) = \Delta E_c - \Delta E_{Fi} \qquad (11a)$$

$$\left(\frac{dE_c}{dz}\right)_{z=0} = 8\pi n_{2D} \qquad (11b)$$

$$E_F(0) = 0 \qquad (11c)$$

The current density is obtained in a self-consistent calculation involving the value of the Fermi level (E_{Fm}) at the semiconductor metal interface. (E_{cg} is the conduction band at the $Al_xGa_{1-x}N$ metal contact, and ϕ_b is the Schottky barrier height of the $Al_xGa_{1-x}N$ metal contact.)

$$E_{Fm} = E_{cg} - q\phi_b \qquad (12)$$

As a result of this calculation, the gate bias needed to achieve the originally given 2DEG concentration is calculated as

$$-qV_G = E_{Fm} \qquad (13)$$

By performing this numerical calculation for different values of the 2DEG concentration, a curve of this concentration as a function of the gate bias is obtained.

d. Drain Current and Transconductance

The current in the channel is given by Eq. (14), which is the usual current continuity equation and where the second term in the bracket is the diffusion term.

$$I_D = qW_D\mu(y)\left[\frac{dV_c}{dy}n_{2D}(V_G - V_c(y)) + \left(\frac{k_BT}{q}\right)\frac{dn_{2D}(V_G - V_c(y))}{dy}\right] \quad (14)$$

where q is the electron charge, y is the direction of the channel length (e.g., the direction from the source to the drain), W_D is the channel width, $V_c(y)$ is the potential in the channel at point y, V_G is the applied gate bias, $n_{2D}(V_G - V_c(y))$ is the 2DEG concentration as a function of $V_G - V_c(y)$, and $\mu(y)$ is the field-dependent (and hence position-dependent) mobility at y. The mobility is given by

$$\mu(y) = \frac{\mu_0}{1 + \frac{1}{F_c}\left(\frac{dV_c}{dy}\right)} \quad (15)$$

The latter term in Eq. (14) will be neglected in the following calculations because its value is small compared to that of the first term at room temperature (k_BT/q being only 0.0259 V).

If we define $V_{GS} = V_G - V_S$ and $V_{GD} = V_G - V_D$, then the total drain-source current is given by integration of Eq. (14) from source to drain (neglecting diffusion)—for example, from $y = 0$ to $y = L$.

$$I_D = \left[\frac{qW_D\mu_0}{L + \frac{V_D}{F_c}}\right]\int_{V_{GD}}^{V_{GS}} n_{2D}(u)du \quad (16a)$$

This formula is valid when the drain current I_D is lower than its saturation value $I_{D,SAT}$, which corresponds to $V_D = V_{D,SAT}$, where $V_{D,SAT}$ is the value of V_D at which $I_D(V_D)$ is maximum. When the saturation is reached, the current is

$$I_D = \left[\frac{qW_D\mu_0}{L + \frac{V_{D,SAT}}{F_c}}\right]\int_{V_G - V_{D,SAT}}^{V_{GS}} n_{2D}(u)du \quad (16b)$$

In the case of a long channel device, this maximum is reached for $V_D = V_G - V_{GT}$ (V_{GT} is the threshold gate bias), leading to the saturation current

$$I_{D,SAT} = \frac{qW_D\mu_0}{L} \int_{V_{GT}}^{V_G} n_{2D}(u)du \qquad (16c)$$

for which the integral is the total surface under the curve of n_{2D} as a function of V_G.

In the case of a short channel device, the drain current for the region where n_{2D} exhibits a constant maximum value $n_{2D,\max}$ as a function of V_G, is

$$I_D = \left[\frac{qW_D\mu_0}{L + \dfrac{V_D}{F_c}}\right] V_D n_{2D,\max} \qquad (16d)$$

Note that Eq. (16d) leads to the well-known expression

$$I_{D,SAT} = qW_D v_{sat} n_{2D,\max} \qquad (16e)$$

when V_D is large as compared to LF_c and where v_{sat} is the saturation velocity of the electrons in GaN.

Using Eq. (16a), the transconductance may be derived as

$$g_m = \frac{1}{W_D}\left(\frac{\partial I_D}{\partial V_G}\right) = \frac{q\mu_0}{L + \dfrac{V_D}{F_c}} [n_{2D}(V_{GS}) - n_{2D}(V_{GD})] \qquad (17a)$$

where $n_{2D}(V_{GS})$ is the 2DEG concentration as a function of V_{GS}, and $n_{2D}(V_{GD})$ is the 2DEG concentration as a function of V_{GD}. This formula is valid for nonsaturation regimes. At saturation, which, in the general case, is caused by both carrier velocity saturation and pinch-off, the transconductance may be calculated by

$$g_m = \frac{q\mu_0}{L + \dfrac{V_{D,SAT}}{F_c}} [n_{2D}(V_{GS}) - n_{2D}(V_G - V_{D,SAT})] \qquad (17b)$$

In the simple case of a long channel device, where the carrier velocity saturation can be neglected, the transconductance in the saturation regime is given by

$$g_m = \frac{q\mu_0}{L} n_{2D}(V_{GS}) \tag{17c}$$

When the velocity saturation is dominant, in the case of a short channel device, Eq. (17b) would reduce to

$$g_m = \frac{qv_{sat}}{V_{D,SAT}} [n_{2D}(V_{GS}) - n_{2D}(V_G - V_{D,SAT})] \tag{17d}$$

The measured transconductance is actually smaller than that given the intrinsic one in that the source resistance, which will be defined shortly, acts as a negative feedback. Through circuit considerations, the measurd extrinsic transconductance is given by

$$g_m^{max}|_{ext} = \frac{g_m^{max}}{1 + R_s g_m^{max}} \tag{18}$$

Equation (18) simply implies that the source access resistance must be lowered for the intrinsic transconductance not to be degraded. Although not transparent this treatment, the source and drain resistances have a deleterious effect on the efficiency and the power level one can achieve.

e. Default Parameters

For our calculations (except in Section G), the default parameters used for both GaN and $Al_xGa_{1-x}N$ (for values of x lower than 0.4) are listed in Table I. The donor level in GaN depends on the impurity atom [69], for the present calculations, it is assumed to be 45 meV. The energy bandgap of $Al_xGa_{1-x}N$ is modeled as

$$E_g(Al_xGa_{1-x}N) = xE_g(AlN) + (1-x)E_g(GaN). \tag{19}$$

f. 2DEG Concentration as a Function of ΔE_{fi} and Calculation Δd

Calculated results for ΔE_{fi} as a function of n_{2D} are shown in Fig. 23. These are compared with a similar curve for GaAs and with a curve obtained by

TABLE I
Default Paramaters Used in the Calculations of 2DEG and MODFET-Related Predictions

Parameter	Symbol for Parameter	Value
Temperature	T	300 K
Boltzmann constant	k_B	1.38066×10^{-23} J/K
Effective electron mass in GaN	m^*	0.2
Dielectric constant (GaN)	ε_r	10.4
Effective density of states (Elec)	N_c (GaN)	2.25×10^{18} cm^{-3}
Energy bandgap (GaN)	E_g	3.43 eV
Energy bandgap (AlN)	E_g	6.0 eV
E_g(AlGaN)-E_g(GaN)	ΔE_g	—
Conductive band discontinuity	ΔE_c	$0.82 \Delta E_g$
Metal/AlGaN barrier height	ϕ_b	1.10 eV
Electron mobility in GaN	μ_0	500 cm^2/Vs
Saturation velocity in GaN	v_{sat}	2.0×10^7 cm/s
AlGaN layer thickness	d	200 Å
Spacer layer thickness	W_{sp}	20 Å
Al mole fraction in AlGaN	x	0.25
Donor concentration in AlGaN	N_d (AlGaN)	5.0×10^{18} cm^{-3}
Channel width	W_D	40 μm
Impurity level in GaN	E_d	45 meV

data fitting. The approximation used for the curve for GaAs is

$$n_{2D} = \left(\frac{\Delta E_{Fi}}{18.38}\right)^{1.5} \quad (20)$$

for which the length units are Bohr radii, and the energy unit is in Rydberg (5.31 meV in GaAs). As a result, the density of 2DEG, n_{2D}, is in 10^{12} cm^{-2} (as the Bohr radius is approximately 100 Å in GaAs). Considering ΔE_{Fi} to be in millielectron volts, and using the value of the Bohr radius (27.51 Å) and of the energy unit of Rydberg (25.15 meV) for GaN, the formula for n_{2D} can be simplified as

$$n_{2D} = 13.21 \times 10^{12} \left(\frac{\Delta E_n}{462}\right)^{1.5}, \text{cm}^{-2} \quad (21)$$

For the sake of convenience and for analytical modeling, the values of n_{2D} for the range of 0.65×10^{12} cm^{-2} to 9.0×10^{12} cm^{-2} may be fitted to

$$n_{2D} = [0.65 + 1.72 \times 10^{-3} \Delta E_{Fi}^{1.42}] \times 10^{12}, \text{cm}^{-2} \quad (22)$$

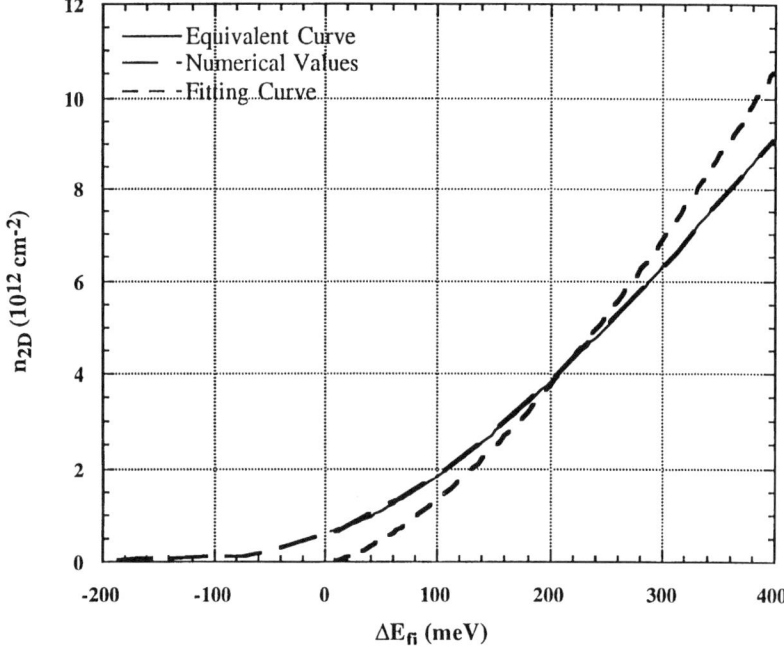

FIG. 23. Position of the Fermi level with respect to the conduction-band edge, ΔE_{fi}, in the GaN side of the heterointerface as a function of the 2DEG concentration. The curves correspond, respectively, to the equivalence with GaAs (equivalent curve), numerical simulation, and simple analytical fitting of this calculation. After Stengel et al. [66].

From Eq. (22) it is evident that, in order to obtain a significantly large 2DEG concentration, the conduction-band discontinuity must also be quite large. For example, this conduction-band discontinuity should be larger than 400 meV for n_{2D} to be about 9×10^{12} cm^{-2}.

The value of Δd, which is the average distance of the 2DEG from the heterointerface, can also be obtained from numerical simulations, the results of which are presented in Fig. 24. A simple analytical formula for Δd obtained as a function of n_{2D} ranging between 0.2×10^{12} cm^{-2} and 10^{13} cm^{-2} may be given by

$$\Delta d = \frac{69}{(n_{2D})^{0.4}}, \text{Å} \qquad (23)$$

For the sake of comparison, the results from Eq. (23) are also presented in

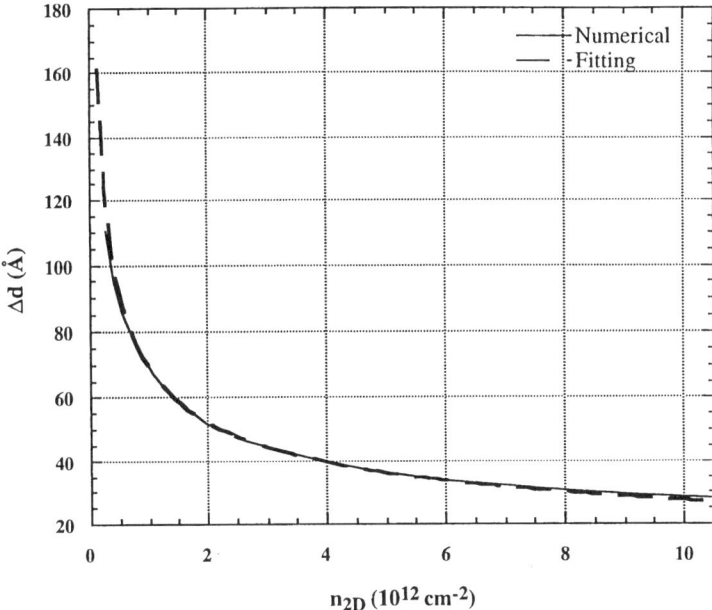

FIG. 24. Average distance (Δd) of electrons of the 2DEG from the heterointerface as a function of the 2DEG concentration. The solid curve corresponds to the numerical simulation, and the dashed curve corresponds to analytical fitting. After Stengel et al. [66].

Fig. 24. From this figure it may be noted that the fitted results compare very well with the numerical ones. This suggests that, for somple analytical models, a value of Δd ranging between 30 and 60 Å would be quite reasonable, which would correspond to n_{2D} varying from 10 to 1×10^{12} cm^{-2}.

g. *Band Diagrams for Normally on and Quasi Normally off MODFETs*

Energy-band diagrams for a quasi normally off (QN-OFF) and a normally on (N-ON) MODFET are shown in Fig. 25 for $x = 0.25$, $N_d = 10^{19}$ cm^{-3}, and $W_{sp} = 20$ Å. For the calculations for QN-OFF MODFET, $V_G = 0.03$ V, and $d = 130$ Å, and for the calculations for N-ON MODFET, $V_G = 0.04$ V, and $d = 200$ Å. Our calculations indicate that for both MODFETs, the 2DEG does not extend to the AlGaN region due to a high $Al_xGa_{1-x}N/GaN$ conduction-band discontinuity (more than

500 meV as compared to 142 meV for $Al_xGa_{1-x}As/GaAs$ at $x = 0.3$). Because of this and the fact that the probability of electron wave functions extending to the $Al_xGa_{1-x}N$ is very low, a thinner spacer would be needed to achieve the optimal mobility in the 2DEG due to a lower alloy scattering [70]. In spite of this, the effect of Coulombic scattering could presumably be opposite, especially because of the lower dielectric constant of GaN and AlGaN, which leads to a higher scattering potential. Very precise calculations or experiments would be needed to resolve this matter. As is apparent from Fig. 25a, for the QN-OFF MODFET, the calculated 2DEG concentration is 2×10^{12} cm^{-2}, and the maximum electron concentration in GaN is about 4×10^{18} cm^{-3}. The energy-band diagram shows that the quasi-Fermi level (origin of the energy axis) in GaN is very close to the first energy level. The donor atoms in $Al_xGa_{1-x}N$ are all ionized, and yet, because the conduction band is far above the Fermi level, there are no electrons present in the $Al_xGa_{1-x}N$ region. The gate bias can thus be raised to increase the 2DEG concentration and be lowered to decrease it.

The situation is, however, different in the case of the N-ON MODFET, for which the density of 2DEG is 3.8×10^{12} cm^{-2} (see Fig. 25b). For this MODFET, the quasi-Fermi level in GaN is far above the lowest energy level, and the peak concentration of electrons in the 2DEG is 10^{19} cm^{-3}. Also, some of the donor atoms (for z between -100 Å and -50 Å) are now neutralized, and some electrons start to appear in the $Al_xGa_{1-x}N$ region. Because of these, a further rise of the gate bias causes not only an increase in the donor neutralization, but also an increase in the electron concentration in $Al_xGa_{1-x}N$. However, the 2DEG concentration remains unaltered.

h. 2DEG Concentration as a Function of V_G

Our calculations employing the default parameters listed in Table I and their variations indicate that the peak value of 2DEG concentration for $Al_xGa_{1-x}N/GaN$ MODFETs is around 2 to 5×10^{12} cm^{-2}. It was noted, however, that a much larger value of n_{2D} can be obtained if the doping level in $Al_xGa_{1-x}N$ is substantially increased. Figure 26a shows the n_{2D} versus V_G curve for different values of the Al mole fraction x. As the conduction-band discontinuity ΔE_c for the $Al_xGa_{1-x}N/GaN$ is not precisely known, it was considered worthwhile to study the dependence of n_{2D} on various values of ΔE_c. This was automatically done in Fig. 26a as the mole fractions $x = 0.20$, 0.25, 0.30, and 0.40 correspond to $\Delta E_c = 0.42$, 0.53, 0.63, and 0.84 eV, respectively. From Fig. 26a it may be noted that the maximum value of n_{2D} is enhanced due to deepening of the quantum well. The value of V_G (or threshold gate bias, V_{GT}) for which the 2DEG becomes negligible

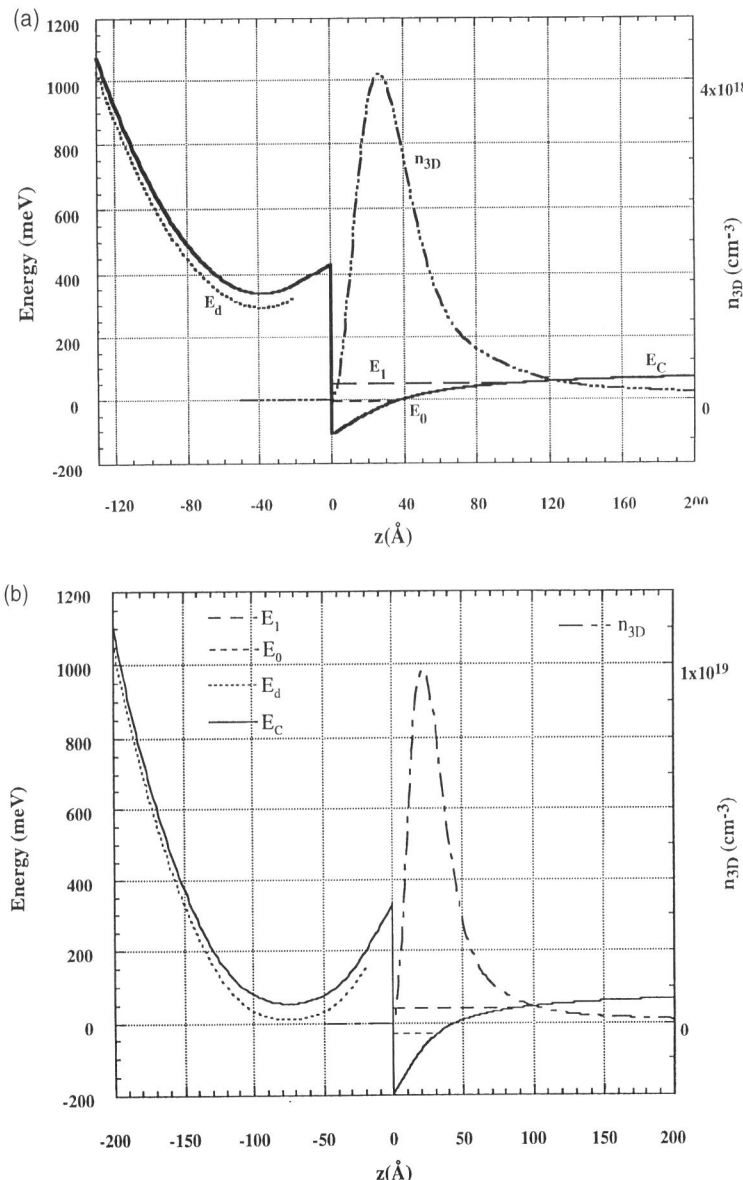

FIG. 25. Band diagram for (a) QN-OFF and (b) an N-ON MODFET. The origin of energy for these band diagrams is the Fermi level. The left side of the $z = 0$ line corresponds to the AlGaN region, and the right side corresponds to the GaN. The condution band is represented on both sides by curve 1, the donor level E_d in AlGaN by curve 2, the energy level E_0 in GaN by curve 3, and the energy level E_1 in GaN by curve 4. After Stengal et al. [66].

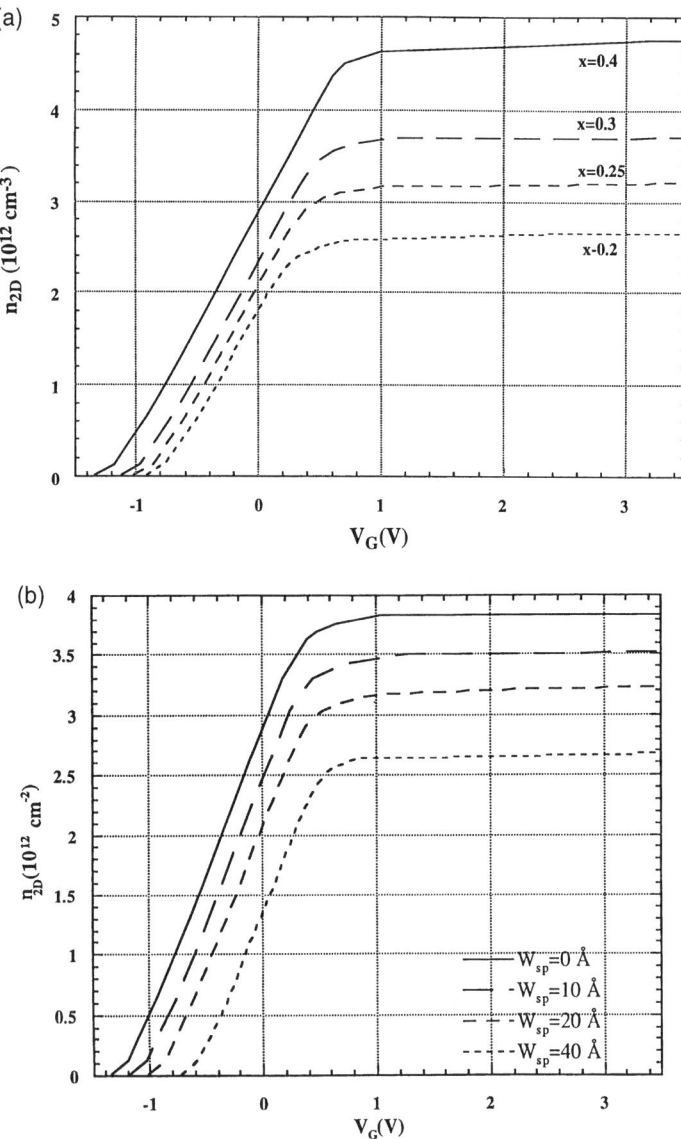

FIG. 26. Plots of the 2DEG concentration as a function of the gate-source bias V_G for various values of the (a) Al mole fraction x in $Al_xGa_{1-x}N$, (b) spacer layer thickness, (c) doping concentration in $Al_xGa_{1-x}N$, and (d) $Al_xGa_{1-x}N$ layer thickness. After Stengel et al. [66].

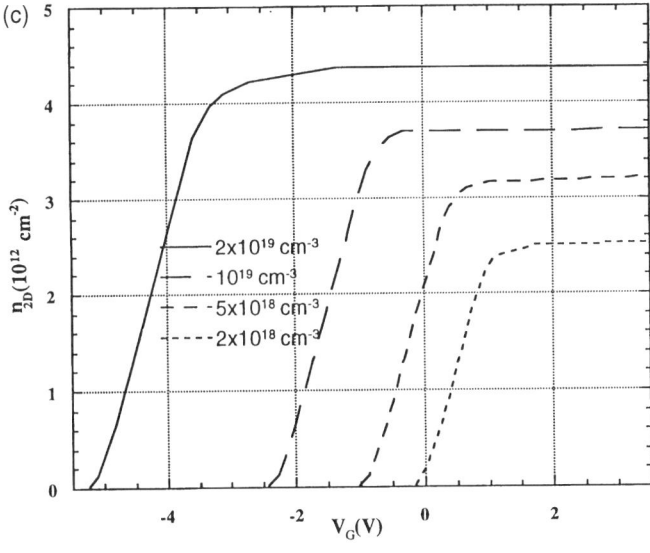

FIG. 26. (Continued).

is also shifted in the negative bias direction. This shift is linear, with -0.1 V for the quantum well being 0.1 eV deeper. From various curves of Fig. 26a, it is also apparent that, in order to achieve the highest value of n_{2D}, the value of x should be as high as possible. However, depending on the depth of the donor level in $Al_xGa_{1-x}N$, a compromise would be required for the optimal value of x, as it was in the case for GaAs/AlGaAs MODFET.

Figure 26b depicts n_{2D} versus V_G plots for various spacer thicknesses. The behavior of various curves of Fig. 26b is observed to be similar to that of the curves of Fig. 26a. It was found that, as long as the electron concentration in the spacer layer is low, the effects of varying x and W_{sp} on n_{2D} are essentially equivalent, because at the end of the spacer layer, the value of $E_c - E_F$ is very close to $E_c(\text{interface}) - W_{sp}(dE_c/dz)(\text{interface})$. Experimental data describing the effect of the spacer layer thickness on the mobility of the electrons in the 2DEG are needed to evaluate optimized values for this spacer. One may, however, predict that these values for the $Al_xGa_{1-x}N/$GaN system would be smaller than those for the $Al_xGa_{1-x}As/GaAs$ system, because the $Al_xGa_{1-x}N/GaN$ system provides a deeper confinement, and the wave function of the 2DEG over a shorter range in $Al_xGa_{1-x}N$ is nonzero.

Figure 26c shows n_{2D} versus V_G plots for various doping levels in $Al_xGa_{1-x}N$. An interesting feature of various curves of this figure is that n_{2D} increases with an increase in doping levels in $Al_xGa_{1-x}N$. Also, as this doping level increases, n_{2D} starts to increase at a more negative value of V_G. More specifically, the higher the value of the doping level in $Al_xGa_{1-x}N$, the larger the shift of the n_{2D} versus V_G curve toward a more negative value of the threshold gate bias V_{GT}. Interestingly, this shift is approximately a linear one, as may be found from a solution of Poisson's equation under the condition that all donors are ionized and the electron concentration in the 2DEG is low; for example,

$$\frac{d^2 E_c}{dz^2} = 8\pi N_d \qquad (24)$$

so that

$$E_{cg} = 4\pi N_d (d - W_{sp})^2 + C_{te} \qquad (25)$$

where C_{te} is a constant of integration. The doping level of the $Al_xGa_{1-x}N$ layer becomes a critical parameter for implementing the design of an N-ON or a QN-OFF device. (A higher donor concentration would lead the MODFET toward the N-ON direction.) It is clear from Fig. 26d, that the total $Al_xGa_{1-x}N$ thickness plays a similar role; for example, a QN-OFF device turns to be an N-ON device due to the increase of $Al_xGa_{1-x}N$ thickness, and vice versa. The slope of the linear region of the curve is also larger for thinner $Al_xGa_{1-x}N$ films, which can be explained by a larger capacitance of the device:

$$C = \frac{dQ}{dV} = \frac{\varepsilon_0 \varepsilon_r}{d + \Delta d} \qquad (26)$$

Further, a larger maximum value of n_{2D} can be obtained for a concentration that could be reached for thicker $Al_xGa_{1-x}N$ layers; it would be possible, particularly, under an extremely large bias. It is also remarkable that all of the $n_{2D}(V_G)$ curves cross at a single point (1.1 V, 3.2×10^{12} cm^{-2}); this is the point at which the conduction band is flat at the metal contact. It also corresponds to the situation in which the 2DEG is at equilibrium (e.g., there is no contact).

i. Transconductance

For the sake of reducing computational time, we have made use of interpolation formulas while trying to model n_{2D}. For this purpose, the n_{2D} versus V_G curves are considered to consist of three parts. The first is a linear part, for which the curves are modeled by straight lines. The second and third are exponential parts, which correspond to a rapid fall of the electrical characteristics in the negative bias and to the saturating characteristics for large positive bias, respectively. For the calculation, the saturation velocity of electrons is taken into account. Also, from a knowledge of the properties of the n_{2D} versus V_G curves, a number of characteristics of the transconductance are deduced. For example, employing Eq. (17), the maximum value of the transconductance is obtained for the gate-source voltage $V_G \equiv V_{Gmax}$ of V_G and corresponding to the point at which n_{2D} attains its maximum value (within $\pm 5\%$). As g_m is proportional to $[n_{2D}(V_{GS}) - n_{2D}(V_{GD})]$, V_D needed to be large enough to provide $V_{GT} = V_G - V_D$, where V_{GT} is the threshold value of $n_{2D}(V_G)$. Once this value is obtained, a larger V_D leads g_m to decrease under the influence of the saturation velocity term. Also, when V_G is lowered, with a starting point at V_{Gmax}, $n_{2D}(V_G)$ decreases down to zero, causing a similar decrease of g_m. On the other hand, when V_G is increased to V_{Gmax}, donors are neutralized, and there occur no change in $n_{2D}(V_G)$; however, $n_{2D}(V_{GD})$ rose gradually and eventually reached a value of $n_{2D}(V_G)$, once again leading g_m to be 0.

The transconductance g_m was studied as a function of a number of different parameters, including the spacer layer thickness, the total $Al_xGa_{1-x}N$ thickness, the doping level in $Al_xGa_{1-x}N$, and the channel length. For each curve, the transconductance was calculated employing Eq. (17) for that value of V_D that corresponds to the highest maximum g_{max}, so that the peak value g_{max} is obtained indeed at the saturation point. The variation of transconductance g_m with spacer layer thickness W_{sp} is shown in Fig. 27a. It was noted that g_m ranged between 60 and 90 mS/mm for W_{sp} varying between 40 and 0 Å, respectively. As the spacer thickness was decreased, the maximum values of g_m were found to correspond to larger V_D and smaller V_G. Due to the lack of relevant information, this result did not, however, take the effect of the spacer thickness on mobility into account. Had it not been the case, and if, for example, a 40-Å-thick spacer layer would bring about an improvement of more than 80% in the 2DEG mobility, it would also increase the peak transconductance of the device.

When the total $Al_xGa_{1-x}N$ thickness was varied from 250 to 120 Å, the calculated peak transconductance ranged between 70 and 85 mS/mm (see Fig. 27b) for a gate-source voltage V_G varying between 0 and 1.6 V, and for

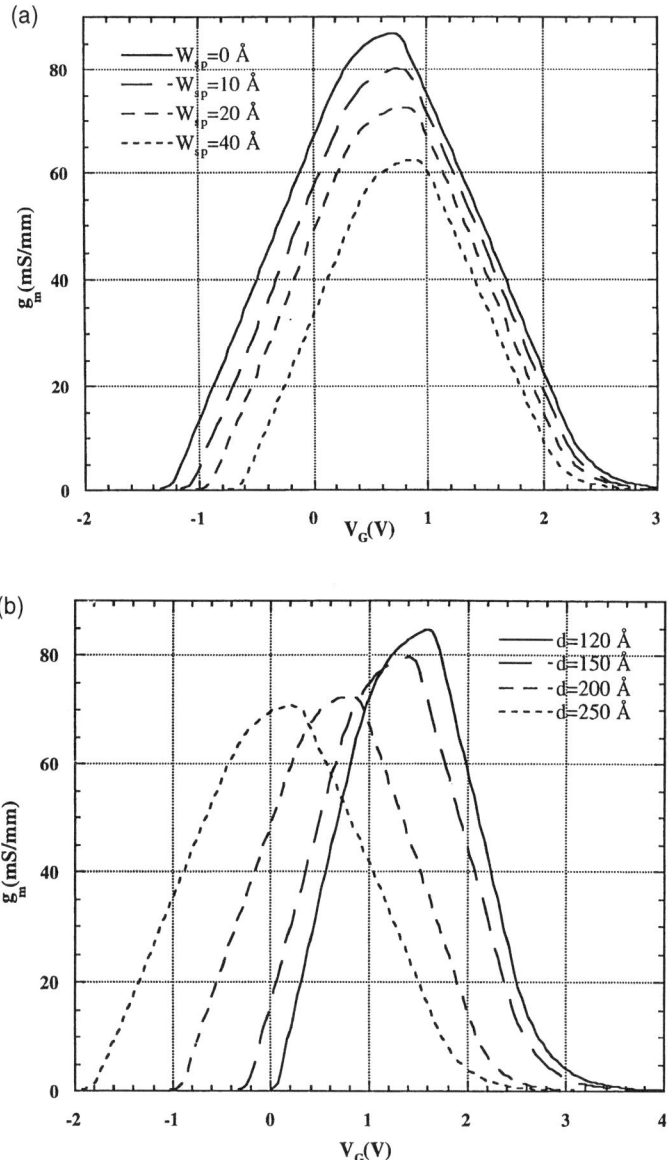

FIG. 27. Variation of the transconductance g_m of MODFETs as a function of the gate-source bias V_G, at the optimal value of the drain-source bias V_D (optimum value of V_D is defined to be the value at which the transconductance peak reaches its maximum value) for each of the plots, for various values of the (a) spacer layer thickness in AlGaN, (b) total thickness of the AlGaN, (c) Si doping concentration in AlGaN, and (d) channel length L. After Stengel et al. [66].

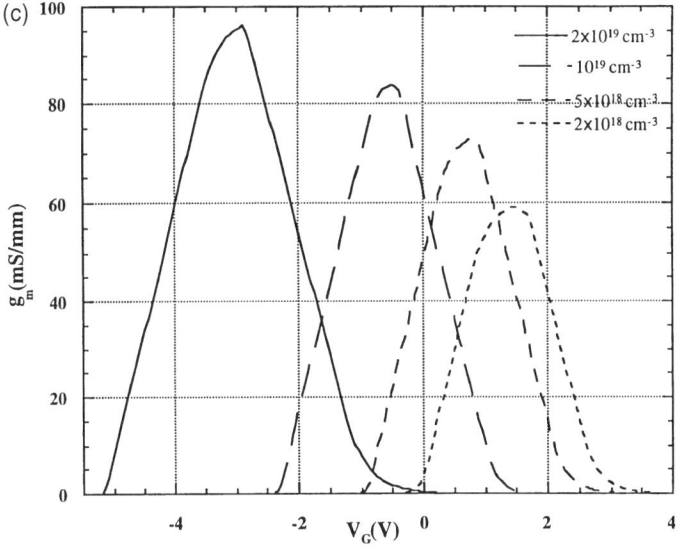

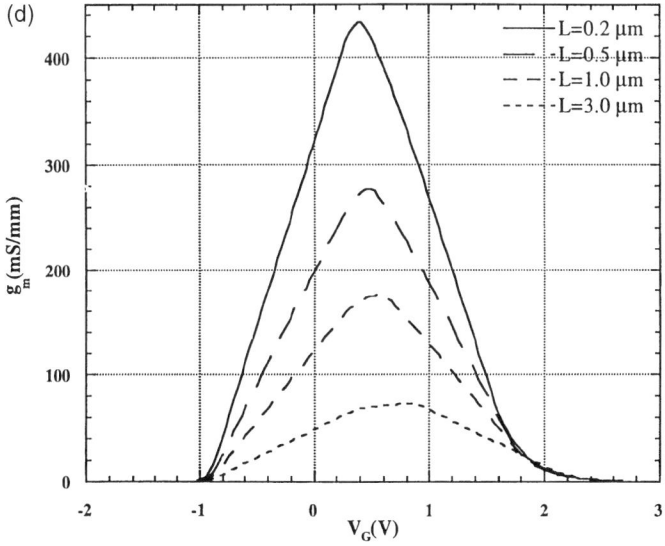

FIG. 27. (*Continued*).

the drain-source voltage V_D lying between 2.1 and 1.6 V. The higher slope of the n_{2D} versus V_G curve led to a higher slope for g_m. Although a higher peak transconductance was obtained for a thinner $Al_xGa_{1-x}N$, there was still an ambiguity regarding whether this layer should be thin or not, depending on the type of device needed (N-ON or N-OFF). After all, these devices also depend strongly on the doping level in $Al_xGa_{1-x}N$ and on the leakage current (within acceptable limit), which is generally larger for a higher applied gate bias (especially when V_G is larger than the Schottky barrier). In view of the fact that a larger value of the peak transconductance corresponds to a lower value of V_D when the spacer layer is thinner, it may be anticipated that, due to higher capacitance of thinner $Al_xGa_{1-x}N$ layers, short-channel MODFETs with thinner $Al_xGa_{1-x}N$ layers would exhibit a significantly larger transconductance.

As the doping level of $Al_xGa_{1-x}N$ layer was varied between 2×10^{18} cm^{-3} and 2×10^{19} cm^{-3}, the calculated peak value of the transconductance was 60 to 120 mS/mm (see Fig. 27c), for a gate-source bias ranging between 1.5 and -3.0 V and a drain-source bias ranging between 2.3 and 1.7 V. In general, the higher the doping level, the higher the peak transconductance. This suggested that, in order to obtain a higher transconductance, one should opt for the highest possible doping level of the $Al_xGa_{1-x}N$ layer. This should, however, be deferred while designing an N-OFF MODFET, for which a very thin layer of $Al_xGa_{1-x}N$ would enhance the electrical performance.

When the channel length was decreased from 3.0 to 0.2 μm, the peak transconductance varied from 70 to 420 mS/mm (see Fig. 27d). A very large value of transconductance was achieved when the values of both V_G and V_D were chosen to be small to avert the velocity saturation of the carriers. This was very encouraging, considering the fact that, for all practical purposes, MODFETs with reduced channel length are very desirable and have lower leakage current at the operating point. Notably, the peak transconductance of these MODFETs is obtained for lower gate biases.

j. Drain Current, Comparison with Experiment

For our experimental investigation [71, 72], $Al_xGa_{1-x}N/GaN$ MODFETs were grown by MBE and fabricated by the conventional self-aligned method. Various parameters for these MODFETs were $d = 250$ Å, $W_{sp} \approx 50$ Å, and $x \approx 0.2$. (Both W_{sp} and x had some uncertainty.) The doping concentration N_d was such that it gave rise to an experimental equilibrium electron concentration of 5×10^{18} cm^{-3} for GaN, which corresponds to a vlaue of $N_d = 2.1 \times 10^{19}$ cm^{-3} when obtained from

$$\frac{N_d}{1 + 2\exp\left(\dfrac{E_d - E_c}{k_B T}\right)} = N_c F_{1/2}\left(\frac{-E_c}{k_B T}\right) \tag{27}$$

The measured gate length of the MODFETs was 3 μm, the channel length was 5 μm, and the channel width was 40 μm. The experimental measurements recorded the channel electron mobility to be 490 cm^2/Vs and the 2DEG concentration to be around 10^{13} cm^{-2}.

Drain current I_D was calculated using Eq. (16), which includes the effect of the saturation velocity of electrons. For the calculations, the mobility was assumed to be 500 cm^2/Vs and the saturation velocity to be 2×10^7 cm/s, which led to a critical electric field, $F_c = 4$ V/μm. To obtain the 2DEG concentration for these calculations, it was observed that the slopes of I_D versus V_D curves for smaller values of V_D may be obtained from the following approximate form of Eq. (16):

$$\int_{V_{GD}}^{V_{GS}} n_{2D}(u)du \approx (V_D - V_S) \times n_{2D}(V_G) \tag{28}$$

It was also assumed that the value of the drain current at saturation for a given V_G is proportional to the area under the n_{2D} versus V_G curve. This is particularly true when the electron velocity does not approach the situation value, which is indeed the case for MODFETs with $L = 5$ μm. Finally, the threshold gate bias V_{GT} was assumed to be around -0.6 to -0.8 V, since the pinch-off phenomenon occurs for values of V_D (0.6–0.8 V) larger than V_G. Based on these assumptions, the parameters $d = 100$ Å, $N_d = 2 \times 10^{19}$ cm^{-3}, $W_{sp} = 20$ Å, and $x = 0.35$ yielded the maximum of n_{2D} to be 7.2×10^{12} cm^{-2}.

Variation of the drain current as a function of drain-source voltage V_D and gate-source voltage V_G is shown in Fig. 28. An inspection of various curves of Fig. 28 would indicate that in the saturation regime the drain current tends to decrease slightly with increasing drain-source voltage V_D. This is more apparent for MODFETs with smaller channel lengths. It is interesting to note that almost similar behavior is observed for Al$_x$Ga$_{1-x}$N/GaN MODFETs fabricated and characterized in our laboratory [71]. As saturation is reached in the current–voltage characteristics due to both pinch-off

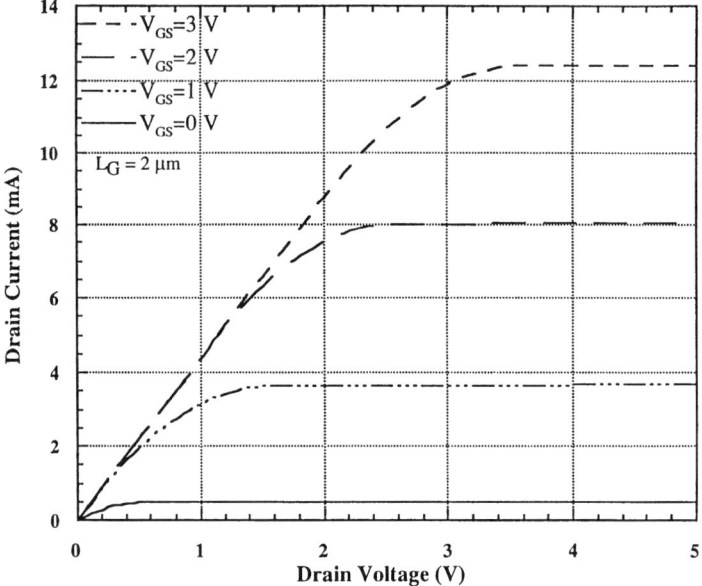

FIG. 28. Comparison of the experimental (solid line) and theoretically calculated results (dashed lines) for drain-source current I_D for AlGaN/GaN MODFETs under various gate-source bias conditions (0, 1, 2, and 3 V). After Stengel *et al.* [66].

and velocity saturation, the electric field and the potential in the channel remain essentially constant, even when a larger drain-source bias is applied (except in a very small region at the edge) [73]. The pinch-off depends on the relative values of the gate-source and drain-source voltages. While the gate-source voltage tends to keep channel electrons stuck to the $Al_xGa_{1-x}N$/GaN surface, the drain-source voltage tends to drag them away to the drain. This competition between the gate-source and the drain-source voltages becomes increasingly imbalanced as the difference between them increases. Consequently, the length of the pinch-off region, and hence the leakage, becomes larger when the drain-source voltage becomes much larger than the gate-source voltage. This causes a slight decrease in drain saturation current, with increasing drain-source voltage V_D.

For the sake of comparison, the experimental results for various gate-source voltages of a MODFET with a channel length of 3 μm are also presented in Fig. 28. It was noted that the theoretical results are in very good agreement with the experimental results. Also, a peak transconductance of 100 mS/mm is obtained if the saturation velocity is taken into

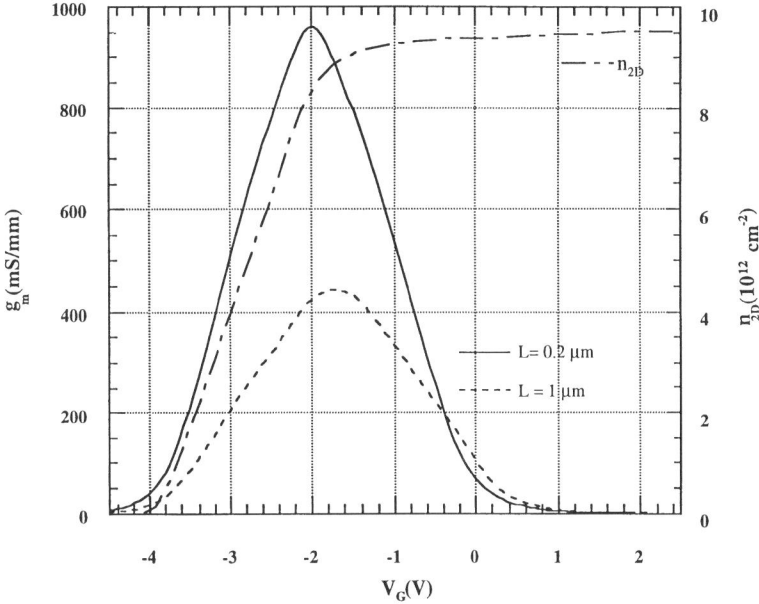

FIG. 29. Variaton of 2DEG concentration and transconductance as functions of the gate-source bias V_G. Curves correspond to transconductance of a 1-μm channel-length MODFET, a 0.2-μm channel-length MODFET, and n_{2D}, for a structure that shows the possibility of achieving $n_{2D} \approx 10^{13}$ cm^{-2}. After Stengel et al. [66].

account [$F_c = 4 \times 10^6$ V/m in Eq. (17)] and $V_G = 2.5$ V; and a peak transconductance of 115 mS/mm is obtained when this saturation velocity is not considered (e.g., if $F_c = +\infty$) and $V_G = 3.0$ V. These data compare rather well with 120 mS/mm found experimentally for $V_D = 3$ V. At this point, it would be worthwhile to bear in mind that the values of various parameters obtained from a fitting of the I_D versus V_D curves could be different, depending, for example, on the channel length and the saturation velocity. This would lead the calculated peak transconductance to be larger.

Using other reasonable parameters (e.g., a value of x lower than that used in the previous calculation), even larger values of the 2DEG (very close to 10^{13} cm^{-2}) could be realized, as is evident from Fig. 29. The values of $d = 100$ Å, $N_d = 5 \times 10^{19}$ cm^{-3}, $W_{sp} = 0$ Å, and $x = 0.3$ would lead to a maximum value of the 2DEG concentration to be as high as 9.5×10^{12} cm^{-2} and the peak transconductance to be 450 mS/mm for $L = 1$ μm and 1000 mS/mm for $L = 0.2$ μm. This is confirmed by calculations employing Eq. (17), which takes the saturation velocity into account.

k. High-Speed Aspects

High-speed devices are commonly analyzed by two port-scattering measurements, which are performed generally in the range of 2 to 26 GHz on the wafer by an on-wafer microwave probe station, although 60 GHz is possible, and this figure is constantly moving upward. From the scattering measurements, one can obtain an equivalent circuit for diagnosis. From the s parameters one deduces the y parameters from which an equivalent circuit is generated. The equivalent circuit model is very useful in the sense that device performance, or lack thereof, can be compartmentalized, with the problem area getting targeted attention.

The transit time under the gate of a submicron MODFET is on the order of a few picoseconds. This being the case, the charging time of the input and the feedback capacitance through the input resistance R_i in the equivalent circuit determine the speed of response. Generally two parameters—the current gain cut-off frequency and maximum oscillation frequency—are figures of merit to gauge the expected high-frequency performance of an FET.

Since the feedback capacitance is negligible compared to the input capacitance, the current gain cut-off frequency, defined as the frequency at which the current gain goes to unity, is given by

$$f_T = \frac{g_m}{2\pi C_{gs}} = \frac{v_s}{2\pi L} \tag{29}$$

As can be seen, and as mentioned, the higher the saturation velocity and the smaller the gate length, the higher the f_T.

The maximum oscillation frequency, defined as the frequency at which the power gain goes to unity, is given by [74]

$$f_{\max} = \frac{f_T}{2\sqrt{r_1 + f_T \tau_3}} \tag{30}$$

where $r_1 = (R_g + R_i + R_s) G_d$, and the feedback time constant $\tau_3 = 2\pi R_g C_{dg}$.

The term r_1 underscores the importance of the need to reduce the output conductance, not just at dc but, of course, microwave frequencies for increased $f_{\max}$ or power gain. Power gain is also aided by the reduction of parasitic resistances and the time constant τ_3, which to a large extent depends on the geometry of the device.

l. Performance of GaN-Based MODFETs

As stated earlier, MODFETs utilize a two-dimensional carrier gas confined at an interface between two layers with an interfacial energy barrier, such as AlGaAs/GaAs [75] and AlGaAs/InGaAs [76]. A GaN MODFET taking advantage of the background donors in the AlGaN layer, which is not a controllable, to say the least, was reported [77]. Congruent with the early stages of development and defect-laden nature of the early GaN and AlGaN layers, the MODFETs exhibited a low-resistance and a high-resistance state before and after the application of a high drain voltage (20 V). As in the case of GaAs/AlGaAs MODFETs [78], hot electron trapping in the larger bandgap material at the drain-side of the gate is primarily responsible for the current collapse. The negative electron charge accumulated due to this trapping causes a significant depletion of the channel layer, more probably a pinch-off, leading to a drastic reduction of the channel conductance and the decrease of the drain current. This continues to be effective until the drain-source bias is substantially increased, leading to a space-charge injection and giving rise to an increased drain-source current.

Paltry performance became a thing of the past with reports of excellent n-channel N-OFF, N-ON normal GaN MODFETs and inverted MODFETs utilizing a doped AlGaN donor layer [71, 79]. For the fabrication of normal MODFETs, the GaN-based heterostructure was grown epitaxially on basal-plane sapphire substrate by MBE.

An artistic view of the cross-sectional diagram of an N-ON MODFET structure is shown in Fig. 30. The Hall measurements of the N-ON MODFET structure yielded an electron mobility over 1000 cm^2/Vs at 40 K, although it was found to decrease first slowly and later rapidly with temperature dropping to about 490 cm^2/Vs at 310 K. The sheet carrier concentration in the layer was 10^{13} cm^{-2}, which was effectively independent of temperature over the temperature range of 40 to 200 K. It rose to about 1.2×10^{13} cm^{-2} at 300 K. These sheet carrier concentrations are at least 1 order of magnitude higher than those measured for GaAs/AlGaAs and InGaAs/InAlAs systems. The inverted MODFET (IMODFET) structure employed in the pursuit of high-power devices is shown in Fig. 31 and is as follows: sapphire substrate (1000 Å) AlN (1 μm) GaN (500 Å) AlGaN (30 Å) AlGaN:Si (30 Å) AlGaN (60 Å) GaN (40 Å) AlGaN. The mole fraction in AlGaN layers is 15%, and the doping level is 5×10^{18} cm^{-3} for the doped AlGaN layer.

Both N-ON and N-OFF devices investigated in the author's laboratory have gate lengths of about 2 μm and two gate fingers with a total width of 78 μm. In some devices with appropriate undercutting, the actual gate length

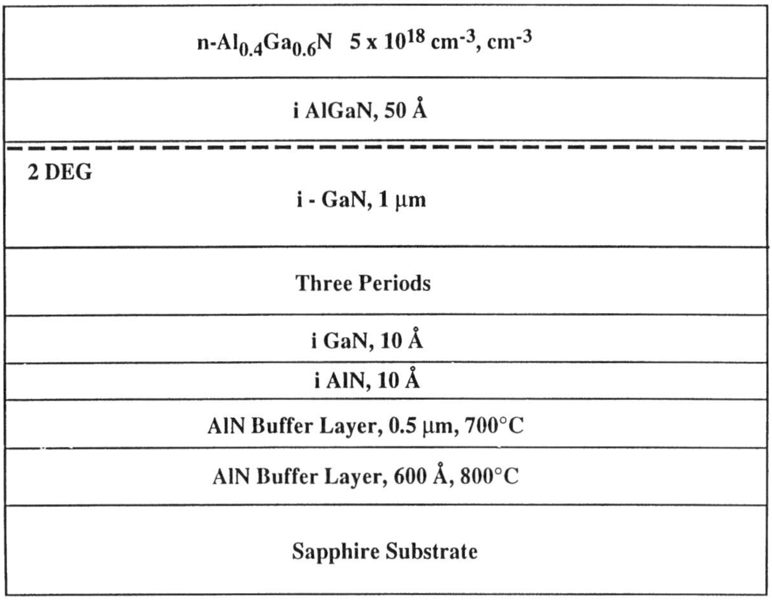

n-Al$_{0.4}$Ga$_{0.6}$N   5 x 10^{18} cm^{-3}, cm^{-3}
i AlGaN, 50 Å
i - GaN, 1 µm
Three Periods
i GaN, 10 Å
i AlN, 10 Å
AlN Buffer Layer, 0.5 µm, 700°C
AlN Buffer Layer, 600 Å, 800°C
Sapphire Substrate

(2 DEG indicated at the AlGaN/GaN interface)

FIG. 30. Cross-sectional diagram of the normal AlGaN/GaN MODFET structure.

was reduced to about 1.5 µm. The transparent nature of the GaN-based structures lacks the contrast for easy optical lithography with contact mask aligner when the dimensions fall below the aforementioned values. For simplicity and convenience, no attempt was made to deposit contact and gate overlay metallization, with the adverse effect of increased access resistances. An electron Hall mobility of 350 cm^2/Vs with a sheet carrier concentration of 1.6×10^{13} cm^{-2} was measured at room temperature. The actual sheet carrier concentration is expected to be somewhat lower because of some degree of parallel conduction in the bulk layers [80]. The measured sheet carrier concentration is larger than can be accounted for by the Si donors in the doped AlGaN layer that may be ascribed to redistribution of charges by the piezo-electric effect. At 77 K, the sheet carrier density was measured as 1.3×10^{13} cm^{-2} with a mobility 1450 cm^2/Vs. These results indicate that, unlike in the GaAs/AlGaAs case [81], the GaN/AlGaN interface with GaN grown on the AlGaN layer is of no worse quality than the normal-order case, at least at the present time.

The dc drain characteristics of MODFETs with a gate length of 2 µm, a gate width of 40 µm, and source-drain separation of 4 µm are presented in

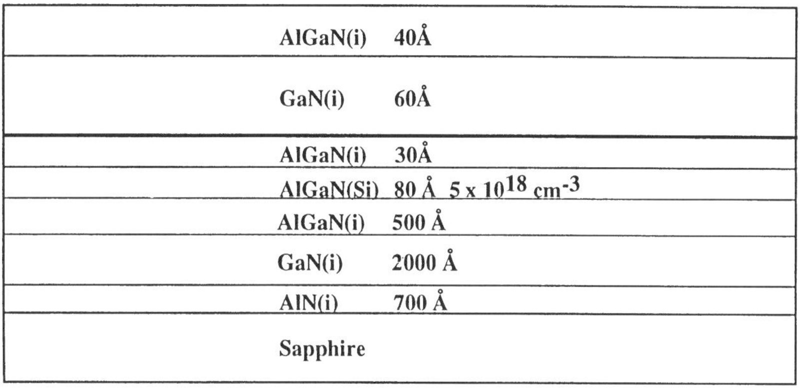

AlGaN(i)	40Å
GaN(i)	60Å
AlGaN(i)	30Å
AlGaN(Si)	80 Å 5×10^{18} cm^{-3}
AlGaN(i)	500 Å
GaN(i)	2000 Å
AlN(i)	700 Å
Sapphire	

FIG. 31. Cross-sectional diagram of the inverted AlGaN/GaN MODFET structure.

Fig. 32. As may be evident from this figure, room-temperature extrinsic transconductance g_m for these MODFETs is about 180 mS/mm. As expected, the extrinsic transconductance of the MODFETs increases with decreasing gate length, as shown in Fig. 33, in which the drain breakdown voltage with respect to drain gate spacing is depicted in the inset. Others reported [82] GaN MODFETs with $L = 0.7\,\mu$m and $W = 150\,\mu$m, with a

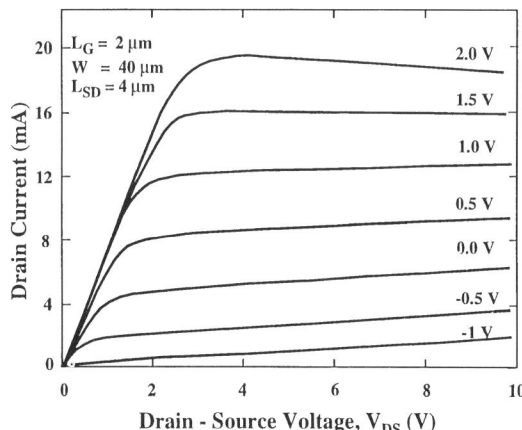

FIG. 32. Output current voltage characteristics of a normal AlGaN/GaN MODFET with a 2-μm gate length exhibiting a transconductance of about 180 mS/mm.

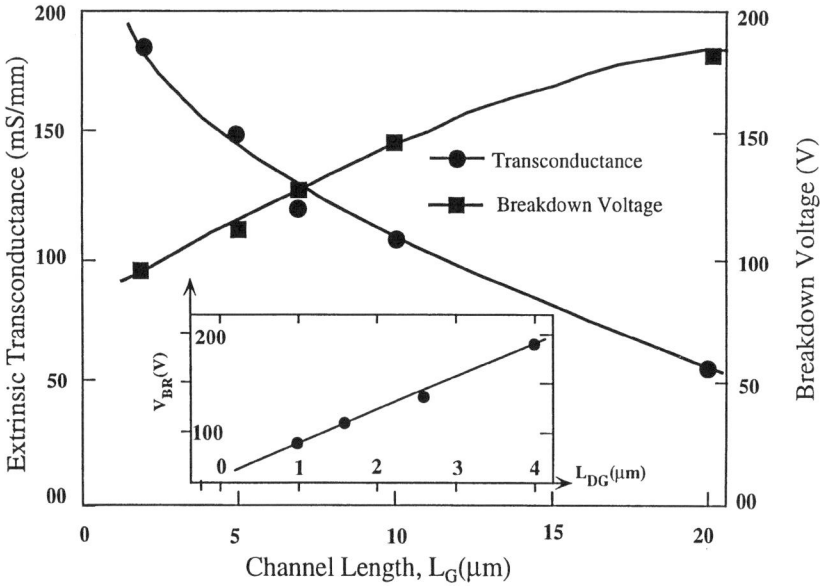

FIG. 33. Variation of transconductance with gate length and drain breakdown voltage with gate drain spacing (inset).

room-temperature extrinsic transconductance $g_m = 45$ mS/mm at $V_{GS} = -2$ V and $V_{DS} = 10$ V.

Similar MODFETs [83] with $L = 0.23$ μm and $W = 100$ μm exhibited, on the other hand, $g_m = 23$ mS/mm at $V_{GS} = -0.6$ V and $V_{DS} = 10$ V. The performance of both of these MODFETs appears to be inferior to the ones fabricated in the author's laboratory. It may be noted from Fig. 32 that the drain-source current I_{DS} of these MODFETs saturates at $V_{DSAT} \approx 3$ V. The large electric field needed to cause velocity saturation and access resistances manifests in the observed dependence of transconductance, g_m, on the gate length L.

Continued effort in the author's laboratory led to much-enhanced performance in terms of transconductance and breakdown voltages, the hallmark of what GaN has to offer. MODFETs with gate lengths of 1.5 μm exhibited transconductances of about 220 mS/mm and drain breakdown voltages of about 100 V for a gate-to-drain spacing of 1 μm [84]. The fact that similar results, in terms of the breakdown voltage [85], are being obtained at other laboratories as well is indicative of the potential of this material system for high-power applications.

Using MODFET structures similar to those employed in author's laboratory, Khan and co-workers [86] at APA optics have also achieved sheet carrier concentrations of 10^{13} cm^{-2}, which led to MODFETs with 1-μm gate lengths exhibiting transconductances between 80 and 120 mS/mm. N-ON devices with 1-μm gate lengths exhibited current gain cut-off frequencies of about 18 GHz, which is larger than that expected even from GaAs devices. Moreover, devices with quarter-micron gate lengths exhibit current gain cut-off and power gain cut-off frequencies of about 40 and 90 GHz, respectively [87]. Preliminary simulations conducted at author's laboratory indicate that GaN should outperform GaAs for short gate length devices [66]. A compilation of the current gain cut-off frequency of GaN-based and SiC-based FET-like devices is shown in Fig. 34.

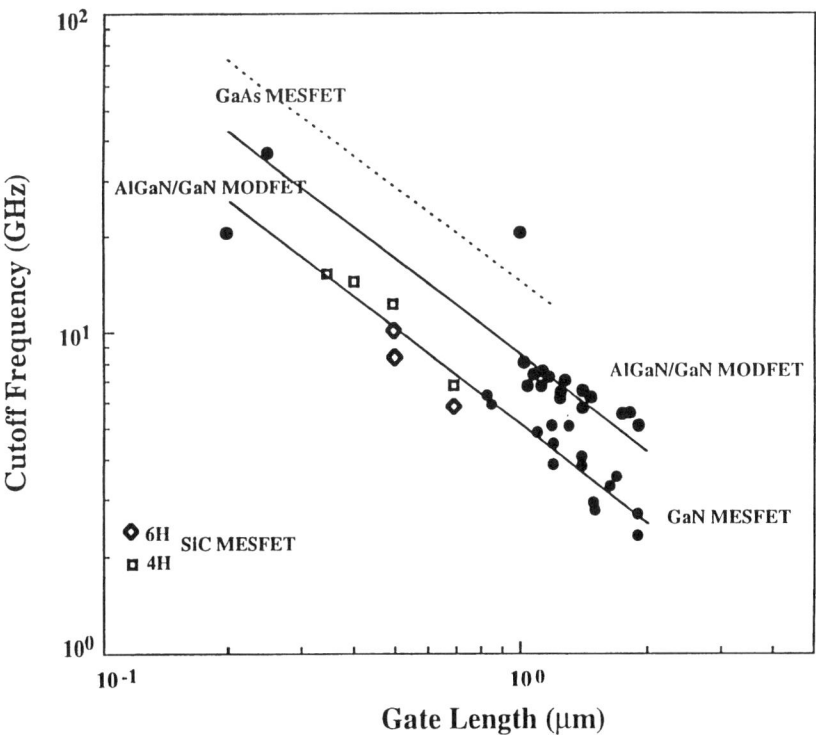

FIG. 34. Current gain cut-off frequency of AlGaN/GaN MODFETs, GaN MESFETs, and SiC MESFETs, as compiled by S. Binari of the Naval Research Laboratory.

The normal MODFET structures grown with reactive MBE and reported previously have exhibited transconductances of 220 mS/mm with low leakage currents [84, 88]. The same devices exhibited breakdown voltages exceeding 90 volts for 1.5-μm gate devices with 1 μm gate-to-drain spacing, which are far from optimum. We have also reported on the characteristics of our devices under UV illumination and presented the 300°C operation [89]. The breakdown voltage, which is a significant parameter in high-power devices, has received some attention. Wu et al. [85] have demonstrated excellent gate-to-drain breakdown voltages of 340 V for 3-μm gate-to-drain spacing in undoped devices, which rely on carrier donation from the unintentionally doped AlGaN, and 170 V for 2-μm gate-to-drain spacing in doped structures where the AlGaN layer is intentionally doped. These results were achieved by decreasing the GaN buffer thickness and employing a larger gate-to-drain distance of 3 and 2 μm in two types of devices. Our devices, however, employ high-resistivity buffer layers of 1 to 3 μm thickness, which is possible due to the particular growth scheme employed.

Inverted AlGaN/GaN MODFETs with 2-μm gate lengths and 78-μm gate widths exhibited a normalized CW output power of 1.5 W/mm and power-added efficiency of 17.5% at 4 GHz without catastrophic failure. These results are extraordinary, considering the low current gain cut-off frequency of 6 GHz, which is a direct result of the long gate length. MODFETs exhibit reduced dc output conductance compared to normal MODFETs, attesting that a part of the negative output conductance observed in nitride-based FETs, which is often attributed to poor thermal conductivity of sapphire substrate, is due to carrier loss to the buffer layer at high fields. The electron mobility and sheet carrier concentrations were 1450 cm^2/Vs (350 at room temperature) and 1.3×10^{13} cm^{-2} at 77 K, respectively, indicating that inverted interfaces in this class of heterostructures, unlike the GaAs/AlGaAs system, are comparable at this stage to the normal interface, with the added advantage of barrier at the buffer layer side reducing the differential output conductance. Moreover, these together with other exciting results reported in the literature attest to the potential of nitride-based heterostructures for high-power applications.

The details of the IMODFET structure employed in this investigation, as shown in Fig. 31, are as follows: sapphire substrate (1000 Å) AlN (1 μm) GaN (500 Å) AlGaN (30 Å) AlGaN:Si (30 Å) AlGaN (60 Å) GaN (40 Å) AlGaN. The mole fraction in AlGaN layers is 15%, and the doping level is 5×10^{18} cm^{-3} for the doped AlGaN layer. The investigated devices have gate lengths of about 2 μm and two gate fingers with a total width of 78 μm. No attempt was made to deposit contact and gate overlay metallization, with the adverse effect of increased access resistances. An electron Hall mobility of 350 cm²/Vs with a sheet carrier concentration of 1.6×10^{13} cm^{-2} was measured at room temperature. The actual sheet carrier concentration is expected to be somewhat lower because of some degree of parallel conduction in the bulk layers [80]. It should be mentioned here that the measured sheet carrier concentration may be upwardly influenced by processes such as the piezo-electric effect.

The 2-μm devices fabricated on this layer exhibited a dc transconductance of 105 mS/mm, peaking at a gate-to-source voltage of -2.5 V. The output characteristics of the IMODFET are shown in Fig. 35. Small-signal S-parameter measurements were performed at bias conditions used for the power measurements—15 V, -2.5 V, and 20 mA for the drain voltage, gate voltage, and drain current, respectively. Short-circuited current gain, maximum available power gain, and the unilateral gain calculated from the small-signal S-parameters are shown in Fig. 36 as a function of frequency under bias conditions of $V_{DS} = 15$ V and $V_{GS} = -2.5$ V. The I_{ds} at this bias was approximately 20 mA, which corresponds to 260 mA/mm. The unity current gain cut-off frequency (f_t) and maximum frequency of oscillation were 6 GHz and 11 GHz, respectively, at both 15- and 30-V bias. The CW microwave power measurement results are presented in Fig. 37. The measurements were taken at 4 GHz, with the input power swept from 5 to

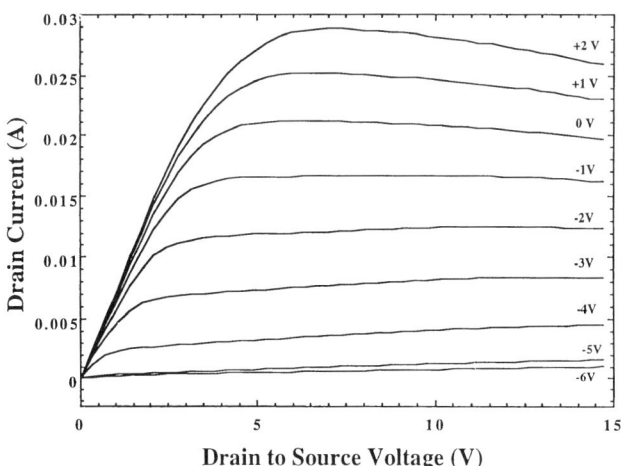

FIG. 35. Output characteristic of the IMODFET, the structue of which is depicted in Fig. 31.

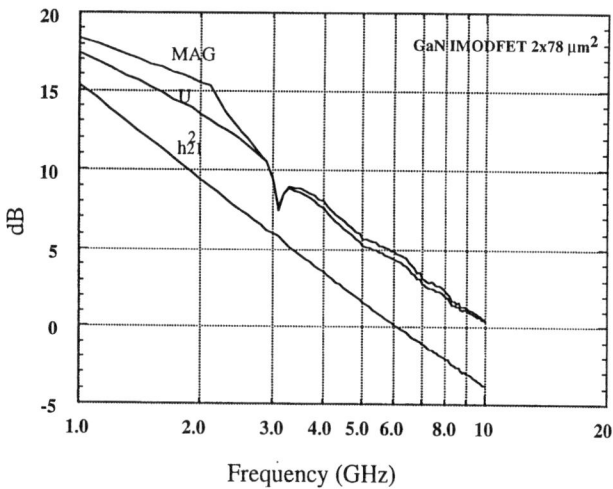

FIG. 36. Short-circuited current gain, maximum available gain, and unilateral gain as a function of frequency under bias conditions of $V_{DS} = 15$ V and $V_{GS} = -2.5$ V. The I_{ds} at this bias was approximately 20 mA, which corresponds to 260 mA/mm.

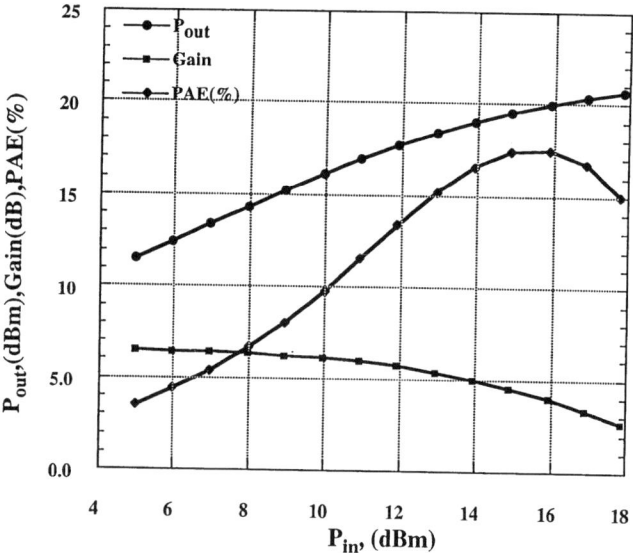

FIG. 37. CW output power (P_{out}), power gain, and the power-added efficiency (PAE) versus input power level.

TABLE II

MICROWAVE PERFORMANCE OF FETs REPORTED BY VARIOUS LABORATORIES

Laboratory	Gate Length (μm)	2DEG Density (cm^{-2})	Mobility (cm^2/Vs)	g_m (mS/mm)	I_{DSS} (mA)	f_t (GHz)
APA and Cornell	~0.25			~30?	~300	31–40
APA Optics groups	1	1×10^{13}	1200	120	600	18?
UCSB	1	7.9×10^{12}	1500	120	340	6
Morkoç IMODFET	2	1.3×10^{13}	350	110	500	6

UCSB, University of California–Santa Barbara.

18 dBm in 14 steps. The input and output matches, which were used during the power sweep, were determined by iterating between source and load pulls. The output match was selected to optimize the output power, and the input match was selected to maximize the delivered power. The devices were biased to $V_{DS} = 15$ V and $V_{GS} = -2.5$ V. The I_{ds} at this bias was approximately 20 mA, which corresponds to 260 mA/mm. From the illustration, we see that the devices exhibited 6-dB gain for various input levels. The maximum output power was 20.6 dB gain and the peak power-added efficiency was 17.5%. This corresponds to a normalized output power density of 1.5 W/mm, which represents the highest value reported in literature to date. This result, together with that of Wu *et al.* [85], attest to the potential of this semiconductor system for power applications. Summary of reported FET-like device performance is tabulated in Table II.

2. BLUE, GREEN, AND YELLOW LEDs

LEDs have undergone a tremendous advancement in performance and are now used in almost every aspect of life. The future of many technologies, including printing, communications, displays, and sensors, depend profoundly on the development of compact, reliable, and inexpensive light sources. Currently, conversion efficiencies of commercial LEDs emitting in the red (650–660 nm) stand at 16%, and those of lasers stand at 75% [90, 91]. A primary goal of GaN research is to efficiently harness its direct energy bandgap for optical emission. Though the band-edge emission in GaN occurs at about 362 nm, which is UV, by appropriately alloying GaN with its cousins AlN and InN, the energy bandgap of the resulting Al(In)GaN can be altered for emission in the range of UV to yellow or may even be red. The first GaN LEDs were reported some 25 years ago by Pankove *et al.* [92]. Due to difficulties in doping GaN *p*-type at the time, these LEDs were metal-insulator semiconductor (MIS) LEDs, rather than *pn* junction LEDs. The electroluminescence of these LEDs could be varied from blue to yellow, depending on the doping of the insulator layer. The measured efficiencies of these preliminary MIS LEDs were not sufficient for competition with the commercially available LEDs of that time.

One of the timely advancements in the nitride effort has been the exploitation of double heterostructures (DHs) for LED [93, 94]. The advantage of DH LEDs over homojunction LEDs is that the entire structure outside of the active region where the light is generated is transparent. This reduces the internal absorption losses. Furthermore, this cladding region serves as an interface for scattering of light, thus minimizing the probability of total internal reflections within the device. These two factors together enhance the probability of escape of the light out of the device.

The first GaN *pn* junction LED was demonstrated by Amano *et al.* [94] and was reported in 1989. The fabricated device consisted of an Mg-doped GaN layer grown on top of an undoped *n*-type (n = 2×10^{17} cm^{-3}) GaN film with the chemical Mg concentration estimated to be 2×10^{20} cm^{-3}. The electroluminescence of the devices was dominated by near-band-edge emission at 375 nm, which was attributed to transitions involving injected electrons and Mg-associated centers in the *p*-GaN region. Additionally, a small shoulder extending to 420 nm, due to defect levels, was also observed.

To achieve other desired colors, InGaN alloys for emission media are required. While the increased InN mole fraction in GaN red shifts the spectrum, this would be at the expense of working against the introducing structural defects unless InGaN is made sufficiently thin so it is not lattice-matched to GaN. Lattice-mismatched films can be grown up to a certain thickness, called "the critical thickness," for a given composition. The larger the composition, the smaller the critical thickness. In view of this, there should be substantial effort devoted to optimization. In this vein, band-edge emissions were also obtained for LEDs employing an Si-doped InGaN quantum well as the active region in a GaN/InGaN DH LED [95]. The InN mole fraction content of the active layer was varied, resulting in a shift of the peak wavelengths of the device's electroluminescent spectra from 411 to 420 nm. Impressively, the researchers at Nichia Chemical [96] were successful in reducing the thickness of InGaN emission layers to about 30 Å. With this achievement, InGaN quantum wells with InN mole fractions up to 70% have been obtained, and LEDs with commercial capabilities are now possible in blue, green, and yellow.

The blue and blue-green LEDs developed by Nichia initially relied on the transitions to Zn centers in InGaN. This was necessitated by the need to extend the wavelength to the desired values, and the amount of In that could be added was limited while maintaining good crystalline quality. These LEDs suffered from wide spectral widths and saturation in the output power with injection current at the intended wavelength of operation. The large spectral width spoiled the color saturation, with the undesirable outcome that not all of the colors could be obtained through color mixing. However, with the quantum well approach, the In mole fraction can be extended to about 70%, paving the way for excellent UV, violet, blue, green, and yellow LEDs. These LEDs exhibit power levels of 5, 3, and 1 mW at 20 mA of injection current for the wavelengths of 450, 525, and 590 nm, respectively.

Very important is the fact that the full width at half maximum (FWHM) of the spectrum was 20, 30, and 80 nm for blue, green, and yellow LEDs, respectively, owing to the fact that these new LEDs take advantage of

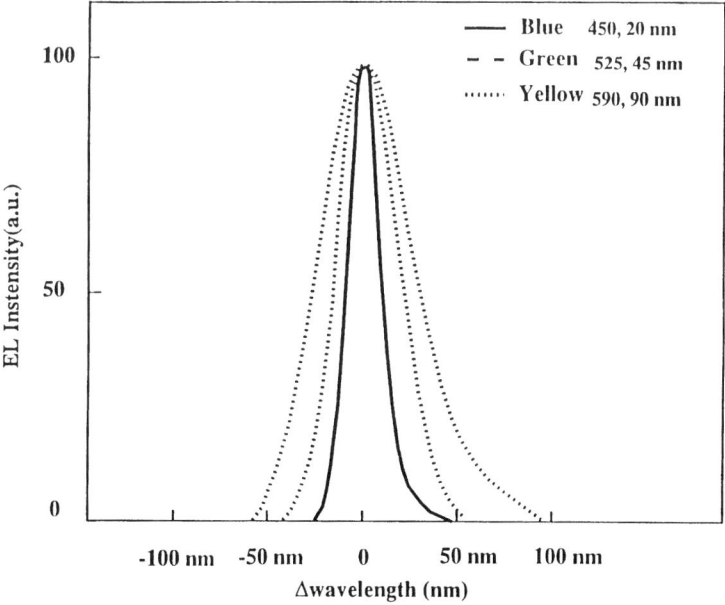

FIG. 38. Spectral width of blue (450 nm), green (525 nm), and yellow (590 nm) Nichia LEDs. FWHM values of 20, 45, and 90 nm have been obtained for 450, 525, and 590 nm emissions respectively. For reference, the Nichia LEDs, which relied on Zn centers, showed an FWHM value of 70 nm. EL, electroluminescence. Courtesy of S. Nakamura of Nichia Chemical Industries, Ltd.

band-to-band transitions (Fig. 38). The In mole fractions used are 20, 43, and 60% for the 450-, 525-, and 590-nm emissions, respectively. The output power as a function of injection current for the blue, green, and yellow LEDs mentioned here are shown in Fig. 39. At the wavelength corresponding to green, the 3-mW power level, which corresponds to 12 cd in a 10-degree viewing cone, was obtained with an efficiency of 6.3%. Most recent results for the GaN and ZnTeSe material systems are over the 20-cd level. In summary, at an injection current of 20 mA, the blue, green, and yellow LEDs produce 5 mW, 2.5 cd (efficiency = 9.1%), 3 mW (efficiency = 6.3%), and 1 mW (efficiency = 2.3%) at blue, green, and yellow wavelengths. The UV LEDs operating at 400 nm exhibit efficiencies of 10% at $I = 20$ mA. At the current injection level of 100 mA, 400- and 450-nm LEDs exhibit power levels of 13 and 12 mW, respectively. A chromaticity diagram in which blue InGaN single quantum well LEDs, green InGaN single quantum well LEDs, green GaP LEDs, AlInGaP LEDs, and red GaAlAs LEDs are indicated is shown in Fig. 40. The outer perimeter in this diagram indicates saturated colors—in other words, emission with negligible broadening. It is evident that addition of nitride-based blue and green along with the previously available red LEDs has paved the way to obtain some 70% of the colors encompassed by the chromaticity diagram. The average human eye would not be able to realize that the entire chromaticity diagram is not accessed, because green LEDs do not quite produce saturated color due to their relatively wide spectra.

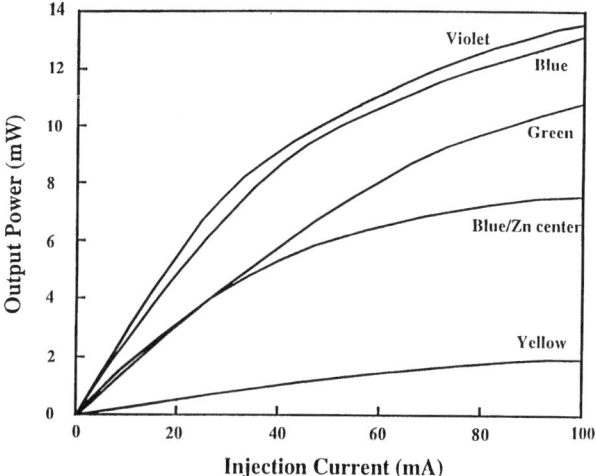

FIG. 39. The output power at an injection current of 20 mA for blue, green, and yellow quantum well LEDs. Courtesy of S. Nakamura of Nichia Chemical Industries, Ltd.

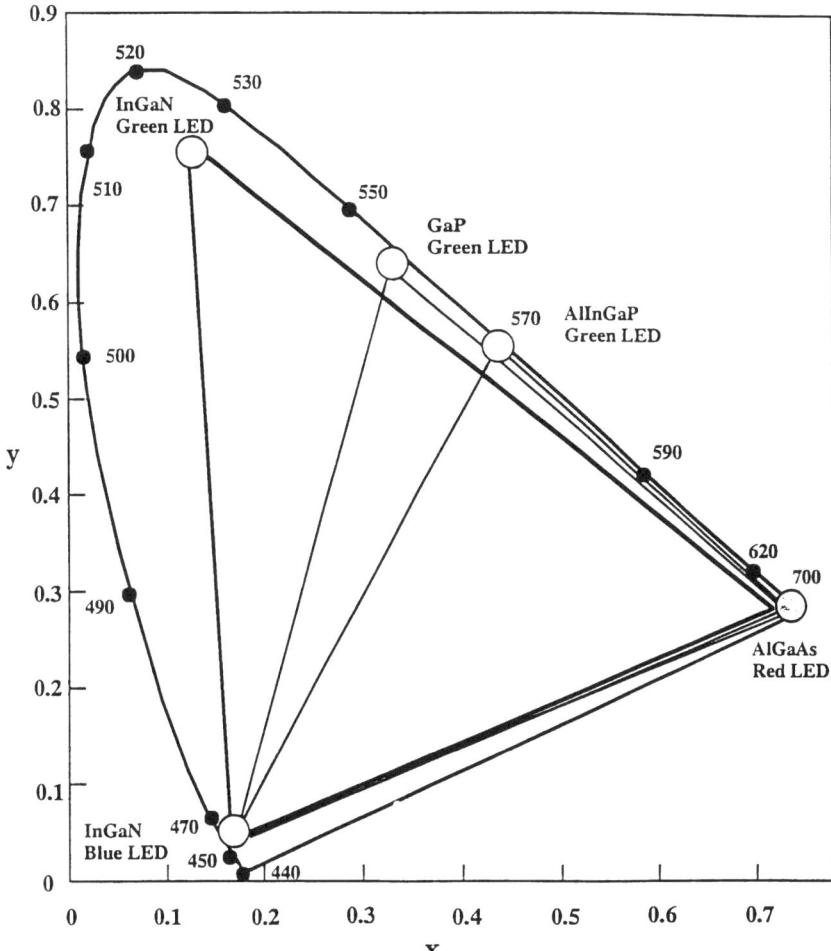

FIG. 40. Chromaticity diagram, in which blue InGaN single quantum well LEDs, green InGaN single quantum well LEDs, green GaP LEDs, AlInGaP LEDs, and red GaAlAs LEDs are shown. Courtesy of S. Nakamura of Nichia Chemical Industries, Ltd.

Nichia is no longer alone in production of the GaN-based LEDs. Continuing on their long-term activity, researchers at Toyota Gosei, using an asymmetric GaN/InGaN/AlGaN DH design, obtained blue InGaN/GaN/AlGaN LEDs with excellent performance. For example, with injection currents of 20 (efficiency = 3.9%) and 100 mA, the output power is 3 and 9 mW, respectively, for the blue LEDs. Since LEDs rely on the Zn centers in InGaN, the spectral width is about 70 nm. This is comparable to the earlier version of Nichia LEDs, which relied on the impurity transitions. Cree Research also has blue GaN-based LEDs grown on SiC, which takes advantage of extrinsic transitions involving Mg-induced complexes in GaN. As of September 1995, the Cree devices exhibit 1.4 mW with an external efficiency of 2.4%. One purported advantage of the Cree device is that it is on conducting SiC substrate in a vertical geometry.

While blue and green InGaN-based LEDs are finding increasing applications in displays and traffic lights, the attention is turning to white-light generation and illumination. One approach is to use the best red, green, and blue LED technologies together in a primary color-mixing configuration. In this scheme, the red device will be either AlGaAs- or InGaAlP-based, while the green and blue will be provided by InGaN, which is complex. Another approach for the production of white light utilizes a bright violet or blue LED that pumps a medium that contains phosphors corresponding to the three primary colors, as was demonstrated and marketed by Nichia Chemical. The medium acting as a host to the phosphors used in this particular case is YAG. The success of this approach will depend on the efficiency of conversion. At present, the violet LEDs are about 10% efficient, and further enhancements can be expected with the development of new substrates.

Nichia LEDs were studied extensively for degradation at various laboratories. To speed the degradation process, LEDs were tested under high injection-curent levels, higher temperatures than the ambient, and in high-humidity environments. Humidity tests indicated slower degradation than that encountered in red LEDs containing AlGaN, as nitride-based heterostructures are resistant toward oxidation. Many investigations concluded that degradation observed, particularly in the early stages of development had to do with metallization-related failures. In more recent LEDs, the metallization-related degradation has been reduced. This bodes well for the III-nitride material system, in that even with a high concentration of structural and point defects, the degradation is minimal under normal operating conditions [97].

3. LEDs BY MBE

Unlike lasers, which sport waveguides, light emitted in LEDs radiates randomly; thus, appropriate measures must be taken to collect as much of the spontaneous emission as is possible. This is done with transparent substrates and contacts, coupled with the epoxy dome, which also increases the angle of collection. LED-like devices not incorporating the aforementioned measures can only be used as research tool as a precursory device to LEDs. It is in this context that LEDs utilizing GaN emission layers that emit in the violet have been produced in several laboratories [98, 99], including ours [100]. There are also others who fabricate InGaN-based LEDs by MBE [101]. Both homojunction and DH LEDs have been explored, the structural details of which are shown in Fig. 41. For both devices, the substrates were subjected to nitridation for about 15 min prior to the deposition of a 600-Å-thick AlN buffer layer. The device structure for

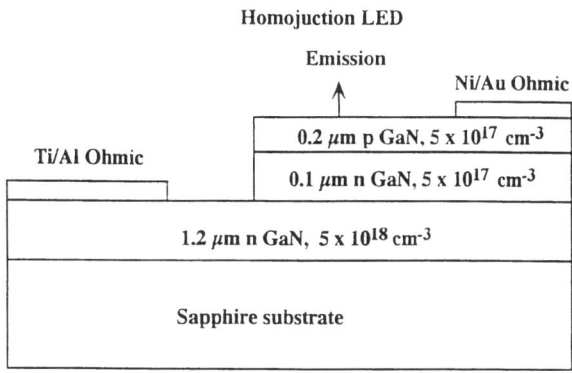

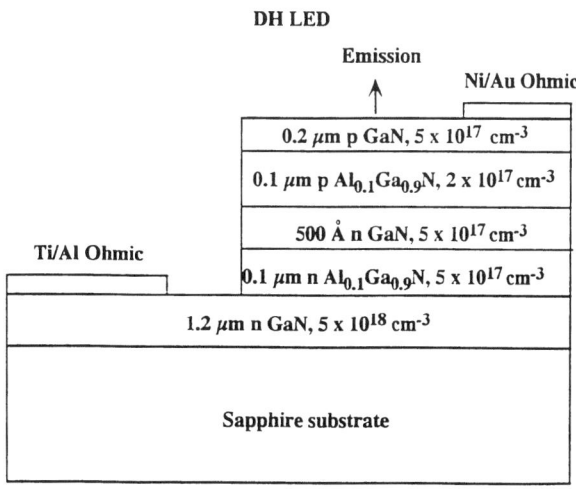

FIG. 41. Cross-sectional diagrams of *pn* junction and DH *pn* junction LEDs fabricated by RMBE. After Salvador *et al.* [100].

the GaN homojunction consisted of a 1.2-mm Si-doped (5×10^{18} cm^{-3}) n^+-GaN layer followed by 0.1-mm-thick Si-doped (5×10^{17} cm^{-3}) layer and capped with a 0.2-mm Mg doped p-GaN layer. For the GaN/ Al$_{0.1}$Ga$_{0.9}$N DH LED, a 1.7-mm Si-doped (5×10^{18} cm^{-3}) GaN layer followed by 0.1-mm Si-doped (5×10^{17} cm^{-3}) Al$_{0.1}$Ga$_{0.9}$N layer and then the 500-Å Si-doped (3×10^{17} cm^{-3}) active layer was grown. A 0.1-mm Mg-doped (2×10^{17} cm^{-3}) Al$_{0.1}$Ga$_{0.9}$N layer was grown on top of active layer and then capped with 500-Å Mg-doped p-GaN, which functioned as a contact layer. The Mg-doped GaN was as-grown p-type; no postgrowth treatment was required to activate the Mg acceptors.

For fabrication, mesas were etched to the Si-doped n^+-GaN. Ti/Al metallization for ohmic contacts on the n^+-layer was achieved in N$_2$ ambient at 900°C for 30 s using an RTA furnace. Ni/Au metallization for ohmic p-contacts on p-GaN was, on the other hand, performed at 600°C for 90 s using a gold image furnace. To ensure that the electroluminescence emission occurs from the front side of the epilayer, only about one-fourth of the top surface area of the mesa was metallized. The electroluminescence spectra of a DH pn diode resulting as collected from the front surface, which really capture a small fraction of the internally generated light, are shown in Fig. 42. The backside electroluminescence spectra of GaN DH LEDs at lower drive currents are similar to those of the pn homojunction diode. On the other hand, the electroluminescence spectra of the GaN DH LEDs was broader on the low-energy side. Amano et al. [102] observed the same feature of GaN/AlGaN DH LEDs grown by MOCVD and attributed it to the blue shift in the energy of the emitted photons from Mg-related states in the p-AlGaN layer. It should be noted that the device structure is such that the emission from the GaN active layer is absorbed in other GaN layers, which undoubtedly reduced the intensity of the light collected and favored the longer wavelength tail of the emission. Properly mounted blue LEDs fabricated in active layers grown by MBE on MOCVD-grown buffer layers, prepared at Meijo University, showed efficiencies of around 0.3%.

4. Lasers in Semiconductor Nitrides

During the past several years, wide-bandgap nitrides have attracted attention in the field of digital information storage. Unlike display and lighting applications, digital information storage and reading require coherent light sources, namely lasers. The output of these coherent light sources can be focused on a diffraction-limited spot, paving the way for an optical system in which bits of information can be recorded and read with ease and uncommon accuracy [103]. As the wavelength of the light gets shorter, the

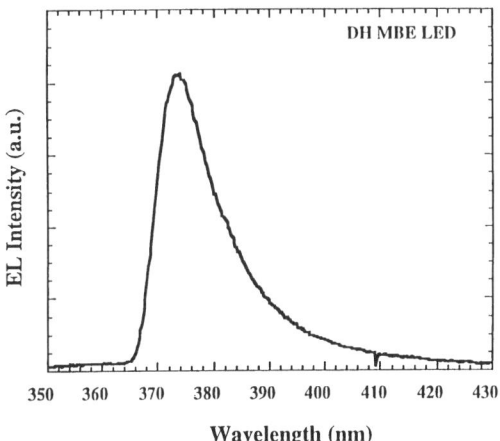

FIG. 42. Electroluminescence (EL) spectrum of a DH LED fabricated by RMBE. After Salvador et al. [100].

focal diameter becomes smaller. The interim approach adopted by the industry relies on red lasers with which pit dimensions of about 0.4 μm can be read. Using a two-layer scheme in what has been dubbed *the digital versatile disk* (DVD) (misnomers such as *digital video disk* are already commonplace), the density can be increased from 1 Gb to about 17 Gb per compact disk [104]. The cycle in consumer electronics is rather short, in that even if the red laser–based DVDs are implemented, the violet or the blue laser can be implemented 2 years after the red lasers. For consumer applications, CW operation lifetimes on the order of 10,000 hr at 60°C are required. Predictions are that DVDs using blue or shorter wavelength injection lasers will be available around the turn of the century.

Nitride-based lasers stand alone as the alternative approach employing II–VI compound semiconductors [105, 106], such as zinc-selenide ($E_g = 2.67$ eV), and ZnSdSe pseudo-morphic ternary and lattice-matched ZnMgCdSe quaternary alloys have been rather slow in attaining the required operation lifetimes needed. In addition, the shortest wavelength that can be obtained with II–VI is limited to well above that which can be obtained with nitrides. Specifically, the ZnSe has low thermal conductivity, poor thermal stability, large ohmic contact resistance, and a low damage threshold. Though tremendous strides have been made at Sony laboratories and elsewhere, operating lifetimes of these devices are still on the order of 100 hr, with informal reports alluding to improved longevities in more recent devices.

The injection level required for achieving the lasing condition or for rendering the active layer nonabsorbing at the wavelength of emission in GaAs is about 1.6×10^{18} cm^{-3}. At this injection level, the separation of the quasi-Fermi levels equals the bandgap of the active layer [76]. The lower the injection level, the lower the injection-current requirement for the lasing condition or transparency. It then follows that in semiconductors with a large density of states (joint density of states), the transparency condition is reached at higher injection levels. Using the hexagonal band parameters, the injection level for transparency in GaN-based lasers with GaN active layers is about 6×10^{18} cm^{-3}. Simply put, all else being equal, the transparency current for GaN lasers is about 3 to 4 times larger than in GaAs-based lasers.

Uenoyama and Suzuki [107] considered the gain in bulk GaN and 40-Å-thick quantum well lasers utilizing the wurtzitic and zinc-blende phases. With the joint density of states, polarization effects and matrix elements, and the occupation probabilities taken into account, they calculated the gain versus normalized injection current. The unique valence-band nature of wurtzitic GaN, where spin orbit and crystal field splitting lead to three bands, results in extremely large polarization effects and large gain, which is much larger than that for the zinc-blende case (Fig. 43). The zinc-blende structure, however, exhibits much-enhanced performance in the quantum well case due to its enhanced matrix elements. Nevertheless, a laser based on the zinc-blende polytype of GaN would still be much inferior to the wurtzitic one. This exercise leads one to conclude that when and if the quality of GaN-based heterostructures is improved to the point where scattering losses and nonradiative recombination losses are lowered and smooth facets are formed, one can expect excellent lasers.

As anxiously anticipated, Nakamura *et al.* [108] reported the observation of laser oscillations at room temperature in InGaN quantum wells utilizing GaN waveguides and AlGaN cladding layers. Although the initial laser reported utilized 26 periods of $In_{0.2}Ga_{0.8}N/In_{0.05}Ga_{0.95}N$ MQW structures consisting of 25-Å-thick $In_{0.2}Ga_{0.8}N$ well layers and 50-Å-thick $In_{0.05}Ga_{0.95}N$ barrier layers, more recent structures incorporate MQWs with as few as 7 periods. A 200-Å-thick *p*-type $Al_{0.2}Ga_{0.8}N$:Mg layer was employed to prevent dissociation of the InGaN layers during the growth of the subsequent GaN and AlGaN layers, which require much higher substrate temperatures. The 0.1-μm-thick layer of *n*-type $In_{0.1}Ga_{0.9}N$ was imbedded in the buffer layer to prevent cracking. Since it is difficult to cleave the *c*-face sapphire substrates, reactive ion etching (RIE) was employed to form the cavity facets. High-reflectivity facet coatings (60–70%) were used to reduce the threshold current. An Ni/Au contact was evaporated onto the entire area of the *p*-type GaN layer, and a Ti/Al contact onto the *n*-type

GaN layer. Akasaki et al. [109] also reported lasing action in a single $In_{0.1}Ga_{0.9}N$ at 376 nm, the shortest of any injection semiconductor laser, in a separate confinement structure with a threshold current of 2.9 kA/cm² at room temperature under a pulsed-operation duty cycle of 1% and a pulse width of 0.3 μs.

The lasers utilizing films grown on the c-plane of sapphire had etched cavities because sapphire does not cleave well. This is complicated further in that the GaN epilayers are rotated with respect to the underlying sapphire, making it impossible to align the cleavage plane, a plane of GaN and sapphire. It is for this reason that Nakamura et al. [110] explored lasers grown on the a-plane of sapphire. Even though the materials quality is barely comparable to that on the c-plane, improved cavity formation along the r-plane ($1\bar{1}02$) outweighed its reduced materials quality. Moreover, laser structures on (111) $MgAl_2O_4$, spinel substrates, which lead to wurtzitic GaN along the c-plane have also been explored for optical pumping [111] and injection laser experiments [112]. In this scheme, spinel cleaves along the (100) plane, inclined to the surface with cleavage following the r plane of GaN about where the epilayers is reached. Even though the facet quality in this scheme is the best among the aforementioned approaches, materials quality degradation is too severe to pull it ahead of the other approaches. It is, however, clear that a substrate with good cleavage characteristics and on which GaN can be grown without rotation is desperately needed. While SiC meets some of these criteria, poor surface quality of SiC (as far as the

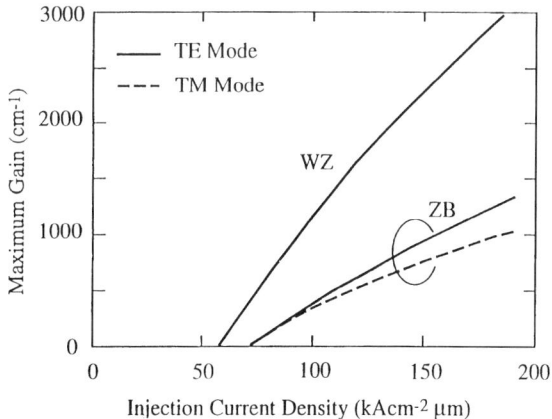

FIG. 43. The gain in wurtzitic and zinc-blende GaN as a function of injection current, which is normalized to 1 μm of the active pumped region. After Uenoyama and Suzuki [107], with permission.

growth is concerned) is making its implementation difficult, although optically pump-stimulated emission has been reportd [113]. Until such time as native substrates are available, ZnO meets the foregoing criteria. Again, this approach also has been hampered by the scarcity of high-quality substrates prepared by the sublimation technique.

Recently, utilizing a four-period $In_{0.15}Ga_{0.85}N(35\,\text{Å})/In_{0.02}Ga_{0.98}N(70\,\text{Å})$ quantum well structure, Nakamura et al. [114] achieved a room-temperature CW operation of about 27 hr. CW light and voltage versus current characteristics of an InGaN MQW laser having four $In_{0.15}Ga_{0.85}N$ quantum well gain media are shown in Fig. 44. The threshold current is slightly over 80 mA, and voltage across the device at threshold is about 5.5 V. Despite the enormous reduction in the voltage drop from about 30 V in earlier devices, this voltage must be reduced further to reduce the thermal load on the device. A good portion of the excess voltage, which stands at about 2 V, is attributed to large contact resistance on p-type GaN and must be reduced for longer CW operation. The operating lifetime of the most recent laser reported by Nakamura et al. [114] is depicted in Fig. 45, in which the evolution of increased current needed to maintain a CW power output of 1.5 mW is shown. The most recent devices were fabricated in structures grown on the c-face of sapphire, and facets were formed by etching. The success of these seemingly defective structures may lie in self-organized dots in the gain medium driven by strain.

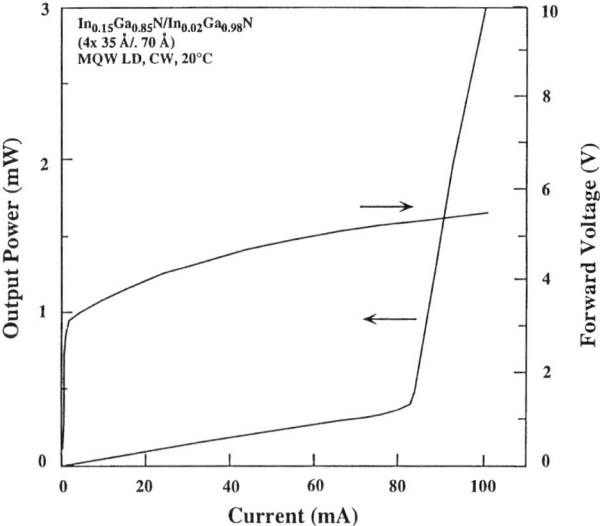

FIG. 44. Light and voltage versus injection-current characteristics of the laser structure reported by Nakamura et al. After Nakamura et al. [114], with permission.

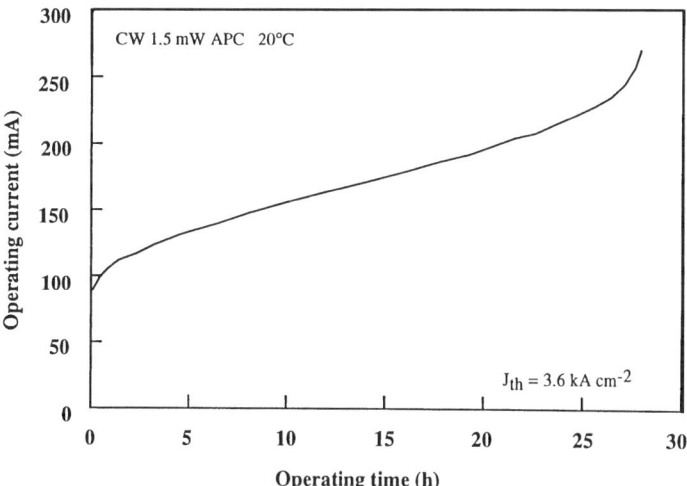

FIG. 45. Evolution of the injection current to maintain a constant output power of 1.5 mW under CW conditions at 20°C. After Nakamura et al. [114], with permission.

5. UV Detectors

Interest in GaN-based detectors stems in part from the fact that the earth's ozone layer absorbs radiation in the band of 180 to 280 nm. Thus, the background radiation–related quantum noise in detectors would be minimized in this wavelength range, and systems that are already pushed to their detection limits would be more accurate. Thus, the term *solar blind detector* was coined. This band should also prove useful in space-to-space systems as well, because of its reflection from the earth's surface; thus, the background radiation is reduced. At present, satellite monitoring of the earth's atmosphere relies on Si photodetectors that require bulky band filters to block the visible solar radiation background. With the use of UV-sensitive, visible blind GaN detectors, the need for these filters will be largely eliminated, simplifying the design of the spectroscopic monitoring equipment. Another area in which GaN-based photodetectors can find application is in the combustion monitoring of gases, wherein UV emission is a normal by-product.

Even though solar blind detection has been the driving force for GaN materials development in the United States, its progress has been dwarfed in comparison to that in the LED and laser front. This is in part because the problem is more formidable, in that defects that are not detrimental to the operation of LEDs at brightness levels accepted by the display society cause havoc in detectors. These defects reduce device speed, increase the dark current, and cause gradual band-edge absorption. Honeywell Inc. and the APA Optics groups [115, 116] have been quite active in this area, with Honeywell interested in application in furnace controllers as well. The APA Optics groups [115] reported its GaN detectors as having responsivity up to a wavelength of 365 nm and an estimated gain of about 6×10^3. The absorbing GaN layer was about 0.8 μm thick and was grown on sapphire following a 0.1-μm-thick AlN buffer layer. Measurements were conducted in the wavelength range of 200 to 365 nm, with a responsivity of 2000 A/W at 365 nm. The sharp absorption edge was characterized with a 3 orders of magnitude drop in the responsivity over just 10 nm, from 365 to 375 nm. The responsivity was observed to be a linear function of the incident optical power over 5 orders of magnitude, attesting to the quality of the GaN absorber.

The APA Optics group [115] described the fabrication and investigations of AlGaN photoresistors sensitive to radiation with $1 < 360$ nm. Layers had a sharp band-edge cut-off and a large responsivity in the UV that was nearly wavelength independent. Devices operating at several hertz performed quite well. The typical values of the dark resistance were about 0.5 MW for the devices with 1-mm^2 area. The dark current increased from 10^{-11} to 10^{-6} A as the temperature increased from 20 to 220°C. The photoresponse was linear in a photocurrent range of 10^{-7} to 10^{-3} A. Photoconductors have also been investigated by Goldenberg *et al.* [116]. In their particular investigation, it was found that the photoconductive response of GaN decreased with increasing frequency. Also, increased intensity lowered the photoconductive gain.

The photoconductive gain of a detector designed at Honeywell [116] decreased rapidly with increased UV intensity (Fig. 46). If reproducible, this indicates that the device has a built-in automatic gain control that allows the detector electronics to function over a narrow range of voltages, even though the UV intensity varies over many orders of magnitude. The device has a very low resistance at a very low light level, which allows it to be insignificantly affected by interference from stray ac electric or magnetic fields. The resistance is low enough to effectively short stray electric fields, but high enough to avoid voltages due to ground loop currents. Unless externally limited, saturation of the gain with input power is an indication that the gain mechanism is dominated by defect participation.

GaN-based detectors have been characterized by very high gains and low bandwidths due to the inordinate concentration of defects. Due to storage of carriers in defects for long periods, the detector response is generally very slow, limiting the exciting potential applications of nitride-based detectors. Efforts are underway to reduce the defect concentration and thus increase the detector speed as well as the responsivity [117]. As previously discussed,

N deficiency is known to lead to defects. As the ammonia flux is increased during growth by a factor of about 3, the responsivity of GaN photoconductive detectors increases by about a factor of 3, reaching the maximum of 3000 A/W. Congruent with the dependence of the responsivity, the response time also showed a similar enhancement, with increased ammonia flow during growth. For example, as the ammonia flow is increased by a factor 3, the response times, both tur*n*-on and tur*n*-off (tur*n*-off being longer due to charge storage) decrease by a factor of about 10.

Increasingly, more exotic detector structures are being explored for enhanced performance, which is made possible partially by the successes in obtaining *p*-type material. A popular detector structure is the pin type, the structure of which is shown in Fig. 47, which can also be a precursor device for eventual avalanche photodiodes when the semiconductor quality is sufficiently improved. Shown in Fig. 48 is the spectral response of a pin detector [118] built entirely in GaN, having a 1-μm *i* region and a 0.2-μm *p*-layer, before and after correction for the spectral dependence of the incident Hg light source and detector response. The detector is characterized with a sharp cut-off near the band edge, a flat response above the band edge, and what appears to be an excitonic absorption near the band edge. The responsivity is about 0.1 A/W. The response was flat up to the maximum tested frequency of 3.3 kHz.

The number of detectors that have been reported and/or experimented with is quite respectable. Among them are the intrinsic GaN and AlGaN

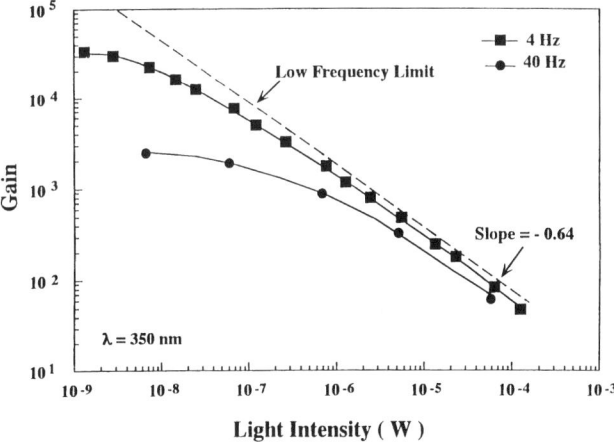

FIG. 46. Variation of photoconductive gain with intensity of light at 350 nm. After Goldenberg *et al.* [116] and courtesy of Honeywell.

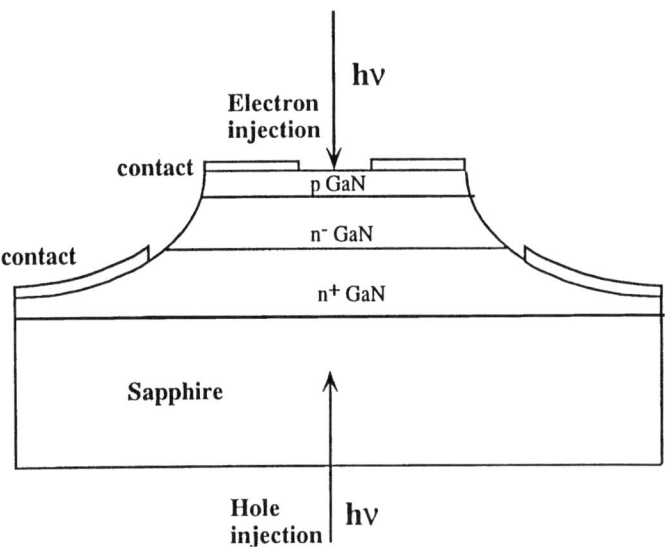

FIG. 47. A schematic representation of the GaN-based pin detector structure employed in the author's laboratory.

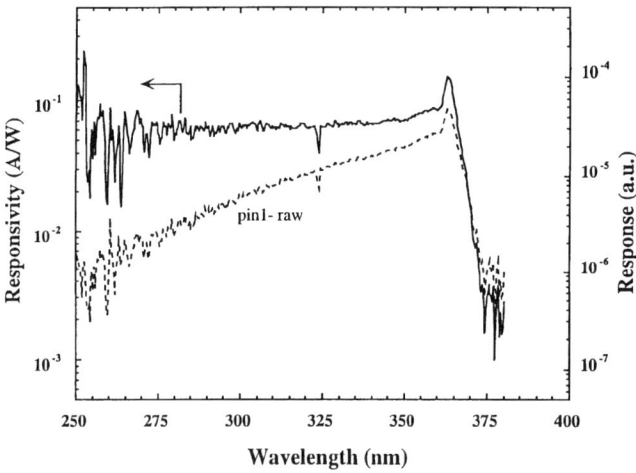

FIG. 48. The spectral response of a pin detector built entirely in GaN, having a 1-μm i region and a 0.2-μm p-layer, before and after correction for the spectral dependence of the incident Hg light source and detector response, from the author's laboratory.

photoconductive detectors, *n*-type GaN and AlGaN photoconductive detectors, *p*-type GaN photoconductive detectors, and GaN and GaN/AlGaN *pn* junction detectors. In addition, the detection response of GaN-based FETs has also been measured in an effort to exploit this class of devices for detectors. Outside of furnace controllers, the solar blind detectors must operate in the wavelength region where the ozone layer absorbs solar radiation. To the knowledge of this author, no such detector has been reported. In fact, attainment of such a detector with sufficiently low noise places more stringent requirements on the material quality and device fabrication procedures than the laser operation and requires a nitride technology that has not been developed. It will be a question of time and effort for this technology to be developed.

V. Conclusions

Owing to their large bandgaps, good thermal conductivities, type I heterostructures, and favorable transport properties, the wide-bandgap nitrides have attracted a great deal of attention in recent years. Nitrides produced the first successful commercial LEDs operative in blue and green, paving the way for full-color outdoor displays. Improved material quality has also led to CW room-temperature operation of lasers with operation lifetimes of about 27 hr. Nitrides produced the shortest wavelength semiconductor lasers demonstrated to this point (376 nm). Applications to high-density storage are awaiting commercialization of lasers, which is expected to occur in a few years. Compact disk industry requirements are that lasers have an average longevity of about 10,000 hr at 60°C of operation. On the electronics side, a large breakdown field combined with good thermal conductivity and transport properties have led to the exploration of these materials for high-power applications. Current gain cut-off frequencies of about 40 GHz and power gain cut-off frequencies of about 90 GHz have already been achieved. Incredibly, MODFETs with 2-μm gate lengths have demonstrated normalized CW output power levels of 1.5 W/mm at 4 GHz in chip form. Considering that the devices lacked the conventional overlay metallization, with the resultant large probe resistances, much improved results can only be imminent. Many more breakthroughs, pending the development of lattice-matched substrates, are expected in the near future and will have an unprecedented impact on the optoelectronics and electronics industries.

ACKNOWLEDGMENTS

The research projects in the author's laboratory are funded by grants from ONR, AFOSR, under the supervision of Mr. M. Yoder and Drs. G. L. Witt, Y. S. Park, and C. Wood. The author acknowledges insightful discussions and information exchange with many active researchers in the field of nitride semiconductors, and contributions by and discussions with his collaborators and members of his groups. The author benefited tremendously from collaborations with Profs. J. J. Song, H. Jiang, J. Lin, B. Gil, D. J. Smith, and K. T. Tsen, and Dr. J. Tsen. The autor is particularly thankful to Drs. A. Garscadden, D. C. Reynolds, C. W. Litton, D. C. Look, P. Hemenger, and W. Mitchel, and Mr. M. Roth, Ms. L. T. Kehias, and Mr. T. Jenkins of Wright Laboratory for their support and/or collaboration. He also acknowledges discussions with, and/or advance copies of papers and figures from, Profs. J. Bernholc, I. Akasaki, S. N. Mohammad, J. Schetzina, and E. Haller, and Drs. C. Van deWalle, J. Northrup, F. Ponce, S. Nakamura, C. Wetzel, W. Walukiewicz, T. Susk, and C. Kisilowski. Last but not least, the author would like to acknowledge valuable discussions with Profs. L. Eastman, U. Mishra and S. DenBaars, and Drs. M. A. Khan, A. Higgins, and C. Weitzel regarding FETs based on GaN.

REFERENCES

1. Morkoç, H., Strite, S., Gao, G. B., Lin, M. E., Sverdlov, B., and Burns, M. (1994). *J. Appl. Phys. Rev.* **76**, 1363.
2. Mohammad, S. N., Salvador, A., and Morkoç, H. (1995). *Proc. IEEE* **83**, 1306.
3. Mohammad, S. N., and Morkoç, H. "Progress and Prospects of Group III–V Nitride Semiconductors," *Progress in Quantum Electronics*, **20**, 321 (1996).
4. Nakamura, S., Senoh, M., Isawa, N., and Nagahama, S. (1995). *Jpn. J. Appl. Phys.* **34**, L797.
5. Leszczynski, M., Gregory, I., and Beckowski, M. (1993). *J. Crystal Growth* **126**, 601.
6. Chu, T. L. (1971). *J. Electrochem. Soc.* **118**, 1200.
7. Grimmeiss, H. G., and Monemar, B. (1970). *J. Appl. Phys.* **41**, 4054.
8. Amano, H., Hiramatsu, K., and Akasaki, I. (1988). *Jpn. J. Appl. Phys.* **27**, L1384.
9. Naniwae, K., Itoh, S., Amano, H., Itoh, K., Hiramatsu, K., and Akasaki, I. (1990). *J. Cryst. Growth* **99**, 381.
10. Gil, B., Briot, O., and Aulombard, R. L. (1995). *Phys. Rev. B* **52**, R17028; Gil, B., Hamdani, F., and Morkoç, H. (1996) *Phys. Rev. B*, **54**, R7678.
11. Yoshida, S., Misawa, S., and Gonda, S. (1983). *J. Vac. Sci. Technol. B* **1**, 250.
12. Yoshida, S., Misawa, S., and Gonda, S. (1983). *Appl. Phys. Lett.* **42**, 427.
13. Amano, H., Sawaki, N., Akasaki, I., and Toyoda, Y. (1986). *Appl. Phys. Lett.* **48**, 353; Koide, Y., Itoh, N., Itoh, K., Sawaki, N., and Akasaki, I. (1988). *Jpn. J. Appl. Phys.* **27**, 1156.
14. Akasaki, I., Amano, H., Koide, Y., Hiramatsu, K., and Sawaki, N. (1989). *J. Cryst. Growth* **98**, 209; Amano, H., Akasaki, I., Hiramatsu, K., Koide, N., and Sawaki, N. (1988). *Thin Solid Films* **163**, 415.

15. Kistenmacher, T. J., and Bryden, W. A. (1991). *Appl. Phys. Lett.* **59**, 1844.
16. Nakamura, S. (1991). *Jpn. J. Appl. Phys.* **30**, L1705.
17. Ponce, F. A., Major, J. S. Jr., Plano, W. E., and Welch, D. F. (1994). *Appl. Phys. Lett.* **65**, 2302.
18. Davis, R. F., Sitar, Z., Williams, B. E., Kong, H. S., Kim, H. J., Palmour, J. W., Edmond, J. A., Ryu, J., Glass, J. T., and Carter, C. H. Jr. (1988). *Mater. Sci. Eng. B* **1**, 77.
19. For a description of observed stacking faults, see, for example, Davis, R. F. (1991). *Proc. IEEE* **79**, 702 (1991).
20. Lei, T., Moustakas, T. D., Graham, R. J., He, Y., and Berkowitz, S. J. (1992). *J. Appl. Phys.* **71**, 4933.
21. Strite, S., Sariel, B., Smith, D. J., Chen, H., and Morkoç, H. (1993). Seventh International MBE Conference, Schwäbisch Gmünd, Germany (1992). *J. Cryst. Growth* **127**, 204–208.
22. Seifert, W., and Tempel, A. (1974). *Phys. Status Solidi A* **23**, K39.
23. Powell, R. C. (1992). Ph.D. Thesis, University of Illinois at Urbana–Champaign.
24. Lei, T., and Moustakas, T. D. (1992). *Mater. Res. Soc. Symp. Proc.* **242**, 433.
25. Maruska, H. P., and Tietjen, J. J. (1969). *Appl. Phys. Lett.* **15**, 327.
26. Neubauer, J., and Van de Walle, C. G. (1996). *Appl. Phys. Lett.* **69**(4), 22 July.
27. Perlin, P., Suzuki, T., Teisseyre, H., Leszczynski, M., Gregory, I., Jun, J., Porowski, S., Boguslawski, P., Bernholc, J., Chervin, J. C., Polian, A., and Moustakas, T. D. (1995). *Phys. Rev. Lett.* **75**, 296.
28. Wetzel, C., Walukiewicz, W., Haller, E. E., and Ager III, J. (1996). *Phys. Rev. B* **53**, 1322.
29. Jenkins, D. W., and Dow, J. D. (1989). *Phys. Rev.* 5 **39**, 3317; Jenkins, D. W., Dow, J. D., and Tsai, M.-H. (1992). *J. Appl. Phys.* **72**, 4130.
30. Neugebauer, J., and Van de Walle, C. G. In Diamond, SiC and Nitride Wide Bandgap Semiconductors (Eds. C. H. Carter, Jr., G. Gildenblatt, S. Nakamura, and R. J. Nemanich), *Materials Res. Soc. Symp. Proc.* Vol. 339, Pittsburgh, PA.
31. Bogusklawski, P., Briggs, E. L., and Bernholc, J. (1995). *Phys. Rev. B* **51**, 17255.
32. Kim, W., Botchkarev, A. E., Salvador, A., Popovici, G., Tang, H., and Morkoç, H. (1997). *J. Appl. Phys.*, **82**, 219.
33. Chung, B-C., and Gershenzon, M. (1992). *J. Appl. Phys.* **72**, 651.
34. Neugebauer, J., and Van de Walle, C. G. (1996). *Appl. Phys. Lett.* **68**, 1829.
35. Brandt, M. S., Johnson, N. M., Molnar, R. J., Singh, R., and Moustakas, T. D. (1994). *Appl. Phys. Lett.* **64**, 2264.
36. Götz, W., Johnson, N. M., Walker, J. Bour, D. P., Amano, H., and Akasaki, I. (1995). *Appl. Phys. Lett.* **67**, 2666.
37. Amano, H., Kito, M., Hiramatsu, K., and Akasaki, I. (1989). *Jpn. J. Appl. Phys.* **28**, L2112.
38. Nakamura, S., Mukai, T., Senoh, M., and Iwasa, N. (1992). *Jpn. J. Appl. Phys.* **31**, L139.
39. Nakamura, S., Iwasa, N., Senoh, M., and Mukai, T. (1992). *Jpn. J. Appl. Phys.* **31**, 1258.
40. Van Vechten, J. A., Zook, J. D., Hornig, R. D., and Goldenberg, B. (1992). *Jpn. J. Appl. Phys.* **31**, 3662.
41. Neugebauer, J., and van de Walle, C. G. (1995). *Phys. Rev. Lett.* **75**, 4452.
42. Okamoto, Y., Saito, M., and Oshiyama, A. (1996). *Jpn. J. Appl. Phys.* **35**, L807.
43. Kim, W., Salvador, A., Botchkarev, A. E., Aktas, Ö, Mohammad, S. N., and Morkoç, H. (1996). *Appl. Phys. Lett.* **69**(4), 559–561.
44. Chevallier, J., Clerjaud, B., and Pajot, B. (1991). In "Hydrogen in Semiconductors" (Eds. J. I. Pankove and N. M. Johnson), *Semiconductors and Semimetals*, Vol. 34, Academic Press, New York.
45. Yuan, S., Salagaj, T., Guray, A., Zawadzki, P., Chern, C. S., Kroll, W., Stall, R. A., Li, Y., Schurman, M., Hwang, C.-Y., Mayo, W. E., Lu, Y., Pearton, S. J., Krishnankutty, S., and Kolbas, R. M. (1995). *J. Electrochem. Soc.* **142**, L163.
46. Götz, W., Johnson, N. M., and Bour, D. P. (1996). *Appl. Phys. Lett.* **68**, 3470.

47. Han, J., Ringle, M. D., Fan, Y., Gunshor, R. L., and Nurmikko, A. V. (1994). *Appl. Phys. Lett.* **65**, 3230.
48. Li, J. Z., Lin, J. Y., Jiang, H. X., Salvador, A., Botchkarev, A., and Morkoç, H. (1996). *Appl. Phys. Lett.* **69**, 1474.
49. Götz, W., Johnson, N. M., and Street, R. A. (1995). *Appl. Phys. Lett.* **66**, 1340.
50. Hacke, P., Detchprohm, T., Hiramatsu, K., and Sasaki, N. (1994). *J. Appl. Phys.* **76**, 304.
51. Götz, W., Johnson, N. M., Amano, H., and Akasaki, I. (1994). *Appl. Phys. Lett.* **65**, 463.
52. Lee, W. I., Huang, T. C., Guo, J. D., and Feng, M. S. (1995). *Appl. Phys. Lett.* **67**, 1721.
53. Gaskill, D. K., Wickenden, A. E., Doverspike, K., Tadayon, B., and Rowland, L. B. (1995). *J. Electronic Mater.* **24**, 1525.
54. Molnar, R. J., Lei, T., and Moustakas, T. D. (1993). *Appl. Phys. Lett.* **62**, 72.
55. Seifert, W., Franzheld, R., Butter, E., Sobotta, H., and Riede, V. (1983). *Cryst. Res. Technol.* **18**, 383.
56. Chung, B.-C., and Gershenzon, M. (1992). *J. Appl. Phys.* **72**, 651.
57. Wang, C., and Davis, R. F. (1993). *Appl. Phys. Lett.* **63**, 990.
58. Look, D. C., Reynolds, D. C., Kim, W., Aktas, O., Botchkarev, A., Salvador, A., and Morkoç, H. (1996). *J. Appl. Phys.* **80**, 2960.
59. Mott, N. F., and Twose, W. D. (1961). *Adv. Phys.* **10**, 107.
60. Kennedy, T. A., Glaser, E. R., Freitas, J. A., Carlos, W. E., Khan, M. A., and Wickenden, D. K. (1995). *J. Electron. Mater.* **24**, 219; Glaser, E. R., Kennedy, T. A., Doverspike, K., Rowland, L. B., Gaskill, D. K., Freitas, J. A., Khan, M. A., Olson, D. T., Kuznia, J. N., and Wickenden, D. K. (1995). *Phys. Rev. B* **51**, 13326.
61. Shan, W., Song, J. J., Goldenberg, B., Wetzel, C., and Haller, E. E., private communication.
62. Reynolds, D. C., Look, D. C., Jogai, B., and Morkoç, H. (1997). *Solid State Com* **101**, 643.
63. Hofmann, D. M., Kovalev, D., Steude, G., Meyer, B. K., Hoffmann, A., Eckey, L., Heitz, R., Detchprom, T., Amano, H., and Akasaki, I. (1995). *Phys. Rev. B* **52**, 16702.
64. Smith, M., Chen, G. D., Lin, J. Y., Jiang, H. X., Salvador, A., Kim, W., Aktas, Ö., Botchkarev, A., Morkoç, H., and Goldenberg, B. (1996) *Appl. Phys. Lett.*, **68**, 1883.
65. Salvador, A., Kim, W., Aktas, Ö., Botchkarev, A., Fan, Z., and Morkoç, H. (1995). *Appl. Phys. Lett.* **67**, 3322.
66. Stengel, F., Mohammad, S. N., and Morkoç, H. (1996). *J. Appl. Phys.* **80**(5), 3031–3042.
67. Masselink, W. T. (1986). Ph.D. Thesis, University of Illinois at Urbana–Champaign.
68. Ponce, F., Masselink, W. T., and Morkoç, H. (1985). "Quasi-Fermi level bending in MODFETs and its effect on FET transfer characteristics," *IEEE Trans. Electron Devices* **ED-32**, 1017.
69. Strite, S., and Morkoç, H. (1992). "GaN, AlN, and InN: A review," *J. Vac. Sci. Technol. B* **10**, 1237.
70. Weisbuch, C., and Vinter, B. (1991). "Quantum Semiconductor Structures," Academic Press, New York.
71. Aktas, Ö., Kim, W., Fan, Z., Botchkarev, A., Salvador, A., Mohammad, S. N., Sverdlov, B., and Morkoç, H. (1995). *Electron. Lett.* **31**, 1389.
72. Aktas, Ö., Kim, W., Fan, Z., Stengel, F., Botchkarev, A., Salvador, A., Mohammad, S. N., and Morkoç, H. (1995). *Tech Dig. IEDM*, Washington, DC.
73. Kim, Y. M., and Roblin, P. (1986). *IEEE Trans. Electron Devices* **ED-**33, 1644; Roblin, P., Rice, L., Bibky, S. B., and Morkoç, H. (1988). *IEEE Trans. Electron Devices* **ED-35**, 1207.
74. Liechti, C. A. (1976). *IEEE Trans. Microwave Theory Technol.* **MTT-24**, 279.
75. Drummond, T. J., Masselink, W. T., and Morkoç, H. (1986). *Proc. IEEE* **74**, 773.
76. Morkoç, H., Sverdlov, B., and Gao, G. B. (1993). *IEEE* **81**, 492.
77. Khan, M. A., Shur, M. S., Chen, Q. C., and Kuznia, J. N. (1994). *Electron. Lett.* **30**, 2175.

78. Fischer, R., Drummond, T. J., Klem, J., Kopp, W., Henderson, T., Perrachione, D., and Morkoç, H. (1984). *IEEE Trans. Electron. Dev.* **ED-31**, 1028.
79. Aktas, Ö., Fan, Z. F., Botchkarev, A., Mohammad, S. N., Roth, M., Jenkins, T., Kehias, L., and Morkoç, H. (1997). "Microwave Performance of AlGaN/GaN Inverted MODFETs," *IEEE EDL*, **18**, 293.
80. Lee, K., Shur, M. S., Klem, J., Drummond, T. J., and Morkoç, H. (1984). *Jpn. J. Appl. Phys. Lett.* **23**, L230; Lee, K., Shur, M. S., Drummond, T. J., and Morkoç, H. (1984). *IEEE Trans. Electron. Dev.* **ED-31**, 29.
81. Witkowski, L. C., Drummond, T. J., Barnett, S. A., Morkoç, H., Cho, A. Y., and Greene, J. E. (1981). *Electron. Lett* **17**, 126.
82. Binari, S. C., Rowland, L. B., Kelner, G., Kruppa, W., Dietrich, H. B., Doverspike, K., and Gaskill, D. K. (1994). 21st Int. Symp. Compound Semicond. Proc.
83. Khan, M. A., Shur, M. S., Kuznia, J. N., Cheng, Q. C., Burm, J., and Schaff, W. (1995). *Appl. Phys. Lett.* **66**, 1083.
84. Fan, Z., Mohammad, S. N., Aktas, Ö., Botchkarev, A., Salvador, A., and Morkoç, H. (1996). *Appl. Phys. Lett.* **69**, 1229.
85. Wu, Y. F., Keller, B. P., Keller, S., Kapolnek, D., Kozodoy, P., Denbaars, S. P., and Mishra, U. K. (1996). *Appl. Phys. Lett.* **69**, 1438; Wu, Y. F., Keller, B. P., Keller, S., Kapolnek, D., Denbaars, S. P., and Mishra, U. K. (1996). *IEEE Elect. Dev. Lett.* **17**, 455.
86. Khan, M. A., Chen, Q., Shur, M. S., Dermott, B. T., Higgins, J. A., Burm, J., Schaff, W., and Eastman, L. F. (1996). *Elect. Lett.* **32**, 357–358; Khan, M. A., Chen, Q., Wang, J. W., Shur, M. S., Dermott, B. T., and Higgins, J. A. (1996). *IEEE Elect. Dev. Lett.* **17**, 325.
87. Eastman, L. F., private communication; Burm, A. U., Schaff, W. J., Eastman, L. F., Amano, H., and Akasaki, I. (1996). *Appl. Phys. Lett.* **68**, 2849.
88. Mohammad, S. N., Fan, Z. F., Salvador, A., Aktas, Ö., Botchkarev, A. E., Kim, W., and Morkoç, H. (1996). *Appl. Phys. Lett.* **69**, 1420.
89. Aktas, Ö., Fan, Z. F., Mohammad, S. N., Botchkarev, A. E., and Morkoç, H. (1996). *Appl. Phys. Lett.* **69**, 1420.
90. Crawford, M. G. (1992). *IEEE Circuits Devices Mag.* **8**, 24.
91. Bhargava, R. M. (1992). *Optoelectronics—Devices and Technol.* **7**, 19–47.
92. Pankove, J. I., Miller, E. I., and Berkeyheiser, J. E. (1971). *RCA Rev.* **32**, 383.
93. Nakamura, S., Mukai, T., and Senoh, M. (1991). *Jpn. J. Appl. Phys.* **30**, L1998.
94. Amano, H., Kito, M., Hiramatsu, K., and Akasaki, I. (1989). *Jpn. J. Appl. Phys.* **28**, L2112; Akasaki, I., and Amano, H. (1994). *J. Electrochem. Soc.* **141**, 2266–2271.
95. Amano, H., Hiramatsu, K., and Akasaki, I. (1988). *Jpn. J. Appl. Phys.* **27**, L1384.
96. Nakamura, S., Senoh, M., Isawa, N., and Nagahama, S. (1995). *Jpn. J. Appl. Phys.* **34**, L797.
97. Osinski, M., Zeller, J., Chiu, P. C., and Phillips, B. S. (1996). *Appl. Phys. Lett.* **59**(7), 898.
98. Moustakas, T. D., Vaudo, R. P., Korakakis, D., Misra, M., Sampath, A., Goepfert, I. D. (1996). *Inst. Phys. Conf. Ser.* **142**, 833.
99. Schetzina, J., private communication.
100. Salvador, A., Botchkarev, A. E., Fan, Z.-F., Kim, W., Aktas, Ö., and Morkoç, H. Topical Workshop for Nitrides '96, March 1996, St. Louis.
101. Reichert, H. and Schetzina, J., private communications.
102. Amano, H., Kito, M., Hiramatsu, K., and Akasaki, I. (1989). *Jpn. J. Appl. Phys.* **28**, L2112; Akasaki, I., and Amano, H. (1994). *J. Electrochem. Soc.* **141**, 2266.
103. See for example, Asthana, P. (1994). *IEEE Spectrum* **31**, 60.
104. Bell, A. (1996). Scientific American, July, 42–46.
105. Haase, M. A., Qui, J., Depuydt, J. M., and Cheng, H. (1991). *Appl. Phys. Lett.* **59**, 1272.

106. Salovatke, A., Jeon, H., Hoviven, M., Kelkar, P., Nurmikko, A. V., Grillo, D. C., He, L., Han, J., Fan, Y., Ringle, M., and Gunshor, R. L. (1993). *Electron. Lett.* **29**, 2041.
107. Uenoyama, T., and Suzuki, M. Proc. of the Internation Symposium on Blue Light Emitting Diodes and Lasers, March 4–7, 1996, Chiba, Japan; and *Inst. Phys. Conf. Ser.* **142**, 915.
108. Nakamura, S., Senoh, M., Nagahama, S., Iwasa, N., Yamada, T., Matsushita, T., Kiyoku, H., and Sugimoto, Y. (1996). *Jpn. J. Appl. Phys.* **35**, L74.
109. Akasaki, I., Sota, S., Sakai, H., Tanaka, T., and Amano, H. (1996). *Electron. Lett.* **32**, 1105.
110. Nakamura, S., Senoh, M., Nagahama, S., Iwasa, N., Yamada, T., Matsushita, T., Kiyoku, H., and Sugimoto, Y. (1996). *Jpn. J. Appl. Phys.* **35**, L217.
111. Kuramata, A., Horino, K., Domen, K., Shinohora, K., and Tanahashi, T. (1995). *Appl. Phys. Lett.* **67**, 2521.
112. Nakamura, S., Senoh, M., Nagahama, S., Iwasa, N., Yamada, T., Matsushita, T., Kiyoku, H., and Sugimoto, Y. (1996). *Appl. Phys. Lett.* **68**, 2105.
113. Zubrilov, A. S., Nikolaev, V. I., Tsevetkov, D. V., Dmitirev, V. A., Irvine, K. G., Edmond, J. A., and Carter, C. H. (1995). *Appl. Phys. Lett.* **67**, 533; Song, J.-J., private communication.
114. Nakamura, S., Senoh, M., Nagahama, S., Yamada, T., Matsushita, T., Sumigoto, Y., and Kiyoku, H. *Proc. IEEE LEOS meeting*, November 18–21, 1996, Boston.
115. Khan, M. A., Kuznia, J. N., Olson, D. T., Van Hove, J. M., Blasingame, M., and Reitz, L. F. (1992). *Appl. Phys. Lett.* **60**, 2917.
116. Goldenberg, B., Zook, J. D., and Ulmer, R. J. (1995). "Fabrication and performance of GaN detectors," Topical Workshop on III–V Nitrides Proc., Nagoya, Japan; and MRS Fall Meeting, Nove. 1995, Boston.
117. Chang, Z. C., Mott, D. B., and Shu, P. K. (1996). Device Research Conference, Santa Barbara, CA, June 24–26.
118. Xu, G., Salvador, A., Kim, W., Fan, Z., Lu, C., Tang, H., Morkoç, H., Smith, G., Estes, M., Goldenberg, B., Yong, W., and Krishnonkutly, S. (1997). *Appl. Phys. Lett.* **71**, 2154.

Index

A

Acheson process, 23
AIM-Spice, 186–188, 187f
AlGaN/GaN, electron velocity-electric field characteristics, 243, 244f
AlN. *See* Aluminum nitride
Aluminum doping, 45–46, 46f, 203–204, 204f
Aluminum nitride (AlN)
  buffer layer thickness, 310
  properties, 166
  SiC chemical vapor deposition on, 49–50
Annealing
  effects
    on Schottky barrier height, 101, 102f
    on Schottky contacts to 6H-SiC, 97, 100–101, 102f, 103–108, 104f–107f
  post deposition, of Al Schottky contacts, 105–106, 107f
ATLAS, 185
Avalanche frequency, 274
Avalanche process, 273
AX centers, 329, 329f

B

Baliga's figure of merit (BFOM), 167
Bardeen limit, 84
Barrier height
  for metal-semiconductor contacts, 81–84
  Schottky. *See* Schottky barrier height
  Schottky contacts to 6H-SiC, 97–100
Base region transit time, for bipolar transistors, 269
BFOM (Baliga's figure of merit), 167
Bipolar junction transistors (BJTs)
  base resistance values, 270
  circuit model, 271f, 271–272
  collector region design, 269, 271
  design considerations, 269–270
  dimensions, 269
  RF active, 267–272, 268f, 271f, 272f
  RF performance, 272, 272f
  structure, 268, 268f
  vs. FETs, 163
BLAZE, 185
Boron-implanted diodes
  4H-SiC, 121, 122f, 123
  6H-SiC, 142, 143f

C

CCDs (charge-coupled devices), 173–174
CFLPE (container-free liquid-phase epitaxy), 53f, 53–55, 54f
CFOM (combined dimensionless figure of merit), 167, 168
Charge-coupled devices (CCDs), 173–174
Chemical vapor deposition (CVD)
  commercial reactors, 59
  3C-SiC
    on 6H-SiC and 15R-SiC substrates, 49
    on silicon substrates, 48
  homoepitaxial, 36
  rapid thermal, 138
  SiC, 63
    on AlN, 49–50
    bulk crystal growth, 34
  SiC epitaxial growth, 36–50, 39f–42f, 45f–47f
    cool-down phase, 41–42
    equipment, 36–37

Chemical vapor deposition, SiC epitaxial growth, (*continued*)
  growth rate, 38–39, 39 *f*
  heteroepitaxy of SiC, 48–50
  homoepitaxy of 3C-SiC, 47–48
  homoepitaxy of α-SiC, 39–47, 40 *f*–42 *f*, 45 *f*–47 *f*
  precursors, 37–38
  reaction chemistry, 37–38
  vertical low-pressure, 203
CMOS transistors (complementary MOS transistors), 168
Combined dimensionless figure of merit (CFOM), 167, 168
Container-free liquid-phase epitaxy (CFLPE), 53 *f*, 53–55, 54 *f*
Crucible
  design, 25 *f*, 26
  for sublimation sandwich method, 56, 57 *f*
Current-density equation, 247–248
CVD. *See* Chemical vapor deposition

D

Deep-level transient spectroscopy (DLTS)
  defect analysis, 330, 331 *f*–333 *f*, 333
  of α-SiC epitaxial layers, 44–45
Defects
  analysis, by deep-level transient spectroscopy, 330, 331 *f*–333 *f*, 333
  intrinsic. *See* Point defects
  Mg vis-à-vis, p-type doping by, 320–330, 322 *f*–324 *f*, 326 *f*–329 *f*
  micropipe. *See* Micropipes
  native. *See* Point defects
  point. *See* Point defects
  from p-type doping in GaN, 339–342, 340 *f*, 341 *f*
  in SiC CVD layers, 42–44
  from SiC physical vapor growth, 201–202
  in sublimation-grown SiC crystals, 29–31, 30 *f*
  surface, in SiC CVD layers, 42
Diamond, properties, 166, 242
Dielectric constant, 242
Digital integrated circuits, SiC, 180 *f*, 180–182, 181 *f*
Digital versatile disk (DVD), 381

DIMOS (double-implant MOS), 172–173, 173 *f*
Diodes
  boron-implanted
    4H-SiC, 121, 122 *f*, 123
    6H-SiC, 142, 143 *f*
  IMPATT. *See* IMPATT diodes
  light-emitting. *See* Light-emitting diodes
  p-n junction. *See* p-n junction diodes
  Schottky. *See* Schottky barrier diodes
  UV photodiodes. *See* Photodiodes, UV
Dipole surface charge barrier (DSCB), 82–83
DLTS. *See* Deep-level transient spectroscopy
DMOS (double-diffused MOS), Si, 172
Doping
  aluminum, 45–46, 46 *f*, 203–204, 204 *f*
  nitrogen, 203–204, 204 *f*
  p-type
    in GaN, defects from, 339–342, 340 *f*, 341 *f*
    by Mg vis-à-vis defects, 320–330, 322 *f*–324 *f*, 326 *f*–329 *f*
    SiC, 60, 239–240
    during crystal growth, 60
    diffusion of impurities in, 61, 61 *f*
    impurities for, 60
    by ion implantation, 61–62
  in situ, 131, 133
Double-implant MOS (DIMOS), 172–173, 173 *f*
DSCB (dipole surface charge barrier), 82–83
DVD (digital versatile disk), 381

E

Edge termination
  B-implanted, 4H-SiC diodes, 121, 122 *f*, 123
  in 6H-SiC Schottky diodes, 113–114, 115 *f*, 116 *f*
Einstein relation, 247
Electrical static discharge survivability, of nitride LEDs, 305
Electron-beam lithography, 216 *f*, 217
Electrotechnical Laboratory, surface preparation technique, 86–87

INDEX 397

Epitaxy
  liquid-phase. *See* Liquid-phase epitaxy
  molecular beam. *See* Molecular beam epitaxy
  step-controlled, 36
  sublimation, 56–58, 57f, 58f
Equipment
  for chemical vapor deposition, 36–37
  for sublimation growth, 25f, 26

F

Faraday's law, 247, 248
Field-effect transistors (FETs)
  GaN-based, 308, 342. *See also* MODFETs, GaN-based
    $Al_xGa_{1-x}N$ layer, 345
    band diagrams, 351–352, 353f
    default parameters, 348
    2DEG concentration, 343–344, 348–351, 350f, 351f, 352, 354f–355f, 355–361
    drain current, 346–347, 360–363, 362f, 363f
    high-speed aspects, 364–365
    theoretical method, 342–343
    transconductance, 347–348, 357, 358f–359f, 360
  junction. *See* Junction field-effect transistors
  junction-gate, 240
  metal semiconductor. *See* MESFETs
  metal-oxide semiconductor. *See* MOSFETs
  modulation-doped. *See* MODFETs
  SiC
    microwave, 177–180, 178f
    principle of operation, 162–168, 163f–165f
    structure, 162, 163f
    threshold voltage, 163
Floating metal field rings (FMRs), 114

G

Gallium arsenide (GaAs)
  bandgap, 242
  electron velocity-electric field characteristics, 243, 244f
  properties, 242
    electronic, 196–198
    physical, 196–198
    vs. other semiconductors, 166
Gallium nitride (GaN)
  band-gap emission, 373–374
  conduction mechanism, 334–335
  crystal structure
    lattice mismatch and, 309–310
    strain effect on, 309–310
  defect analysis, by deep-level transient spectroscopy, 330, 331f–333f, 333
  devices, 307–309. *See also specific devices*
  electron velocity-electric field characteristics, 243, 244f
  GaN:SiC blue LEDs, 302f–304f, 302–305
  point defects, 314f–317f, 314–316
  properties, 242
    electronic, 196–198
    physical, 196–198
    vs. other semiconductors, 166
  p-type
    doping, defects from, 339–342, 340f, 341f
    persistent photoconductivity and, 327f–329f, 327–330,
  stacking fault defects, 311–313, 312f
  unintentionally doped, defect-aided current in, 333–335
Gated-off thyristors (GTOs), 229, 229f, 230f
Graphite boat, liquid-phase epitaxy growth from, 51–52
Growth pits, in SiC CVD layers, 42–43
GTOs (gated-off thyristors), 229, 229f, 230f

H

Hall effect, 298–299, 299f
HBT, 273, 280
High-power RF systems, SiC, 204–205
Huang clean procedure, 86
Hydrofluoric acid, for oxide removal, 86

I

ICTS (isothermal capacitance transient spectroscopy), 330, 331

IGBT (insulated-gate bipolar transistor), 170, 171f, 183
IMPATT diodes
　advantages, 273
　applications, 272–273
　double-drift structure, 273, 273f, 275–276
　electrical properties, 273–275, 274f, 275f
　fabrication, 273–274
　optimal operation, 276
　simulations, 276–279, 278f
　thermal limitations, 273
Impurities, for SiC doping, 60, 61, 61f
Indium phosphide (InP), properties, 242
Insulated-gate bipolar transistor (IGBT), 170, 171f, 183
Integrated circuits, 188, 195–196
Intrinsic defects. *See* Point defects
Ion implantation
　mesa structure junction diodes, 136, 136f, 137f
　p-n junction diodes, 133, 136
　for SiC doping, 61–62
　for SiC MOSFET fabrication, 169–170
　surface coping concentration, for 3C-SiC ohmic contacts, 126, 128 f
Isothermal capacitance transient spectroscopy (ICTS), 330, 331

J

JFETs. *See* Junction field-effect transistors
JFM (Johnson's figure of merit), 165, 167
JGFETs (junction-gate field-effect transistors), 240
Johnson's figure of merit (JFM), 165, 167
Junction field-effect transistors (JFETs), 170
　principle of operation, 162–168, 163f–165f
　SiC, 175f–177f, 175–177
Junction-gate field-effect transistors (JGFETs), 240

K

Keye's figure of merit (KFM), 167
Kyoto clean procedure, 86

L

Lasers, nitride-based, 380–384, 382f–384f

LEDs. *See* Light-emitting diodes
Lely process, 23–24, 25, 239
　modified or seeded. *See* Sublimation growth
Light-emitting diodes (LEDs), 283–286, 305, 308
　asymmetric double heterostructures, 285
　double heterostructures, 374
　GaN-based
　　blue, green and yellow, 373–378, 375f–378f
　　critical thickness, 374
　　by MBE, 379–380, 379f
　　p-n junction, 374
　GaN:SiC blue, 302f–304f, 302–305
　6H-SiC blue, development of, 284
　InGaN SQW, 285
　nitride, electrical static discharge survivability of, 305
　sapphire substrate, 285
　SiC blue
　　epitaxy, 286–287
　　fabrication, 286–287
　　performance, 287–288, 288f
　　SEM micrograph, 286f, 287
　SiC green
　　epitaxy, 288–289
　　ion implantation, 288–289
　　performance, 289f, 289–290
Liquid-phase epitaxy (LPE)
　container-free, 53f, 54f, 53–55
　of 3C-SiC on 6H-SiC, 55–56
　doping, 51
　endpoint, 50–51
　low-temperature, 55
　micropipe defects, 50f, 51, 52
　p-n junction diode fabrication, 144
　from Si melt, 51–52
　by vertical dipping, 52f, 52–53
Low-temperature liquid-phase epitaxy (LTLPE), 55
LPE. *See* Liquid-phase epitaxy
LTLPE (low-temperature liquid-phase epitaxy), 55

M

Magnesium vis-à-vis defects, p-type doping by, 320–330, 322f–324f, 326f–329f

MBE. *See* Molecular beam epitaxy
MDD (micropipe defect density), 199–200, 201*f*
MESFETs, 231, 279
  3C-SiC, 240
  design
    considerations for, 205–207, 206*f*
    parameters for, 259
  drain-gate breakdown, 206
  fabrication, 207, 208*f*, 209*f*, 243
  for high-power RF systems, 205
  4H-SiC
    dc and RF performance, 257–258, 258*f*, 259*f*, 261, 262, 263*f*, 263*f*
    fabrication results, 211–212, 212*f*
  6H-SiC, 240–241
    dc and RF performance, 254*f*–256*f*, 254–257, 259, 260*f*–261*f*, 261, 263–264
    fabrication results, 209, 210*f*, 211
  ionization energy, 207
  microwave
    performance of, 206–207
    power densities for, 188
  principle of operation, 162–168, 163*f*–165*f*
  SiC, 174*f*, 174–175, 176
  structure, 253*f*, 253–254, 254*f*
Metal contacts
  to 3C-SiC, 125
    ohmic, 125–126, 127*f*, 128*f*
    Schottky type, 126, 128–130
  to 4H-SiC
    ohmic, 118, 119
    Schottky-type, 118, 120*f*–124*f*, 120–125
  to 6H-SiC, ohmic, 87–96, 90*f*–95*f*
Metal-insulator-semiconductor (MIS), 87, 90*f*
Metal-semiconductor contacts
  electrical properties, 81–86, 82*f*, 83*f*
  for high-power applications, 84–86
  ideal, Schottky model for, 81–83, 82*f*, 83*f*
Metal-SiC systems
  electrical properties, 81–86, 82*f*, 83*f*
  barrier height, 81–82
  for high-power applications, 84–86
Micropipe defect density (MDD), 199–200, 201*f*
Micropipes

bulk growth techniques and, 199
  characteristics, 29–30
  in liquid-phase epitaxy, 50*f*, 51, 52
Microwave devices, SiC, 237–238, 279–280
  contact properties, 241–246
  high-power, 204–205
  models, 247–250, 249*f*
  RF active, 252*f*, 252–253, 253*f*
    bipolar transistors, 267–272, 268*f*, 271*f*, 272*f*
    IMPATT diodes, 272–279, 273*f*–275*f*, 278*f*
    MESFETs, 252*f*–263*f*, 253–259, 261, 263–264
    SITs, 264*f*–266*f*, 264–267
  semiconductor material properties, 241–246
  temperature effects, 250–252, 251*f*
MIS (metal-insulator-semiconductor), 87, 90*f*
MODFETs
  band diagrams, 351–352, 353*f*
  GaN-based, performance of, 365–373, 366*f*–369*f*, 371*f*, 372*f*, 373
  thermal methods, 342–343
  transconductance, 357, 358*f*–359*f*, 360
Molecular beam epitaxy (MBE)
  LED fabrication, 378–380, 379*f*
  SiC, 58–60, 310
MOSFETs
  buried-channel, 169*f*, 170
  3C-SiC, 240
  design, 169*f*, 169–170
  n-type channel or NMOS, 168
  operation principles, 162–168, 163*f*–165*f*
  p-type channel or PMOS, 168
  SiC, 168–174, 169*f*, 171*f*, 172*f*, 173*f*, 176
    double-implant, 172–173, 173*f*
    limitations of, 170–171
  UMOSFETs, 170–171, 171*f*, 172*f*

N

Native defects. *See* Point defects
Nickel contacts, 100–101, 102*f*
Nitride-based heterostructures, 308–309. *See also* Gallium nitride
  crystal structure
    lattice mismatch and, 309–310

400 INDEX

Nitride-based heterostructures, crystal structure, (*continued*)
  strain effect on, 309–310
  defects
    deep-level transient spectroscopy, 330, 331f–333f, 333
    Mg vis-à-vis, p-type doping by, 320–330, 322f–324f, 326f–329f
    optical manifestations of, 335–342, 336f–341f
    point. *See* Point defects
    stacking fault, 311–313, 312f
  dislocations, 310–311
  in field-effect transistors. *See* Field-effect transistors, GaN-based
Nitrides. *See also* Nitride-based heterostructures; *specific nitrides*
  III, on 6H-SiC
    blue LEDs, 302f–304f, 302–305
    characterization, 298
    electrical static discharge survivability, 305
    epitaxial growth, 297–298
    fabrication, 297–298
    Hall effect and, 298–299, 299f
    photoluminescence, 300, 300f, 301f, 302
  III-V, 308
  wide-bandgap, in lasers, 380–384, 382f–384f
Nitrogen doping, 45, 45f, 203–204, 204f
NMOS transistors, 168, 176, 180f, 180–182, 181f
Nonvolatile random-access memories (NVRAMs), 183, 184 f

O

Ohmic contacts, 246
  Al/p-SiC, 96
  Al-Ti, 90–91, 91f, 96
  Al/Ti/Al, contact resistivity of, 94f, 94–95
  3C-SiC, 125–126, 127f, 128f
  to 4H-SiC, 118, 119
  to 6H-SiC, 87–96, 90f–95f
  silicide, 87, 90
  Ti, 95
  W/Au-based, 90
Open-circuit voltage, UV photodiode, 293–294, 294f

P

Photodiodes, UV
  applications, 296–297
  electrical characteristics
    dark current, 292, 293f
    power generation, 293–294, 294f
  fabrication, 290–292, 291f
  optical responsivity
    epilayer thickness effects, 294–296, 295f
    operating temperature effect, 296, 297f
  SEM micrograph, 291f, 292
  structure, 290–291, 291f
Photoluminescence, of group III nitrides on 6H-SiC, 300, 300f, 301f, 302
Physical vapor growth, SiC
  defects, 201–202
  techniques, 22–23
    Acheson process, 23
    Lely process, 23–24, 24f, 25f
    modified Lely process. *See* Sublimation growth
PMOS transistors, 168, 176
p-n junction
  3C-SiC, 133–138
  fabrication, double charge method for, 139
  grown in SiC
    solution growth technique for, 131
    traveling solvent method for, 131
  6H-SiC, 138–140, 141f–146f, 142, 144–146
p-n junction diodes
  fabrication
    high-temperature implantation, 142, 143f
    liquid-phase epitaxy, 144
  high-voltage
    on 3C-SiC, 134–135
    on 4H-SiC, 134–135
    on 6H-SiC, 134–135
  4H-SiC, 147–149, 148f
  6H-SiC, 138–140, 141f–146f, 142, 144–146
  mesa structure
    3C-SiC, 136f, 137f, 136–138
    6H-SiC, 139–140, 142
  planar, on 6H-SiC, 145–146, 146f
p-n junction rectifiers, on-resistance, 85

INDEX 401

Point defects
  antisites, 313
  in GaN, 314f–317f, 314–316
  oxygen donors and, 319–320
  secondary ion mass spectroscopy, 316–317, 318f–319f
  self-interstitials, 313
  vacancies, 313–314, 314f
Poisson's equation, 247, 248
Polytype inclusions, in SiC CVD layers, 43–44
Postgrowth diffusion processes, 133
Power rectifiers
  on-resistance, 85–86
  power losses in, 85–86
Power switching devices, SiC, 226–229, 228f–230f, 231

R

Rapid thermal annealing (RTA), 207, 321–320
Reactive ion etching (RIE), 207, 382
Rectifiers, 85–86, 131, 133
Resistive Schottky barrier field plates (RESPs), 114
RIE (reactive ion etching), 207, 382
RTA (rapid thermal annealing), 207, 321–320

S

Sapphire substrate, for LEDs, 285
S-band SITs, 223–224, 224f
SBDs. *See* Schottky barrier diodes
SBH. *See* Schottky barrier height
Schottky barrier diodes (SBDs)
  Al-Ti, 114, 116f
  high-voltage on 4H-SiC, 118, 120f–124f, 120–125,
  6H-SiC
    edge termination in, 113–114, 115f, 116f
    excess leakage current in, 114–115, 117f, 118
    Ni/6H-SiC, 112, 113f
    PT/3C, current-voltage characteristics of, 130, 132f

SiC
  excess leakage current in, 114–115, 117f, 118
  performance of, 149–151, 150f
Schottky barrier height (SBH), 82, 85
  annealing effects, 101, 102f, 105–106
  3C-SiC, metal work function, 129–130, 131f
  diffusion potential, temperature and, 107–108, 108f
  etch depth and, 101, 103f
  4H-SiC, 118, 120f
  of metal contacts to n-type 6H-SiC, 100, 101f
  postdeposition annealing conditions and, 108
  Pt-Schottky contacts on 3C-SiC, 129, 130f
Schottky contacts, 85. *See also* Schottky barrier diodes
  high-voltage on 6H-SiC, 109f, 109–112, 111f–113f
  to 4H-SiC, 118, 120f–124f, 120–125
  to 6H-SiC, 96–108, 101f–108f
    annealing effects, 97, 100–101, 102f, 103–104, 104f, 105f
Schottky model, for ideal metal-semiconductor contact, 81–83, 82f, 83f
Schottky rectifiers, on-resistance, 85
Schottky-Mott limit, 82
Schottky-Mott theory, 83
Secondary ion mass spectroscopy (SIMS)
  Al-doped epitaxial layer, 46f, 46–47
  of point defects, 316–317, 318f–319f
Short-circuit current, UV photodiode, 293
SiC. *See* Silicon carbide
Silicon (Si)
  DMOS, 172
  electron velocity-electric field characteristics, 243, 244f
  melt, liquid-phase epitaxy from, 51–52, 53, 53f
  properties, 242
    electronic, 196–198
    physical, 196–198
    vs. other semiconductors, 166
  thermal conductivity, 165, 166
Silicon carbide (SiC)
  applications

402 INDEX

Silicon carbide, applications (*continued*)
   in high-power electronics, 195–196, 197f, 198, 231–232
   in optoelectronics, 285
  bandgap, 242, 284
  bulk crystal growth, 22
   by chemical vapor deposition, 34
   from liquid phase, 34–35
   by physical vapor transport, 22–34, 24f, 25f, 27f
  3C-SiC (cubic; β-SiC), 3, 78
   chemical vapor deposition, 48–49
   electron mobility, 78, 239
   high-voltage p-n junction diodes, 134–135
   homoepitaxy of, 47–48
   JGFETs, 240
   LPE on 6H-SiC, 55–56
   MESFETs, 240
   MOSFETs, 240
   ohmic contacts, 125–126, 127f, 128f
   p-n junctions, 133–138
   properties, 166
   structure, 239
  devices, 149–151, 150f. *See also specific devices*
   advantages, 77–78
   p-n junctions diode rectifiers, 131, 133
  doping. *See* Doping, SiC
  electrical contacts. *See* Metal-SiC systems
  epitaxial growth, 35, 63
   by chemical vapor deposition. *See* Chemical vapor deposition, SiC epitaxial growth
   doping during, 240
   liquid-phase. *See* Liquid-phase epitaxy
   molecular beam, 58–60
   sublimation epitaxy. *See* Sublimation epitaxy
   α-epitaxial layers
    electrical properties, 44–47, 45f–47f
    homoepitaxial growth, 39–47, 40f–42f, 45f–47f
    structural properties, 42–44
  growth
   bulk, 198–202, 199f–201f
   epitaxial, 202–204, 203f, 204f
  heteroepitaxy, 48–50
  for high-power RF systems, advantages of, 204–205

  historical aspects, 240–241
  4H-SiC (noncubic; α-SiC), 3–4, 78
   electrical contact formation properties, 78–80
   electron mobility in, 80
   electron velocity-electric field characteristics, 243, 244f
   high-voltage p-n junction diodes, 134–135
   high-voltage Schottky contacts on, 110–111
   p-n junction diodes, 147–149, 148f
   properties, 166, 242
  6H-SiC (noncubic; α-SiC), 3–4, 78
   electrical contact formation properties, 78–80
   electron mobility anisotropy ratio, 80
   electron velocity-electric field characteristics, 243, 244f
   fabrication, 239
   high-voltage p-n junction diodes, 134–135
   high-voltage Schottky contacts on, 109f, 109–112, 111f–113f
   homoepitaxial growth, 203
   ionization rates, 245, 246
   LPE growth of 3C-SiC on, 55–56
   MESFETs, 240–241
   n-type doping, dependence of contact resistance on, 91–95, 92f–95f
   ohmic contacts, 87–96, 90f–95f
   p-n junction diodes, 138–140, 141f–146f, 142, 144–146
   p-n junctions, 138–140, 141f–146f, 142, 144–146,
   properties, 166, 242
   Schottky contacts, 96–108, 101f–108f
  interest in, 21
  ion implantation, doping during, 240
  lattice parameters, 7–12
  microwave devices. *See* Microwave devices, SiC
  polytypes, 2f–8f, 2–7, 17–18, 239. *See also specific polytypes*
   3C, bulk growth by physical vapor transport, 33–34
   control, in sublimation growth techniques, 31–32, 32f, 58
   electrical properties of, 16

hexagonal or rhombohedral or α, 3–4
inclusions, in SiC CVD layers, 43–44
mechanical properties of, 12–13
optical properties of, 14–15
symbolic notations, 2–7, 9
thermal properties of, 12–14
properties
electronic, 196–198
physical, 196–198
source material, for sublimation growth, 26
stacking notations, 2–3
structural unit, 3, 3f
sublimation growth. See Sublimation growth
synthesis, 239
transistors. See Transistors, SiC
Silicon oxide (SiO$_2$), for MOSFETs, 170
SIMS. See Secondary ion mass spectroscopy
SITs. See Static induction transistors
Solution growth technique (SG), 131
Solvents, for surface preparation, 86–87
Space-charge limited current (SCLC), 213–214
Specific contact resistivity, 84
Stacking fault defects, III-V nitride-based heterostructures, 311–313, 312f
Static induction transistors (SITs)
applications, 279–280
current-voltage characteristic, 213–214, 214f
design considerations, 212–215, 213f, 214f
fabrication, 215, 216f, 217, 243
4H-SiC, 219, 219f, 220f, 221
6H-SiC, 217f, 218f, 217–219
gate-to-drain spacing, 215
microwave performance, 177, 179
potential diagram, 264, 265f
power density, 231
RF active, 264f–266f, 264–267
saturated dc I-V characteristics, 265–266, 266f
S-band, 223–224, 224f
S-band scale-up, 224–226, 225f
structure, 178, 178f, 264, 264f, 180
450-W UHF, 221f, 222f, 221–222
Step bunching, in SiC CVD layers, 43

Sublimation epitaxy
polytype control, 58
sandwich method, 56–58, 57f, 58f
Sublimation growth, 35
electrical characteristics, 32–33, 33f
equipment, 25f, 26
growth rate, 28f, 28–29
polytype control, 31–32, 32f
problems, 198–199
source material, 26
source temperature, 24
structural defects, 29–31, 30f
vapor phase equilibrium, 26–28, 27f
Surface defects, in SiC CVD layers, 42
Surface preparation techniques, ohmic contact and, 86–87

T

Tairov-Tsvetkov method. See Sublimation growth
Temperature
Hall effect and, 298–299, 299f
Schottky contact function and, 105–106, 106f
SiC CVD growth rate and, 38–39, 39f
SiC microwave devices and, 250–252, 251f
UV photodiode optical responsivity and, 296, 297f
Thermal conductivity, 242
Thyristors, SiC, 228
gated-off, 229, 229f, 230f
structure, 182–183, 183f
Titanium contacts, 100–101, 102f
Transistors, SiC, 161–162
applications, 188
bipolar, 182–183, 183f, 184f, 185
circuit simulation, 186–188, 187f
digital integrated circuits, 180f, 181f, 180–182
field-effect
principle of operation, 162–168, 163f–165f
structure, 162, 163f
modeling
analytical, 186–188, 187f
two-dimensional, 185
potential performance, 188

Traveling solvent method (TS), 131

U

UCCM (unified charge-control model), 186–187
UHF TV module, 2.0-kW, 222–223, 223f
UMOSFETs, SiC, 170–171, 171f, 172f
Unified charge-control model (UCCM), 186–187
UV detectors, GaN-based, 385–389, 386f–388f
UV photodiodes. *See* Photodiodes, UV

V

Vertical dipping, for SiC liquid-phase epitaxy, 52f, 52–53

W

W deposition, on 3C-SiC ohmic contacts, 125–126, 127f

Y

Yellow-band emission, in wide-bandgap semiconductors, 335–336, 336f–339f, 338–339

Contents of Volumes in This Series

Volume 1 Physics of III–V Compounds

C. Hilsum, Some Key Features of III–V Compounds
Franco Bassani, Methods of Band Calculations Applicable to III–V Compounds
E. O. Kane, The k-p Method
V. L. Bonch-Bruevich, Effect of Heavy Doping on the Semiconductor Band Structure
Donald Long, Energy Band Structures of Mixed Crystals of III–V Compounds
Laura M. Roth and Petros N. Argyres, Magnetic Quantum Effects
S. M. Puri and T. H. Geballe, Thermomagnetic Effects in the Quantum Region
W. M. Becker, Band Characteristics near Principal Minima from Magnetoresistance
E. H. Putley, Freeze-Out Effects, Hot Electron Effects, and Submillimeter Photoconductivity in InSb
H. Weiss, Magnetoresistance
Betsy Ancker-Johnson, Plasma in Semiconductors and Semimetals

Volume 2 Physics of III–V Compounds

M. G. Holland, Thermal Conductivity
S. I. Novkova, Thermal Expansion
U. Piesbergen, Heat Capacity and Debye Temperatures
G. Giesecke, Lattice Constants
J. R. Drabble, Elastic Properties
A. U. Mac Rae and G. W. Gobeli, Low Energy Electron Diffraction Studies
Robert Lee Mieher, Nuclear Magnetic Resonance
Bernard Goldstein, Electron Paramagnetic Resonance
T. S. Moss, Photoconduction in III–V Compounds
E. Antoncik ad J. Tauc, Quantum Efficiency of the Internal Photoelectric Effect in InSb
G. W. Gobeli and I. G. Allen, Photoelectric Threshold and Work Function
P. S. Pershan, Nonlinear Optics in III–V Compounds
M. Gershenzon, Radiative Recombination in the III–V Compounds
Frank Stern, Stimulated Emission in Semiconductors

Volume 3 Optical of Properties III–V Compounds

Marvin Hass, Lattice Reflection
William G. Spitzer, Multiphonon Lattice Absorption
D. L. Stierwalt and R. F. Potter, Emittance Studies
H. R. Philipp and H. Ehrenveich, Ultraviolet Optical Properties
Manuel Cardona, Optical Absorption above the Fundamental Edge
Earnest J. Johnson, Absorption near the Fundamental Edge
John O. Dimmock, Introduction to the Theory of Exciton States in Semiconductors
B. Lax and J. G. Mavroides, Interband Magnetooptical Effects
H. Y. Fan, Effects of Free Carries on Optical Properties
Edward D. Palik and George B. Wright, Free-Carrier Magnetooptical Effects
Richard H. Bube, Photoelectronic Analysis
B. O. Seraphin and H. E. Bennett, Optical Constants

Volume 4 Physics of III–V Compounds

N. A. Goryunova, A. S. Borschevskii, and D. N. Tretiakov, Hardness
N. N. Sirota, Heats of Formation and Temperatures and Heats of Fusion of Compounds $A^{III}B^V$
Don L. Kendall, Diffusion
A. G. Chynoweth, Charge Multiplication Phenomena
Robert W. Keyes, The Effects of Hydrostatic Pressure on the Properties of III–V Semiconductors
L. W. Aukerman, Radiation Effects
N. A. Goryunova, F. P. Kesamanly, and D. N. Nasledov, Phenomena in Solid Solutions
R. T. Bate, Electrical Properties of Nonuniform Crystals

Volume 5 Infrared Detectors

Henry Levinstein, Characterization of Infrared Detectors
Paul W. Kruse, Indium Antimonide Photoconductive and Photoelectromagnetic Detectors
M. B. Prince, Narrowband Self-Filtering Detectors
Ivars Melngalis and T. C. Harman, Single-Crystal Lead-Tin Chalcogenides
Donald Long and Joseph L. Schmidt, Mercury-Cadmium Telluride and Closely Related Alloys
E. H. Putley, The Pyroelectric Detector
Norman B. Stevens, Radiation Thermopiles
R. J. Keyes and T. M. Quist, Low Level Coherent and Incoherent Detection in the Infrared
M. C. Teich, Coherent Detection in the Infrared
F. R. Arams, E. W. Sard, B. J. Peyton, and F. P. Pace, Infrared Heterodyne Detection with Gigahertz IF Response
H. S. Sommers, Jr., Macrowave-Based Photoconductive Detector
Robert Sehr and Rainer Zuleeg, Imaging and Display

Volume 6 Injection Phenomena

Murray A. Lampert and Ronald B. Schilling, Current Injection in Solids: The Regional Approximation Method
Richard Williams, Injection by Internal Photoemission
Allen M. Barnett, Current Filament Formation

R. Baron and J. W. Mayer, Double Injection in Semiconductors
W. Ruppel, The Photoconductor-Metal Contact

Volume 7   Application and Devices
Part A

John A. Copeland and Stephen Knight, Applications Utilizing Bulk Negative Resistance
F. A. Padovani, The Voltage-Current Characteristics of Metal-Semiconductor Contacts
P. L. Hower, W. W. Hooper, B. R. Cairns, R. D. Fairman, and D. A. Tremere, The GaAs Field-Effect Transistor
Marvin H. White, MOS Transistors
G. R. Antell, Gallium Arsenide Transistors
T. L. Tansley, Heterojunction Properties

Part B

T. Misawa, IMPATT Diodes
H. C. Okean, Tunnel Diodes
Robert B. Campbell and Hung-Chi Chang, Silicon Junction Carbide Devices
R. E. Enstrom, H. Kressel, and L. Krassner, High-Temperature Power Rectifiers of $GaAs_{1-x}P_x$

Volume 8   Transport and Optical Phenomena

Richard J. Stirn, Band Structure and Galvanomagnetic Effects in III–V Compounds with Indirect Band Gaps
Roland W. Ure, Jr., Thermoelectric Effects in III–V Compounds
Herbert Piller, Faraday Rotation
H. Barry Bebb and E. W. Williams, Photoluminescence I: Theory
E. W. Williams and H. Barry Bebb, Photoluminescence II: Gallium Arsenide

Volume 9   Modulation Techniques

B. O. Seraphin, Electroreflectance
R. L. Aggarwal, Modulated Interband Magnetooptics
Daniel F. Blossey and Paul Handler, Electroabsorption
Bruno Batz, Thermal and Wavelength Modulation Spectroscopy
Ivar Balslev, Piezopptical Effects
D. E. Aspnes and N. Bottka, Electric-Field Effects on the Dielectric Function of Semiconductors and Insulators

Volume 10   Transport Phenomena

R. L. Rhode, Low-Field Electron Transport
J. D. Wiley, Mobility of Holes in III–V Compounds
C. M. Wolfe and G. E. Stillman, Apparent Mobility Enhancement in Inhomogeneous Crystals
Robert L. Petersen, The Magnetophonon Effect

Volume 11 Solar Cells

Harold J. Hovel, Introduction; Carrier Collection, Spectral Response, and Photocurrent; Solar Cell Electrical Characteristics; Efficiency; Thickness; Other Solar Cell Devices; Radiation Effects; Temperature and Intensity; Solar Cell Technology

Volume 12 Infrared Detectors (II)

W. L. Eiseman, J. D. Merriam, and R. F. Potter, Operational Characteristics of Infrared Photodetectors
Peter R. Bratt, Impurity Germanium and Silicon Infrared Detectors
E. H. Putley, InSb Submillimeter Photoconductive Detectors
G. E. Stillman, C. M. Wolfe, and J. O. Dimmock, Far-Infrared Photoconductivity in High Purity GaAs
G. E. Stillman and C. M. Wolfe, Avalanche Photodiodes
P. L. Richards, The Josephson Junction as a Detector of Microwave and Far-Infrared Radiation
E. H. Putley, The Pyroelectric Detector—An Update

Volume 13 Cadmium Telluride

Kenneth Zanio, Materials Preparations; Physics; Defects; Applications

Volume 14 Lasers, Junctions, Transport

N. Holonyak, Jr. and M. H. Lee, Photopumped III–V Semiconductor Lasers
Henry Kressel and Jerome K. Butler, Heterojunction Laser Diodes
A Van der Ziel, Space-Charge-Limited Solid-State Diodes
Peter J. Price, Monte Carlo Calculation of Electron Transport in Solids

Volume 15 Contacts, Junctions, Emitters

B. L. Sharma, Ohmic Contacts to III–V Compounds Semiconductors
Allen Nussbaum, The Theory of Semiconducting Junctions
John S. Escher, NEA Semiconductor Photoemitters

Volume 16 Defects, (HgCd)Se, (HgCd)Te

Henry Kressel, The Effect of Crystal Defects on Optoelectronic Devices
C. R. Whitsett, J. G. Broerman, and C. J. Summers, Crystal Growth and Properties of $Hg_{1-x}Cd_xSe$ alloys
M. H. Weiler, Magnetooptical Properties of $Hg_{1-x}Cd_xTe$ Alloys
Paul W. Kruse and John G. Ready, Nonlinear Optical Effects in $Hg_{1-x}Cd_xTe$

Volume 17 CW Processing of Silicon and Other Semiconductors

James F. Gibbons, Beam Processing of Silicon
Arto Lietoila, Richard B. Gold, James F. Gibbons, and Lee A. Christel, Temperature Distribu-

tions and Solid Phase Reaction Rates Produced by Scanning CW Beams
Arto Leitoila and James F. Gibbons, Applications of CW Beam Processing to Ion Implanted Crystalline Silicon
N. M. Johnson, Electronic Defects in CW Transient Thermal Processed Silicon
K. F. Lee, T. J. Stultz, and James F. Gibbons, Beam Recrystallized Polycrystalline Silicon: Properties, Applications, and Techniques
T. Shibata, A. Wakita, T. W. Sigmon, and James F. Gibbons, Metal-Silicon Reactions and Silicide
Yves I. Nissim and James F. Gibbons, CW Beam Processing of Gallium Arsenide

Volume 18 Mercury Cadmium Telluride

Paul W. Kruse, The Emergence of $(Hg_{1-x}Cd_x)Te$ as a Modern Infrared Sensitive Material
H. E. Hirsch, S. C. Liang, and A. G. White, Preparation of High-Purity Cadmium, Mercury, and Tellurium
W. F. H. Micklethwaite, The Crystal Growth of Cadmium Mercury Telluride
Paul E. Petersen, Auger Recombination in Mercury Cadmium Telluride
R. M. Broudy and V. J. Mazurczyck, (HgCd)Te Photoconductive Detectors
M. B. Reine, A. K. Soad, and T. J. Tredwell, Photovoltaic Infrared Detectors
M. A. Kinch, Metal-Insulator-Semiconductor Infrared Detectors

Volume 19 Deep Levels, GaAs, Alloys, Photochemistry

G. F. Neumark and K. Kosai, Deep Levels in Wide Band-Gap III–V Semiconductors
David C. Look, The Electrical and Photoelectronic Properties of Semi-Insulating GaAs
R. F. Brebrick, Ching-Hua Su, and Pok-Kai Liao, Associated Solution Model for Ga-In-Sb and Hg-Cd-Te
Yu. Ya. Gurevich and Yu. V. Pleskon, Photoelectrochemistry of Semiconductors

Volume 20 Semi-Insulating GaAs

R. N. Thomas, H. M. Hobgood, G. W. Eldridge, D. L. Barrett, T. T. Braggins, L. B. Ta, and S. K. Wang, High-Purity LEC Growth and Direct Implantation of GaAs for Monolithic Microwave Circuits
C. A. Stolte, Ion Implantation and Materials for GaAs Integrated Circuits
C. G. Kirkpatrick, R. T. Chen, D. E. Holmes, P. M. Asbeck, K. R. Elliott, R. D. Fairman, and J. R. Oliver, LEC GaAs for Integrated Circuit Applications
J. S. Blakemore and S. Rahimi, Models for Mid-Gap Centers in Gallium Arsenide

Volume 21 Hydrogenated Amorphous Silicon
Part A

Jacques I. Pankove, Introduction
Masataka Hirose, Glow Discharge; Chemical Vapor Deposition
Yoshiyuki Uchida, di Glow Discharge
T. D. Moustakas, Sputtering
Isao Yamada, Ionized-Cluster Beam Deposition
Bruce A. Scott, Homogeneous Chemical Vapor Deposition

Frank J. Kampas, Chemical Reactions in Plasma Deposition
Paul A. Longeway, Plasma Kinetics
Herbert A. Weakliem, Diagnostics of Silane Glow Discharges Using Probes and Mass Spectroscopy
Lester Gluttman, Relation between the Atomic and the Electronic Structures
A. Chenevas-Paule, Experiment Determination of Structure
S. Minomura, Pressure Effects on the Local Atomic Structure
David Adler, Defects and Density of Localized States

Part B

Jacques I. Pankove, Introduction
G. D. Cody, The Optical Absorption Edge of a-Si:H
Nabil M. Amer and Warren B. Jackson, Optical Properties of Defect States in a-Si:H
P. J. Zanzucchi, The Vibrational Spectra of a-Si:H
Yoshihiro Hamakawa, Electroreflectance and Electroabsorption
Jeffrey S. Lannin, Raman Scattering of Amorphous Si, Ge, and Their Alloys
R. A. Street, Luminescence in a-Si:H
Richard S. Crandall, Photoconductivity
J. Tauc, Time-Resolved Spectroscopy of Electronic Relaxation Processes
P. E. Vanier, IR-Induced Quenching and Enhancement of Photoconductivity and Photoluminescence
H. Schade, Irradiation-Induced Metastable Effects
L. Ley, Photoelectron Emission Studies

Part C

Jacques I. Pankove, Introduction
J. David Cohen, Density of States from Junction Measurements in Hydrogenated Amorphous Silicon
P. C. Taylor, Magnetic Resonance Measurements in a-Si:H
K. Morigaki, Optically Detected Magnetic Resonance
J. Dresner, Carrier Mobility in a-Si:H
T. Tiedje, Information about band-Tail States from Time-of-Flight Experiments
Arnold R. Moore, Diffusion Length in Undoped a-Si:H
W. Beyer and J. Overhof, Doping Effects in a-Si:H
H. Fritzche, Electronic Properties of Surfaces in a-Si:H
C. R. Wronski, The Staebler-Wronski Effect
R. J. Nemanich, Schottky Barriers on a-Si:H
B. Abeles and T. Tiedje, Amorphous Semiconductor Superlattices

Part D

Jacques I. Pankove, Introduction
D. E. Carlson, Solar Cells
G. A. Swartz, Closed-Form Solution of I–V Characteristic for a a-Si:H Solar Cells
Isamu Shimizu, Electrophotography
Sachio Ishioka, Image Pickup Tubes

P. G. LeComber and W. E. Spear, The Development of the a-Si:H Field-Effect Transistor and Its Possible Applications
D. G. Ast, a-Si:H FET-Addressed LCD Panel
S. Kaneko, Solid-State Image Sensor
Masakiyo Matsumura, Charge-Coupled Devices
M. A. Bosch, Optical Recording
A. D'Amico and G. Fortunato, Ambient Sensors
Hiroshi Kukimoto, Amorphous Light-Emitting Devices
Robert J. Phelan, Jr., Fast Detectors and Modulators
Jacques I. Pankove, Hybrid Structures
P. G. LeComber, A. E. Owen, W. E. Spear, J. Hajto, and W. K. Choi, Electronic Switching in Amorphous Silicon Junction Devices

Volume 22 Lightwave Communications Technology
Part A

Kazuo Nakajima, The Liquid-Phase Epitaxial Growth of IngaAsp
W. T. Tsang, Molecular Beam Epitaxy for III–V Compound Semiconductors
G. B. Stringfellow, Organometallic Vapor-Phase Epitaxial Growth of III–V Semiconductors
G. Beuchet, Halide and Chloride Transport Vapor-Phase Deposition of InGaAsP and GaAs
Manijeh Razeghi, Low-Pressure Metallo-Organic Chemical Vapor Deposition of $Ga_x in_{1-x} As P_{1-y}$ Alloys
P. M. Petroff, Defects in III–V Compound Semiconductors

Part B

J. P. van der Ziel, Mode Locking of Semiconductor Lasers
Kam Y. Lau and Ammon Yariv, High-Frequency Current Modulation of Semiconductor Injection Lasers
Charles H. Henry, Special Properties of Semiconductor Lasers
Yasuharu Suematsu, Katsumi Kishino, Shigehisa Arai, and Fumio Koyama, Dynamic Single-Mode Semiconductor Lasers with a Distributed Reflector
W. T. Tsang, The Cleaved-Coupled-Cavity (C^3) Laser

Part C

R. J. Nelson and N. K. Dutta, Review of InGaAsP InP Laser Structures and Comparison of Their Performance
N. Chinone and M. Nakamura, Mode-Stabilized Semiconductor Lasers for 0.7–0.8- and 1.1–1.6-μm Regions
Yoshiji Horikoshi, Semiconductor Lasers with Wavelengths Exceeding 2 μm
B. A. Dean and M. Dixon, The Functional Reliability of Semiconductor Lasers as Optical Transmitters
R. H. Saul, T. P. Lee, and C. A. Burus, Light-Emitting Device Design
C. L. Zipfel, Light-Emitting Diode-Reliability
Tien Pei Lee and Tingye Li, LED-Based Multimode Lightwave Systems
Kinichiro Ogawa, Semiconductor Noise-Mode Partition Noise

Part D

Federico Capasso, The Physics of Avalanche Photodiodes
T. P. Pearsall and M. A. Pollack, Compound Semiconductor Photodiodes
Takao Kaneda, Silicon and Germanium Avalanche Photodiodes
S. R. Forrest, Sensitivity of Avalanche Photodetector Receivers for High-Bit-Rate Long-Wavelength Optical Communication Systems
J. C. Campbell, Phototransistors for Lightwave Communications

Part E

Shyh Wang, Principles and Characteristics of Integrable Active and Passive Optical Devices
Shlomo Margalit and Amnon Yariv, Integrated Electronic and Photonic Devices
Takaoki Mukai, Yoshihisa Yamamoto, and Tatsuya Kimura, Optical Amplification by Semiconductor Lasers

Volume 23 Pulsed Laser Processing of Semiconductors

R. F. Wood, C. W. White, and R. T. Young, Laser Processing of Semiconductors: An Overview
C. W. White, Segregation, Solute Trapping, and Supersaturated Alloys
G. E. Jellison, Jr., Optical and Electrical Properties of Pulsed Laser-Annealed Silicon
R. F. Wood and G. E. Jellison, Jr., Melting Model of Pulsed Laser Processing
R. F. Wood and F. W. Young, Jr., Nonequilibrium Solidification Following Pulsed Laser Melting
D. H. Lowndes and G. E. Jellison, Jr., Time-Resolved Measurement During Pulsed Laser Irradiation of Silicon
D. M. Zebner, Surface Studies of Pulsed Laser Irradiated Semiconductors
D. H. Lowndes, Pulsed Beam Processing of Gallium Arsenide
R. B. James, Pulsed CO_2 Laser Annealing of Semiconductors
R. T. Young and R. F. Wood, Applications of Pulsed Laser Processing

Volume 24 Applications of Multiquantum Wells, Selective Doping, and Superlattices

C. Weisbuch, Fundamental Properties of III–V Semiconductor Two-Dimensional Quantized Structures: The Basis for Optical and Electronic Device Applications
H. Morkoc and H. Unlu, Factors Affecting the Performance of (Al, Ga)As/GaAs and (Al, Ga)As/InGaAs Modulation-Doped Field-Effect Transistors: Microwave and Digital Applications
N. T. Linh, Two-Dimensional Electron Gas FETs: Microwave Applications
M. Abe et al., Ultra-High-Speed HEMT Integrated Circuits
D. S. Chemla, D. A. B. Miller, and P. W. Smith, Nonlinear Optical Properties of Multiple Quantum Well Structures for Optical Signal Processing
F. Capasso, Graded-Gap and Superlattice Devices by Band-Gap Engineering
W. T. Tsang, Quantum Confinement Heterostructure Semiconductor Lasers
G. C. Osbourn et al., Principles and Applications of Semiconductor Strained-Layer Superlattices

Volume 25 Diluted Magnetic Semiconductors

W. Giriat and J. K. Furdyna, Crystal Structure, Composition, and Materials Preparation of Diluted Magnetic Semiconductors

W. M. Becker, Band Structure and Optical Properties of Wide-Gap $A^{II}_{1-x}Mn_xB^{IV}$ Alloys at Zero Magnetic Field

Saul Oseroff and Pieter H. Keesom, Magnetic Properties: Macroscopic Studies

Giebultowicz and T. M. Holden, Neutron Scattering Studies of the Magnetic Structure and Dynamics of Diluted Magnetic Semiconductors

J. Kossut, Band Structure and Quantum Transport Phenomena in Narrow-Gap Diluted Magnetic Semiconductors

C. Riquaux, Magnetooptical Properties of Large-Gap Diluted Magnetic Semiconductors

J. A. Gaj, Magnetooptical Properties of Large-Gap Diluted Magnetic Semiconductors

J. Mycielski, Shallow Acceptors in Diluted Magnetic Semiconductors: Splitting, Boil-off, Giant Negative Magnetoresistance

A. K. Ramadas and R. Rodriquez, Raman Scattering in Diluted Magnetic Semiconductors

P. A. Wolff, Theory of Bound Magnetic Polarons in Semimagnetic Semiconductors

Volume 26 III–V Compound Semiconductors and Semiconductor Properties of Superionic Materials

Zou Yuanxi, III–V Compounds

H. V. Winston, A. T. Hunter, H. Kimura, and R. E. Lee, InAs-Alloyed GaAs Substrates for Direct Implantation

P. K. Bhattachary and S. Dhar, Deep Levels in III–V Compound Semiconductors Grown by MBE

Yu. Yu. Gurevich and A. K. Ivanov-Shits, Semiconductor Properties of Supersonic Materials

Volume 27 High Conducting Quasi-One-Dimensional Organic Crystals

E. M. Conwell, Introduction to Highly Conducting Quasi-One-Dimensional Organic Crystals

I. A. Howard, A Reference Guide to the Conducting Quasi-One-Dimensional Organic Molecular Crystals

J. P. Pouquet, Structural Instabilities

E. M. Conwell, Transport Properties

C. S. Jacobsen, Optical Properties

J. C. Scott, Magnetic Properties

L. Zuppiroli, Irradiation Effects: Perfect Crystals and Real Crystals

Volume 28 Measurement of High-Speed Signals in Solid State Devices

J. Frey and D. Ioannou, Materials and Devices for High-Speed and Optoelectronic Applications

H. Schumacher and E. Strid, Electronic Wafer Probing Techniques

D. H. Auston, Picosecond Photoconductivity: High-Speed Measurements of Devices and Materials

J. A. Valdmanis, Electro-Optic Measurement Techniques for Picosecond Materials, Devices, and Integrated Circuits.

J. M. Wiesenfeld and R. K. Jain, Direct Optical Probing of Integrated Circuits and High-Speed Devices

G. Plows, Electron-Beam Probing

A. M. Weiner and R. B. Marcus, Photoemissive Probing

Volume 29 Very High Speed Integrated Circuits: Gallium Arsenide LSI

M. Kuzuhara and T. Nazaki, Active Layer Formation by Ion Implantation
H. Hasimoto, Focused Ion Beam Implantation Technology
T. Nozaki and A. Higashisaka, Device Fabrication Process Technology
M. Ino and T. Takada, GaAs LSI Circuit Design
M. Hirayama, M. Ohmori, and K. Yamasaki, GaAs LSI Fabrication and Performance

Volume 30 Very High Speed Integrated Circuits: Heterostructure

H. Watanabe, T. Mizutani, and A. Usui, Fundamentals of Epitaxial Growth and Atomic Layer Epitaxy
S. Hiyamizu, Characteristics of Two-Dimensional Electron Gas in III–V Compound Heterostructures Grown by MBE
T. Nakanisi, Metalorganic Vapor Phase Epitaxy for High-Quality Active Layers
T. Nimura, High Electron Mobility Transistor and LSI Applications
T. Sugeta and T. Ishibashi, Hetero-Bipolar Transistor and LSI Application
H. Matsueda, T. Tanaka, and M. Nakamura, Optoelectronic Integrated Circuits

Volume 31 Indium Phosphide: Crystal Growth and Characterization

J. P. Farges, Growth of Discoloration-free InP
M. J. McCollum and G. E. Stillman, High Purity InP Grown by Hydride Vapor Phase Epitaxy
T. Inada and T. Fukuda, Direct Synthesis and Growth of Indium Phosphide by the Liquid Phosphorous Encapsulated Czochralski Method
O. Oda, K. Katagiri, K. Shinohara, S. Katsura, Y. Takahashi, K. Kainosho, K. Kohiro, and R. Hirano, InP Crystal Growth, Substrate Preparation and Evaluation
K. Tada, M. Tatsumi, M. Morioka, T. Araki, and T. Kawase, InP Substrates: Production and Quality Control
M. Razeghi, LP-MOCVD Growth, Characterization, and Application of InP Material
T. A. Kennedy and P. J. Lin-Chung, Stoichiometric Defects in InP

Volme 32 Strained-Layer Superlattices: Physics

T. P. Pearsall, Strained-Layer Superlattices
Fred H. Pollack, Effects of Homogeneous Strain on the Electronic and Vibrational Levels in Semiconductors
J. Y. Marzin, J. M. Gerárd, P. Voisin, and J. A. Brum, Optical Studies of Strained III–V Heterolayers
R. People and S. A. Jackson, Structurally Induced States from Strain and Confinement
M. Jaros, Microscopic Phenomena in Ordered Suprlattices

Volume 33 Strained-Layer Superlattices: Materials Science and Technology

R. Hull and J. C. Bean, Principles and Concepts of Strained-Layer Epitaxy
William J. Schaff, Paul J. Tasker, Marc C. Foisy, and Lester F. Eastman, Device Applications of Strained-Layer Epitaxy

S. T. Picraux, B. L. Doyle, and J. Y. Tsao, Structure and Characterization of Strained-Layer Superlattices
E. Kasper and F. Schaffer, Group IV Compounds
Dale L. Martin, Molecular Beam Epitaxy of IV–VI Compounds Heterojunction
Robert L. Gunshor, Leslie A. Kolodziejski, Arto V. Nurmikko, and Nobuo Otsuka, Molecular Beam Epitaxy of II–VI Semiconductor Microstructures

Volume 34 Hydrogen in Semiconductors

J. I. Pankove and N. M. Johnson, Introduction to Hydrogen in Semiconductors
C. H. Seager, Hydrogenation Methods
J. I. Pankove, Hydrogenation of Defects in Crystalline Silicon
J. W. Corbett, P. Deák, U. V. Desnica, and S. J. Pearton, Hydrogen Passivation of Damage Centers in Semiconductors
S. J. Pearton, Neutralization of Deep Levels in Silicon
J. I. Pankove, Neutralization of Shallow Acceptors in Silicon
N. M. Johnson, Neutralization of Donor Dopants and Formation of Hydrogen-Induced Defects in n-Type Silicon
M. Stavola and S. J. Pearton, Vibrational Spectroscopy of Hydrogen-Related Defects in Silicon
A. D. Marwick, Hydrogen in Semiconductors: Ion Beam Techniques
C. Herring and N. M. Johnson, Hydrogen Migration and Solubility in Silicon
E. E. Haller, Hydrogen-Related Phenomena in Crystalline Germanium
J. Kakalios, Hydrogen Diffusion in Amorphous Silicon
J. Chevalier, B. Clerjaud, and B. Pajot, Neutralization of Defects and Dopants in III–V Semiconductors
G. G. DeLeo and W. B. Fowler, Computational Studies of Hydrogen-Containing Complexes in Semiconductors
R. F. Kiefl and T. L. Estle, Muonium in Semiconductors
C. G. Van de Walle, Theory of Isolated Interstitial Hydrogen and Muonium in Crystalline Semiconductors

Volume 35 Nanostructured Systems

Mark Reed, Introduction
H. van Houten, C. W. J. Beenakker, and B. J. van Wees, Quantum Point Contacts
G. Timp, When Does a Wire Become an Electron Waveguide?
M. Büttiker, The Quantum Hall Effects in Open Conductors
W. Hansen, J. P. Kotthaus, and U. Merkt, Electrons in Laterally Periodic Nanostructures

Volume 36 The Spectroscopy of Semiconductors

D. Heiman, Spectroscopy of Semiconductors at Low Temperatures and High Magnetic Fields
Arto V. Nurmikko, Transient Spectroscopy by Ultrashort Laser Pulse Techniques
A. K. Ramdas and S. Rodriguez, Piezospectroscopy of Semiconductors
Orest J. Glembocki and Benjamin V. Shanabrook, Photoreflectance Spectroscopy of Microstructures
David G. Seiler, Christopher L. Littler, and Margaret H. Wiler, One- and Two-Photon Magneto-Optical Spectroscopy of InSb and $Hg_{1-x}Cd_xTe$

Volume 37 The Mechanical Properties of Semiconductors

A.-B. Chen, Arden Sher and W. T. Yost, Elastic Constants and Related Properties of Semiconductor Compounds and Their Alloys
David R. Clarke, Fracture of Silicon and Other Semiconductors
Hans Siethoff, The Plasticity of Elemental and Compound Semiconductors
Sivaraman Guruswamy, Katherine T. Faber and John P. Hirth, Mechanical Behavior of Compound Semiconductors
Subhanh Mahajan, Deformation Behavior of Compound Semiconductors
John P. Hirth, Injection of Dislocations into Strained Multilayer Structures
Don Kendall, Charles B. Fleddermann, and Kevin J. Malloy, Critical Technologies for the Micromachining of Silicon
Ikuo Matsuba and Kinji Mokuya, Processing and Semiconductor Thermoelastic Behavior

Volume 38 Imperfections in III/V Materials

Udo Scherz and Matthias Scheffler, Density-Functional Theory of sp-Bonded Defects in III/V Semiconductors
Maria Kaminska and Eicke R. Weber, El2 Defect in GaAs
David C. Look, Defects Relevant for Compensation in Semi-Insulating GaAs
R. C. Newman, Local Vibrational Mode Spectroscopy of Defects in III/V Compounds
Andrzej M. Hennel, Transition Metals in III/V Compounds
Kevin J. Malloy and Ken Khachaturyan, DX and Related Defects in Semiconductors
V. Swaminathan and Andrew S. Jordan, Dislocations in III/V Compounds
Krzysztof W. Nauka, Deep Level Defects in the Epitaxial III/V Materials

Volume 39 Minority Carriers in III–V Semiconductors: Physics and Applications

Niloy K. Dutta, Radiative Transitions in GaAs and Other III–V Compounds
Richard K. Ahrenkiel, Minority-Carrier Lifetime in III–V Semiconductors
Tomofumi Furuta, High Field Minority Electron Transport in p-GaAs
Mark S. Lundstrom, Minority-Carrier Transport in III–V Semiconductors
Richard A. Abram, Effects of Heavy Doping and High Excitation on the Band Structure of GaAs
David Yevick and Witold Bardyszewski, An Introduction to Non-Equilibrium Many-Body Analyses of Optical Processes in III–V Semiconductors

Volume 40 Epitaxial Microstructures

E. F. Schubert, Delta-Doping of Semiconductors: Electronic, Optical, and Structural Properties of Materials and Devices
A. Gossard, M. Sundaram, and P. Hopkins, Wide Graded Potential Wells
P. Petroff, Direct Growth of Nanometer-Size Quantum Wire Superlattices
E. Kapon, Lateral Patterning of Quantum Well Heterostructures by Growth of Nonplanar Substrates
H. Temkin, D. Gershoni, and M. Panish, Optical Properties of Ga$1-x$In$_x$As/InP Quantum Wells

Volume 41 High Speed Heterostructure Devices

F. Capasso, F. Beltram, S. Sen, A. Pahlevi, and A. Y. Cho, Quantum Electron Devices: Physics and Applications
P. Solomon, D. J. Frank, S. L. Wright, and F. Canora, GaAs-Gate Semiconductor–Insulator–Semiconductor FET
M. H. Hashemi and U. K. Mishra, Unipolar InP-Based Transistors
R. Kiehl, Complementary Heterostructure FET Integrated Circuits
T. Ishibashi, GaAs-Based and InP-Based Heterostructure Bipolar Transistors
H. C. Liu and T. C. L. G. Sollner, High-Frequency-Tunneling Devices
H. Ohnishi, T. More, M. Takatsu, K. Imamura, and N. Yokoyama, Resonant-Tunneling Hot-Electron Transistors and Circuits

Volume 42 Oxygen in Silicon

F. Shimura, Introduction to Oxygen in Silicon
W. Lin, The Incorporation of Oxygen into Silicon Crystals
T. J. Schaffner and D. K. Schroder, Characterization Techniques for Oxygen in Silicon
W. M. Bullis, Oxygen Concentration Measurement
S. M. Hu, Intrinsic Point Defects in Silicon
B. Pajot, Some Atomic Configurations of Oxygen
J. Michel and L. C. Kimerling, Electical Properties of Oxygen in Silicon
R. C. Newman and R. Jones, Diffusion of Oxygen in Silicon
T. Y. Tan and W. J. Taylor, Mechanisms of Oxygen Precipitation: Some Quantitative Aspects
M. Schrems, Simulation of Oxygen Precipitation
K. Simino and I. Yonenaga, Oxygen Effect on Mechanical Properties
W. Bergholz, Grown-in and Process-Induced Effects
F. Shimura, Intrinsic/Internal Gettering
H. Tsuya, Oxygen Effect on Electronic Device Performance

Volume 43 Semiconductors for Room Temperature Nuclear Detector Applications

R. B. James and T. E. Schlesinger, Introduction and Overview
L. S. Darken and C. E. Cox, High-Purity Germanium Detectors
A. Burger, D. Nason, L. Van den Berg, and M. Schieber, Growth of Mercuric Iodide
X. J. Bao, T. E. Schlesinger, and R. B. James, Electrical Properties of Mercuric Iodide
X. J. Bao, R. B. James, and T. E. Schlesinger, Optical Properties of Red Mercuric Iodide
M. Hage-Ali and P. Siffert, Growth Methods of CdTe Nuclear Detector Materials
M. Hage-Ali and P Siffert, Characterization of CdTe Nuclear Detector Materials
M. Hage-Ali and P. Siffert, CdTe Nuclear Detectors and Applications
R. B. James, T. E. Schlesinger, J. Lund, and M. Schieber, $Cd_{1-x}Zn_xTe$ Spectrometers for Gamma and X-Ray Applications
D. S. McGregor, J. E. Kammeraad, Gallium Arsenide Radiation Detectors and Spectrometers
J. C. Lund, F. Olschner, and A. Burger, Lead Iodide
M. R. Squillante, and K. S. Shah, Other Materials: Status and Prospects
V. M. Gerrish, Characterization and Quantification of Detector Performance
J. S. Iwanczyk and B. E. Patt, Electronics for X-ray and Gamma Ray Spectrometers
M. Schieber, R. B. James, and T. E. Schlesinger, Summary and Remaining Issues for Room Temperature Radiation Spectrometers

Volume 44 II–IV Blue/Green Light Emitters: Device Physics and Epitaxial Growth

J. Han and R. L. Gunshor, MBE Growth and Electrical Properties of Wide Bandgap ZnSe-based II–VI Semiconductors
Shizuo Fujita and Shigeo Fujita, Growth and Characterization of ZnSe-based II–VI Semiconductors by MOVPE
Easen Ho and Leslie A. Kolodziejski, Gaseous Source UHV Epitaxy Technologies for Wide Bandgap II–VI Semiconductors
Chris G. Van de Walle, Doping of Wide-Band-Gap II–VI Compounds — Theory
Roberto Cingolani, Optical Properties of Excitons in ZnSe-Based Quantum Well Heterostructures
A. Ishibashi and A. V. Nurmikko, II–VI Diode Lasers: A Current View of Device Performance and Issues
Supratik Guha and John Petruzello, Defects and Degradation in Wide-Gap II–VI-based Structures and Light Emitting Devices

Volume 45 Effect of Disorder and Defects in Ion-Implanted Semiconductors: Electrical and Physiochemical Characterization

Heiner Ryssel, Ion Implantation into Semiconductors: Historical Perspectives
You-Nian Wang and Teng-Cai Ma, Electronic Stopping Power for Energetic Ions in Solids
Sachiko T. Nakagawa, Solid Effect on the Electronic Stopping of Crystalline Target and Application to Range Estimation
G. Müller, S. Kalbitzer and G. N. Greaves, Ion Beams in Amorphous Semiconductor Research
Jumana Boussey-Said, Sheet and Spreading Resistance Analysis of Ion Implanted and Annealed Semiconductors
M. L. Polignano and G. Queirolo, Studies of the Stripping Hall Effect in Ion-Implanted Silicon
J. Stoemenos, Transmission Electron Microscopy Analyses
Roberta Nipoti and Marco Servidori, Rutherford Backscattering Studies of Ion Implanted Semiconductors
P. Zaumseil, X-ray Diffraction Techniques

Volume 46 Effect of Disorder and Defects in Ion-Implanted Semiconductors: Optical and Photothermal Characterization

M. Fried, T. Lohner and J. Gyulai, Ellipsometric Analysis
Antonios Seas and Constantinos Christofides, Transmission and Reflection Spectroscopy on Ion Implanted Semiconductors
Andreas Othonos and Constantinos Christofides, Photoluminescence and Raman Scattering of Ion Implanted Semiconductors. Influence of Annealing
Constantinos Christofides, Photomodulated Thermoreflectance Investigation of Implanted Wafers. Annealing Kinetics of Defects
U. Zammit, Photothermal Deflection Spectroscopy Characterization of Ion-Implanted and Annealed Silicon Films
Andreas Mandelis, Arief Budiman and Miguel Vargas, Photothermal Deep-Level Transient Spectroscopy of Impurities and Defects in Semiconductors
R. Kalish and S. Charbonneau, Ion Implantation into Quantum-Well Structures
Alexandre M. Myasnikov and Nikolay N. Gerasimenko, Ion Implantation and Thermal Annealing of III-V Compound Semiconducting Systems: Some Problems of III-V Narrow Gap Semiconductors

Volume 47 Uncooled Infrared Imaging Arrays and Systems

R. G. Buser and M. P. Tompsett, Historical Overview
P. W. Kruse, Principles of Uncooled Infrared Focal Plane Arrays
R. A. Wood, Monolithic Silicon Microbolometer Arrays
C. M. Hanson, Hybrid Pyroelectric-Ferroelectric Bolometer Arrays
D. L. Polla and J. R. Choi, Monolithic Pyroelectric Bolometer Arrays
N. Teranishi, Thermoelectric Uncooled Infrared Focal Plane Arrays
M. F. Tompsett, Pyroelectric Vidicon
T. W. Kenny, Tunneling Infrared Sensors
J. R. Vig, R. L. Filler and Y. Kim, Application of Quartz Microresonators to Uncooled Infrared Imaging Arrays
P. W. Kruse, Application of Uncooled Monolithic Thermoelectric Linear Arrays to Imaging Radiometers

Volume 48 High Brightness Light Emitting Diodes

G. B. Stringfellow, Materials Issues in High-Brightness Light-Emitting Diodes
M. G. Craford, Overview of Device issues in High-Brightness Light-Emitting Diodes
F. M. Steranka, AlGaAs Red Light Emitting Diodes
C. H. Chen, S. A. Stockman, M. J. Peanasky, and C. P. Kuo, OMVPE Growth of AlGaInP for High Efficiency Visible Light-Emitting Diodes
F. A. Kish and R. M. Fletcher, AlGaInP Light-Emitting Diodes
M. W. Hodapp, Applications for High Brightness Light-Emitting Diodes
I. Akasaki and H. Amano, Organometallic Vapor Epitaxy of GaN for High Brightness Blue Light Emitting Diodes
S. Nakamura, Group III-V Nitride Based Ultraviolet-Blue-Green-Yellow Light-Emitting Diodes and Laser Diodes

Volume 49 Light Emission in Silicon: from Physics to Devices

David J. Lockwood, Light Emission in Silicon
Gerhard Abstreiter, Band Gaps and Light Emission in Si/SiGe Atomic Layer Structures
Thomas G. Brown and Dennis G. Hall, Radiative Isoelectronic Impurities in Silicon and Silicon-Germanium Alloys and Superlattices
J. Michel, L. V. C. Assali, M. T. Morse, and L. C. Kimerling, Erbium in Silicon
Yoshihiko Kanemitsu, Silicon and Germanium Nanoparticles
Philippe M. Fauchet, Porous Silicon: Photoluminescence and Electroluminescent Devices
C. Delerue, G. Allan, and M. Lannoo, Theory of Radiative and Nonradiative Processes in Silicon Nanocrystallites
Louis Brus, Silicon Polymers and Nanocrystals

Volume 50 Gallium Nitride (GaN)

J. I. Pankove and T. D. Moustakas, Introduction
S. P. DenBaars and S. Keller, Metalorganic Chemical Vapor Deposition (MOCVD) of Group III Nitrides
W. A. Bryden and T. J. Kistenmacher, Growth of Group III-A Nitrides by Reactive Sputtering
N. Newman, Thermochemistry of III-N Semiconductors
S. J. Pearton and R. J. Shul, Etching of III Nitrides
S. M. Bedair, Indium-based Nitride Compounds
A. Trampert, O. Brandt, and K. H. Ploog, Crystal Structure of Group III Nitrides

H. Morkoc, F. Hamdani, and A. Salvador, Electronic and Optical Properties of III–V Nitride based Quantum Wells and Superlattices
K. Doverspike and J. I. Pankove, Doping in the III-Nitrides
T. Suski and P. Perlin, High Pressure Studies of Defects and Impurities in Gallium Nitride
B. Monemar, Optical Properties of GaN
W. R. L. Lambrecht, Band Structure of the Group III Nitrides
N. E. Christensen and P. Perlin, Phonons and Phase Transitions in GaN
S. Nakamura, Applications of LEDs and LDs
I. Akasaki and H. Amano, Lasers
J. A. Cooper, Jr., Nonvolatile Random Access Memories in Wide Bandgap Semiconductors

Volume 51A Identification of Defects in Semiconductors

George D. Watkins, EPR and ENDOR Studies of Defects in Semiconductors
J.-M. Spaeth, Magneto-Optical and Electrical Detection of Paramagnetic Resonance in Semiconductors
T. A. Kennedy and E. R. Glaser, Magnetic Resonance of Epitaxial Layers Detected by Photoluminescence
K. H. Chow, B. Hitti, and R. F. Kiefl, μSR on Muonium in Semiconductors and Its Relation to Hydrogen
Kimmo Saarinen, Pekka Hautojärvi, and Catherine Corbel, Positron Annihilation Spectroscopy of Defects in Semiconductors
R. Jones and P. R. Briddon, The Ab Initio Cluster Method and the Dynamics of Defects in Semiconductors

Volume 51B Identification of Defects in Semiconductors

Gordon Davies, Optical Measurements of Point Defects
P. M. Mooney, Defect Identification Using Capacitance Spectroscopy
Michael Stavola, Vibrational Spectroscopy of Light Element Impurities in Semiconductors
P. Schwander, W. D. Rau, C. Kisielowski, M. Gribelyuk, and A. Ourmazd, Defect Processes in Semiconductors Studied at the Atomic Level by Transmission Electron Microscopy
Nikos D. Jager and Eicke R. Weber, Scanning Tunneling Microscopy of Defects in Semiconductors

ISBN 0-12-752160-7